NEW DIRECTIONS in ORGANIC and BIOLOGICAL CHEMISTRY

Series Editor : C.W. Rees, FRS
Imperial College of Science, Technology and Medicine
London, UK

New and Forthcoming Titles

Chirality and the Biological Activity of Drugs
Roger J. Crossley

Enzyme-Assisted Organic Synthesis
Manfred Schneider and Stefano Servi

C-Glycoside Synthesis
Maarten Postema

Organozinc Reagents in Organic Synthesis
Ender Erdik

Activated Metals in Organic Synthesis
Pedro Cintas

Capillary Electrophoresis: Theory and Practice
Patrick Camilleri

Cyclization Reactions
C. Thebtaranonth and Y. Thebtaranonth

Mannich Bases: Chemistry and Uses
Maurilio Tramontini and Luigi Angiolini

Vicarious Nucleophilic Substitution and Related Processes in Organic Synthesis
Mieczyslaw Makosza

Radical Cations and Anions
M. Chanon, S. Fukuzumi, and F. Chanon

Chlorosulfonic Acid: A Versatile Reagent
R. J. Cremlyn and J. P. Bassin

Aromatic Fluorination
James H. Clark and Tony W. Bastock

Selectivity in Lewis Acid Promoted Reactions
M. Santelli and J.-M. Pons

Dianion Chemistry
Charles M. Thompson

Asymmetric Methodology in Organic Synthesis
David J. Ager and Michael B. East

Synthesis Using Vilsmeier Reagents
C. M. Marson and P. R. Giles

The Anomeric Effect
Eusebio Juaristi

Chiral Sulfur Reagents
M. Mikołajczyk, J. Drabowicz, and P. Kiełbasiński

MANNICH
BASES
Chemistry and Uses

Maurilio Tramontini
Luigi Angiolini

CRC Press
Boca Raton   Ann Arbor   London   Tokyo

Library of Congress Cataloging-in-Publication Data

Tramontini, Maurilio.
    Mannich bases : chemistry and uses / Maurilio Tramontini and Luigi
Angiolini.
       p.   cm. — (New directions in organic and biological
chemistry)
       Includes bibliographical references and index.
       ISBN 0-8493-4430-1
       1. Mannich bases. I. Angiolini, Luigi. II. Title. III. Series.
QD305.A8T73 1994
547'.442—dc20   94-1113
                                                                          CIP

No claim to original U.S. Government works
International Standard Book Number 0-8493-4430-1
Library of Congress Card Number 94- 1113
Printed in the United States of America   2 3 4 5 6 7 8 9 0
Printed on acid-free paper

Table of Contents

Introduction

The first "Mannich reaction" took place accidentally in 1912, when Carl Mannich, at that time a young professor in the pharmaceutical laboratory at Göttingen University, was treating with acid the solution of a pharmaceutical preparation based on salicyl antipyrine and urotropine (hexamethylenetetramine). He obtained a crystalline precipitate, which was later identified, in collaboration with W. Krögen Kröshe, as having the structure shown in Fig. 1.[1]

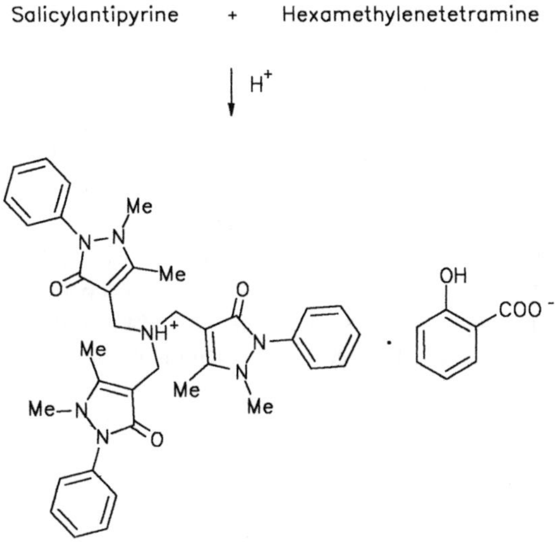

Salicylantipyrine   +   Hexamethylenetetramine

Fig. 1.   The first Mannich reaction.

Having observed that the same condensation product was also formed by mixing antipyrine, formaldehyde, and ammonium chloride, regardless of the order of addition, Mannich realized the great synthetic relevance of the reaction, of which only a few examples were known in the literature of the time. He saw that it allowed the linkage of two different chemical moieties in one step by means of a methylenic bridge. He then studied the reaction in considerable depth, assisted by a number of collaborators and scholars, and demonstrated its general applicability as a method for obtaining aminomethylated products. The main research efforts were of course devoted to the synthesis of pharmaceuticals, and some of them also entered into use as patented products.

Mannich published more than 60 papers on this topic, one fifth of his wide scientific production, which ended with his death, in 1947, at age 70.[2,3]

Since then, Mannich bases have been subjects of growing interest, as evidenced by the persistent number of books and review papers[4–12] in which the relevance of these compounds in synthetic organic chemistry is particularly stressed. This interest derives from both the versatility of the aminomethylation reaction and the possibility of obtaining a large variety of derivatives by further conversion of Mannich bases.

More recently, some review articles have appeared in which several applications of Mannich bases in the pharmaceutical field[13–16] and in other industries,[17,18] such as those connected with macromolecular chemistry,[19] are described. However, no general, complete overview of the widespread practical applications of Mannich bases has been published so far, despite their connections with important branches of industrial organic chemistry. Although the most relevant contribution of the Mannich reaction is still in pharmaceutical research (over 30% of the scientific papers published on this subject are found in journals dealing with pharmaceutical chemistry), Mannich bases have been found to have important uses in the manufacture of polymeric materials (resins and, in particular, surface coatings) and in the production of various additives and auxiliaries (for lubricants, textiles, paper, etc.) as well as, for example, in the production of water-treatment agents.

The aim of the present book is therefore to describe and discuss the developments that have occurred in the last 30 years in the chemistry of Mannich bases, the preceding period being excellently covered by Blicke's overview[4] and by the books of Hellmann and Opitz[5] and Reichert[6] (a scholar of Mannich). The wide range of practical applications made possible by using the chemistry of Mannich bases is also treated here.

We have tried to introduce a range of references as wide as possible, but in view of the huge amount of literature available and the need to keep the book to a reasonable size, we have been obliged to be selective in our presentation, though not without stressing unusual or novel results achieved by scientific research into this topic.

To Rosanna, Angela, Andrea, Luisa, Ingrid, Federico.

Chapter I

Synthesis of Mannich Bases

The Mannich reaction is a three-component condensation in which a compound containing an active hydrogen atom (hereafter called the "substrate") is allowed to react with formaldehyde and an NH-amine derivative (Fig. 2):

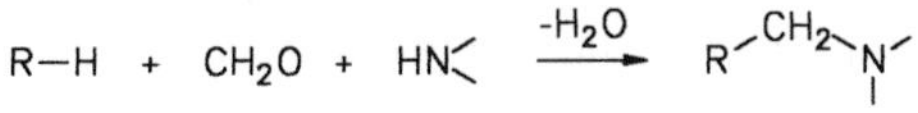

$$R{-}H \ + \ CH_2O \ + \ HN{\textstyle<} \ \xrightarrow{-H_2O} \ R{\sim}^{CH_2}{\sim}N{\sim}$$

Fig. 2.   The general Mannich condensation.

Hence, the resulting product (the Mannich base) is an amine compound having the N atom linked to the R substrate through a methylene group. The substrate may belong to a number of different classes of compounds.

The synthetic relevance of the reaction is clearly evidenced in Fig. 3, which gives a picture of some of the opportunities as well as the drawbacks connected with this method.

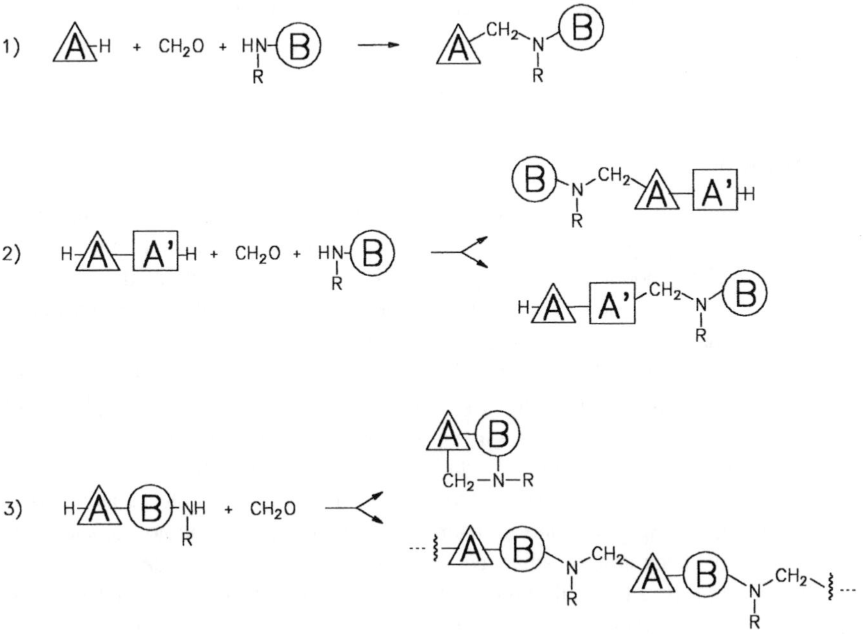

etc.

Fig. 3.   Examples of possible derivatives obtainable by Mannich reaction.

The principal advantage of the Mannich reaction is that it enables two different molecules to be bonded together in one step, as depicted in Eq. 1 (Fig. 3), whereas Eq. 2 stresses the alternatives connected with regioselectivity in polyfunctional substrates, which, on the other hand, also allow interesting cyclization or polymerization reactions to be performed. When the amino group is present in the substrate (Eq. 3), intramolecular Mannich reactions leading to cyclic products or to polymeric derivatives may occur.

Hence, this versatility of the Mannich reaction, along with the remarkable possibilities of exploiting the reactivity of Mannich bases in producing further derivatives (Chap. II), makes it possible to readily attain the most varied chemical structures in conformity with the practical requirements and applications needed in industry (Chap. V).

The substrates suitable for Mannich synthesis are widely available among a number of different compounds, and very few limitations are found in the choice of the amine reactant, except for the unreactive tertiary amine derivatives. As a consequence, a large number of substrate/amine combinations are possible, which may be increased further if one takes into account the possibility of using multifunctional reactants. The most significant chemical moieties that enable Mannich reaction to be performed are listed in Fig. 4 according to the reactive function present in the substrate. Frequently employed amines are also featured in the figure. It can be noted that almost every class of organic compound is covered by the potential substrates and that the amine reactant includes ammonia as well as common aliphatic and aromatic primary or secondary amines. Among these last reagents, dimethylamine and the cyclic secondary amines piperidine and morpholine are by far the most commonly used.

Unless otherwise required, the usual amines reported in Fig. 4 will be, from here on, generically represented in the chemical formulas accompanying the text.

A Aminomethylation Reactants

A.1 Substrate

Substrates usefully employed in Mannich reactions are, in general, XH compounds having nucleophilic properties, with X being equal to C, N, or other heteroatoms (Fig. 4). In particular, CH compounds are suitably activated saturated and unsaturated derivatives, and NH substrates may be amines, amides, heterocycles, etc. Out of OH substrates, alcohols are mainly able to give stable Mannich products. Sulfur- and phosphorus-containing substrates are XH derivatives having the H atom bonded to the heteroatom in the lower oxidation state, i.e., thiols, sulfinic acids, and, respectively, phosphine and phosphorous acid derivatives. As and Se compounds have also been successfully used. All these substrates are listed in more detail in Sec. D of this chapter.

The Mannich reaction is often carried out on ''unusual'' substrates, characterized by molecules having very complex structures, such as hormones, antibiotics, and alkaloids, which have been aminomethylated in the positions indicated by arrows in Table 1. Further examples are found throughout the book and particularly in Chap. IV, which is dedicated to the synthesis and functionalization of natural products. Macromolecular compounds, also, represent a particularly important group of substrates, which are thoroughly treated in Chap. III.

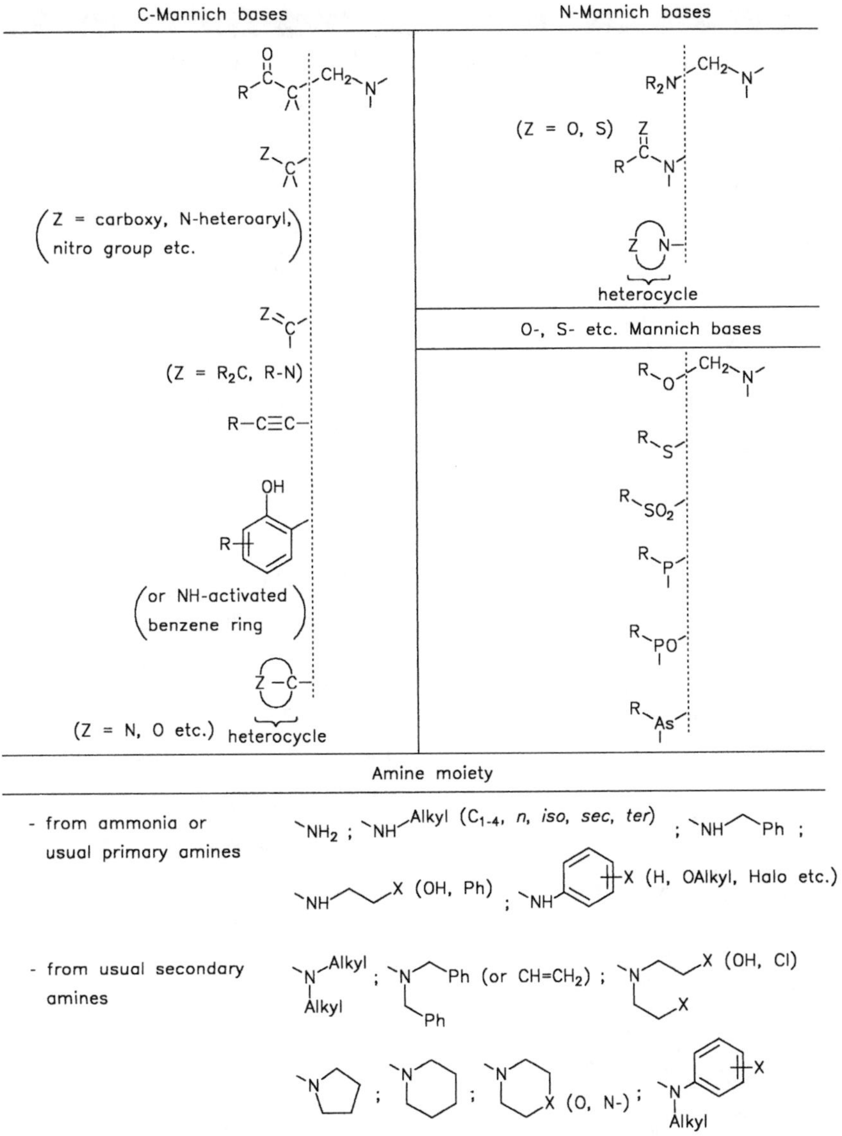

Fig. 4. General classification of Mannich derivatives according to substrate and amine type.

The presence of different reactive centers in the substrate, steric constraints, etc., may originate anomalous or unexpected results in Mannich synthesis. Thus, multifunctional substrates, or substrates containing reactive prochiral centers, give rise to the possibility of chemo-, regio-, and stereoselectivity (Secs. C.1–3), cyclizations and/or polymerizations (Sec. C.4 and Chap. III).

In certain cases, when the selected substrate appears unsuitable to Mannich reaction as it undergoes a hardly controllable reaction giving unacceptably contaminated products, or when its reactivity is unsufficient, the results can be improved by appropriately modifying or, respectively, activating the substrate itself. Thus, the Mannich bases of phosphorous acid (**2**, Fig. 5),[35–37] useful as herbicides and corrosion inhibitors, are better

Table 1

Selected Examples of Particular Substrates Subjected to Aminomethylation[a]

CH-Substrates	Ref.	CH-Substrates (continued)	Ref.
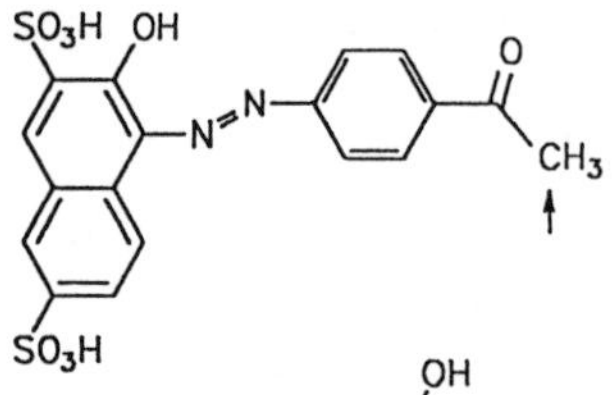	20	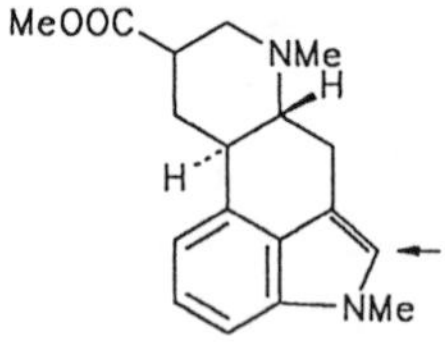	28
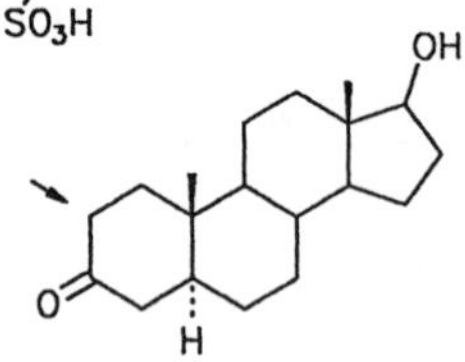	21	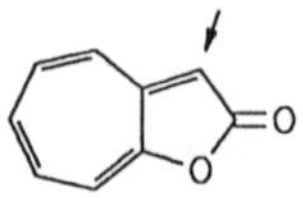	29
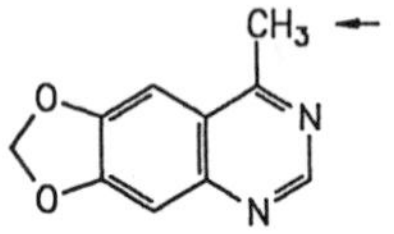	22	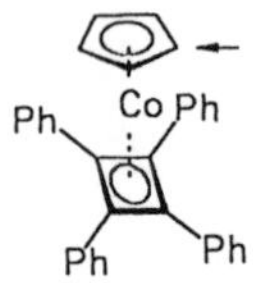	30
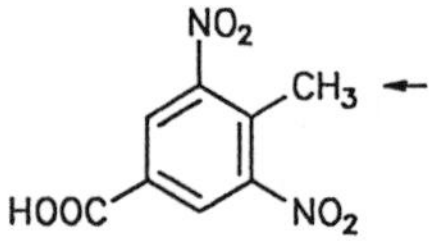	23	**NH-Substrates**	
(Alkyl)$_3$Ge—≡—H ←	24	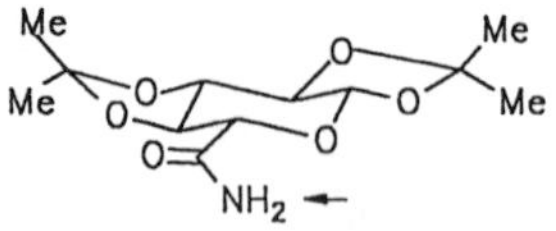	31
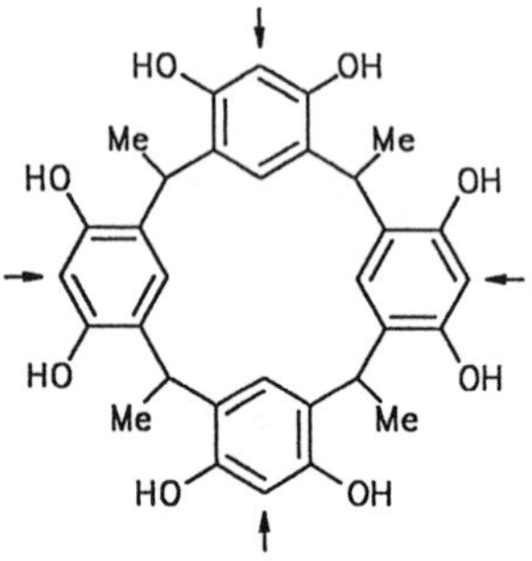	25	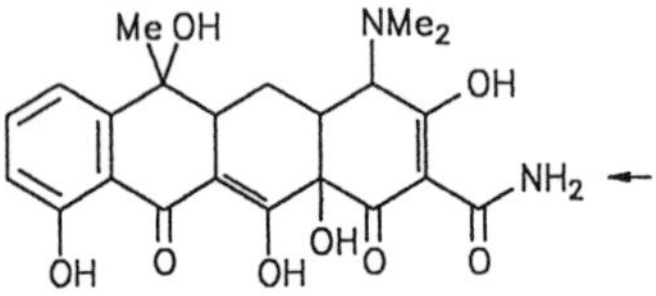	32
		OH-, PH-Substrates	
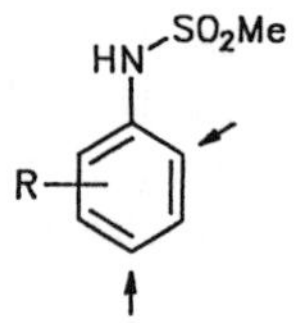	26,27	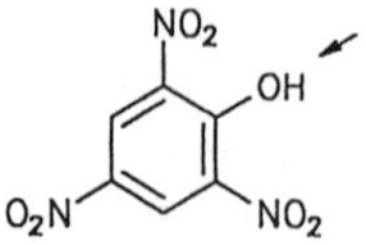	33
		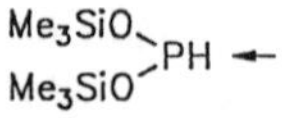 Me$_3$SiO–PH ← Me$_3$SiO	34

[a] Arrow indicates the position of attack

obtained by aminomethylation of the dimethyl ester **1** followed by hydrolysis of the ester groups.

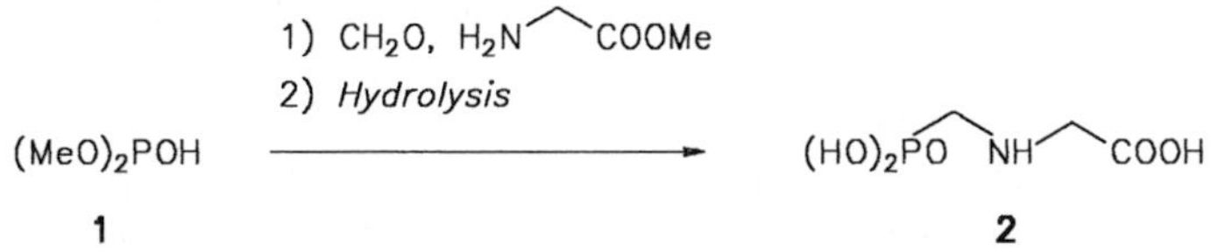

Fig. 5. Synthesis of a Mannich derivative from a phosphorous ester.

Similarly, the palladium complex of nicotinamide **3** (Fig. 6) is aminomethylated in much better yields than those given by the uncomplexed substrate.[38]

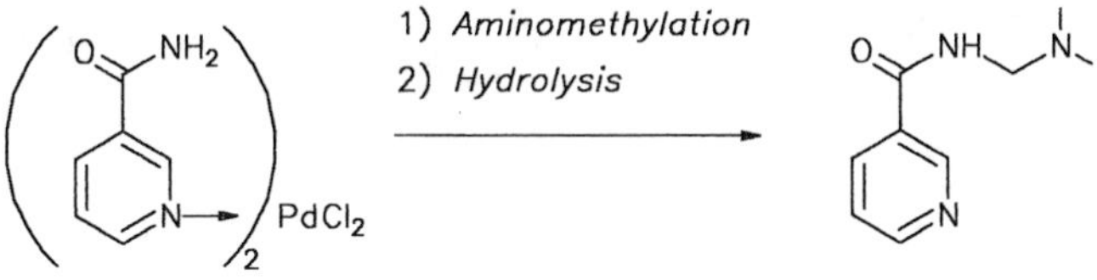

Fig. 6. Synthesis of *N*-aminomethyl nicotinamides.

Activation of a substrate involves a modification in the reactive center in order to enhance its nucleophilic character and make possible, or favor, the attack by the aminomethylating reagent. This is more frequently required by aliphatic carbonyl and carboxyl derivatives (ketones, aldehydes, and, respectively, esters, and lactones). The latter compounds, in particular, would not even be able to give a Mannich reaction in the absence of any activation, as the ester group, unlike the keto group, usually does not render the hydrogen atom in the α position sufficiently acidic. Activation of aliphatic ketones, on the other hand, is carried out mainly when chemo- or regioselective reactions are involved in Mannich synthesis.

Alkyl carbonyl substrates turn out to be more reactive when stabilized in the enolic form either as silyl enolethers **4**, obtained by reaction with $(CH_3)_3SiCl$,[39–58] or as metal enolates **5**, usually produced by lithium alkyls, lithium amides, or hydrides starting, for instance, from enolethers **4**.[39,40,42,59–61]

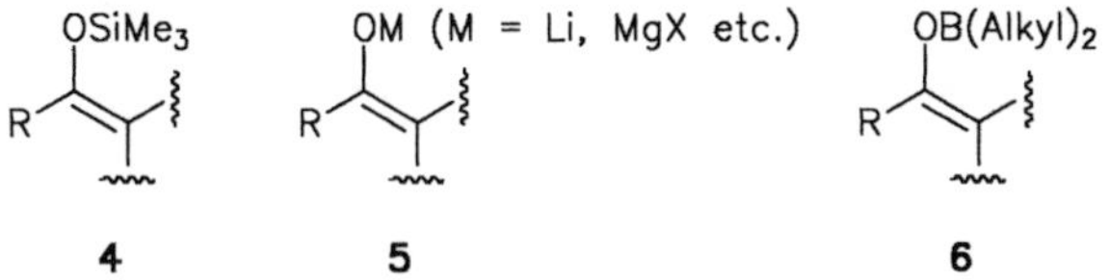

Boron enolates **6**, which are also used as activated derivatives of carbonyl substrates,[62–64] can be similarly obtained from silyl enolethers **4**,[62] or from diazo-carbonyl compounds **7** (Fig. 7), as well as from α, β-unsaturated ketones.[65]

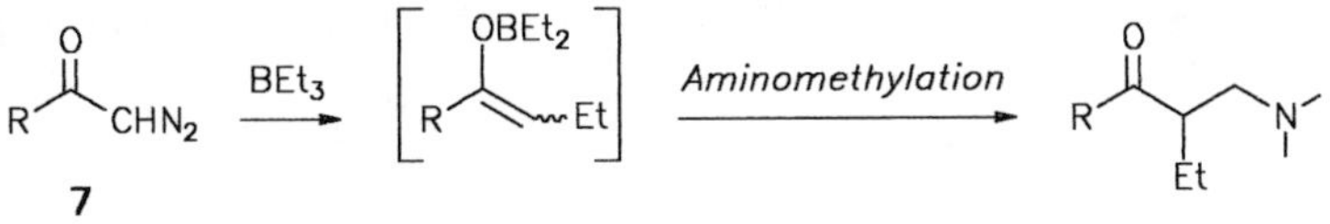

Fig. 7. Diazo-carbonyl compounds as starting material for Mannich synthesis.

In addition to the carbonyl derivatives, substrates such as NH-heterocyclic compounds[66,67] or nitroalkanes[68] have been activated as silyl derivatives.

Cyano- and nitroalkyls,[69,70] nitroamides,[71] and hydroxyphthalimides[72] are activated as alkali salts, whereas alkynes, as is well known, require the presence of copper salts to undergo reaction (see Table 6).

Concurrently with the Mannich reaction, structural modifications of the substrate, such as double bond migration, cyclization, rearrangement, and elimination accompanying the aminomethyl group attachment, may also occur. These are treated in Sec. C.5 and summarized in Fig. 8.

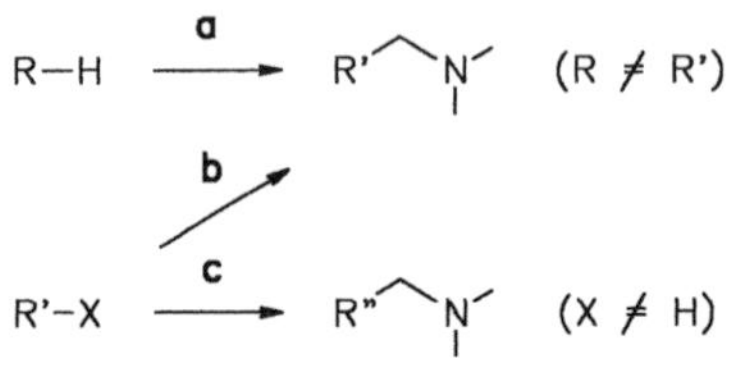

Fig. 8.   Substrate modifications occurring concurrently with Mannich reactions.

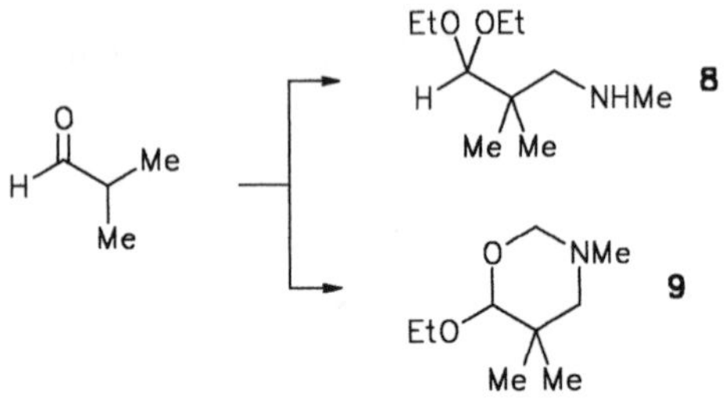

Fig. 9.   Reactions given by an aldehyde substrate during aminomethylation.

Route **a** in Fig. 8 represents the case of modification in the substrate moiety caused by the presence of another reactive species, such as the solvent, participating in the reaction. An example of such an occurrence is afforded by aldehydes acting as substrates, which give the corresponding acetals **8** and **9** (Fig. 9) by reaction with ethanol, the solvent mainly used in Mannich synthesis. The expected Mannich base can however be obtained by further treatment of the acetal **9**.[73]

When the substrate does not possess any hydrogen atom liable to be replaced by the aminomethyl group, aminomethylation can take place equally through the substitution of a sufficiently reactive X group (route **b** in Fig. 8), without modification of the R' moiety in the substrate. This is the case of the anthraquinone dye **10** (Fig. 10), which gives a Mannich base derived by the replacement of the sulfonic group.[74]

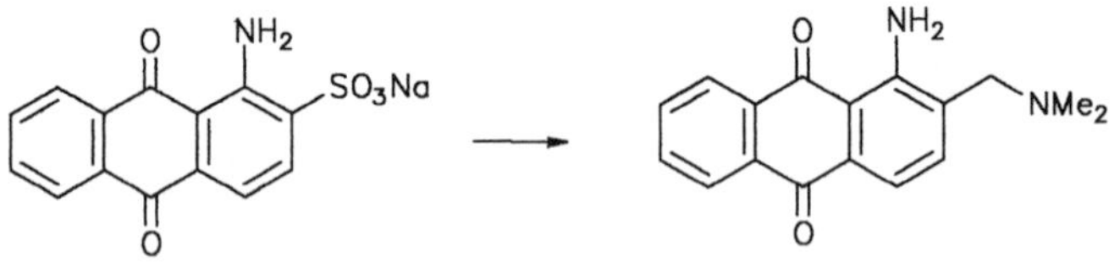

10

Fig. 10.   Mannich reaction on 1-amino-2-anthraquinone sulfonic acid.

Finally, Mannich bases can be obtained from substrates that, although devoid of active hydrogen atoms, may however undergo structural modifications leading to the aminomethylated product (route **c** in Fig. 8), as shown by the triethylester **11** (Fig. 11).[75]

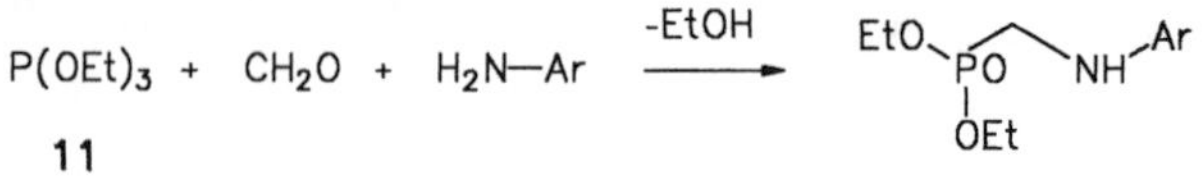

11

Fig. 11.   Mannich reaction on triethyl phosphite.

It is worth noting here that different types of R-XH substrates (Fig. 12), due to their nucleophilicity, can easily react with formaldehyde to give methylol derivatives (**12**), which behave, in turn, as X-methylating agents toward other analogous substrates R'—YH.

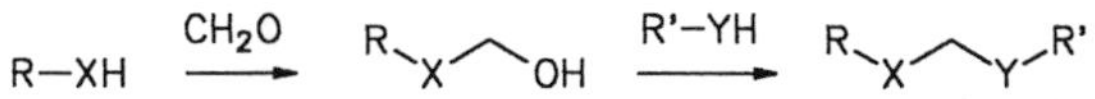

$$R-XH \xrightarrow{CH_2O} \underset{\textbf{12}}{R_{\diagdown X}\diagup{}^{OH}} \xrightarrow{R'-YH} R_{\diagdown X}\diagup{}_{Y}\diagup{}^{R'}$$

Fig. 12. Substrates affording X-methylation reactions.

These reactions have a general relevance in organic synthesis, as they allow the performance of X-methylation reactions such as cyano-, thio-, amidomethylations, etc. starting from hydrogen cyanide, amides, etc.[76–81] When, in particular, the abovementioned methylol derivatives are allowed to react with amine (R'—YH = primary or secondary amine), a Mannich base is produced.

A.2   Amine

The amine reagent of a Mannich reaction, as outlined at the beginning of this chapter, must have at least one reactive hydrogen atom, thus including ammonia, hydrazine, or hydroxylamine derivatives and, much more frequently, aliphatic as well as aromatic primary and secondary amines. Heteroaromatic NH derivatives are also employed. In addition to the usual amines featured in Fig. 4, a huge number of particular amines have been used in Mannich synthesis for specific purposes. A group of these "unusual" reagents, listed in Table 2, is mainly composed of molecules useful for their pharmacological activity, or characterized by peculiar steric hindrance, or containing chemical functions suitable for further reactions. Other important amines are mentioned in the appropriate sections, such as those concerning natural and biological compounds and macromolecular derivatives (Chaps. III and IV). Moreover, a number of amines having a specific role in products used for practical applications are cited in Chap. V.

Some general criteria are usually applied to the selection of the amine reagent to be employed. When an evaluation of the overall feasibility of Mannich reaction on a particular substrate is required, the most relevant features to be considered are steric hindrance and basicity of amine. Thus, the results obtained by alkylamines bearing differently branched and/or extended hydrocarbon chains are checked. In some cases it is in fact sufficient to replace dimethylamine with diethylamine for observing a lower reaction yield. Similarly, if the basic behavior of the amine reagent is to be investigated, the comparison among the products obtained by using, for example, isosteric piperidine or morpholine, is made.

Ethanolamines are employed for improving the hydrophilic properties of Mannich bases. Carboxyacids, such as glycine **13**, along with its precursor aminoacetonitrile,[35] dicarboxymethyl-amine **14**, and other α-amino acids[120] can be used for the same purpose.

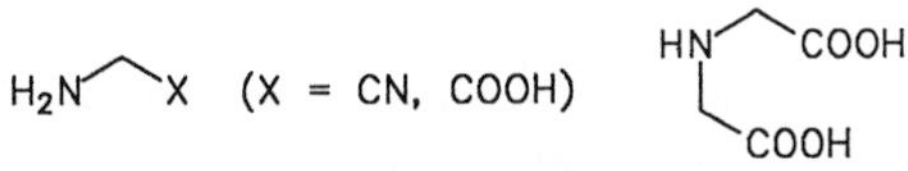

$$H_2N{\diagup}^{X} \quad (X = CN, COOH)$$

13

$$\underset{\textbf{14}}{\overset{HN{\diagup}^{COOH}}{\diagdown_{COOH}}}$$

The above compounds are preferred also for the possibility of obtaining metal complexes of the Mannich base produced by aminomethylation (Chap. V, A.2).

When a Mannich base containing a primary or secondary aminomethyl group is desired, it may be found convenient to make use of amines bearing substituents readily

Table 2

Selected Examples of Particular Amines Used for Aminomethylation

Primary Amines	Ref.	Bifunctional sec. Amines	Ref.
H₂N⟨chain⟩	82	⟨COOH / HN / NH / COOH⟩	94,95
⟨adamantane *⟩	82-85	**Heterocyclic sec. Amines**	
⟨F, OCOMe, OCOMe⟩	86	HN⟨(CH₂)₁₋₂⟩	96
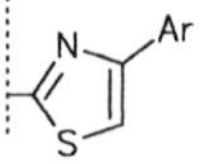 ⟨thiazole-Ar⟩	87	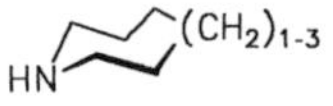 ⟨(CH₂)₁₋₃ ring, HN⟩	97-100
Bifunctional prim. Amines		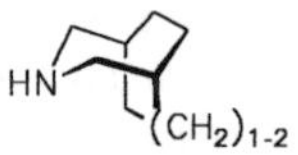 HN⟨(CH₂)₁₋₂ bicyclic⟩	97,98 101-103
H₂N⌒⌒O⌒⌒NH₂	88	HN⟨bicyclic⟩	104
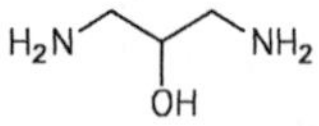 H₂N⌒⌒NH₂ (OH)	89	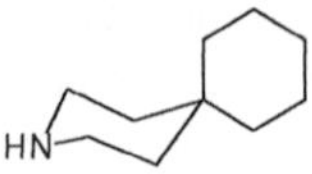 HN⟨spiro cyclohexane⟩	97
Acyclic sec. Amines		HN⟨pyrrolidine-Ar⟩	105
⟨Me, OH, HN, Me, Ph⟩	90,91	HN⟨tricyclic-Me⟩	106
⟨HN, Alkyl, OMe, OMe⟩	92	⟨(3-Indolyl), HN, OMe, OMe⟩	107
HN⟨Me, tetrahydrofuran⟩	93		

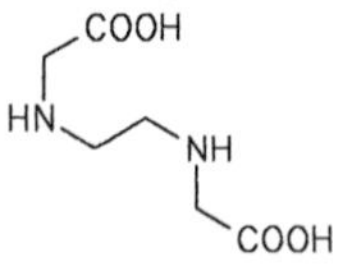

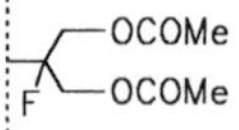

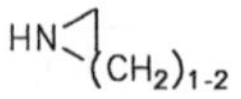

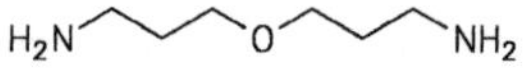

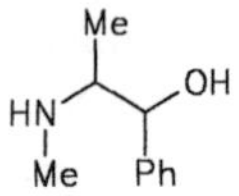

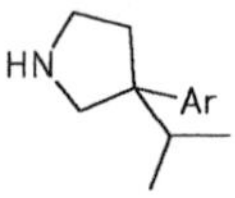

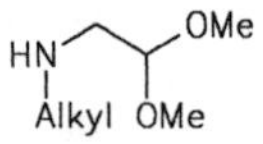

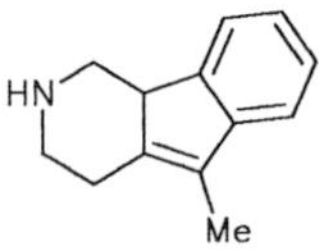

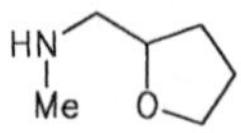

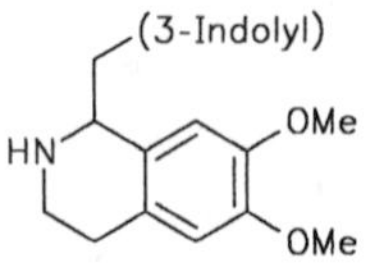

eliminable after Mannich synthesis, rather than to operate directly with ammonia or primary amine. In this case the Mannich base initially produced, having primary or secondary amino groups, will very likely compete for further aminomethylation reactions. For this purpose, benzylamines are mostly used, the benzyl group being easily removable by hydrogenolysis after Mannich reaction.[121-123] The steric hindrance of amine substituents in **15** (Fig. 13) is also exploited in order to prevent further reactions involving the amine hydrogen atom of Mannich base **16**. The bulky groups are then

Table 2

(continued)

Heteroc. sec. Amines (continued)		Heteroc. sec. Amines (continued)	
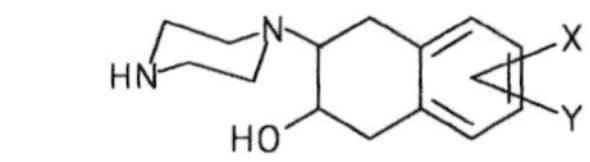	108	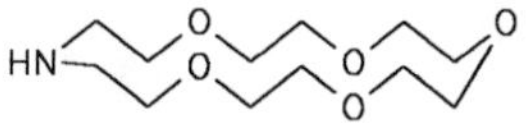	113
	109	X = B-Aryl: = Si:	114,115 54
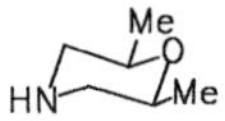	98,110	**Bifunctional Heteroc. sec. Amines**	
(R¹-R³ = H, Alkyl, Ph, fused rings)	111,112	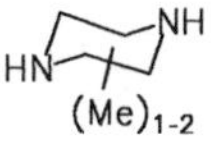 (Me)₁₋₂	116
			117,118
			113,119

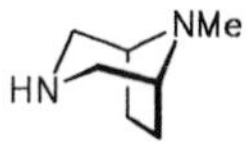

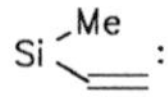

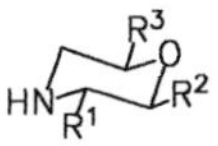

removed to give the aminoketone **17**, which otherwise would not be cleanly obtainable directly from acetophenone, ammonia, and formaldehyde.[124]

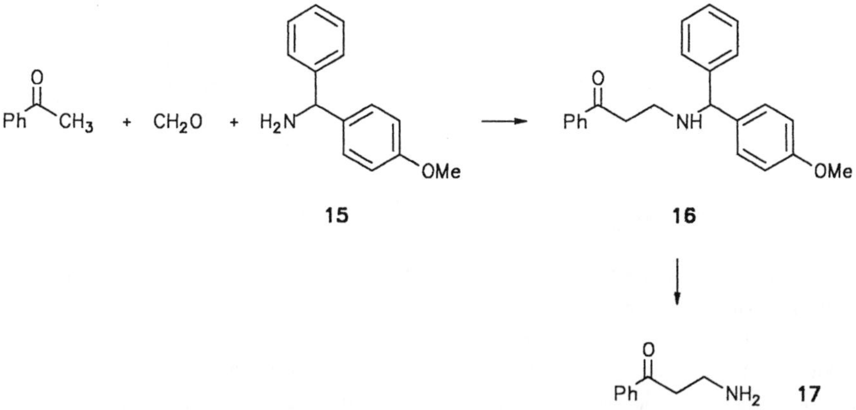

Fig. 13. Synthesis of primary ketoamine by using a hindered amine reagent.

Disilylamines, also, are useful starting materials for the preparation of Mannich bases having a primary amino group.[125] Disilylamine, by reaction with chloromethyl methylether, gives the aminomethylating reagent **18**, used for obtaining the Mannich base **19**, which is then easily converted into the corresponding primary amine **20** (Fig. 14).

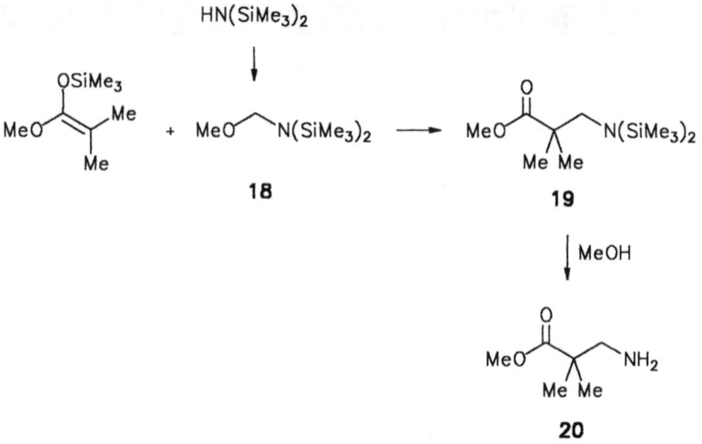

Fig. 14.   Synthesis of primary ketoamine by using a disilylamine reagent.

The amidomethylation reaction is suitable for the same purpose, as NH amides, which are well-known substrates of Mannich synthesis, can also behave like amine reagents. Thus, the resultant amidomethyl derivative **21** (Fig. 15) can be then subjected to hydrolysis by means of hydrazine to give the primary amine product.[126–128] By this method the 6-aminomethyl indole derivative **22** has been obtained.[126]

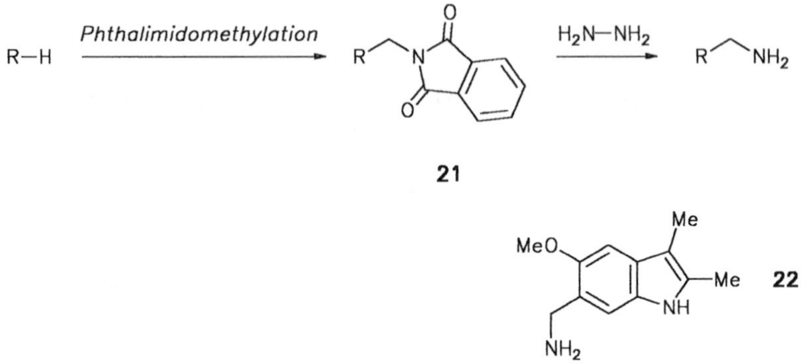

Fig. 15.   Synthesis of primary aminomethyl derivatives by means of phthalimidomethylation.

However, some drawbacks and limitations are connected with the use of particular amines in Mannich synthesis.

First of all, the steric hindrance may seriously affect yield and/or stability of the product, when bulky substituents are bound to the amine reagent.[129,130] Second, complications may arise, as we have seen before, with polyfunctional amines, mainly ammonia and primary amines, due to the possibility that the unreacted hydrogen atoms of the amine may undergo further reaction with formaldehyde, thus producing undesired by-products. Similarly, the use of secondary bifunctional amines, such as piperazine, always leads to a bis-Mannich base, due to reaction of both amino groups. Attempts to limit the reaction to only one amine function, as well as hydrolysis of the Mannich product obtained from aminomethylation of mono-N-acylpiperazines, invariably gives the disubstituted piperazine **23**.[131]

In addition, the polyfunctional reagents bearing two vicinal amino groups—for example, ethylenedi- amines[132] or *o*-arylendiamines—may produce cyclic

$$R \frown N \frown N \frown R \quad \mathbf{23}$$

derivatives upon reaction with formaldehyde. Other vicinal groups may be able to react similarly with formaldehyde, or with the aminomethyl derivative, even if they do not belong to functional groups characterizing the usual substrates capable of giving a Mannich reaction. A notable example is anthranilic acid,[133] which can be used as an amine reagent for the Mannich reaction on phenol (Fig. 16) only in the presence of two equivalents of formaldehyde, as one of them is required for reaction with the vicinal carboxyl group to give the unreactive cyclic intermediate **25**. The expected Mannich base can be therefore obtained only by hydrolysis of the dihydrooxazinone **24**.

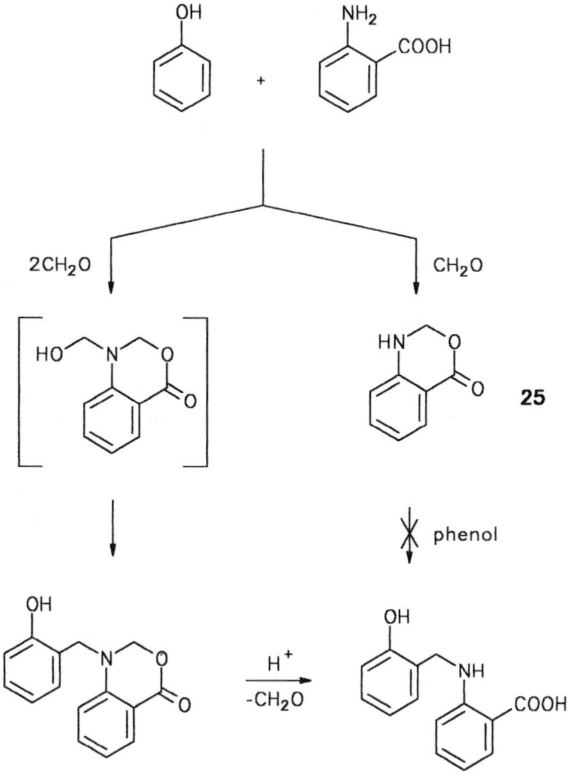

Fig. 16.   Mannich reaction of phenol, formaldehyde, and anthranilic acid.

A number of compounds containing the NH residue may behave as either substrates or amine reagents in the Mannich reaction, depending on the nature of the co-reagent. Among them, NH-heteroaromatic compounds and hydrazine[134] or hydroxylamine[135,136] derivatives may be considered as amine reagents when allowed to react with more acidic substrates, such as alkylthiols affording S-Mannich bases **26**.

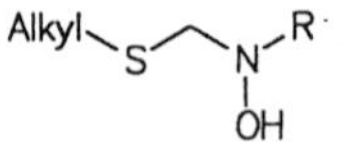

$$\text{Alkyl} \diagup_S \diagdown \diagup_N \diagdown^{R} \\ \qquad\quad \underset{OH}{|}$$

$$R^1 \diagdown_{NH} \overset{\overset{\displaystyle X}{\|}}{\diagdown} R^2 \quad (X = O, S)$$

26 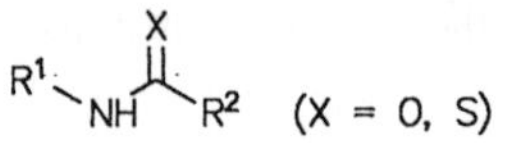   **27**

Compounds of type **27** and analogous derivatives such as sulfonamides[136] are frequently used in condensation reactions (defined as amidomethylations) with formaldehyde and substrates of the same type as those subjected to Mannich synthesis. Yet, the products obtained cannot be strictly considered as Mannich bases, due to their negligible basicity.

On the other hand, amidomethylations play a very important synthetic role, particularly in the production of urea resins and similar derivatives, and of bis-methylolamides used as cross-linking agents. The topic has been accurately investigated and reviewed by several authors as far as the chemistry and technological applications of the reaction are concerned.[137–140] These reactions are therefore mentioned throughout this book whenever analogies and correlations with aminomethylations can be usefully made; an example is seen in the case of **22**, when amidomethylation is used as a tool for obtaining Mannich bases with a primary aminomethyl group.

A.3   Aldehyde

Although formaldehyde is usually employed in Mannich aminomethylation, other aldehydes may be used, and in this case the reaction should be more generally defined as an aminoalkylation.

As is well known, commercial formaldehyde exists in three forms, all of them in a polymeric association, which readily produce the reagent molecule CH_2O.[141,142] In aqueous formaldehyde, or "formalin" (35 to 40% w/w), which is probably the most frequently used, the quantity of molecules in the monomeric state is less than 27%, the remainder consisting of oligomers having a polymerization degree below 10. Trioxymethylene, or 1,3,5-trioxane, is a solid cyclic trimer easily decomposed to formaldehyde by aqueous acid as well as by heating in organic solvent. Finally, paraformaldehyde is a crystalline linear polymer with a polymerization degree below 50, which becomes water soluble after depolymerization on heating.

In some circumstances, formaldehyde is replaced in Mannich synthesis by methylene dihalogenides CH_2XY (X = or $\neq$ Y = Cl, I)[143,144] or by ether derivatives such as chloromethylether.[125]

Aminoalkylation with aldehydes other than formaldehyde (**28**, R′ $\neq$ H in Fig. 17), or even ketones (e.g., acetone[145,146] or cyclopropylketone[147]), has been successfully carried out on several occasions. Table 3 lists a series of reactions reported in the literature, which have been performed in a way very similar to those performed using formaldehyde. Most of them are aminoalkylations leading to ring formation.

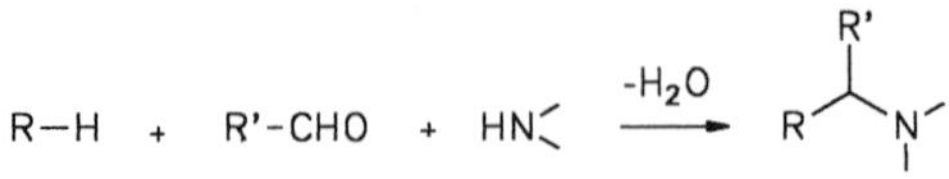

$$R{-}H \;+\; R'{-}CHO \;+\; HN\!\!\big\langle \;\; \xrightarrow{-H_2O} \;\; R \diagup \underset{|}{\overset{R'}{\diagup}} N \diagdown$$

28

Fig. 17.   Mannich reaction with aldehydes other than formaldehyde.

The presence of aldehyde in a Mannich reaction, as outlined above, makes it possible to connect substrate and amine moieties through a methylene group. However, when aldehydes other than formaldehyde are used, some other implications have to be

Table 3

Aminoalkylation Reactions

Type	Aldehyde R'-CHO (28)		
Substrate R-H	REFS for: R' = Alkyl	Aryl	Carboxy and others
C-Aminoalkylation			
Alkyl-ketones and Esters	60,64	60,64,148-151, 152 (c)	153,154
Phenols	155	156	157
Heterocyclics	158,159 (c)	159 (c),160 (c)	
N-Aminoalkylation			
Alkylene-bis-amines or *ortho*-Amino arylamides		161 (c),162 (c)	
Benzotriazole		61	163 [R' = CH(OEt)$_2$]
O-Aminoalkylation			
Aminoalcohols	164 (c)	145 (c)	
P-Aminoalkylation			
Diethylphosphite	122,146,165	122,165	

(c) = Aminoalkylation leading to cyclic products

considered. First, a reactivity decrease, due to enhanced steric requirements of the aminoalkylating agent related to the increased hindrance of the R' group in **28**, is to be expected. Electronic effects originated by R' may also contribute to the lowering of reactivity. Second, the stereochemistry of the resultant product is affected by the creation of a chiral center in the molecule; if other chiral centers were present in the substrate or amine, or if the reactive center of the substrate was prochiral, the formation of diastereomeric Mannich bases takes place (see Refs. 60 and 64). Such an opportunity has been elegantly exploited in the asymmetric synthesis of amino acids[56,166] and other compounds (Sec. C.3). Among the possibilities offered by several aldehydes, the use of glyoxylic acid and derivatives (**28**, R' = COOR″) is particularly interesting,[167] as it makes it possible to obtain α-aminoacids (Refs. 153, 154, and 157 in Table 3).

The selection of aldehyde reagent in Mannich synthesis may be also dictated by the presence in the aldehyde moiety of groups available for subsequent reactions, such as haloalkyl (chloromethyl, etc.) or acetal groups (Refs. 158 and 163 in Table 3).

It is finally worth mentioning that several different synthetic strategies (Secs. C.4,5 and Chap. IV, A) involve aldehydes other than formaldehyde in a number of reactions leading to ring formation; an example is given below, which provides evidence that these aldehydes may also be used to yield sufficiently stable preformed aminoalkylating agents having the imine structure.[168,169] Thus, imine **29**, obtained by reaction between aryl aldehyde and amine, gives intramolecular aminoalkylation producing the γ-piperidone precursor **30** (Fig. 18).

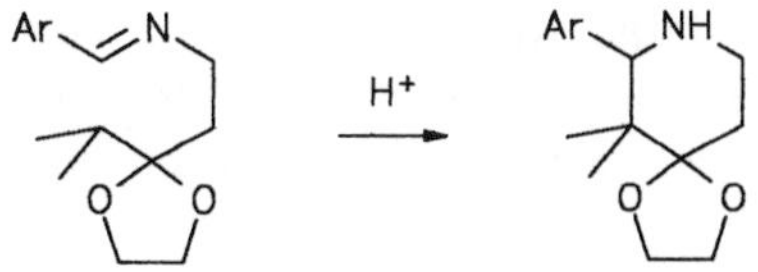

29　　　　**30**

Fig. 18. Aminomethylation reaction by imine reagent.

Interestingly, instead of the usual aliphatic ketone, the "substrate" undergoing aminoalkylation is in this case the corresponding ketal derivative.[168]

A.4 Preformed Aminomethylating Reagents

Although the usual practice still involves mixing the reactants of Mannich synthesis (substrate, amine, and aldehyde) with or without following a particular order of addition, the use of preformed aminomethylating reagents is becoming more and more frequent. Some of them, particularly those deriving from common amines, are available commercially.

Apart from rare exceptions, preformed reagents (Fig. 19) are intermediate derivatives having the structure of methyleneimmonium salts (**31**), X-aminomethyl compounds (**32**, X = heteroatom; see Table 5, below) and imines (**33**), produced by different routes and capable of readily giving Mannich reactions with the appropriate substrate.

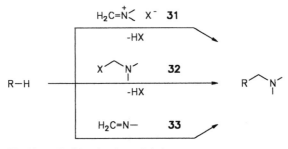

Fig. 19.   Preformed aminomethylating reagents.

Methyleneimmonium salts **31** are conveniently prepared from methylene-bis-amines,[170–174] formaldehyde-N,O-acetals,[175] or methylene halogenides.[176] Electrochemical reactions on various alkylamines[47,177,178] have also been used to generate methyleneimmonium salts, and trimethylamine N-oxide **34** (Fig. 20) is reported as a source of the corresponding methyleneimmonium salt by reaction with trifluoroacetic anhydride.[179]

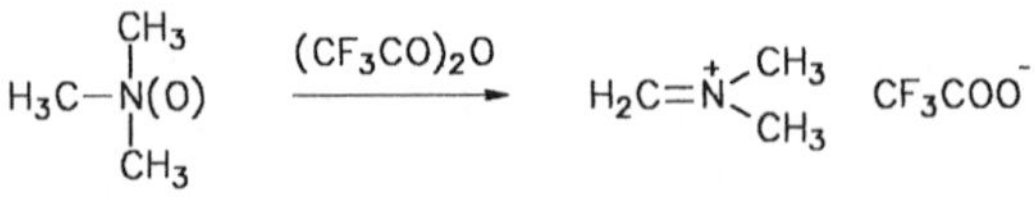

$$H_3C-\underset{\underset{CH_3}{|}}{\overset{\overset{CH_3}{|}}{N}}(O) \quad \xrightarrow{\;(CF_3CO)_2O\;} \quad H_2C=\overset{+}{N}\overset{CH_3}{\underset{CH_3}{\diagup}} \quad CF_3COO^-$$

34

Fig. 20.   Methyleneimmonium salt from trimethylamine N-oxide.

The oxidation of tertiary amines other than trimethylamine with organic perchlorates, aimed at producing salts **35**, has also been carried out.[180]

Analogous alkyleneimmonium salts of type **36**,[167,181] frequently involved in syntheses leading to ring formation, for example, in alkaloid chemistry, are treated in the appropriate chapters.

The nature of the counterion in methyleneimmonium salts derives from the preparation method, as in the abovementioned examples; however chloride or iodide anions are actually mostly reported. It is worth noting that morpholinomethyl fluoride, as distinct from the other halogenides, is a distillable liquid that does not exhibit salt behavior, due

Table 4
Aminomethylation Reactions with Preformed Methyleneimmonium Salts 31

Type Substrate	References
C-Aminomethylation	
Aliphatic Aldehydes	40,170
Alkylketones	40,63,65,170-172,180,183-185
Cyclic Ketones	39-42,57,59,170,171,173,179,186,187
Alkyl Esters and Lactones	39,40,59
Phenols	188-190
Heterocyclics	69,191-196
N-Activated Alkenes	176,197-200
N-Aminomethylation	
Cyanamide	201
Nitroamides	71
Hydrazine derivatives	202
P- and As-Aminomethylation	
Phosphine and Arsine derivatives	34,203

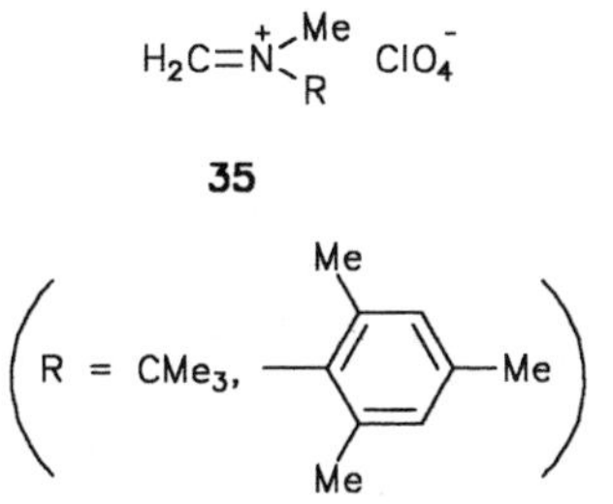

$$H_2C=\overset{+}{N}\underset{R}{\overset{Me}{\diagup}} \quad ClO_4^{-}$$

35

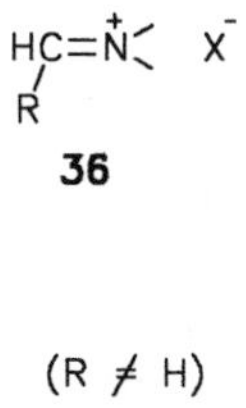

$$HC=\overset{+}{N}\overset{<}{\underset{R}{\diagdown}} \quad X^{-}$$

36

$$\left(R = CMe_3, \quad \underset{Me}{\overset{Me}{\diagdown}}\hspace{-0.3cm}-\hspace{-0.3cm}\diagdown\hspace{-0.3cm}-Me \right) \quad (R \neq H)$$

to the covalent character of the C—F bond. Nevertheless, its reactivity is comparable to that of the other terms of the halogenide series.[182]

The anion of methyleneimmonium salts may assume a remarkable relevance, as in some cases it affects the manipulation of the reagent and product as well as the chemo- or regioselectivity of the Mannich reaction.[40,170,171,173]

Methyleneimmonium salts have been successfully used in Mannich reactions on a high number of different substrates, as reported in Table 4. However, the reactions are frequently performed on substrates such as, in particular, alkyl-carbonyl derivatives, esters and lactones included, previously activated as described above (Refs. 39–42, 59, 65, 69, and 71 in Table 4).

In addition to their use in Mannich reactions, methyleneimmonium salts can be employed as aminomethylating agents of organometallic derivatives,[174,180,204,205] especially organo-tin compounds,[206–208] and as dienophiles in aqueous solution.[209]

The X-aminomethyl reagents **32a–h** are reported in Table 5. Several derivatives obtained from aldehydes other than formaldehyde are also included (see the starred references in the table), which thus allow the range of products achievable by Mannich

Table 5

X-Aminomethyl Derivatives 32 Used as Aminomethylating Reagents

Symmetrical Methylene-bis-Amines		Ref.[a]	Aminomethyl Alkyl Ethers		Ref.[a]
 R_2N ⌢ NR_2	**32a**	62,64*, 210,211, 212*,213	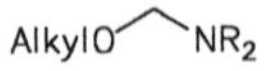 AlkylO ⌢ NR_2 **32d**		52,113, 210, 218-220
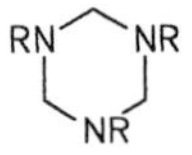 RN ⌢ NR (with NR ring)	**32b**	35,51,84 214-216	 O ⌢ NR (ring) **32e** and similar compds.		53,121*, 221*, 222-224
			Others		
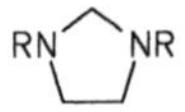 RN ⌢ NR (5-ring)	**32c**	217	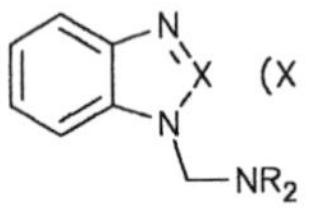 benzotriazole X (X = CH, N) —NR_2	**32f**	61*,225
			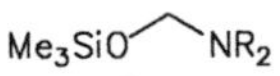 Me_3SiO ⌢ NR_2	**32g**	66,67
			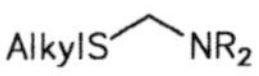 AlkylS ⌢ NR_2	**32h**	226

[a] Starred references include analogous compounds deriving from aldehydes other than formaldehyde

synthesis to be widened. Usually, the X group represents the N atom of a methylene-bis-amine or the ethereal O atom of N,O-acetals. Most of these reagents are cyclic derivatives, such as the hexahydrotriazines **32b**, that are actually trimers of the amino-methylating species, particularly suitable for introducing a secondary aminomethyl group when allowed to react with monofunctional substrates in the molar ratio 1:3.

Reagents **32c** and **32e** are imidazolidines and oxazolidines characterized by sufficient reactivity of the X—CH_2—N moiety for producing substrate aminomethylation. Indeed, when analogous, more stable cyclic derivatives are formed (e.g., **25**), the Mannich reaction does not take place. Compounds **32c** and **32e** are used for synthesizing Mannich bases bearing a further amino or hydroxy functional group in the amine moiety, as depicted in the example reported below for product **37** (Fig. 21).[53]

(X = N, O)   **37**

Fig. 21.   Mannich reaction with a cyclic preformed reagent.

In general, X-aminomethyl reagents have proved their efficacy with the most widely differing substrates. Besides the above reaction, it is worth noting the reactions of tropolones with methylene-bis-amines,[227] the S-[214,215] and P-aminomethylations[35] with hexahydrotriazines, as well as the great versatility of N,O-acetals derived from crown-ethers in several C- and N-aminomethylations.[113]

Certain types of Mannich bases (**32f–32h**) may also behave as aminomethylating reagents, in particular those able to undergo deaminomethylation more readily than deamination (Chap. II, A.2). Thus, the synthesis of Mannich bases carried out using the above compounds may be defined as *trans*-aminomethylation reactions. Cyanomethylamines **38** (Fig. 22), for example, which can be formally considered Mannich bases derived from hydrogen cyanide, give an intermediate methyleneimmonium salt, which is subsequently involved in intramolecular aminomethylation leading to cyclic Mannich derivatives.[228,229]

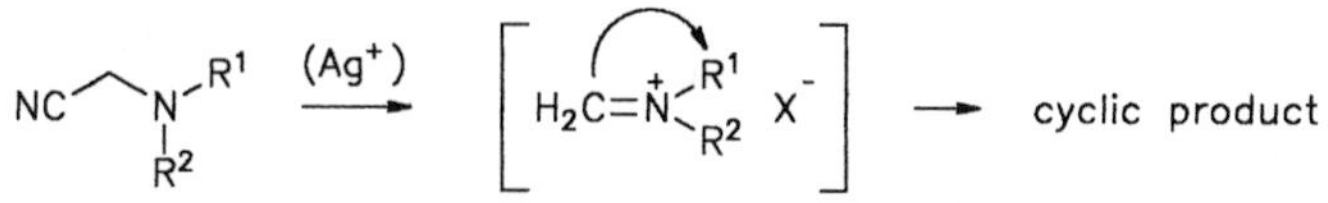

38

Fig. 22. *In situ* generation of methyleneimmonium reagents.

The above reaction is frequently adopted in alkaloid chemistry (Chap. IV, A.1).

The imine derivatives **33** (Fig. 19), used as preformed reagents, include methyleneimines produced by condensation of formaldehyde with particularly hindered amines (adamantanamine, etc.),[84,85,230] as well as the derivatives of aldehydes other than formaldehyde, such as the aryl derivatives **39**, which are most frequently used.[50,56,122,169]

Reagents **33** are employed, in particular, in order to obtain Mannich bases having a secondary amino group in the molecule, as in the case of the tetrahydropyridine **40**, which allows the preparation of Mannich bases having the interesting 2-substituted piperidine structure.[231] Aliphatic ketones are preferred substrates with this type of reagent, although examples of N-aminomethylations and P-aminoalkylations are reported in the literature.[85,122,230]

B Aminomethylation Reaction

B.1 Reaction Conditions

Formaldehyde is employed in Mannich reactions either as an aqueous solution ("formalin") or in the form of paraformaldehyde or trioxane. The amine reactant is used as a free base or hydrochloride. The most commonly adopted solvents for the reaction are alcohols (ethanol mainly, methanol, and isopropanol), water, or acetic acid. Aprotic solvents or neat conditions are also occasionally used.

The reaction is usually carried out by mixing the reactants in equimolar amounts. In some cases, however, amine and aldehyde are allowed to react first and then combined with the substrate. No general rule concerning the choice of reagents and reaction conditions exists; however, the very large number of experimental methods reported in the literature can be divided into a number of main groups, as summarized in Table 6, where the synthetic methods predominantly applied to each individual substrate class are listed. Only monofunctional reactants are taken into consideration here, as the presence of different reactive centers involves the possibility of selective reactions that have to be

Table 6

Commonly Adopted Reaction Conditions for Aminomethylation

Type Substrate	React. cond.	Reference
C-Aminomethylation		
Activated Alkyl Substrates:		
Alkyl Ketones, Aliphatic Aldehydes	A	232,233
Alkyl Esters, Lactones	E	39,40
Nitroalkanes	C	234
N-Heteroaromatic Alkyl Derivatives	A,B C	22,235,236 237
CH-Unsaturated Substrates:		
Alkenes	A	238
Alkynes	D	220,239
Hydrogen Cyanide	B,C	240
Aromatic, Heteroaromatic Substrates:		
Phenols	C	241,242
N-Heteroaryls	B	243,244
Furans, Tiophenes	C E	245 196
Ferrocenes	A	246,247
X-Aminomethylation		
Imines, Amides	B C	248 249,250
NH-Heterocyclics	C	251,252,225
Thiols	B,C	253,254
Sulfinic Acids	C	255
Phosphines	C	256
Phosphorous Acid Derivatives	B	257

Reaction Conditions

	A	B	C	D	E
Formaldehyde	Aq. soln. or Paraformald. or Trioxane	Aq. soln.	Aq. soln.	Same as A	Methylene immonium salt
Amine	Hydrochloride	Hydrochloride or free base in acetic ac.	Free base	Free base	
Solvent	Alcohol, Alcohol/Water (Acetic acid or neat)*		Alcohol/Water (Benzene, Dioxane)*	Dioxane	Anhydrous CH_3CN (CH_2Cl_2, CF_3COOH)*
pH/Catalyst	Acidic	Acidic	Neutral-Basic	Same as C + Cu Salts	Activated substrate

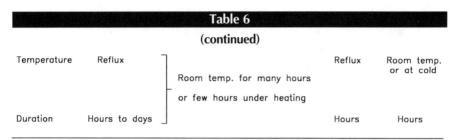

Table 6					
(continued)					
Temperature	Reflux			Reflux	Room temp. or at cold
		Room temp. for many hours			
		or few hours under heating			
Duration	Hours to days			Hours	Hours

* Less frequently adopted conditions

examined one by one. This point is better illustrated in Sec. C.1, which deals with chemoselectivity of the Mannich reaction. A more comprehensive survey of synthetic methods, including a description of the experimental conditions adopted, can be found in Ref. 258.

Table 6 confirms that the Mannich reaction is usually quite simple to achieve. The use of methyleneimmonium salts (condition E in Table 6), however, requires rather accurately controlled conditions, such as anhydrous solvents or a very low reaction temperature.[39,59] These reagents can be prepared *in situ* with good results[43,259] employing, for example, formaldehyde N,O-acetals in the presence of trimethylsilyl halide.[44] The most frequently used solvent is acetonitrile, followed by dichloromethane, but it is important to remember that the type of solvent[40,43,65,171,259] as well as the concentration[171] of reactants may be critical for reaction yield or selectivity.

The reaction conditions reported in Table 6 become increasingly mild on going from A (the most severe, particularly for pH, temperature, and reaction time) to E; these last, however, are characterized by the high concentration of the aminomethylating agent.

Aminomethylation of heteroatoms (X-aminomethylation in Table 6) is generally the easiest, whereas C-aminomethylation requires a great variety of experimental conditions. Indeed, the C—H moiety bound to a carboxyl group is among the less reactive systems, so that alkyl ketones require conditions A; alkyl esters, which would be unstable under such conditions, have to be allowed to react with methyleneimmonium salts. Alkenes, as well as ferrocenyl derivatives, are also barely reactive. By contrast, the nitro group of nitroalkanes behaves as a very good activator of aminomethylation, and phenols are readily aminomethylated.

The reaction solvent plays a particularly important role with methyleneimmonium salts, as observed in their preparation from N,O-acetals[175] and in the aza-Cope rearrangement,[260] frequently applied to the synthesis of alkaloids. However, under "classical" reaction conditions, with formaldehyde and amine, also, the selection of a suitable type of solvent can produce important consequences, as found in the aminomethylation of steroid hormones.[261] Moreover, the use of dry solvents may be a determining factor,[262] for instance in the reaction on a quite unusual substrate such as phthalic anhydride **41** (Fig. 23) with an oxazolidine reagent.[263] In anhydrous conditions, the oligomeric polyester **42**, deriving from aminomethylation of the carbonyl group with ring opening of anhydride, is in fact produced, whereas the monoester **43** is readily formed in the presence of water.

Acidic conditions are usually adopted in Mannich synthesis in order to favor the reaction, although in some cases, such as in the aminomethylation of thiols,[253] they only serve to enhance the stability of the product formed.

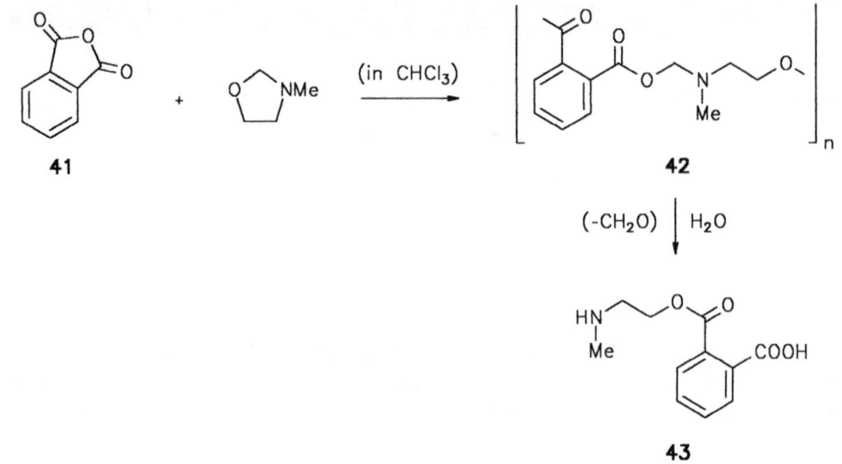

Fig. 23.   Aminomethylation of phthalic anhydride.

An interesting consideration relating to the influence of acidity in the reaction medium is derived from the comparison between the conditions adopted for the synthesis of the β-aminoketones **44** and **45**, and those required for the analogous derivatives that lack the carboxy group. Whereas the latter compounds are prepared under the severe conditions of type A (Table 6), both the syntheses of **44**, employing glyoxal as aldehyde reagent,[154] and that of **45**, which is prepared from 3-benzoyl propionic acid,[264] take place readily under mild conditions (type B or C).

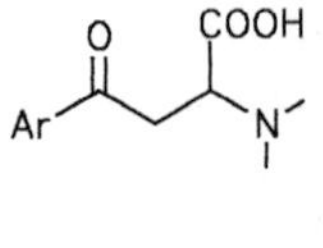

In both cases the carboxy group of the substrate is located in proximity to the reactive center. For this reason it is reasonable to assume that this group is involved in the reaction through stabilization of a methyleneimmonium form of the reagent in a way that favors the attack on the substrate. A similar interpretation may also hold for the remarkable catalytic action exhibited by amino acids in the hydroxymethylation of polyadenilic acid with formaldehyde.[265]

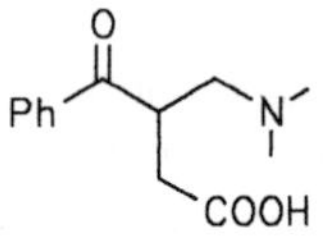

The presence of a catalyst is required in the aminomethylation of alkynes (column D in Table 6) and also in the reaction between imines **33** (Fig. 19) and aryl alkyl ketones.[169]

The common procedure for adding reactants consists either in simultaneously mixing all the chemical species involved in the reaction, or in allowing the amine and the aldehyde to react first and then adding the substrate. In some cases, however, the condensation of substrate and formaldehyde is carried out first in order to isolate the corresponding methylol derivative, which is subsequently submitted to react with the amine (X-methylation of amino derivatives; see Fig. 12): this is advantageous with several substrates, such as nitroalkanes,[266] ferrocenes,[267] sulfonic acids,[268] and phosphines.[269]

B.2   Mechanism

Since it would hardly be feasible to give a concise, detailed discussion of the mechanisms related to syntheses involving so many different classes of substrates, the main paths of the Mannich reaction are considered in only a general way here. The topic is treated in Refs. 258 and 270 and the mechanism of the analogous amidomethylation reaction is discussed in Refs. 79 and 139.

Since the Mannich reaction is a condensation involving three reactants (substrate, aldehyde, and amine), pathways **a** or **b** (Fig. 24) can be followed, if one excludes a rather unlikely trimolecular mechanism:

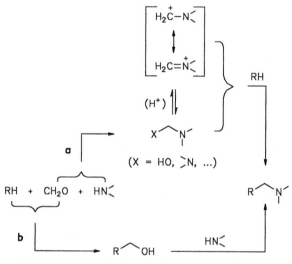

Fig. 24.   Mechanistic routes of the Mannich reaction.

In the case where formaldehyde reacts initially with the amine (path **a**), a condensation product having the structure of X-aminomethyl derivative or methyleneimmonium salt is formed, which is then able to attack the substrate RH. Alternatively (path **b**), a hydroxymethyl derivative is generated, which gives the Mannich base by reaction with the amine. Thus, the main questions concerning the reaction mechanism are

- The relative importance of paths **a** and **b**
- The nature of the aminomethylating species in path **a**
- The manner of attack by the reactive species on the substrate

The Relative Importance of Paths a and b

Path **a** is generally considered the preferred one, at least when the amine is the most nucleophilic species present in the reaction medium. Accordingly, preformed aminomethylating reagents are quite active, as is well known, in performing Mannich synthesis. Indeed, the presence of aminomethylating intermediates has been observed in some cases, mainly by spectroscopic methods. Moreover, the results of several investigations on specific matters favor the predominance of path **a** over path **b**. For instance, in the intramolecular aminomethylation of γ-carboxyglutamic acid with formaldehyde, a detailed study[271] has made it possible to exclude initial attack by formaldehyde on the substrate, and in the C-aminomethylation of indole with glutamic derivatives, it has been confirmed[272] that the hydroxymethylated substrate 3-hydroxymethylindole is unable to give a condensation reaction with the amine reactant. Analogously, in the synthesis of N-aminomethyl-2-pyrrolidone **46** (Fig. 25), the reaction involving N-hydroxymethyl pyrrolidone occurs much more slowly than that of simple pyrrolidone, due to the need for demethylolation of the former reactant.[273]

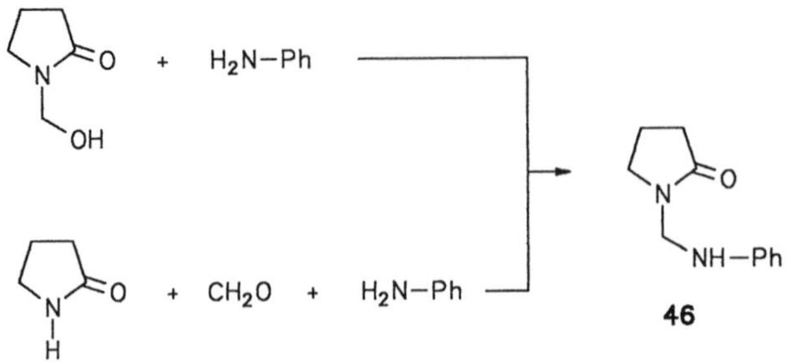

Fig. 25.   Synthesis of the N-Mannich base of 2-pyrrolidone.

On the other hand, some experimental results are in agreement with the predominance of path **b** (Fig. 24); they are due in particular to the aptitude of a very large number of substrates to react readily with aldehydes (see references reported for **12** and Ref. 274). Indeed, several successful syntheses of Mannich bases have been carried out starting from the hydroxymethyl derivative of the substrate, as reported for C-Mannich bases obtained from ferrocenyl derivatives,[267,275] nitroalkanes,[266,276] and hydrogen cyanide[77] as well as for N-, S-, P-Mannich bases of benzimidazoles,[277] sulfonic acids,[268] phosphines,[269] etc.

In addition, some reactions on α-nitroso ketones[278] and 3-mercapto coumarin,[279] as well as the aminoalkylation of benzotriazole with aldehydes and arylamines,[280] can be interpreted only on the basis of aldehyde attack on the substrate as the first reaction step.

Although path **b** is assumed not to represent a mechanism of general validity, experimental findings induce us to look upon Mannich synthesis as the result of a complex series of equilibria, related to the nature of the reactants and to the reaction conditions, which determine the predominant reaction pathway. In this connection, the preceding comments on the synthesis of **46**, in conjunction with some reported studies of the hydrolysis of N-hydroxymethyl phthalimide[281] and the following example (Fig. 26) on N-hydroxymethyl pyrrole **47**,[282] significantly indicate the occurrence of equilibria affecting type and yield of product given by the reaction:

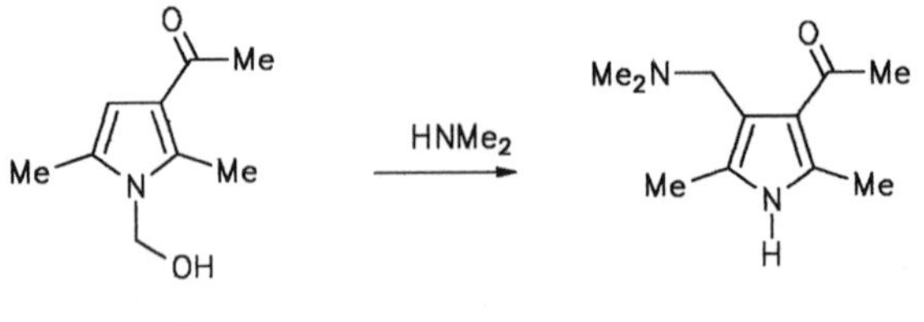

47

Fig. 26.   C-Aminomethyl derivative of pyrrole starting from the corresponding N-hydroxymethyl compound and amine.

The reported observations of the cleavage (deaminomethylation and deamination; see Chap. II, A.1 and 2) as well as the *trans*-aminomethylation reactions (see **32f–h** in Table 5 and Ref. 283) of Mannich bases also give useful support to this argument.

Finally, some important studies of the influence of pH in the reaction between formaldehyde and amine in biological systems,[284,285] in the aminomethylation of poly-acrylamide,[286] and in the use of aldehydes other than formaldehyde[287,288] lead to the conclusion that under acidic conditions, aldehyde attack by the amine is the rate-deter-mining step. Under neutral or basic conditions the rate-determining step is hydroxyl elimination from the methylolamine $HO—CH_2—NR_2$ with formation of the methyl-eneimmonium cation.[285,287] Polarographic methods demonstrate[289] that the maximum concentration of this last species occurs at a pH near 10.

The Aminomethylating Species in Path a

Knowledge of the actual aminomethylating agent resulting from the equilibrium mixture of Fig. 24, constituted mainly by the methyleneimmonium ion, methylene-bis-amine, hydroxymethylamine, and in some cases, the ether derivative of this last species, is an aspect of the Mannich reaction that has aroused much scientific interest.[221,284–287,290–292]

Because the methyleneimmonium ion is considered to be the most reactive species present in the system, it is usually invoked as the actual reagent (see references related to **31** in Sec. A.4). This ion is stabilized by resonance (Fig. 24), and its presence has been ascertained[270] in the reaction medium under acidic conditions and, of course, when preformed reagents **31** are employed. The immonium ion derives primarily from meth-ylene-bis-amine and secondarily from hydroxymethylamine, both of which are formed by the initial attack of formaldehyde on the amine. It can also be produced by preformed reagents of the X-aminomethyl type (compounds **32** in Sec. A.4) in an acidic aqueous solution as well as in an organic medium by reaction with trifluoroacetic acid.[221,291,292] However, strongly acidic conditions are not strictly required for the presence of the immonium ion,[285,287] as it can also be generated by catalytic amounts of chlorosilanes, zinc chloride, etc.[43,52,121,175,210] The bond undergoing cleavage in the precursor (Fig. 27) has been particularly investigated in open chain[175] and cyclic[221,291,292] N,O-formalde-hyde-acetals, as only cleavage of type **a** affords the immonium cation leading to the Mannich derivative.

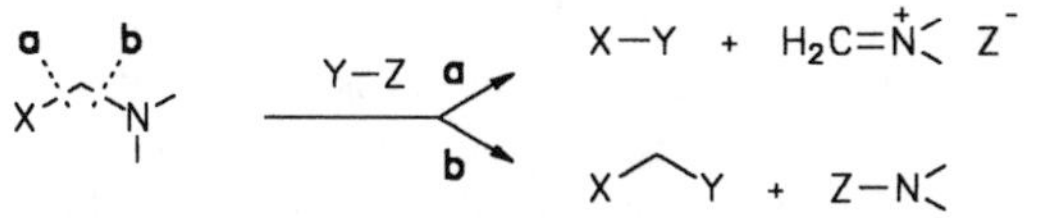

Fig. 27.   Bond cleavage required for the formation of methyleneimmonium ion.

Aminomethylation with primary amines (Fig. 28) also involves the alkyleneim-monium ion **49** as the reactive species. This is produced, for instance, from imine **48**,[50] or hexahydrotriazine derivatives,[49,51] by reaction with trimethylsilyl trifluoromethane sulfonate:

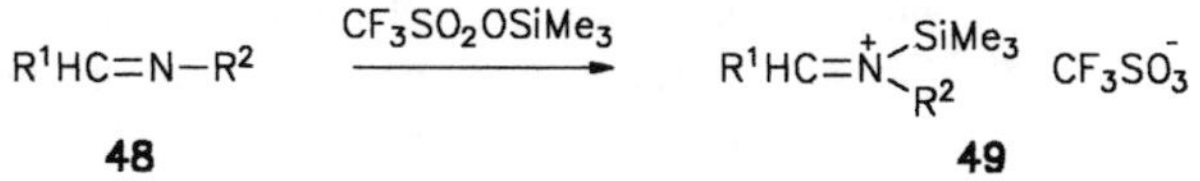

Fig. 28.   Preparation of silylated alkyleneimmonium salts.

In this case, the trimethylsilyl group linked to the nitrogen atom allows the imine reagent to assume the required cationic nature for giving the Mannich reaction. The imine group, however, also can be activated by protonation.[293]

Other cationic structures have been proposed for the aminomethylating species. In the reactions of N,O-acetals catalyzed by trimethylsilyl derivatives Me₃SiX, the reactive agent is claimed[44] to be the oxonium cation **50** (see also Refs. 210 and 211). Moreover, the intermediate hydrochloride **51** could be formed by the action of anhydrous hydrogen chloride[214] on hexahydrotriazines **32b**, although more recent studies[294] suggest the formation, under the same conditions, of equimolecular amounts of methyleneimmonium chloride and the halogenated salt **52**.

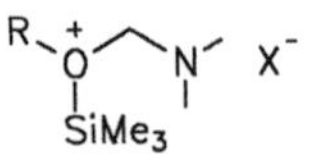

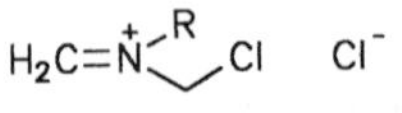

In the alkaline medium, the reagent is postulated to be hydroxymethylamine or methylene-bis-amine,[154,189,286,295–301] although the participation of the N,O-acetal (Alkyl-O—CH₂—N<), when the reaction is carried out in alcoholic solvents, cannot be excluded, and relevant amounts of methyleneimmonium cation are observed.[289] The ^{13}C-NMR measurements[286] indicate that the methylolamine concentration is maximum when the molar ratio amine/aldehyde is *ca.* 1, whereas the formation of methylene-bis-amine is favored when this ratio increases.

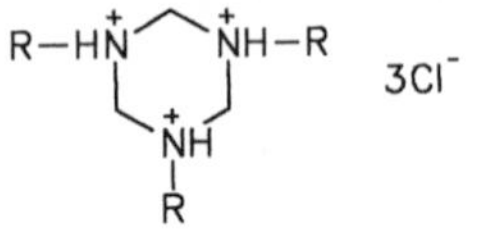

The reactants present in the equilibrium mixture, however, do not display the same reactivity; indeed, studies carried out in an aprotic medium with acetylenic substrates[296] and in an aqueous medium with polyacrylamide[286] show that unlike methylolamine, methylene-bis-amine exhibits poor reactivity or none at all toward the above substrates. By contrast, methylene-bis-amine is an active intermediate in the aminomethylation of alkylphenols.[298]

The Attack by the Reactive Species on the Substrate

The manner of attack on the substrate by the reactive species appears to be strictly related to the nature of the aminomethylating reagent. In an acidic medium, or when a consistent amount of methyleneimmonium ion is present, the electrophilic reaction by the reagent on the substrate takes place (**53**, Fig. 29).[258,300] For instance, in C-aminomethylation of the Me group of methyl-nitrooxazoles, kinetically investigated in hydroalcoholic solution, an S_{E2} attack by the methyleneimmonium cation upon the substrate has been postulated.[301]

The enolic form of alkylcarbonyl substrates is usually considered when these substrates undergo attack by the reagent.[225] The use of enolates and enolethers is thoroughly described in Sec. A.1 (see also Refs. 49, 51, and 259).

When the aminomethylating species is an X-aminomethyl reagent, a mechanism involving the hydrogen-bonded complex **54** is generally proposed for carbonyl,[154,288,297] phenolic,[189] and other[295,299] substrates. Alternatively, an S_{N2} mechanism, involving attack by the carbanion derived from the substrate, has been hypothesized.[154,288,297,299,301]

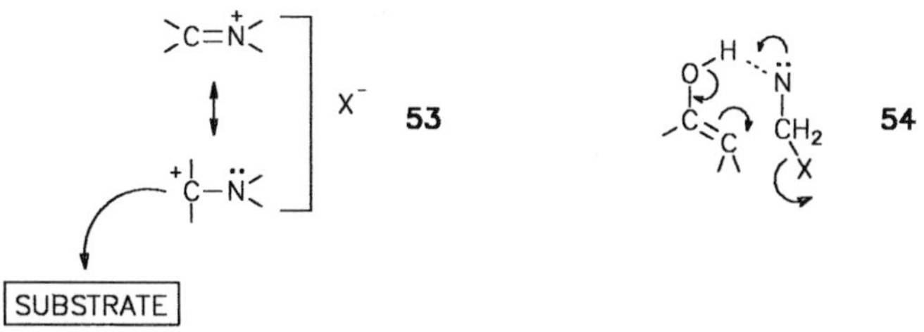

Fig. 29. Types of attack on the substrate by aminomethylating agents.

In intermediate **54** the substrate is considered to be in the enolic form or in analogous tautomeric forms. This assumption usefully clarifies the role played by vicinal groups capable of giving hydrogen bonding with the reagent.[171,189,295,298,299,302] In particular, the preference for ortho attack observed in the Mannich reaction on phenolic substrates can be explained according to this hypothesis.

Enolization of the substrate can be rate determining, as it occurs at a rate comparable with the reagent attack. A demonstration comes from studies on the regioselective synthesis of Mannich bases from nonsymmetric dialkylketones,[171] in which the possibility of kinetic or thermodynamic control of the reaction is thoroughly discussed and the relevance of steric hindrance (see also Refs. 129, 297, and 303) as regards the reaction pathway is considered. A correct choice of reagent concentration, reaction temperature and duration, as well as an effective activation of the substrate as enolate or enolether, can be usefully applied to the solution of problems of chemo- and regioselectivity.[40,42,259,281]

B.3 By-Products

An important aspect of the studies of Mannich reactions is the accurate optimization of reaction conditions so as to minimize the formation of undesired derivatives that could cause difficulties in purification of the final product. In this context, a thorough knowledge of the compounds accompanying the main product is particularly relevant to the manufacture of industrial products such as, for example, pharmaceuticals, where the availability of efficient methods of analysis and dosage of contaminants is of fundamental importance.[304]

Apart from the by-products originated by particular conditions occurring in individual syntheses, the predictable main reactions are reported in Fig. 30, where paths **a–d** represent the routes leading to products other than the expected Mannich base.

Route **a** represents the case of particularly high stability of the aminomethylating reagent (methylene-bis-amine, N,O-acetal derivatives, etc.), which may in turn derive from poor reactivity of the substrate. This actually results in the formation of by-products exclusively from the reagent, without any involvement of the substrate.[305] It is moreover worth mentioning that the reducing properties of formaldehyde may modify the aminomethylating agent, through the formation, for example, of an N-methyl derivative, thus preventing the possibility of further reaction.[306,307] In addition, some undesired reactions may take place during activation of the reagent, as indicated in Fig. 27. When particular amines are employed, such as ethylenediamine[132] or bis-(2-haloethyl)amine, widely used in the synthesis of cytostatic drugs,[308,309] the formation of cyclic derivatives of type **56** has been observed.

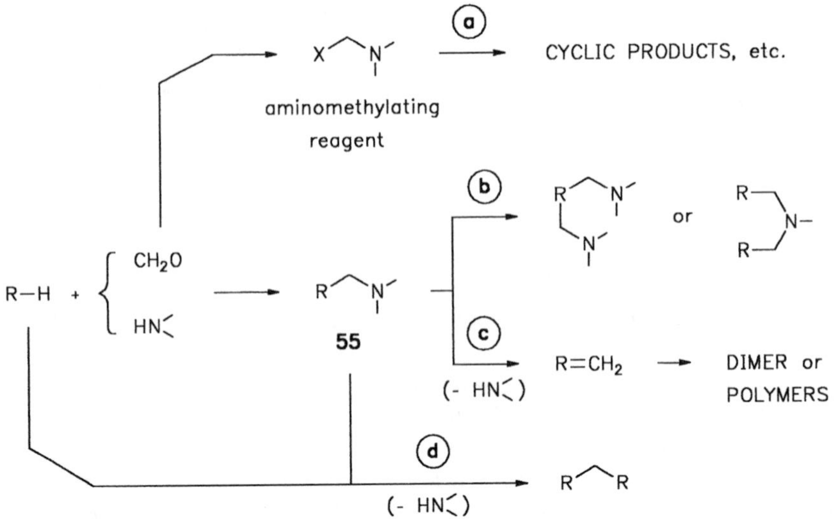

Fig. 30.   By-products accompanying Mannich synthesis.

Route **b** in Fig. 30 represents a typical reaction of the freshly formed Mannich base, consisting of a further aminomethylation, with the Mannich base behaving either as substrate or as amine, depending on the functionalities present in the starting products.[310]

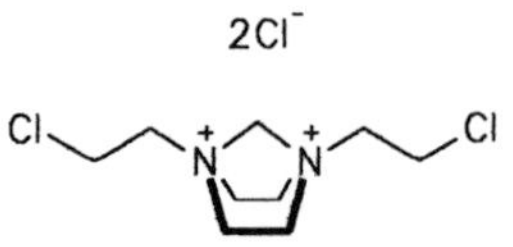

This drawback is not easily surmountable in certain cases, even if an accurate stoichiometric dosage of the reactants is made. Some possible solutions to this problem can be found in Sec. A.2 and A.4, dealing with amine and preformed reagents, respectively. The synthesis of Mannich bases containing a secondary amino group may be also elegantly performed by allowing the primary amine to react as an oxalic acid salt.[311]

When by-products may be originated as a consequence of multiple functionalities in the substrate (see, e.g., Refs. 211 and 312 to 314), the knowledge of chemo- and regioselectivity of the aminomethylation reaction is convenient. This aspect will be treated in Secs. C.1 and C.2.

Details of products derived from the further reaction of Mannich bases (paths **b**–**d**) mostly refer to aminomethylated alkyl carbonyl compounds, particularly alkyl ketones. In Fig. 31 a comprehensive survey of the possible by-products deriving from a typical Mannich base, the β-aminoketone **57**, is depicted.

An excessively high tendency to deaminomethylation can lead to unsaturated products **58** and **62**.[43,44,150,283] It may in fact be sufficient in some cases to attempt to obtain the free base from the hydrochloride salt for producing vinyl ketone **58**.[184] This last compound, which is known to behave at the same time as a diene and a dienophile, frequently dimerizes,[173,198,315,316] thus affording the dihydropyranyl derivative **60**, or polymerizes. When the pristine substrate is very reactive toward Michael-type additions,[173,177,198,199,262,317,318] by-products having structure **59**, that is, the methylene-bis-derivative of the substrate, are formed. In other cases, more than one aminomethyl group may be introduced into the same substrate,[43,44,319] thus producing **61**, which in turn may undergo deamination[44,183,320] with formation of **62**.

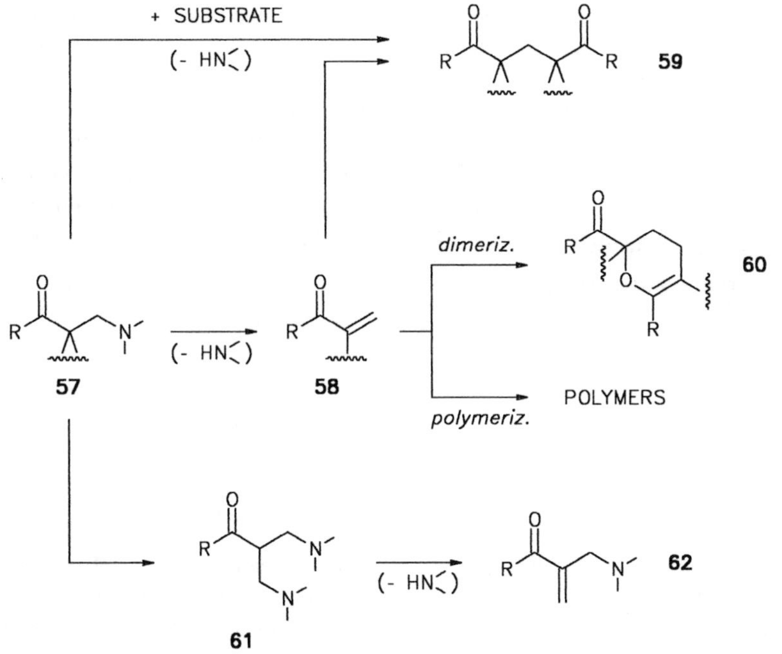

Fig. 31. By-products in the synthesis of β-aminoketones.

Only a few examples of side reactions involving other types of Mannich bases are reported in the literature. In particular, the formation of methylene-bis-derivatives of the substrate (path **d** in Fig. 30) has been observed in phenolic,[321,322] C-heterocyclic,[211,323] and alkylsulfone[324] Mannich bases.

Finally, a pronounced tendency of Mannich bases to undergo cyclization (Sec. C.4) may also produce additional derivatives.[192,220,318,325]

C Synthesis and Chemical Structure of Mannich Bases

The synthetic conditions that make it possible to obtain a Mannich base, starting from a substrate, amine, and aldehyde, have been highlighted in the previous section, with particular attention to the possibility of linking together, through a methylene bridge, various substrate and amine moieties. In the present section the structural features characterizing the molecule of the Mannich base are discussed from several viewpoints, ranging from chemo-, regio-, or stereoselectivity of the reaction to the use of polyfunctional reagents affording cyclic products and to the possibility of concurrent reactions leading to relevant modifications of substrate and/or amine moiety. Stereochemical aspects as well as the presence of tautomeric equilibria in Mannich bases are also considered. An exhaustive treatment of the relationships between structure and properties of Mannich bases is thoroughly developed in Chap. V.

C.1 Chemoselectivity

When chemically different reactive sites capable of reacting independently with the aminomethylating agent are present in the substrate molecule (Table 7), the selectivity of the reaction is substantially determined by the relative reactivity of each reactive center. This also holds for substrates possessing more than one functional group of the same type, as in bis-alkylcarbonyl derivatives **63–66**, exhibiting increasing mutual distance between the activating groups. The α-diketones **63** are the least selective, as they show a tendency to give bis-aminomethylation in the α,α' positions with respect to carbonyls even in the presence of an equimolecular amount of aminomethylating agent.[326,327] A mono-aminomethylated product that derives from a reaction on the less substituted α carbon atom has in fact been isolated in low yield in only one case.

R^1 = Me, Ph	R^1 = Me, Ph, OR	R^1, R^2 = H, alkyl	
R^2, R^3 = H or cycle	R^2, R^3 = H, alkyl		

| **63** | **64** | **65** | **66** |

When the activating carbonyl groups have the 1–3 relative position (**64**), a consistent activation of the C-2 atom occurs, this position being, with rare exceptions,[328] the sole site of attack. The same result is observed in analogous substrates having one of the carbonyl groups replaced by SO_2, which in the absence of other activating groups, is known to be unable to promote aminomethylation.[329,330]

When the carbonyl groups are removed further apart, as in ketoesters **65**, the reaction product derives from aminomethylation on the less alkyl-substituted carbon atom in the α position with respect to the keto group.[331–333] The carboxyl is in fact unable to activate the adjacent carbon atom under the usual reaction conditions.

The case of diketone **66** is more difficult to interpret, as this substrate undergoes a Mannich reaction initially on the acetyl group at position 4 of the 1-naphthol ring and is then further aminomethylated at the acetyl group in position 2.[334]

The examples listed in Table 7 concern the selective aminomethylation of alkyl-ketopyrroles, N-propargyl anilines, thiophene amides, etc., which are better treated on the basis of the relative reactivities induced by the activating groups present in each individual substrate. Thus, the alkyl keto group readily reacts under acidic conditions, phenols in neutral or basic medium, alkynes in the presence of copper salts, etc. according to the indications reported in Table 6. In some cases,[184,326,334,342] the adoption of suitable reaction conditions and an accurate stoichiometric dosage of the reagent make it possible to aminomethylate with good selectivity only one of the two active sites in the substrate. A second aminomethyl group can be then introduced by using an excess of reagent.

The selective aminomethylation of either reactive center has been also achieved.[337,338,340,342] In tetracycline it has been shown[347] that aminomethylation occurs at the amide N atom and not at the phenolic D ring, in accordance with the higher reactivity of the amide N—H group as compared with that of the C atom in ortho position to the phenolic hydroxy group.[352]

A second group of substrates displaying chemoselectivity is schematically listed in Table 8 and includes compounds having active sites capable of mutual interaction in the course of the Mannich reaction, as they are constituted by functional moieties, usually unsaturated heterocycles, in which the possibility of equilibrium between different forms of the substrate (tautomers, etc.) may lead to correspondingly different products.

Table 7

Chemoselective Aminomethylation Reactions on Polyfunctional Substrates[a]

Polyfunctional Substrate HX—R—ZH		Preferentially aminomethylated group	References
—XH	—ZH		
    O ‖ —C—CH —	OH (phenol ring)	XH XH or ZH ZH	184,334-336 337,338 184,339
	—C≡CH ←	XH or ZH	340
	pyrrole ring N H (or Me)	XH ZH	341,342 92,282
	 furan/thiophene ring X (X = O, S)	XH	232,343
	NH ← COR	XH	344
OH (phenol ring)	NH ← COMe	XH ZH	345 346,347
	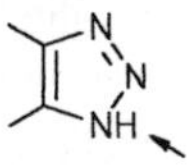 triazole ring N N N NH	XH	241
—C≡CH ←	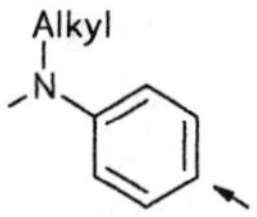 Alkyl N (aniline ring)	XH or ZH	239
	>P(O)H ←	ZH	348

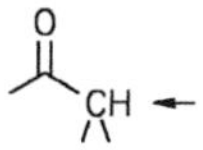

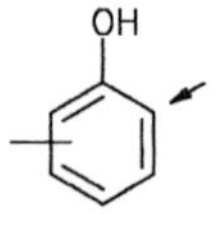

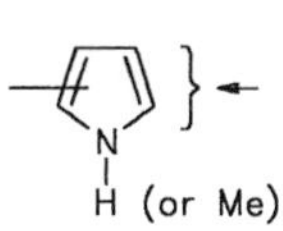

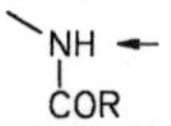

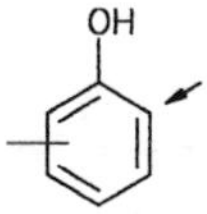

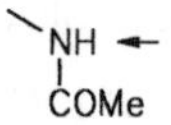

Typical examples of this class of reaction are the C- or N-aminomethylations of pyrrole and indole derivatives, the N- or S-aminomethylation of thioamides, etc. Aminomethylated products may also easily undergo interconversions, as described below in Sec. C.6.

Alkyl vinyl ketones having the general structure **67** (X = H, Table 8) exhibit different reactive sites, depending on the nature of the substituents and the steric requirements. Vinyl ketobases (for the analogous styryl derivatives see Refs. 353 and 354) are obtained by attack of type A on acyclic or exocyclic unsaturated substrates in acidic medium, or with preformed methyleneimmonium salts. Vinylogous Mannich bases are

Table 7		
(continued)		

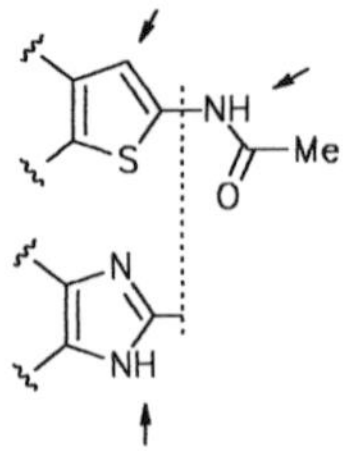

XH   349,350

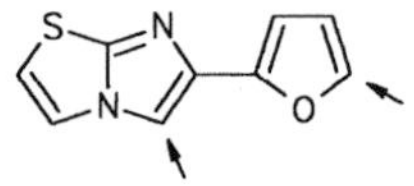

XH   351

[a] Arrow indicates the position of attack

produced under the same conditions from cyclic α,β-unsaturated ketones (steroid derivatives included) through attack of type C. Other vinylogous Mannich bases are reported.[373,374] When aminomethylation is carried out with free amine and formaldehyde, the reaction involves the unsaturated C-α atom (type B attack). This occurs with acyclic and cyclic substrates, the "push-pull olefines", bearing an NH group on the unsaturated C-β atom.[200,356,357] When the substrate bears an N,N-dimethylamino group, a vinylogous Mannich base is produced (type C attack).[358]

Heterocyclic substrates, such as pyrrole and imidazole derivatives **68**, may undergo selective Mannich reactions. C-Aminomethylation is favored by acidic conditions, whereas N-Mannich bases are produced when free amine and formaldehyde, or N,O-acetals in anhydrous solvents, are employed. Heterocyclic N-Mannich bases, however, are not particularly stable and may therefore behave as aminomethylation agents (see, e.g., **32f**, Table 5). Thus, the corresponding heterocyclic C-Mannich bases can be obtained when reaction duration and temperature are increased.[282,375] In aminomethylation of 5-phenyl-hydantoin **69**, the site of attack by the second entering aminomethyl group is affected by the nature of the reacting amine as well as, probably, by the extent of the basicity of the medium. With morpholine, the second aminomethyl group is linked to the N-1 atom, whereas with piperidine it is linked to the C-5 position.

The analogous six-membered heterocyclic substrates also exhibit a complex behavior. Aminomethylation of the uracils **71** occurs on the N atom in position 3, although the reaction with formaldehyde alone shows an equilibrium constant for hydroxymethylation in position 1 which is about twice as much as the value obtained for the reaction at position 3.[376] By contrast, in barbituric acid derivatives **72** (X = O), position 5 is the most reactive when unsubstituted. Otherwise, the reaction is directed toward the 1 or the 3 position.[377–379] When present, the imine NH group (**72**, X = NH) is the last moiety to undergo attack.

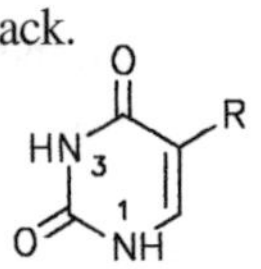

71 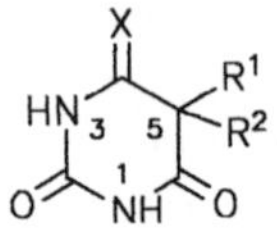   **72**   X = O, NH; R[1], R[2] = H, alkyl

Table 8

Chemoselective Aminomethylation Reactions on Substrates Having Interacting Reactive Sites[a]

Substrate	Selective attack type	Reference
67 (X = H, NHR, NR$_2$, rings included)	A	355
	B	198,200,356-360
	C	41,358,361,362
triazole structure	A	192
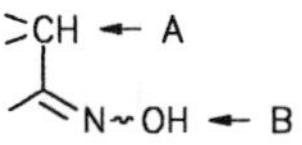 >CH ← A / =N~OH ← B	A or B	363
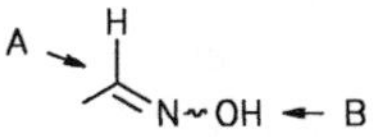 A → structure / =N~OH ← B	A	364
68 (X = CH, N)	A	303
	A or B	365,366
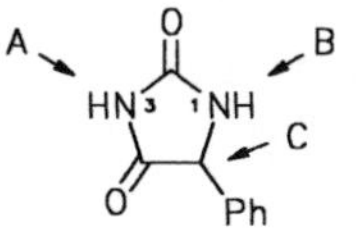 **69**	A,B or A,C	367
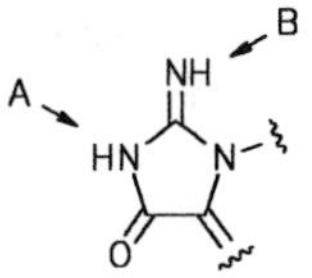 structure	B	368
70	A	369-371
	A or B	371,372

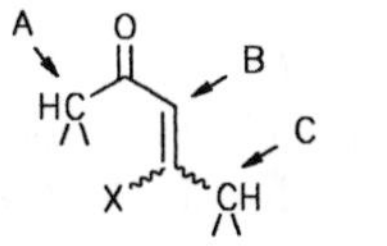

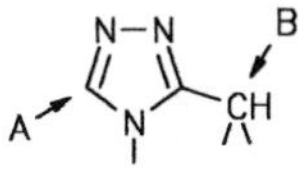

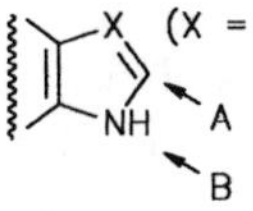

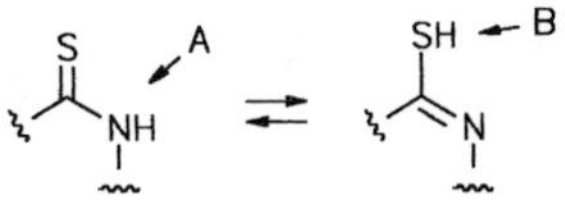

[a] Arrow indicates the position of attack

The behavior of thioamides (**70**, Table 8) is influenced by tautomeric effects. In the case of bis-aminomethylation of 2-imidazolidinothione,[371,372] for example, the mobile H atom can be selectively substituted in both the forms by reaction with secondary dialkylamines, or primary arylamines, to give, respectively, N,N- (A attack) or N,S- (B attack) bis-Mannich bases.

C.2 Regioselectivity

The possibility of regioselective reaction arises when aminomethylation may occur on different sites of a substrate, such as, for example, the unequivalent α or α' positions of a nonsymmetrical ketone, the ortho or para positions of phenol, the different sites of a heterocyclic compound, and so on.

Aliphatic Ketones and Other Activated Alkyl Substrates

Several studies are reported of the aminomethylation of dialkylketones of the type $R^1R^2CHCOCHR^3R^4$ (R^1, $R^2 \neq R^3$, R^4). Both open-chain and cyclic derivatives offer two unequivalent positions suitable to Mannich reaction. These are indicated by A (less substituted C atom) and B (more substituted C atom) in Table 9. Accurately made investigations[40,41,171,380] concerning the preferred site of reaction lead to the conclusion that enolization of a carbonyl group can play a determining role as it may occur at a comparable rate with respect to attack by the reagent.[171] The possibility of kinetic or thermodynamic control of the reaction has been also extensively discussed and the relevance of the steric factors affecting the reaction path highlighted.[40,380]

The results[258] (Table 9) indicate that the Mannich bases deriving from attack on the less substituted C atom are obtained more frequently, the B attack being preferred only under particularly selected reaction conditions, such as the use of methyleneimmonium salts in dilute trifluoroacetic acid solution for very long reaction times[171] or with ethanolic primary arylamine hydrochlorides.[381] However, few exceptions to the prevailing behavior have been observed.[48,335,381]

Besides the carbonyl, other residues, such as N-heteroaromatic rings bearing methyl or methylene groups, are also capable of activating the alkyl C—H moieties of the substrate. Dimethyl-nitrooxazine **73** reacts only on the methyl group in the α position with respect to the O atom.[393] The triazole-benzodiazepine **74** behaves similarly, and it has been found[259] that mono- or bis-aminomethylation as well as aminomethylation of the methyl and the methylene group are notably affected by the type of solvent used and even by the order of the addition of reactants.

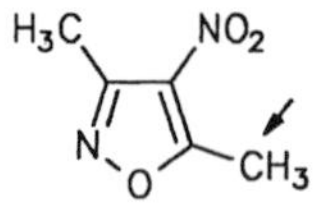

$R^1 = Ar$
$R^2 = H$, Halo, etc.

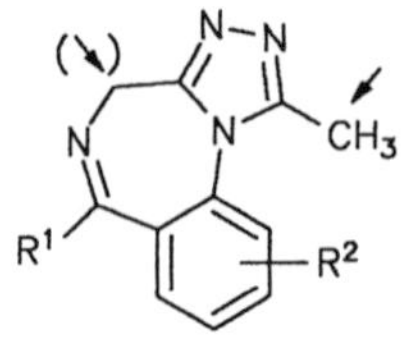

73   **74**

Phenols and NH-Activated Aryl Substrates

When the aromatic ring is suitably activated by electron-donor substituents (NR_2, OH, or alkoxy groups) for electrophilic attack, it can be easily subjected to Mannich reaction.

Table 9

Regioselective Aminomethylation Reactions on Nonsymmetrical Dialkyl Ketones

Dialkyl ketone	Predominant attack type	Reference
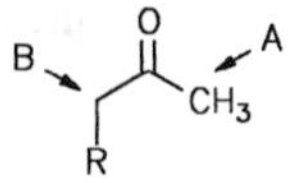 B⟶C(=O)–CH₃ (A), with R substituent (R = Alkyl, Ph)	A[a]	40,48,143,171,331
	B[a]	40,43,171,335,381
B⟶C(=O)–CH₃ (A), with R₂ (R₂ = Alkyl₂, ring)	A	40,48,171,320[b],335,382-385
	B	171,183
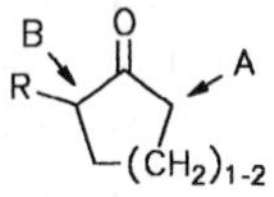 R–C(B)⟶C(=O)⟶(A), –(CH₂)₁₋₂– (R = Alkyl, Ph)	A	41,43,45[b],171,386,387,388[b]
	B[a]	41,48,170,171
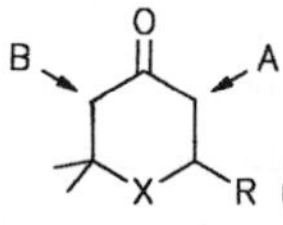 B⟶C(=O)⟶A ring with X and R (R = H, Alkyl) (X = CH₂, O, S)	A	389-391,392[b]

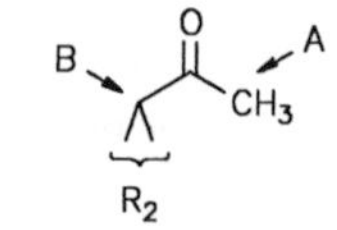

[a] Slightly selective in some cases

[b] Same results reported with analogous substrates

Phenolic derivatives are by far the most widely studied substrates, as is proved by the considerable number of papers dealing with this class of compounds. Some of them[214,394–397] also report accurate evaluations of the predominating position of aminomethylation in the aromatic ring, thus enabling (Fig. 32) the following general considerations concerning the regioselectivity of the reaction to be postulated:[258,270,322,398–400]

- Phenols are aminomethylated preferably in the ortho position, if available (**75**), the para position being alternatively attacked (**76**).
- Between two unequivalent ortho positions, the less sterically hindered one is preferred. In the case of fused aromatic nuclei (e.g., β-naphthols,[398] 5- and 6-hydroxyindoles or -hydroxy benzothiophenes,[322] 6- and 7-hydroxycoumarins and analogous derivatives[399,400]), however, the ortho/α position (**77**) is still predominantly aminomethylated, even when it is appreciably hindered.[401,402] In this connection, it has been discovered[397,403,404] that the ortho/β position of 5- and 6-hydroxyindoles **78** is not active at all,

as the aminomethylation takes place at position 3, if available, or at the heterocyclic
NH group when the ortho/α position is occupied by a substituent.

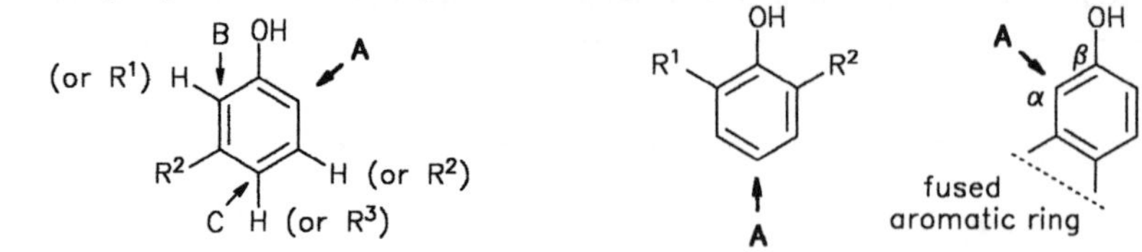

[included R¹, R² = fused aromatic ring]

75   **76**   **77**

A ⟶ : preferred attack; B, C ⟶ : other possible positions of attack

Fig. 32.   The preferred aminomethylation site in phenolic substrates.

The observation that the ortho position is preferred by the aminomethylating agent
has led several authors[189,397,405,406] to propose a concerted mechanism involving the
occurrence of hydrogen bonding between the phenolic hy-
droxy group and the reagent, with steric and electronic effects playing a decisive role in determining the site of
attack. It is in fact reasonable to expect that under the neutral or alkaline conditions usually adopted for this reaction,
the aminomethylating agent is in the X—CH_2—NR_2 form
(**54**, Sec. B.2), well suited to the above behavior, even if the
activating effect by the hydroxy group is also still sufficiently present at the para position.

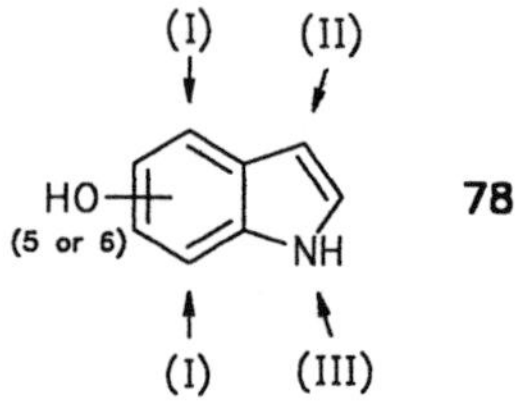

In some cases, however, the selectivity of the reaction is not particularly relevant,
and the production of regioisomers negatively affecting the purification of the main
product has been reported.[214,294,394,407] Attempts to render the reaction more selective
have also been made.[190,312,408] In particular, the influence of the solvent[408] and the possibility of complex formation of the type ''guest-host'' of phenol with β-cyclodextrins[312]
have been investigated.

A consistent number of exceptions to the usual behavior of phenolic substrates
include halogen-substituted phenols, a few nitro derivatives, naphthols, and a series of
methyl-substituted derivatives, as reported in Table 10.

Aminomethylation of the above substrates has however been carried out in conditions that differ from the usual practice, employing methyleneimmonium salts (on α-
naphthol) or hexahydrotriazine reagent in the presence of acid (on alkylphenols). In
these cases, in fact, the aminomethylating agent is in a cationic form (**53**, Sec. B.2),
which is unsuitable for producing hydrogen bonding with the phenolic substrate in the
transition state.[189] On the other hand, in the case of β-naphthol, the unexpected result
can be attributed to the particularly cumbersome reagent used, that is, methylene-bis-
dicyclohexylamine.[213]

In conclusion, besides steric and electronic effects, the nature of the aminomethylating agent also has a notable influence on the regioselectivity of the Mannich reaction
on phenolic substrates, with the obvious consequence that a correct interpretation of the
experimental results calls for consideration of all the factors involved.

Table 10

Anomalous Aminomethylation Sites on Phenolic Substrates[a]

Substrate	Ref.	Substrate	Ref.
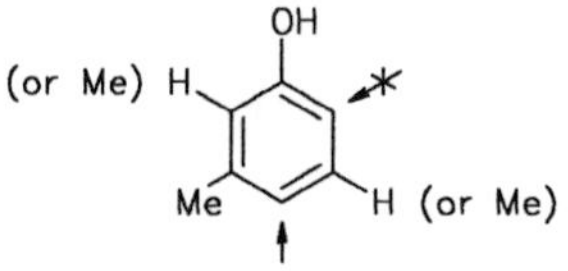	214	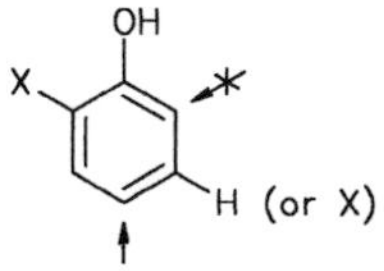 (X = Cl, Br)	396,409
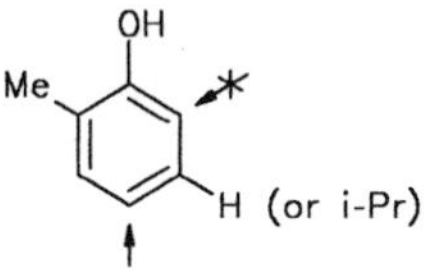	214,394 409,410	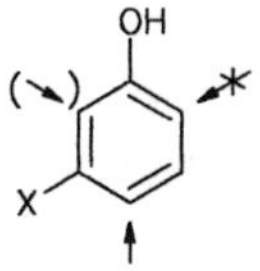 (X = Halogen)	396,409
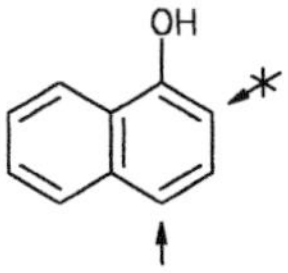	189	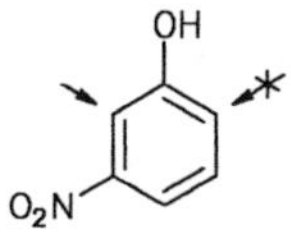	395
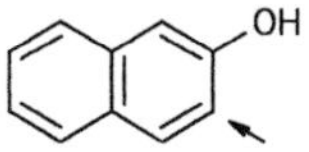	213		

[a] Arrow and crossed arrow indicate experimental and expected position of attack, respectively

The introduction of a second aminomethyl group into aminomethylated phenolic substrates, by distinct steps or by using an excess of reagent, still follows the general rules mentioned above, the reaction taking place at the positions marked B and C (see **75**, Fig. 32).

So far as the aminomethylation of di- or poly-hydroxy phenolic substrates is concerned, it has been discovered[214,411–413] that the reaction always occurs in the ortho position with respect to one hydroxy group, in accordance with the orienting properties of the substituents.

By contrast, aromatic rings activated by secondary or tertiary aminic or amidic groups give the Mannich reaction at the para position, as observed in the aminomethylation of arylamines **79**, **80**[414,415] and sulfonamides **81**. In this latter case the regioselectivity of the reaction is attributed[27] to the dissociation in alkaline conditions of the NH-sulfonamide group which produces a para-quinonoid intermediate leading to para attack by the reagent. The corresponding NCH$_3$ derivative, in fact, does not give any reaction.

Arylamines **79** are allowed to react under acidic conditions, in the presence of acetic acid, which affects even the chemoselectivity of the reaction with formation of N-Mannich base when the molar ratio acid/substrate is very low (1:25).[414]

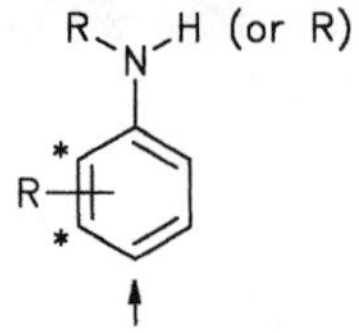

(R = Alkyls)
79

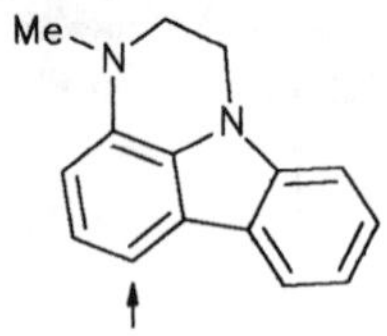

80

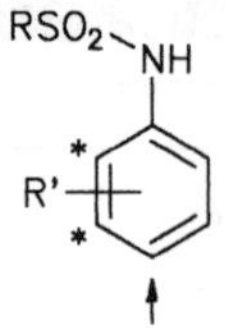

$$\begin{pmatrix} R = Me, Ph \\ R' = various\ subst. \end{pmatrix}$$
81

Aromatic Heterocyclic Substrates

This group of substrates includes a number of compounds, particularly the polyhetero-atomic terms, which undergo Mannich reactions often exhibiting a hardly distinguishable regioselectivity from chemoselectivity. However, the five-membered rings, which, like those depicted in Fig. 33, contain only one heteroatom, make it easy to interpret the experimental results.

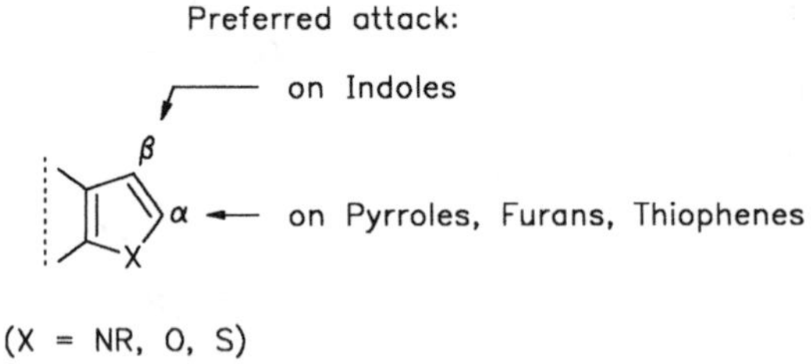

Preferred attack:

(X = NR, O, S)

Fig. 33.   The preferred aminomethylation site in five-membered heteroaromatic substrates.

N-Heterocyclic substrates are the most used for Mannich reactions, usually in acetic acid as a solvent, but methyleneimmonium salts in acetonitrile are also frequently employed as reagents. Such conditions appear adequate to overcome the chemoselectivity problem posed by the formation of the undesired N-aminomethylated product. Apart from rare exceptions, the rules generally followed are as follows:

- Pyrrole is aminomethylated in position 2, that is, in the α position with respect to the heteroatom.[92,303,416] When the pyrrole ring is fused to another aromatic ring (e.g., indole d   e   -
rivatives), the reaction takes place at C-3, that is, in the β position with respect to the N atom.[416-422] If the abovementioned positions are occupied by substituents, the Mannich reaction occurs anyway at the β and α positions, respectively, to the heteroatom.[365]

    When the indole moiety is inserted into the indole[1,7-a,b][1]benzazepine structure **82**, an exception is observed, as the Mannich reaction takes place at the α position with respect to the heteroatom.[236] The orientation of attack by the Mannich reagent to either the α or β position (C-6 or C-5, respectively) of the pyrimidino-pyrrole systems of type **83** is strongly affected by different variables, such as reaction conditions, steric hindrance of substituents, and electronic effects mainly related to the nature of substituent R at C-2.[243,423,424]

    Indolizines **84** are again aminomethylated on the five-membered ring, preferably in the α position to N, the remaining C atoms of the ring giving reaction when a substituent is present in such a position.[194,425,426]

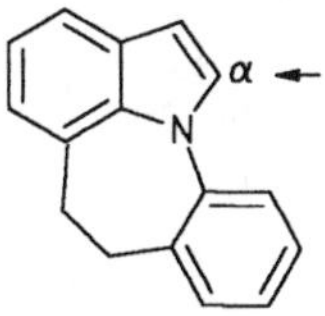

82 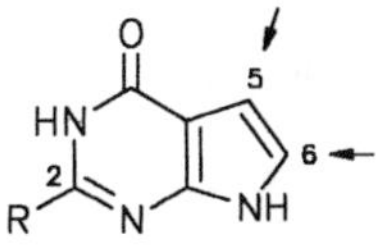   **83** 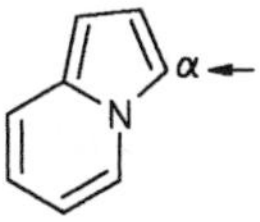   **84**

- Furane and thiophene derivatives are poorly reactive substrates; thiophenes in particular require the presence in the ring of electron donors such as, for example, the methoxy group, in order to undergo the Mannich reaction successfully. The preferred site of attack is again in the α position with respect to the heteroatom.[196,245,351,416,417,427] Interestingly, the reaction takes place at position α-2 in the 3-substituted derivatives, rather than at the less hindered α-5 position.[196,245]

Five-membered rings containing more than one heteroatom behave in a more complex way as regards aminomethylation. Imidazoles and thiazoles **85** react at 4 and 5 positions,[416,417,428,429] although reactions of thiazole with arylamines occurring at position 2 are reported.[430]

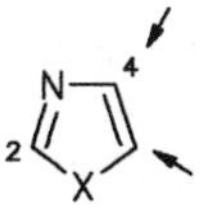

(X = NH, S)

85   **86** 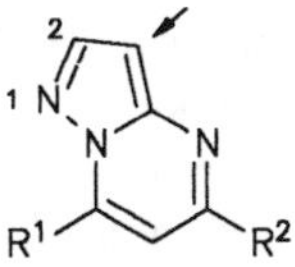   **87**

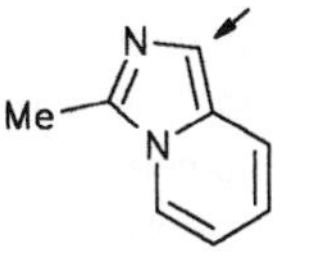

In pyrazole derivatives **87** the preferred reaction site turns out to be the C-3;[431] this example, along with those represented by **84**, **86**[244] and in Table 7 (Ref. 351), usefully demonstrates that the five-membered rings constitute a quite suitable structure for C-aminomethylation. Moreover, when the substrate is a fused polycyclic derivative, the N-containing ring is preferred by the aminomethylating agent.

The N-aminoalkylation of pyrazole and triazole (**88**, Fig. 34), also, can take place regioselectively in a fashion similar to the abovementioned N-aminoalkylations that occur on 5-phenyl idantoin **69**.[367] This is related to the tautomeric equilibrium present in the heteroaromatic ring, with the prevailing site of the reaction of type A (see also Fig. 79, below).[251,252,310,432]

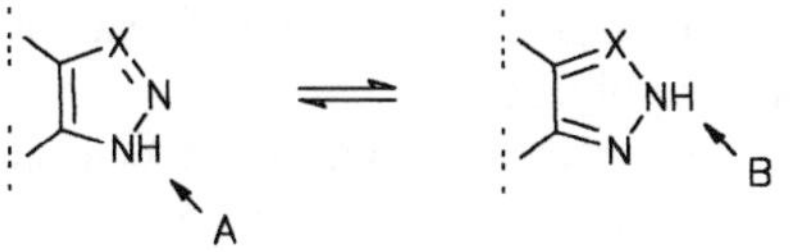

(X = CR, N)

Fig. 34.   Regioselective aminomethylation on pyrazole and triazole substrates.

Six-membered heterocyclic substrates showing regioselective aminomethylation include a substantial number of hydroxy-pyridine, -pyrimidine, and -pyridazine derivatives as well as their N-oxide derivatives (**89–91**, Fig. 35).

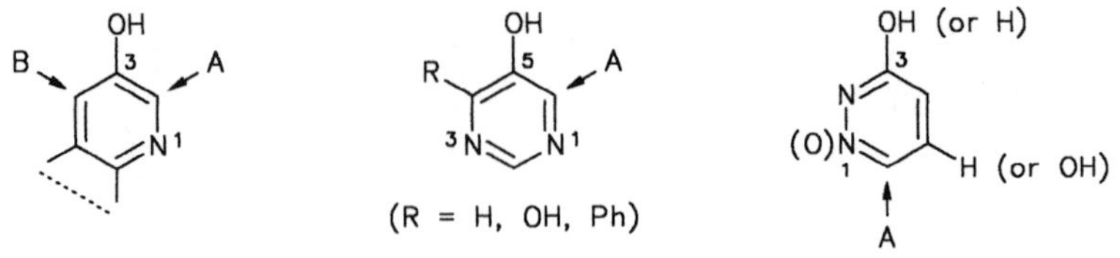

89 (refs. 395, 433-435)   **90** (refs. 436-438)   **91** (refs. 439,440)

Fig. 35. Regioselective aminomethylation on hydroxy-pyridine, -pyrimidine, and -pyridazine substrates.

All these substrates behave very similarly to phenols, as the hydroxy group usually drives the reaction toward the ortho position. When two unequivalent ortho positions are present, the preferred reaction site is at the C atom vicinal to N (A attack), unless the substrate has a fused ring structure.[433,435] In this latter case, as in quinoline derivatives, for example, the prevailing attack is of type B (**89**), similar to that observed in the case of naphthols.

Aminomethylation in the para position with respect to the hydroxy group is observed only in the 3-hydroxy pyridazine derivative **91**.

Uracil derivatives **92** are usually aminomethylated at position 3, although the reaction with formaldehyde alone shows an equilibrium constant for hydroxymethylation in position 1, which is about twice the value obtained for the reaction at position 3.[376] The C-aminomethylation reaction takes place at position 5, when unsubstituted, as these substrates resemble the structure of "push-pull olefines" (**67**, Table 8).[441-444]

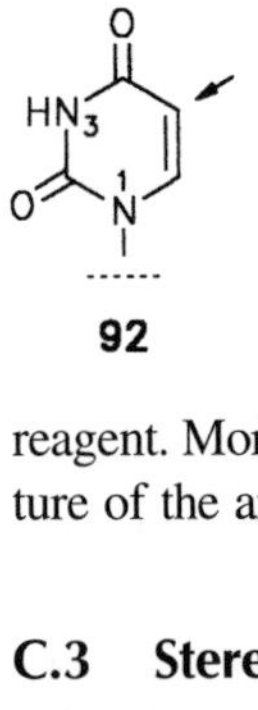

92

In conclusion, the experimental results reported in the literature indicate that the regioselectivity of the Mannich reaction is essentially determined by the possibility of hydrogen bonding between substrate and reagent. Moreover, it may be considerably affected by steric effects related to the structure of the amine reagent.

C.3   Stereoselectivity

Chiral derivatives or compounds having a prochiral group in the reactive center are frequently present among substrates, amines, and aldehydes used as reagents in Mannich synthesis. When at least two out of the three reagents are chiral or prochiral, the resulting Mannich base will be made up of a diastereomeric mixture of products. The main possible combinations of chiral or prochiral couples of reactants leading to diastereomeric derivatives are reported in Fig. 36.

A remarkable number of papers dealing with the stereochemical aspects of Mannich synthesis are present in the literature, some of them reporting the use of optically active enantiomeric reactants, instead of the commonly employed racemic pairs. Interesting asymmetric syntheses are also described.

Among chiral or prochiral substrates, alkyl ketones **93** are those most widely studied from the stereochemical point of view. The analog nitroalkanes have also been investigated.[276]

The possibility that the aminomethylating reagent will preferentially replace H_A or H_B in **93** is essentially determined by the presence of asymmetry in the substrate or in

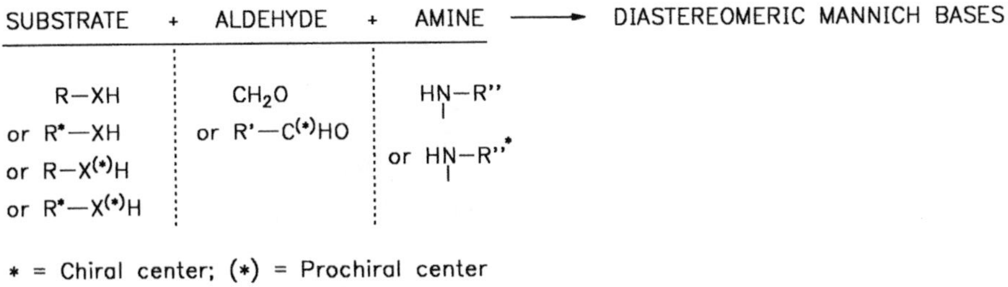

SUBSTRATE + ALDEHYDE + AMINE ⟶ DIASTEREOMERIC MANNICH BASES

R—XH   CH$_2$O   HN—R''
                                    |
or R*—XH   or R'—C$^{(*)}$HO

or R—X$^{(*)}$H   or HN—R''*
                                    |
or R*—X$^{(*)}$H

* = Chiral center; (*) = Prochiral center

Fig. 36.   Reactants combinations affording diastereomeric Mannich bases.

the reagent, so as to induce a predominant reaction at either hydrogen atom. Thus, the synthesis of α- and β-asymmetric β-aminoketones employing prochiral alkyl or aryl aldehyde has been investigated, and the determination of relative configurations as well as diastereomeric ratios in the products has made it possible to demonstrate the presence of low stereoselectivity in this type of system.[64,445]

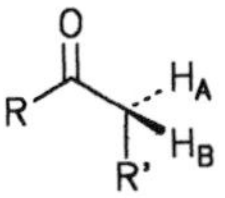

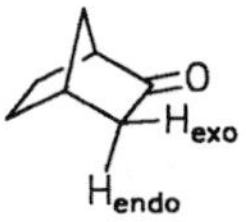

93 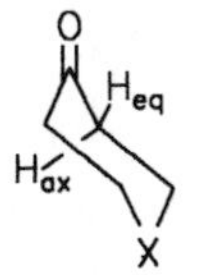   (X = CH$_2$, S, Se)   **95**
                        94

A much higher stereoselectivity has been observed in the analogous reactions on the more rigid cyclic ketones,[60,315] the reactive hydrogen atom to be replaced being in this case the H$_{ax}$ or H$_{eq}$ of cyclohexanone or analogous heterocyclic system **94**, or the H$_{endo}$ or H$_{exo}$ of polycyclic ketones **95** of the norbornanone series.

Substrates **94**, in the presence of ammonium acetate and two equivalents of aryl aldehyde, afford aza-bicyclononanones of type **96** with remarkable stereoselectivity, the formation of only one diastereomeric product being usually observed.[151,446,447] Unexpected factors such as, for instance, the substituents of aryl aldehyde, can however strongly affect the stereochemistry of the reaction, as *p*-chlorobenzaldehyde leads only to product **96B**, whereas simple benzaldehyde produces a mixture of the diastereomers **96A** and **B**.[151]

A   B
          96

The bicycloheptanone **95** undergoes aminomethylation with predominant replacement of the less hindered *exo*-H atom,[186] unlike *d*-camphor, which affords the more stable *endo*-aminomethyl derivative.[57,448] Bicycloheptanedione **97** (Fig. 37), also, gives the unexpected bis-*endo,endo*-aminomethyl derivative **98**.[186]

The prochirality of aldehydes other than formaldehyde is frequently exploited in order to obtain asymmetric Mannich bases **99A** and **B**, as the reaction with substrate may occur from either side of the double bond plane of the aminomethylating agent (Fig. 38).

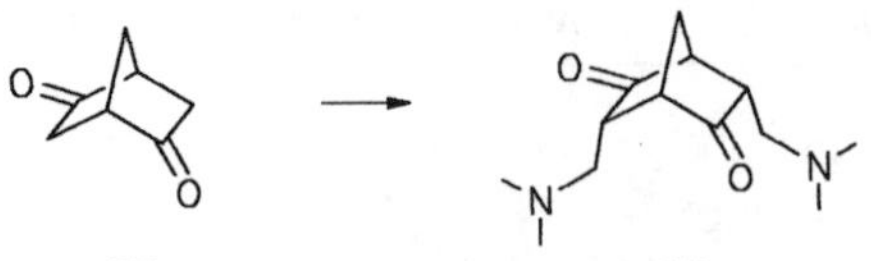

97   **98**

Fig. 37.   Bis-aminomethylation of bicycloheptanedione.

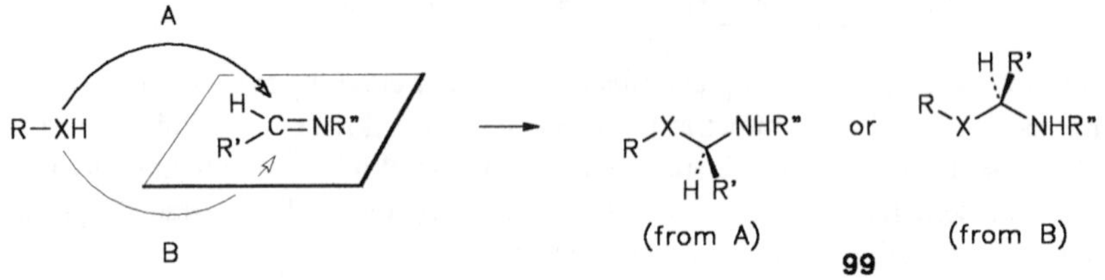

Fig. 38.   Stereochemistry of the reaction between a substrate and a prochiral imine reagent.

In addition to the aza-bicycloheptanones of type **96** described above, some examples of stereospecific cyclization (Sec. C.4) leading to dihydro-1,3-benzoxazine derivatives starting from phenolic substrate, primary amine, and two equivalents of aryl aldehyde are reported in the literature.[449]

The interesting asymmetric synthesis of the optically active β-amino acid **101** (Fig. 39), as well as of intermediates in alkaloid chemistry,[56,58] may also be included in the same group of reactions.

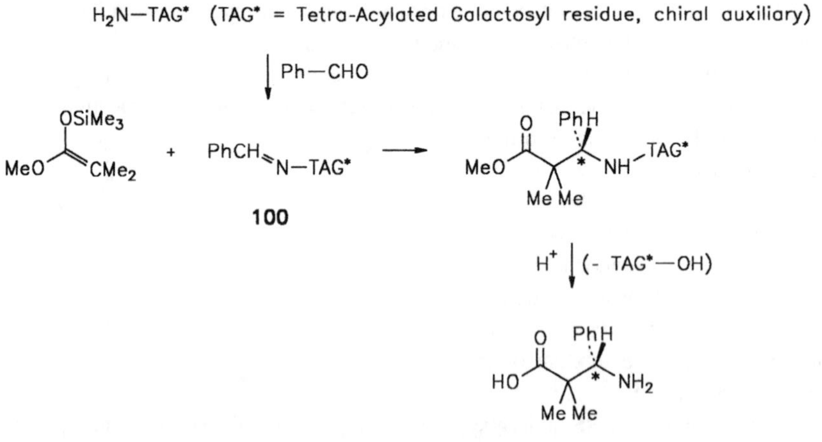

H_2N-TAG^* (TAG* = Tetra-Acylated Galactosyl residue, chiral auxiliary)

101

Fig. 39.   Asymmetric synthesis of optically active β-aminoacid.

In this case the imine aminoalkylating agent **100** is prepared from optically active galactosylamino derivative. The success of the synthesis is determined by the good yield of the aminomethylation step in conjunction with the remarkable stereoselectivity of the reaction.

Chiral amines, also employed in the synthesis of perhydro-oxazepines and γ-diaze-pines,[223] have been used for the preparation of asymmetric P-Mannich bases such as the derivatives **102**, which exhibit the contemporary presence of optically active P and amino group. The synthesis involves reaction of the hydroxymethyl derivative of the substrate with the chiral amine.[269]

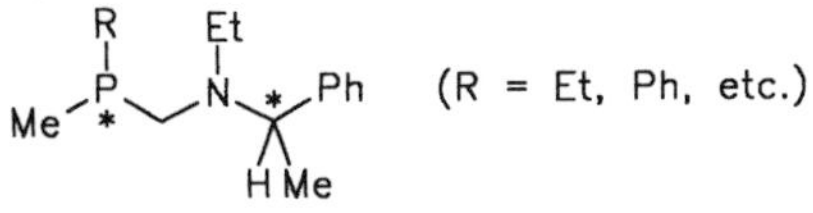

(R = Et, Ph, etc.)

102

C.4   Synthesis of Cyclic Mannich Bases

The possibility of obtaining cyclic derivatives by Mannich synthesis is one of the most significant aspects of the versatility of this reaction. Indeed, a large number of nitrogen-containing heterocyclic compounds can be produced by this method, six-membered rings being the most frequently synthesized, due to their stability. Pentaatomic rings are also reported; larger rings, mainly seven-membered, as well as polycyclic systems, particularly aza-adamantanone derivatives, are rather frequently described in the literature (see also Sec. C.5 and Chap. IV). Further references to several other ring-forming Mannich reactions with aldehydes other than formaldehyde can be found in Table 3.

Two main strategies are usually adopted for the synthesis of N-heterocyclic Mannich bases:

- Normal Mannich reaction employing three components (bi- or polyfunctional substrate, amine, and formaldehyde)
- Condensation by formaldehyde (ring closure) of compounds containing both active hydrogen atom and amine moiety

Cyclization of Polyfunctional Substrate with Amine and Aldehyde

Several types of substrate may be subjected to cyclization. In particular, compounds having two or more hydrogen atoms on the same reactive center (activated methylene or methyl group, phosphinic P atom, etc.) undergo reaction with formaldehyde and primary amine in the molar ratios 1:3:1 and 1:3:2 or with bis-amine in the ratio 1:2:1, in that order, affording products **103–105** and **106, 107**, respectively (Fig. 40).

In addition to the common six-membered rings, a large number of aza-adamantane derivatives **105** and seven-membered rings **106** are produced by this method. Larger rings such as diphospha-diaza-cyclooctanes (**108**, Fig. 41), possessing attractive complexing properties, are prepared from hydroxymethylated phosphine and primary amine.[269]

Most of the polyfunctional substrates belonging to this group of compounds have two (or more) active hydrogen atoms located on different positions of the molecule. Thus, in the case of symmetrical substrate with two identical reactive XH groups **109**, the Mannich bases **110–115** (Fig. 42) are obtained by reaction of the substrate with formaldehyde and amine in the molar ratio 1:2:1.

Aza-bicyclononanones **111** are undoubtedly among the products more widely studied, particularly from the stereochemical point of view, as they exhibit interesting

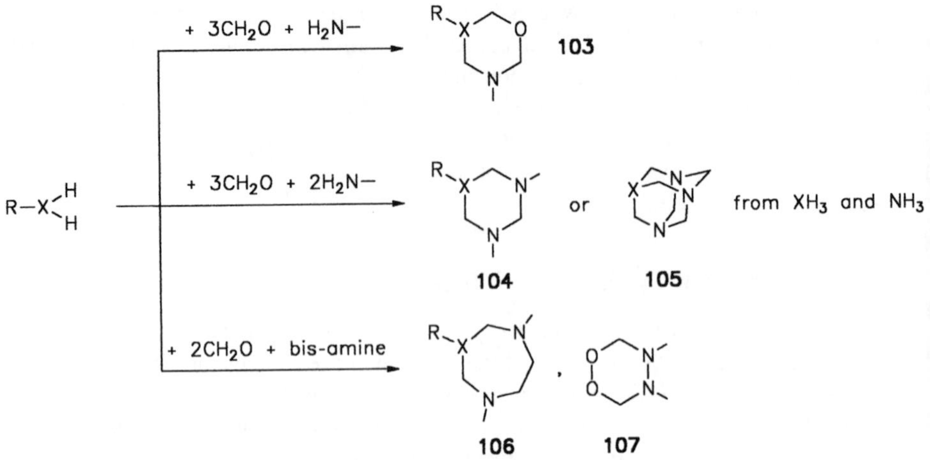

103,104; $R-X< = $ Alkyl$(O_2N)C<$, $(MeCO)_2C<$, $HCON<$. (Refs 230,234,450,451)

105; $X< = O_2N-C<$, $P<$. (Refs 269,452,453)

106; $R-X< = $ Alkyl$(O_2N)C<$, $ArCOCH<$, $ArS-CS-N<$. (Refs 223,234,454)

107; (Ref. 455)

Fig. 40. Mannich cyclizations of bifunctional substrates of the type R—XH$_2$.

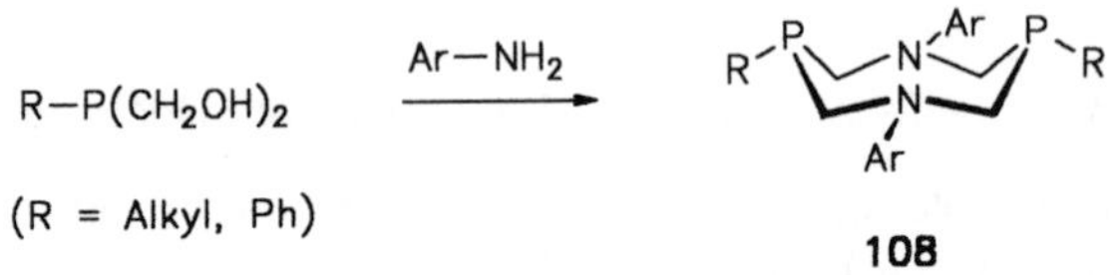

Fig. 41. Synthesis of an eight-membered heterocyclic Mannich base.

conformational and configurational features (see Sec. C.5). In addition, these compounds are strictly connected with polycyclic derivatives such as aza-adamantanones,[456,468] which can be similarly prepared starting from cyclohexanediones and ammonia (**116**),[469] for example. Analogous reactions have been carried out on trinitro cyclohexanes[470] or even metal complexes, to give the so-called sepulcrates.[471]

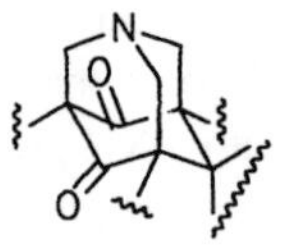

116 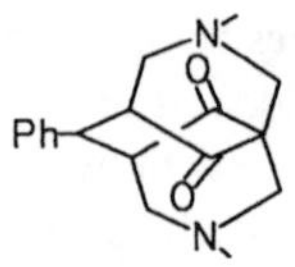   **117** 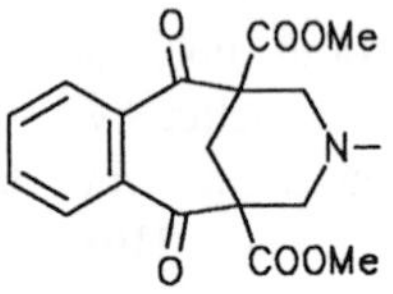   **118**

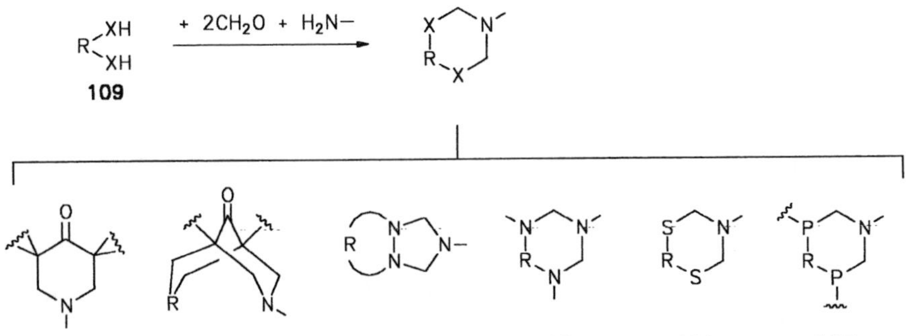

110; (Refs 258,456)

111; $R\zeta$ = —, $-(CH_2)_{1-2}-$, Alkyl-Nζ , Oζ , Sζ , Seζ . (Refs. 149,446,447,457-460)

112; $R\zeta$ = Me$_2$Cζ , $-CO-CH=CH-CO-$. (Refs 461,462)

113; $R\zeta$ = $-(CH_2)_2-$, COζ , CSζ , SO$_2\zeta$. (Refs 88,463,464)

　　　= Ni(complex). (Ref. 465)

114; $R\zeta$ = $-(CH_2)_2-$, MeCO(CN)C=Cζ . (Refs. 135,466)

115; $R\zeta$ = ortho-C$_6$H$_4\zeta$. (Ref. 467)

Fig. 42.　Mannich cyclizations of bifunctional substrates of the type R(XH)$_2$.

Cyclic diketones and primary amines also yield other interesting polycyclic derivatives of types **117** and **118**.[472,473]

When the bifunctional reactant is aldehyde (glutaraldehyde), cyanopiperidines **119** (Fig. 43) are obtained by reaction with hydrogen cyanide and primary amine, although one of the nitrile groups may be partially hydrolyzed to amide during the synthesis.[474]

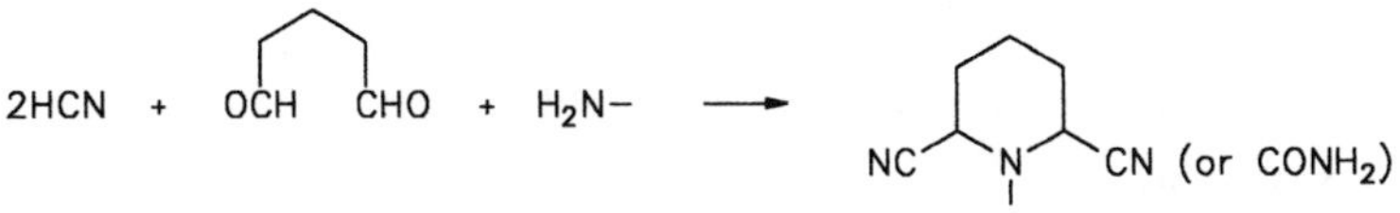

119

Fig. 43.　Mannich cyclization leading to a 2,6-disubstituted piperidine ring.

In the presence of two different functional groups in the substrate (**120**, Fig. 44), the resulting cyclic Mannich bases are γ-piperidones **121**, diazines **122**, **123**, triazines **124**, **125**, oxazines **126**, and thiadiazines **127**.

Phenolic substrates leading to dihydrobenzoxazines **126** are the most used;[258] however, diazines **123**, deriving from enaminones (considered as push-pull olefines) or enaminothiones, are also frequently synthesized.[488] Chromone derivatives **128** (Fig. 45), in which the CH$_2$Ph group constitutes the second reactive center required for cyclization, due to the activation induced by the carbonyl group through a vinylogous system, react similarly.[489]

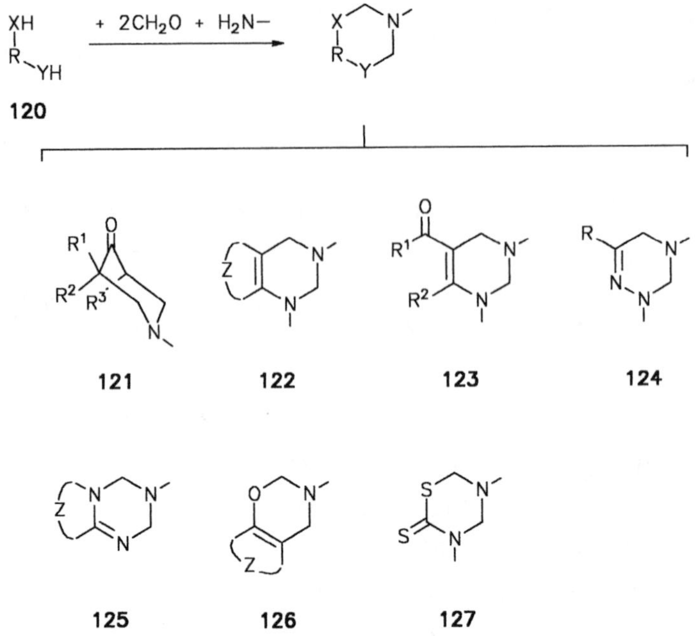

120

121   122   123   124

125   126   127

(Z = Aromatic or heteroaromatic fused ring)

121; R^1-R^3 = H, Ph, COOEt, ring. (Refs 460,475,476)

122; (Refs 323,349)

123; R^1,R^2 = Alkyl, OEt, ring. (Refs 199,318,357,477,478)

124; R = Ar, NO_2. (Refs 479-481)

125; (Refs 379,482,483)

126; (Refs 214,258,322,398,438,484,485)

127; (Refs 486,487)

Fig. 44.   Mannich cyclizations of bifunctional substrates of the type HX—R—YH.

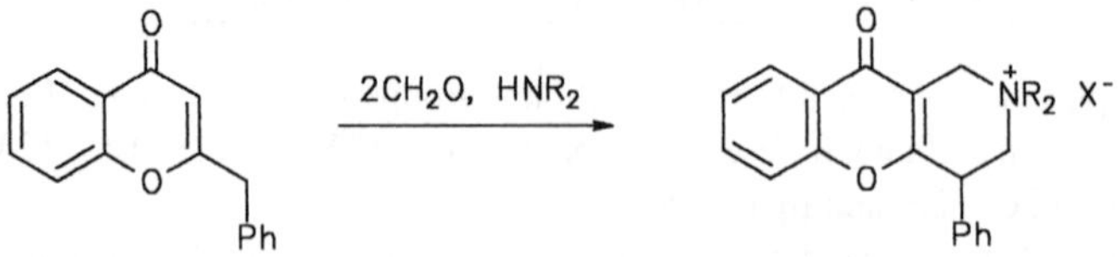

128   129

Fig. 45.   Mannich cyclization on 2-benzyl-chromone.

Product **129** also represents one of the rather infrequent cases of Mannich reaction directly yielding a quaternary ammonium salt.

The tendency to give the most stable cyclic derivative, usually a six-membered ring, as seen before, is evidenced by the Mannich reaction on 3-ethylindole,[365] which takes place in positions 1 and 2 of the substrate. Although the formation of a pentaatomic ring would be expected, a less-strained larger ring (**130**) is preferentially produced through the involvement of an additional aminomethyl group.

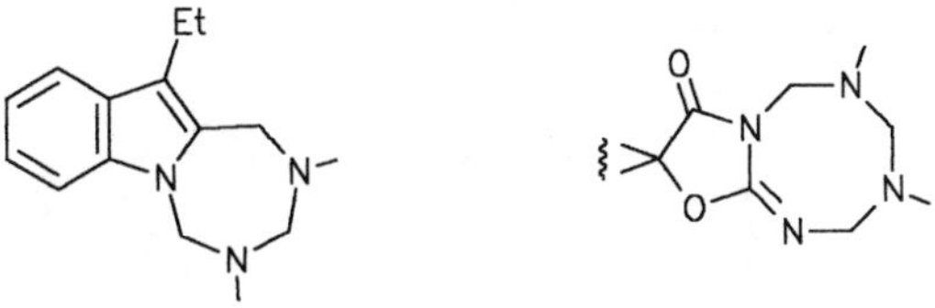

130   **131**

An analogous behavior probably also gives rise to products **131**, deriving from the Mannich reaction on pemoline.[314]

Cyclization of Aminic Substrate by Formaldehyde

Substrates of type **132** (Fig. 46) may produce, by reaction with formaldehyde, a variety of cyclic compounds (A–C), depending on the nature and size of the R group.

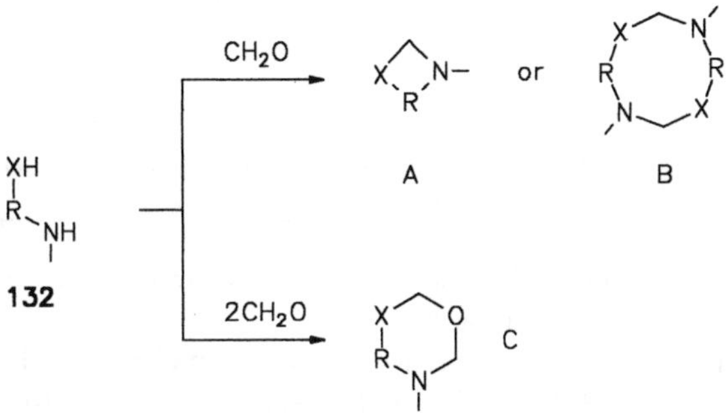

Fig. 46. Possible Mannich cyclizations of substrates containing the amino group.

The thermodynamic stability of the ring to be formed also plays a very important role in this group of Mannich derivatives, as formaldehyde in some cases may enter in the final cyclic product in the dimeric form, thus yielding the oxa-aza-heterocyclic structures of type C.

When XH in **132** is an amino group identical to the NH residue, the symmetrically substituted cyclic methylene-bis-amines **133–135** (Fig. 47) are obtained. These products derive from substrates having ethylenediamine (**133**),[132,490] propylenediamine (**134**),[89,491,492] or a more complex bis-amine structure, such as methylene-bis-hydrazone or methylene-bis-arylamine (**135**).[480,493]

In the presence of a suitable substrate, the formation of the diaza-heterocyclic product from primary bis-amine is accompanied by the normal Mannich reaction.[89,132] Thus, the hexahydropyrimidine derivative **137** (Fig. 48) is obtained from the nitroalkanols **136**, formaldehyde and the bis-amine 1,3-diamino-2-hydroxypropane.[89]

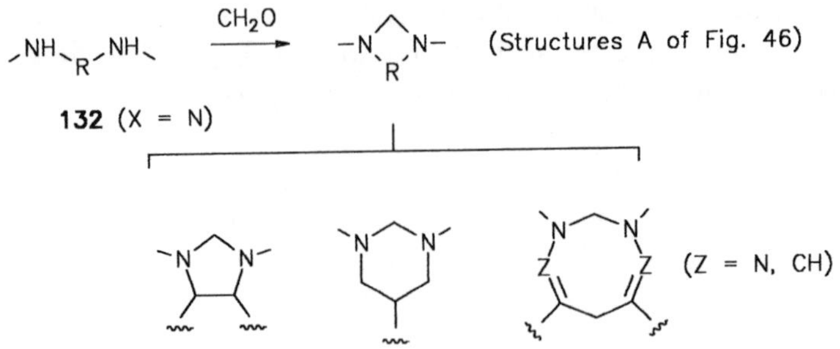

Fig. 47.   Mannich cyclization of symmetrical bis-amines with formaldehyde.

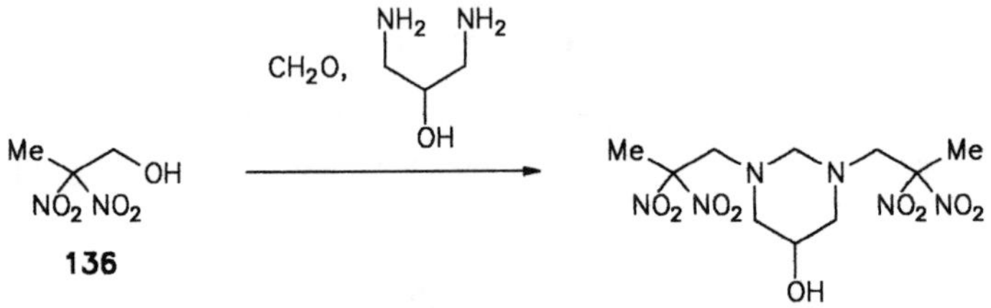

Fig. 48.   Simultaneous cyclization-aminomethylation reaction involving
1,3-diamino-2-hydroxypropane.

Polycyclic derivatives, such as diaza-adamantane **138**[492] and tetraaza-tricyclodo-decane **139**,[132] may also be synthesized.

The formation of product **139** occurs and is even preferred, under particular conditions, in place of **133** (Fig. 47) in the reaction between formaldehyde and ethylenediamine.[132]

When XH in **132** (Fig. 46) is other than an amino group, the symmetrical Mannich bases B may again be obtained by condensation of two molecules of substrate and two molecules of formaldehyde.[494] Hexa- (**140, 141**) and octaatomic (**142**) rings are thus prepared, starting from imidazole,[495] hydrazine,[494,496] and aniline[497] derivatives.

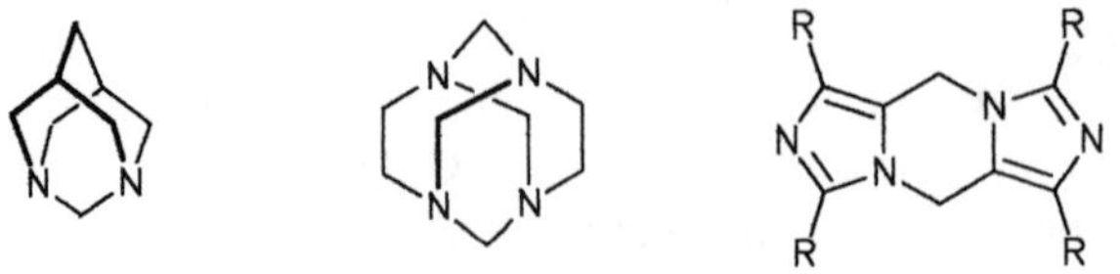

138   139   140

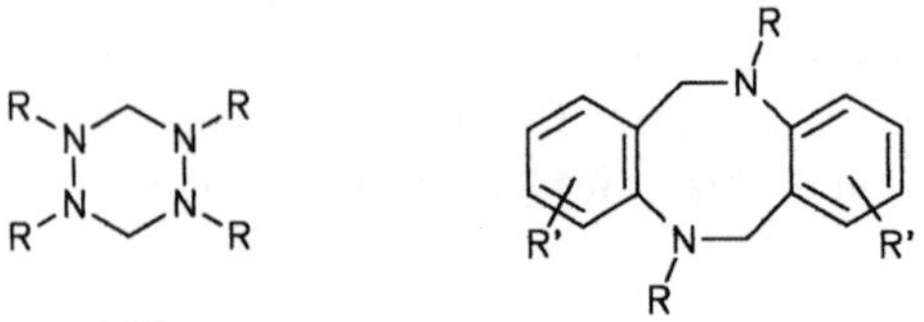

141   142

In most cases, however, the substrate molecule reacts with one molecule of formaldehyde to give five- to seven-membered cyclic derivatives (**143–146**, Fig. 49). Eight-membered rings containing phosphorus have also been obtained.[498]

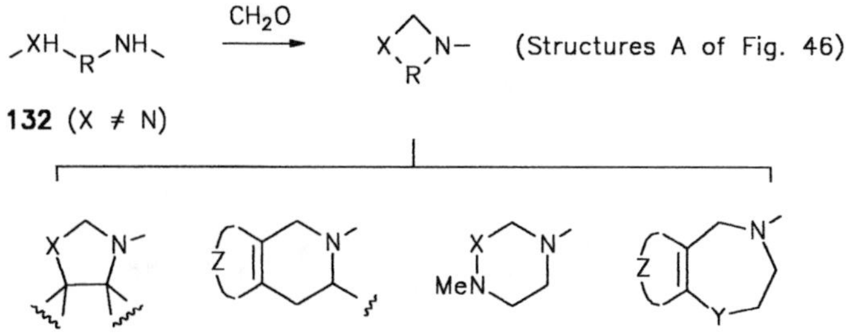

(Z = Aromatic or heteroaromatic fused ring)

143; X = N, O. (Refs 499-501)

    X = S, P(O)OEt. (Refs 325,498,502,503)

144; (Refs. 504-508)

145; X = MeN, O. (Refs 509,510)

146; Y = N, S. (Refs 152,511-513)

Fig. 49. Mannich cyclization of aminic substrates with formaldehyde (1:1 molar ratio).

The above synthetic routes may be accompanied by side reactions, such as, for example, nitrosation or P-methylation of the heterocyclic nitrogen atom.[325,500] Similarly, the Mannich reaction on 2-amino-ethanethiol (Fig. 50), in the presence of phosphorous acid, produces the quaternary ammonium salt **147**:

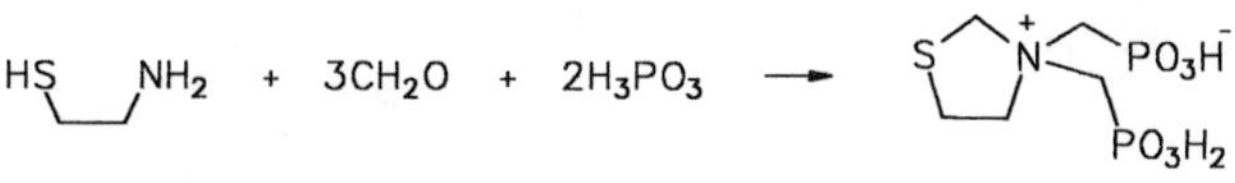

147

Fig. 50. Simultaneous cyclization-aminomethylation reaction involving 2-amino-ethanethiol.

Compound **147** can be considered the product of bis-phosphonomethylation of thiazolidine and represents a quite uncommon example of the direct formation of a quaternary ammonium salt in X-alkylation reactions.[325]

The analogous *trans*-amino cycloalkanethiols **148** (Fig. 51) exhibit, upon treatment with formaldehyde, some interesting features connected with the stability of the heterocyclic structures produced.[499]

The expected reaction path leading to products **143** is not always followed, as it is affected by the substrate structure. In particular, a normal behavior is observed when $n = 2$, but when $n = 3$ further reaction with formaldehyde of the thiazolidine NH takes place to give N,N'-bis-methylene derivative **150**. When $n = 1$, the polycyclic product **149** is formed.

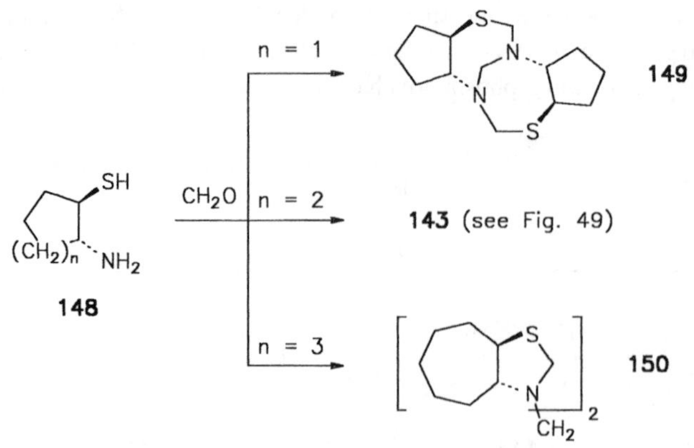

Fig. 51.   Mannich cyclization of 1-amino-cycloalkane-2-thiols.

Some of the Mannich derivatives reported in Fig. 49 are relevant in alkaloid chemistry. In particular, obtaining the tetrahydroisoquinoline ring **144** is a fundamental step in the synthesis of protoberberines (Chap. IV, A).

Phenolic compounds of type **151** (Fig. 52) display different regioselective reaction pathways determined by the presence of substituents bonded to the oxygen atom. Route **a** is usually preferred when R^2 is H,[506] whereas route **b** is exclusively followed when R^2 is an alkyl group. This is observed in the cyclization of protoberberines[506] and similar substrates.[504,507] However, when no alternative way of condensation is offered, route **a** is followed.[505]

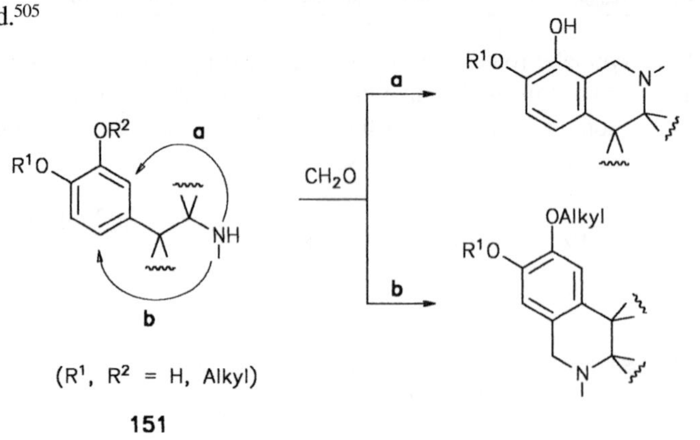

Fig. 52.   Mannich cyclization of 2-aminoethyl-phenols or -phenolethers.

The cyclization of diacetylene derivatives **152** is carried out in two steps through the synthesis of a methoxymethyl intermediate, which then undergoes cyclization in the presence of a cuprous salt as catalyst (Fig. 53). In addition to the main product **153**, macrocycle **154** is formed in a relative amount of about 1/5.[220]

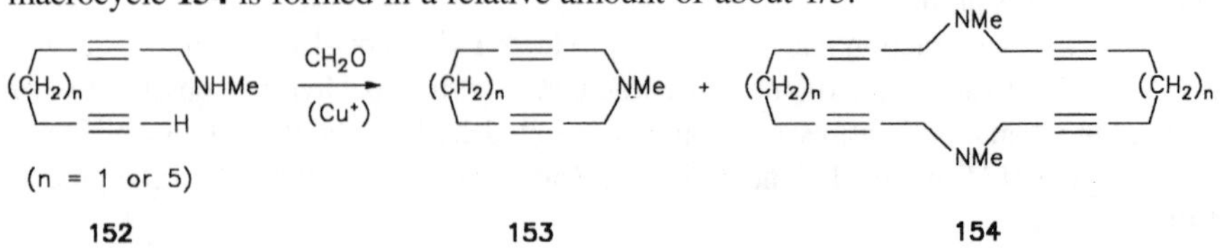

Fig. 53.   Synthesis of cyclic amines containing alkyne groups.

When ring closure by one molecule of formaldehyde would lead to an excessively strained structure, relief is produced by the reaction of the dimeric form of aldehyde, which thus affords oxa-aza-heterocyclic derivatives (**155–158**, Fig. 54), constituted by more stable five- to seven-membered rings.

Acceptably stable metal complexes also may be prepared, such as the cobalt-ethylenediamine product **156**. The reaction on benzimidazole derivatives affording seven-membered rings (**158**) confirms the tendency to produce cycles larger than the pentaatomic ones, as a molecule containing two fused five-membered rings would in principle be expected from this type of substrate (see also the derivatives **130, 131, 140, 149** described above).

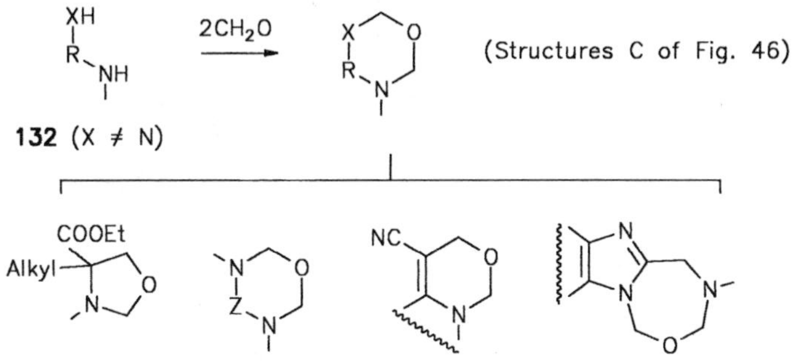

132 (X ≠ N)

155   **156**   **157**   **158**

155; (Ref. 514)

156; Z = CH$_2$, Co(complex). (Refs 471,515,516)

157; (Ref. 517)

158; (Ref. 518)

Fig. 54.   Mannich cyclization of aminic substrates with formaldehyde (1:2 molar ratio).

Some particular examples of the synthesis of cyclic Mannich bases are reported in the literature which, although not included in the general schemes described in this section, are however of relevance in the study of large sized molecules which may exhibit biological activity.

First, 2-pyrrol-1-yl-benzaldehyde **159** (Fig. 55), which combines in the same molecule the functions of both substrate and aldehyde, gives aminoalkylation in position 2 of the pyrrole ring by reaction with the secondary amine, thus yielding an interesting precursor (**160**) of polycyclic derivatives with fused nuclei.[519]

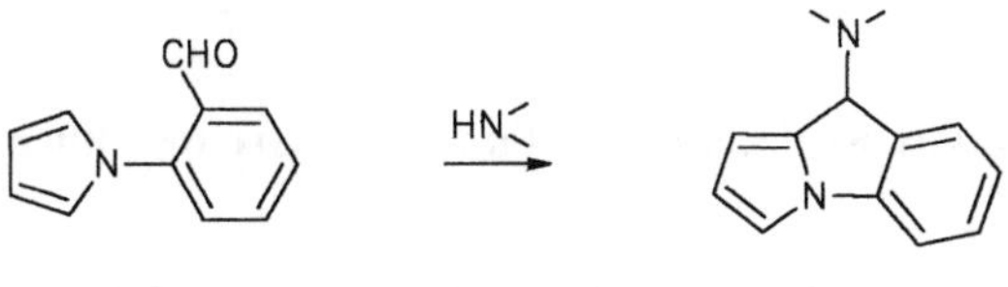

159   **160**

Fig. 55.   Mannich cyclization of a substrate containing the aldehyde moiety.

Substrates containing all the three reactive moieties of Mannich synthesis, that is, substrate, amine, and aldehyde, such as **161** (Fig. 56) also have been subjected to cyclization.[520,521] Two carbonyl functions are present in these molecules, which may produce intramolecular aminoalkylation.

As expected, the structural features of the molecules involved, particularly the size of the cycles that are formed, exert an important influence on the success of this type of synthesis.

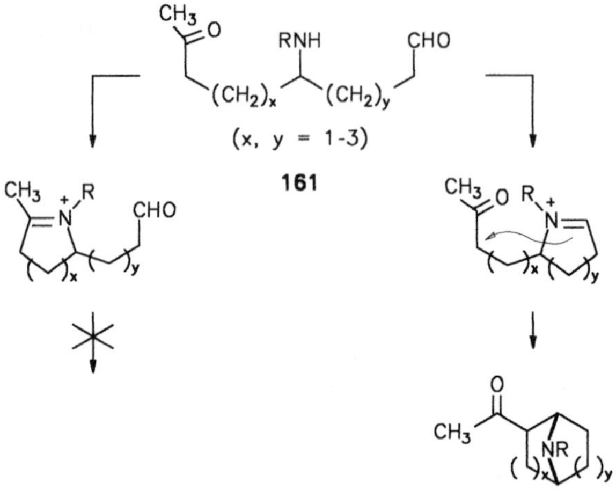

Fig. 56.   Mannich cyclization of trifunctional compounds.

C.5   Concurrent Reactions Involved in Mannich Synthesis

Reactions leading to more or less relevant changes in the structure of the final product, usually involving the substrate moiety, may accompany Mannich synthesis. Some examples have appeared in the preceding sections; for instance, the formation of methyleneimmonium salts from cyanomethylamines (Fig. 22) is frequently adopted in intramolecular cyclization in order to prepare products, which are then deprived of the nitrile group. Although a very large number of cases embracing wide sectors of organic chemistry are reported in the literature, the main types of concurrent reactions taking place more frequently can be summarized as follows:

- Replacement of functional groups in the substrate other than the reactive hydrogen atom
- Further condensation reaction with formaldehyde or other aldehydes
- Hydrolysis, alcoholysis, etc.
- Double bond migration
- Cyclization

Aminomethylation by Replacement of Functional Groups Other Than Hydrogen

Sulfonic, carboxylic, etc., groups present in the substrate may easily undergo substitution (Fig. 8) during Mannich synthesis, due to their reactivity toward electrophilic reagents;

the same groups may also favorably affect the reaction through the stabilization, for example, of enolic structures or carbanions. In other cases the entering aminomethyl group can directly determine the elimination of the above moieties, as suggested for some readily occurring decarboxylation reactions.[258,522]

The carboxyl group is frequently involved in replacement reactions (path **a**, Fig. 57), along with bis-aminomethylation (path **b**), followed or not by deamination to give the methylene derivative **165** (path **c**). These last compounds are treated in the following section, which deals with further condensations of aldehyde.

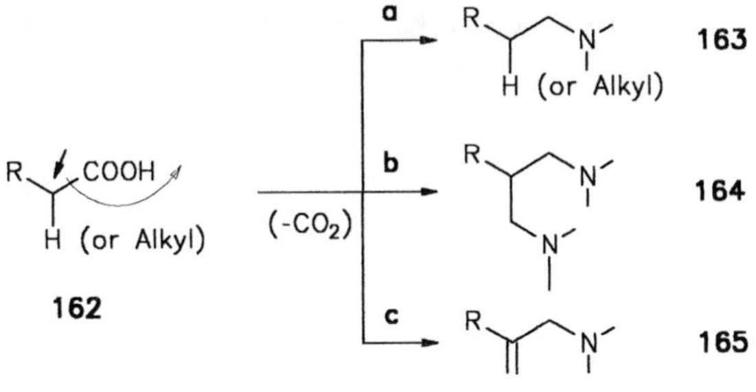

163: R = MeCO, ArSO$_2$. (Refs 104,522,523)

164: R = (O$_2$N)$_2$C$_6$H$_3$, ArCH$_2$SO$_2$. (Refs 524-526)

165: R = CN, COOH. (Refs 525,527)

Fig. 57. Aminomethylation with decarboxylation of carboxyl substrates.

Substituent R in substrates **162** (Fig. 57) is usually an electron-withdrawing group capable of favoring decarboxylation. Other derivatives, such as 3-carboxyl-dihydro-1,4-pyridine **166**, can also provide aminomethylation with displacement of the carboxyl group.[195]

The acetyl group behaves similarly to the carboxy group when it is located in a position adjacent to carbonyl moieties (**167** and **168**).[528-530] In hydrazone derivatives **167** (R = OH), elimination of the acetyl group appears to be favored even over that of the carboxyl.

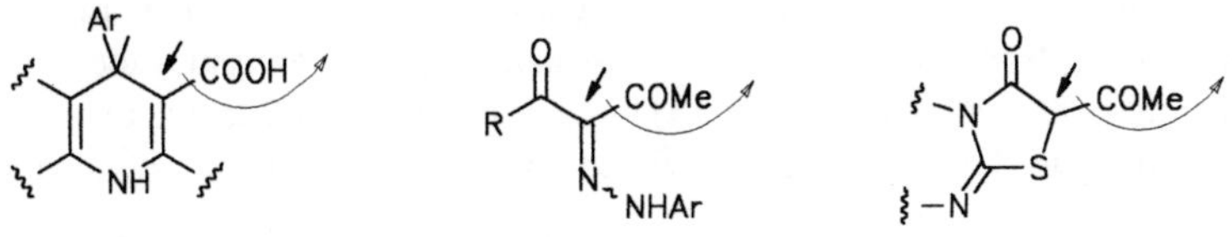

166　　　**167**　　　**168**

The sulfonic group present in anthraquinone dyes is readily replaced by the Mannich reagent, as has been mentioned previously[531] (see **10** in Sec. A.1); in particular, 1-amino- and 1-hydroxy-2-anthraquinone-sulfonic acids may occasionally be involved in the production of cyclic Mannich bases. Finally, the methyl group of dimethylsulfite (CH$_3$)$_2$SO$_3$ may be similarly replaced by the dimethylaminomethyl group giving $^-$O$_3$S—CH$_2$—N$^+$(CH$_3$)$_2$H.[532]

Further Condensation Reaction with Formaldehyde or Other Aldehydes

Besides its usual behavior in Mannich synthesis, formaldehyde may react further with the substrate R—H, thus yielding by-products (Figs. 30 and 31), which may even occur in a preponderating amount (Fig. 58).

The Mannich ketobases, which have two aminomethyl groups, one of which undergoes deamination, are frequently involved. As a consequence, products of type **169**, containing a methylene group, are formed (see also **165** in Fig. 57).[44,183,223] The derivatives of chromanone **171**, produced by simultaneous formation of the cyclic Mannich base (see Fig. 45) and introduction of the methylene group in the vinylogous position with respect to the activating carbonyl group, are likewise formed.[533]

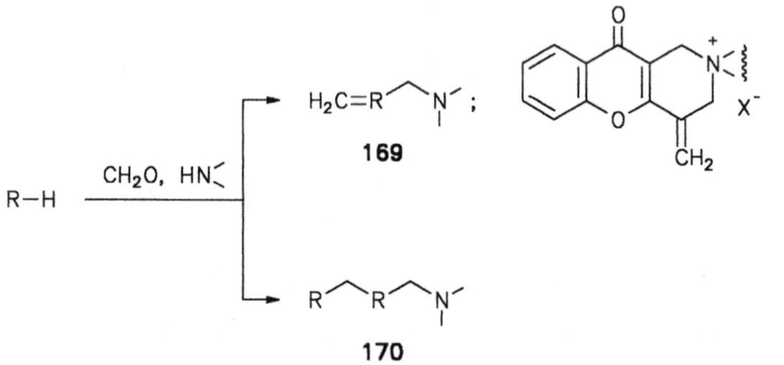

Fig. 58. Further reactions of formaldehyde in Mannich synthesis.

The product of the Mannich reaction may occasionally behave as a substrate undergoing aminomethylation. Thus, when several mobile hydrogen atoms are present in the substrate, it is possible that formaldehyde reacts to give Mannich bases **170**, with a methylene bridge placed between two substrate moieties.[534]

Bond Cleavage by Hydrolysis, Alcoholysis, etc.

Bonds connecting carbon to heteroatoms in the substrate may be frequently broken by the hydrolytic action of the solvent molecules, as observed, for example, in the case of the C—N bond of enamines,[197] which may give a ketonic Mannich base in acid medium.

Acetal formation by the solvent occurs during the aminomethylation of isobutyraldehyde (see **8, 9** in Sec. A.1), and a transesterification accompanies the reaction of dithiobutyric esters with methylolamine ethers (Fig. 59). Thus, Mannich bases **172**, in which the thioester bond has been converted into O-ester bond, are obtained in good yields.[535]

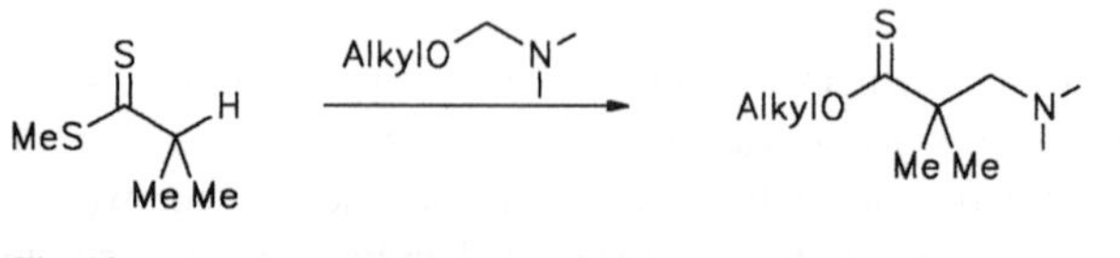

Fig. 59. Simultaneous transesterification-aminomethylation reaction involving a dithiocarboxy ester.

The cleavage of heterocyclic rings, such as the alcoholysis of 1,3-oxazolidinones,[536] affording Mannich bases aminomethylated in the α position with respect to the carboxy-ester group, has been reported as well. Thiazolium salts **173** (Fig. 60) undergo a similar ring opening, but aminomethylation occurs on the thiol group.[537]

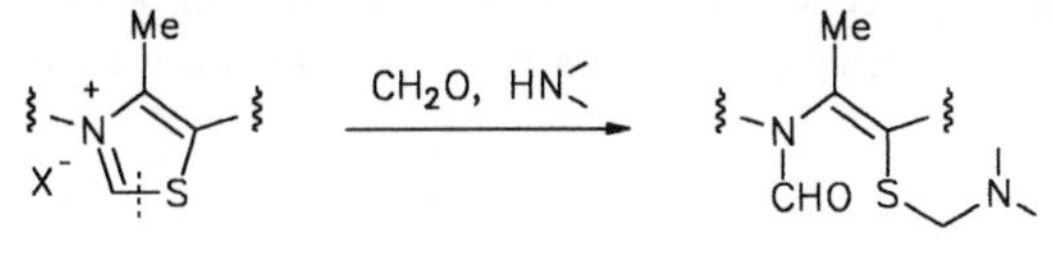

173

Fig. 60. Aminomethylation with ring opening of thiazolium salts.

A large number of Mannich reactions leading to P-aminomethyl derivatives involves hydrolytic steps in the process, the resulting products being mainly the aminomethyl phosphonic or aminomethyl alkylphosphinic acid derivatives **174** and **175** (Fig. 61).

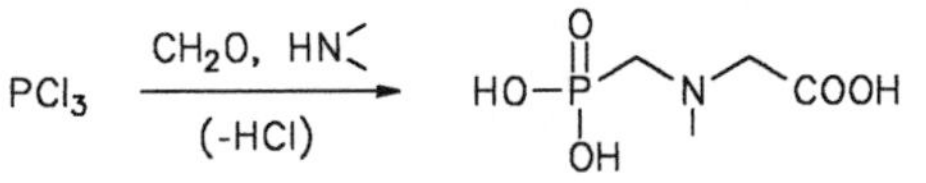

	X	R¹	OR²	Ref.
174				538
175	O	Alkyl	OAlkyl	539
	O	OAlkyl, OPh	OAlkyl, OPh	75,305,540
	C(COOEt)₂, NPh	OEt	OEt	541

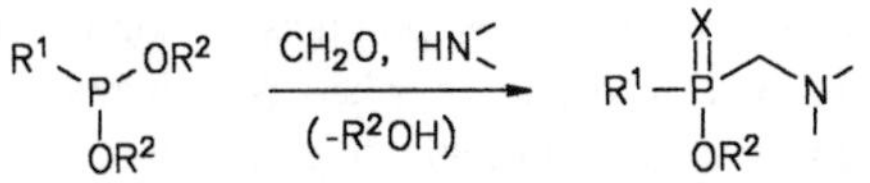

Fig. 61. Aminomethylation of substrates lacking a mobile hydrogen atom.

The above syntheses are carried out in aqueous solution on phosphorus trichloride, thus causing hydrolysis of the P—Cl bonds, or on dialkyl alkylphosphonites or trialkyl phosphites (see also Fig. 11). The process is widely employed in the production of pesticides, flame-proofing agents, etc. (Chap. V).

Double Bond Migration

The Mannich reaction on unsaturated compounds containing C—C double bonds lacking particular activating groups usually involves double bond migration accompanied by other concurrent rearrangements.

Aminomethylation of alkenes[270] affords a quite large range of products, the aminomethylated ones being reported in Fig. 62, as substantial amounts of nonbasic materials are also formed. Alkenes used as substrates (Table 11) are usually allowed to react with dimethylamine; other amines[542,543] tend to complicate the reaction further. Cyclic derivatives obtained by Mannich synthesis are not included in the table, as they are reported in the following section.

The reaction mechanism[542,547,548] schematically depicted in Fig. 62, involves the methyleneimmonium ion as aminomethylating agent. This is generated by acidic reaction conditions (usually acetic acid is used as solvent). Attack by this reagent on the substrate leads to an intermediate carbocation, which then can follow various reaction

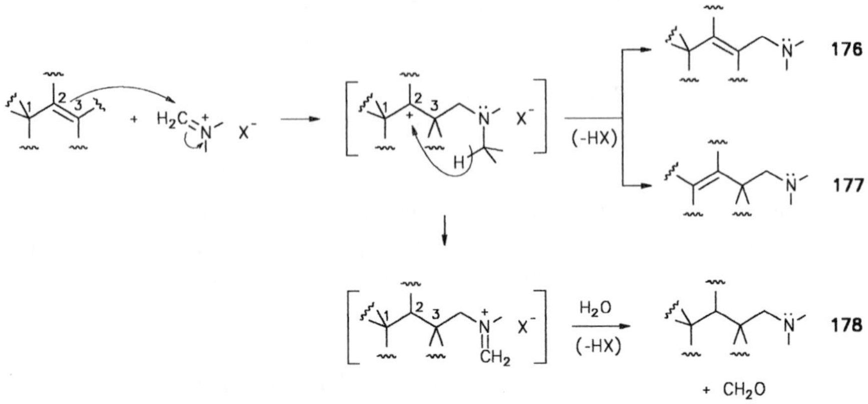

Fig. 62. The Mannich reaction on alkenes.

paths, depending on the substrate structure. Thus, products **176** or **177** are produced by loss of a proton from position 3 or 1, respectively, whereas derivatives with hexocyclic double bonds tend to be given by cyclic substrates (e.g., santene, 2-pinene). Products **178** are formed by intramolecular hydride transfer followed by hydrolysis and loss of formaldehyde. This last reaction path also appears to occur with amines other than dimethylamine, and accounts for the considerable modifications observed in the amine moiety of the final product.[542]

The results reported in Table 11, although indicative of the complexity of the reaction, permit the following general considerations to be put forward:

- Aminomethylation may also occur in the absence of hydrogen atoms bonded to the unsaturated carbon atom.
- The aminomethyl group is preferably linked to the less hindered position.
- Double bond migration is a general feature in this type of reaction, as products **177**, **178** and their derivatives are more frequently generated. Products **176** and analogs appear to prevail when monomethylamine is employed.[549]

Cyclization

The reactions described here follow schemes similar to those described in Sec. C.4, but are accompanied by relevant modification of the substrate moiety. In particular, this relates to the number and position of unsaturated bonds as well as to the molecular skeleton as a whole.

A first group of reactions deals with unsaturated substrates, aromatic derivatives included, giving cyclic products, according to Figs. 44 and 46 of Sec. C.4; this is done by condensation with amino derivatives and one equivalent of formaldehyde or by

Table 11

Aminomethylation Sites on Alkenes[a]

Substrate	Predominant product	Reference
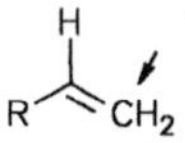 (structure) R—CH₂ with H (R = Alkyl, Ph)	**178 (176,177)**	544
(structure) Me₂C=CMe₂	**177**	183
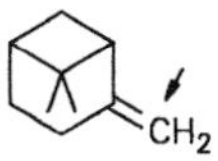 (structure) and similar derivs	**176,177**	545-548
(structure cyclohexene)	**178**	544
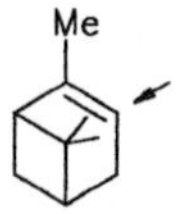 (structure) and similar derivs	**177,178**	543,545,547-549

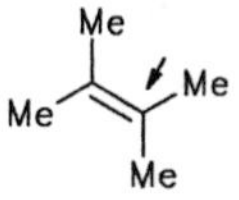

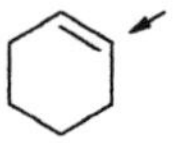

[a] Arrow indicates the position of attack

reaction with formaldehyde and primary amines. In particular, activated alkyl substrates of type **179** (Fig. 63), having the Z-crotonic structure, undergo vinylogous aminomethylation on the methyl group,[550] but the reaction with primary amines R^2—NH_2 also involves tautomeric effects producing the tricyclic derivatives **180**.

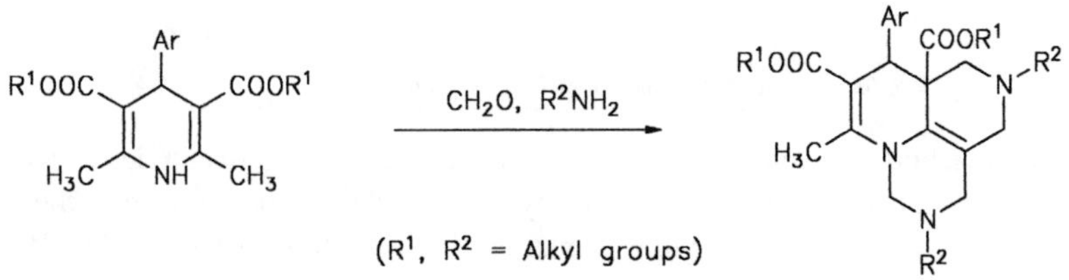

$(R^1, R^2 = Alkyl\ groups)$

179

180

Fig. 63. Simultaneous cyclization-vinylogous aminomethylation reaction involving dihydropyridine derivatives.

Formaldehyde condensation of phenolamines **181** (Fig. 64) interferes even with the aromaticity of the phenol ring to give the spiro-derivatives **182**;[551] this reaction occurs also with analogous arylamines formed by rearrangement of unstable phenolic Mannich bases.[321] Upon treatment with formaldehyde the methyleneimmonium salt **183** is formed, which then similarly yields a spiro-derivative. Analogous reactions on enaminones are cited in Ref. 552.

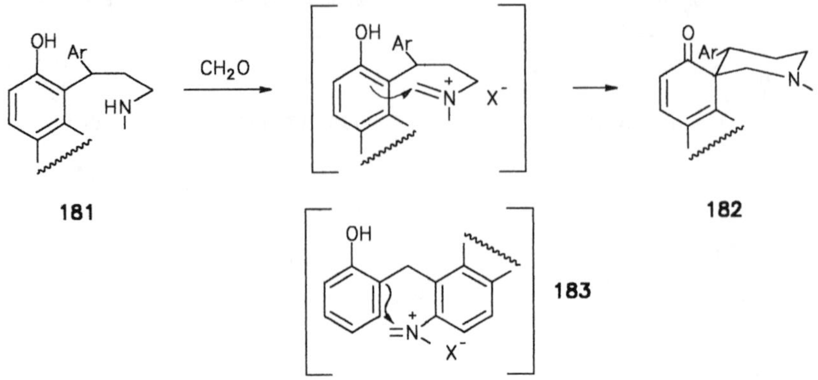

Fig. 64.   Mannich cyclization leading to spiro-derivatives.

Except in a few cases,[553] cyclic Mannich bases derived from alkenes are produced by the reaction of unsaturated amino derivatives with formaldehyde.[554-556] Intramolecular aminomethylation (Fig. 65) takes place with aminoalkenes (**184**, X = H; path **a**) or with analogous silyl allyl derivatives (**184**, X = SiMe$_3$; path **b**):[556]

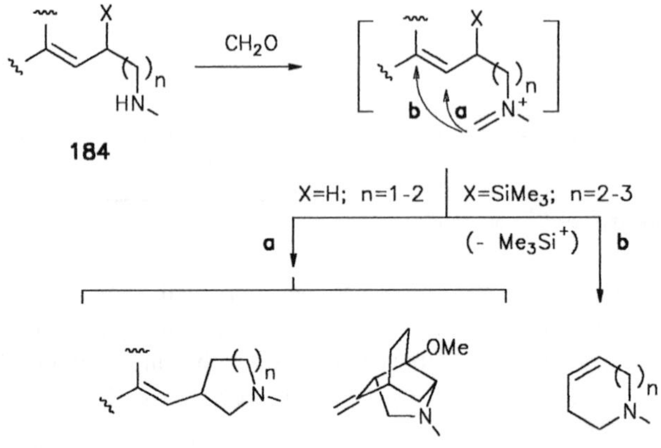

Fig. 65.   Intramolecular Mannich cyclization of alkenes.

The reaction is frequently applied in alkaloid chemistry to obtain derivatives with complex structure (e.g., **186**).[554,555] Attack by the methyleneimmonium ion on the double bond, followed by cyclization with formation of hexo- (**185**) or endocyclic (**187**) unsaturated product, is suggested.

In contrast to the above reaction, the C—C double bond may undergo intramolecular addition to produce ring closure with the formation of a saturated derivative. An example (Fig. 66) is provided by the synthesis of oxazine derivatives **188**[549,557] and similar compounds.[121]

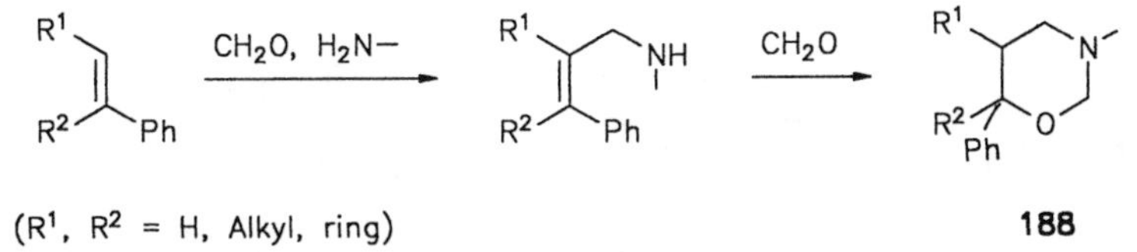

(R¹, R² = H, Alkyl, ring)   **188**

Fig. 66.   Aminomethylation of alkenes followed by ring closure by formaldehyde.

When the unsaturated substrate is an acetylene derivative, furans[558,559] or benzo-furans[560] may be produced. For instance, 2-aminoethyl-benzofuran **190** (Fig. 67) is obtained at pH 7 from the acetylenic alcohol **189** instead of the expected Mannich base, which is produced at a higher pH value.

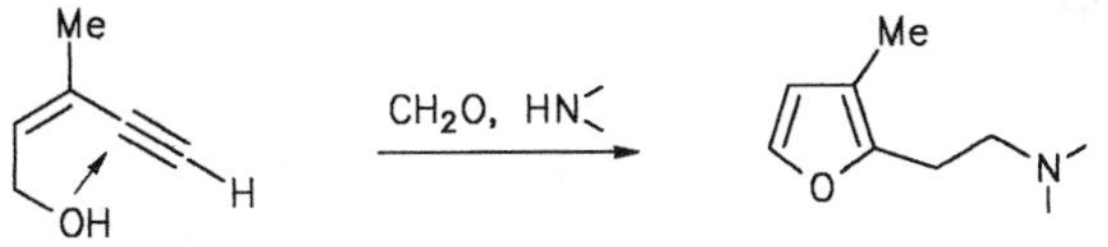

189   **190**

Fig. 67.   Simultaneous cyclization-aminomethylation reaction involving hydroxy and alkyne groups.

A second group of reactions concerns rearrangements, occurring mostly in the cyclic Mannich base, which may generate deep structural modifications of the molecule. Due to its relevance in the synthesis of alkaloids, the tandem cationic aza-Cope rearrangement–Mannich cyclization (Fig. 68) is the most widely studied of this class of reactions.[260,561,562]

Unsaturated amino derivatives containing an allylic hydroxy group, or a precursor of it, undergo the above reaction when the amino group is converted into the methyleneimmonium ion **191** by any of the well-known methods (Sec. A.4), ranging from the common condensation reaction with aldehyde to the elimination of the CN group of cyanomethylamines in the presence of silver ion. As the methyleneimmonium ion attacks the double bond, the C—C single bond near the OX group is broken and the intermediate **192** is formed. Aminoalkylation of the enolic double bond then occurs, leading to the final cyclic amine **193**, bearing a carbonyl substituent. Studies of the stereochemistry of the process have ascertained that a chair conformation is preferred in the transition state, as is shown for the E-alkene **194**. Racemization would be expected in these reactions as long as the initially formed rearranged immonium ion underwent C—C single bond rotation more rapidly than intramolecular Mannich cyclization. This limitation, observed in the synthesis of **195**, should not extend to the ring-enlarging pyrrolidine annulation version of this rearrangement, where racemization of the sigmatropic rearrangement product would be much less likely. A good example of this is provided by the formation of the optically active product **196**.

Studies of alkaloid synthesis provide another example of aminomethylation with concurrent molecular rearrangement (Fig. 69). In this case the methyleneimmonium ion is inserted into a tetrahydroisoquinoline ring, and the cyclization occurs by reaction with a brominated aromatic ring activated by alkoxy groups.[563]

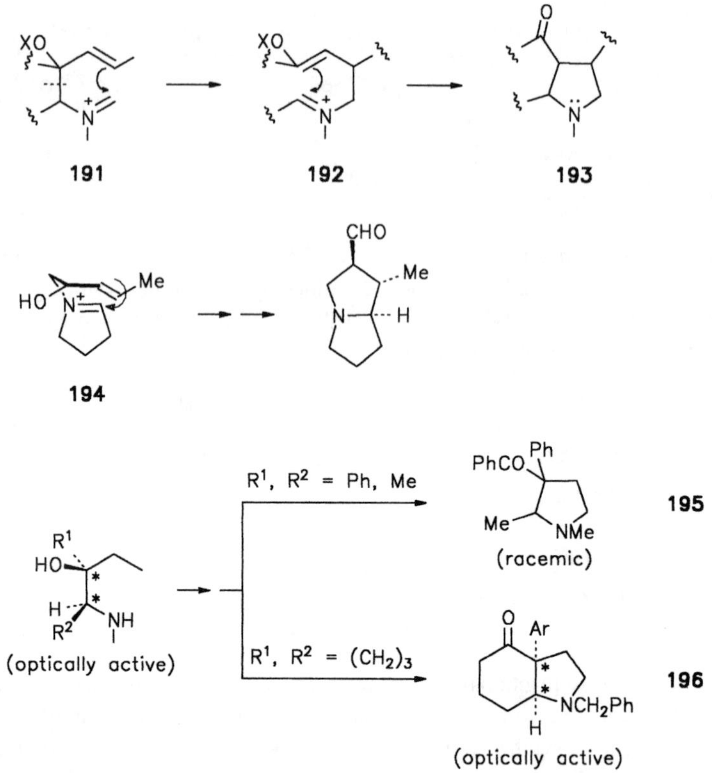

191   **192**   **193**

194

Fig. 68.   Tandem cationic aza-Cope rearrangement–Mannich cyclization.

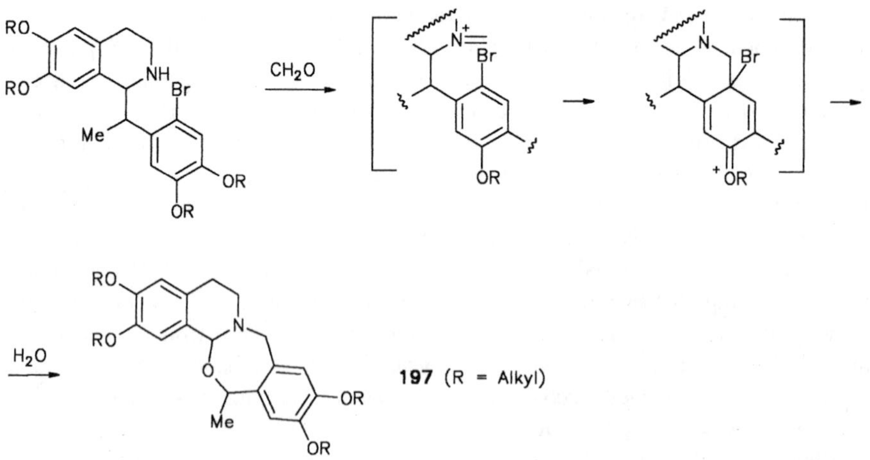

Fig. 69.   Simultaneous rearrangement-ring forming aminomethylation reaction leading to polycyclic derivatives.

The resulting cyclic product **197** contains an unexpected seven-membered ring due to the insertion of an oxygen atom attributed to the opening of the intermediate six-membered ring, initially formed by aminomethylation, and elimination of the halogen atom.

Another example of ring enlargement is provided by the aminomethylation of the potassium salt of N-hydroxy-phthalimide (Fig. 70), which affords the Mannich base **198** of dihydro-benzoxazine-dione, instead of the O-Mannich base of N-hydroxy-phthalimide, as previously proposed.[72]

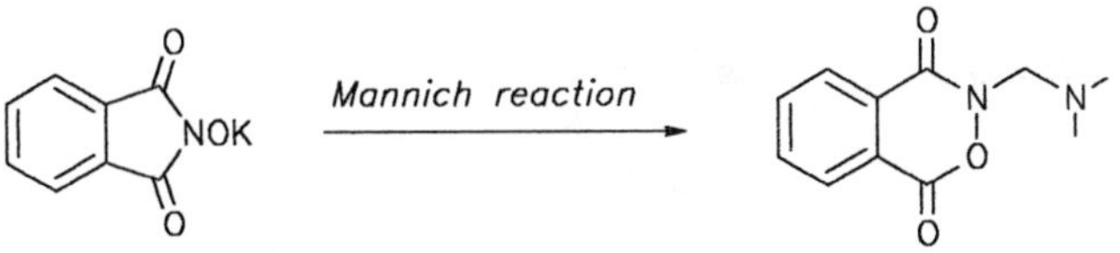

Fig. 70. Simultaneous ring enlargement-aminomethylation reaction of N-hydroxy-phthalimide salt.

Finally, it is worth mentioning a heterogeneous series of papers covering syntheses unrelated to the schemes illustrated in the present section. They deal mainly with condensation reactions, for example, the formation of cyclic acetals that were described in Sec. A.1.[564] Moreover, cyclic Mannich bases may be formed by mutual condensation of functional groups present in the aminomethylation product.[70,565] In other cases the participation of additional molecules of formaldehyde is necessary in order to produce a sufficiently stable ring derivative.[566]

This last series of cyclization reactions has been thoroughly reviewed.[270]

C.6   Stereoisomerism and Tautomeric Equilibria in Mannich Bases

An important matter connected with Mannich synthesis is the possibility of obtaining stereoisomeric products and/or derivatives exhibiting tautomerism. The preparation of enantiomeric products as well as their racemization, along with some uses of optically active Mannich bases in research, are here described. The main results of the studies of conformational and tautomeric equilibria in Mannich bases are also highlighted.

Optically Active Mannich Bases

Enantiomeric Mannich bases may be obtained either by using optically active starting materials or by optical resolution of racemic derivatives. In the former case, the reactants are mostly provided by natural products, such as components of essential oils (e.g., camphor[448]), hormones,[261] nucleic acids,[567] employed as substrates, or α-amino acids[501,568,569] mainly used as amine reagents, etc. A list of optically active reactants reported in the literature is summarized in Table 12.

As shown in the table, chiral substrates, amines, or aldehydes are employed in synthesis. Cyclic derivatives can also be obtained according to the reaction schemes reported in Sec. C.4.

Table 12

Reactants Employed for the Synthesis of Optically Active Mannich Bases[a]

Substrate	Amine	Aldehyde	Reference
β-Ketoesters*	Secondary Amines	CH_2O	570
α-Hydroxyethylbenzimidazole*	Primary Amines	CH_2O (C)	571
Galacturonic acid amide*	Secondary Amines	CH_2O	31
Alkyl Thiol*	Secondary Amines	CH_2O	572
Aralkyl Ketones, Furan	Ephedrines*	CH_2O	53,573
Alkyl Ketones, Esters	O-Acyl Glycosamines*	Ar-CHO	56,58
Phosphines	Phenylethylamine*	CH_2O	269
(R,S)-Cysteine		Galactose*, Glucose*, etc. (C)	166
Dihydroxyphenylalanine*		CH_2O (C)	574
Cu- and Co-Amino complexes*		CH_2O (C)	471,575
Cyanomethylamino	hydroxyalkenes*	(C)	260

[a] If not otherwise stated, usual amines (Fig. 4) are employed. * = Optically active. (C): cyclic Mannich base produced

The optical resolution of racemic Mannich bases has been applied mainly to ketobases (Table 13), preferably employing (−)-dibenzoyltartaric acid as a resolving agent.

A considerable number of compounds have been subjected to this type of optical resolution, the absolute configuration of the chiral centers having also been established for some of them.[576,580,584] Some unsuccessful attempts at resolution, due to not readily explainable reasons, are also reported, with the tendency to racemize being a peculiar feature of the ketonic Mannich bases. As a consequence, the goal of obtaining appreciably high optical purity appears to be hardly achievable by this method.[258] Optically active dibenzoyltartaric acid is also effective in the resolution of Mannich bases derived from indole (Table 13), whereas the cyclic derivatives **199** (Fig. 71), produced from racemic cysteine (see Table 12) and D-(+)-galactose as an optically active aldehyde reagent, are used for the resolution of cysteine (**200**).[166]

Racemization of the Mannich base may be caused by intrinsic instability[575,580] of the optically active final product, or it may occur during the synthesis or the optical resolution. The former case is frequently observed in the preparation of cyclic derivatives of natural products and concerns Mannich reactions of different types, including the tandem aza-Cope–Mannich rearrangement, which affords more or less extensively racemized products starting from optically active materials.[55,585,586] This finding is explained on the basis of the equilibrium involved in the 3,3-rearrangement leading to ketones **201** (Fig. 72), key intermediates for the synthesis of alkaloids.

Table 13

Optical Resolution of Mannich Bases by (−)- or (+)-Dibenzoyltartaric Acid

Mannich base	R[1]	-NR[2]$_2$[a]	Reference
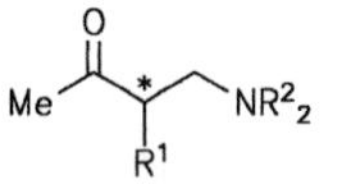	Me	DMA, PIP	576,577
	CH$_2$Ph	DMA, PIP [b]	578
	Ph	DMA	579
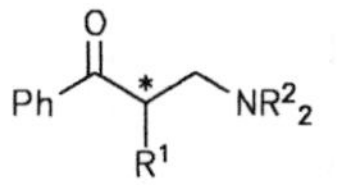	Me	DMA, PIP, MOR	579,580
	n-Pr, n-Bu	PYR	581
	CH$_2$Ph	DMA[b], PIP	264
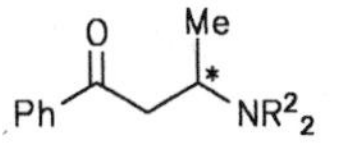		DMA[b], PIP	264,582
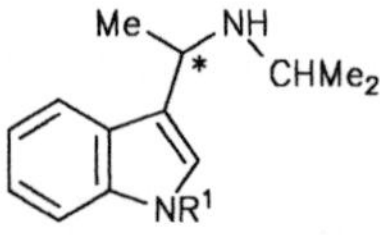	H, Me		583

[a] DMA: dimethylamine; PYR: pyrrolidine; PIP: piperidine; MOR: morpholine
[b] Attempts to obtain optical resolution of this base were unsuccessful (ref. 578, pp. 237, 356, 394)

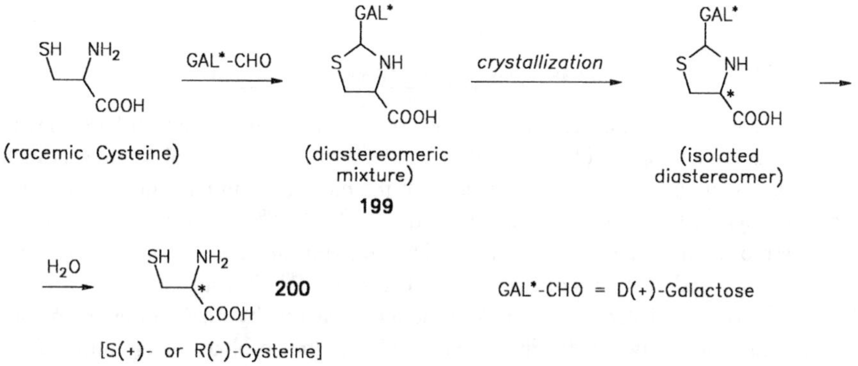

Fig. 71. Optical resolution of (±)-cysteine.

The optically active β-aminoketones reported in Table 13 may be involved in racemization processes occurring in the course of the optical resolution carried out with dibenzoyltartaric acid.[586] Indeed, the diastereomeric salt which initially precipitates (optically active)   **201**

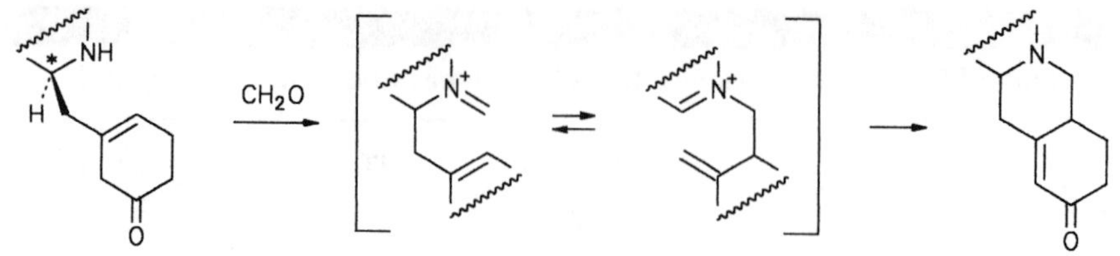

Fig. 72.   Racemization of an optically active aminic substrate upon Mannich cyclization with formaldehyde.

readily, corresponding to the amount of about 50% of the starting racemic aminoketone, proves to be identical to the product that is formed after longer deposition times from the remaining part of the Mannich base. The reason for this is that aminoketone racemizes in solution and the same enantiomer always reacts to give the less soluble salt, thus producing only one diastereomeric derivative. However, the resulting final free aminoketone is not optically pure, due to inhomogeneity of the salt and/or the alkaline treatment required in order to free the base. Recrystallization of the salt usually does not improve optical purity and may even lead to decomposition of aminoketone. This behavior is attributed to easy enolization of the carbonyl group, as depicted for the keto-enolic equilibrium **a/b** of the ketobase **202** (Fig. 73). However, the participation in the equilibrium of forms of type **c,** justifying the behavior of β-substituted ketobases or some deaminoalkylation processes observed in the synthesis of alkaloids[587] cannot be excluded. Moreover, a role may also be played by tartaric acid, as the carboxyl group appears to have a notable effect on the mechanism of synthesis, decomposition, and racemization of Mannich bases (see **44, 45,** Sec. B.1).

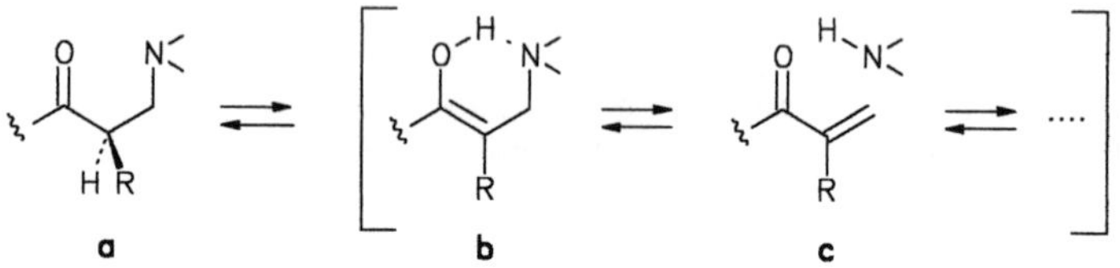

a   b   c

202

Fig. 73.   Equilibrium species afforded by the β-aminoketone molecule.

For fundamental research, the relevance of optically active Mannich bases lies in the stereochemical aspects of their synthesis and the study of structural parameters such as absolute configuration and conformation of the molecules. In particular, besides interesting asymmetric syntheses of amino acids, etc.[56,58,588,589] (see also Fig. 39) and the resolution of both the enantiomers of cysteine **200**, optically active Mannich bases are usefully employed in studies of reaction mechanisms;[260,590] they are employed particularly in the catalyzed stereoselective de-deuteration reaction by means of an aminomethylated camphor derivative.[448] Finally, knowledge of the absolute configuration of the starting ketobases has made it possible to investigate the stereochemistry of the synthesis of diastereomeric aminoalcohols;[591] also, the possibility of running circular dichroism and optical rotatory dispersion spectra has made it possible to study the configuration and conformation of various chiral derivatives ranging from metal complexes to biological molecules.[471,567,577] Other applications, for example, in the field of compounds possessing peculiar biological activity, are described in the appropriate chapters.

Conformational Stereoisomerism of Mannich Bases

Stereoisomeric conformational equilibria have been investigated on alkyl ketonic and, for the most part, ortho-phenolic Mannich bases. In particular, it has been found that the acyclic ketobases with a chiral center in the α position with respect to the carbonyl group behave differently, depending on whether they are in the form of salt or of free base; the preferred conformation of hydrochlorides is determined mainly by the repulsion among polar groups rather than by the steric hindrance of the substituents.[577] Cyclic ketobases, particularly oxa-azabicyclononanones **203**, exhibit exceptionally high conformational stability of both the **a** and **b** forms, as is demonstrated by the possibility of their physical separation.[592]

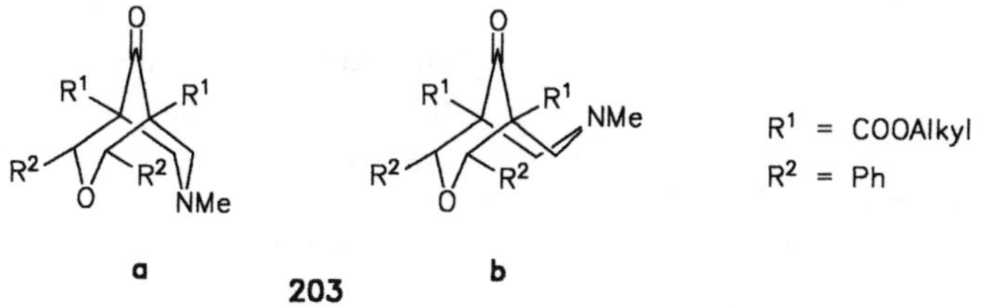

R^1 = COOAlkyl
R^2 = Ph

a   **203**   **b**

Phenolic Mannich bases having the aminomethyl group in the ortho position (**204**, Fig. 74) and, similarly, tropolone derivatives[227] show prevailing closeness between amino and hydroxy groups. This is due to stabilization of the equilibria **a** and **b** by the presence of intramolecular hydrogen bonding; the occurrence of intermolecular hydrogen bonding has been also observed.[593]

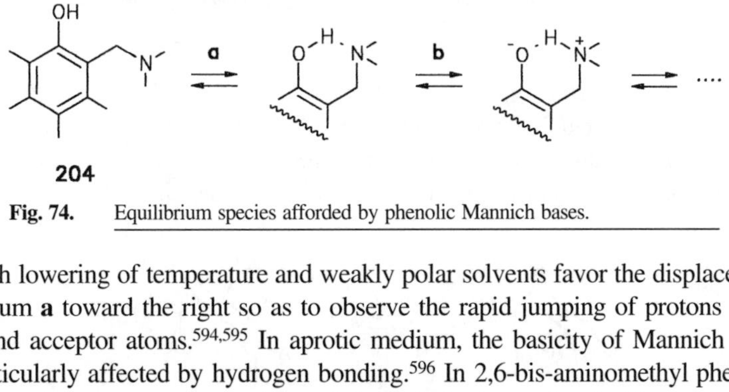

204

Fig. 74.   Equilibrium species afforded by phenolic Mannich bases.

Both lowering of temperature and weakly polar solvents favor the displacement of equilibrium **a** toward the right so as to observe the rapid jumping of protons between donor and acceptor atoms.[594,595] In aprotic medium, the basicity of Mannich bases is also particularly affected by hydrogen bonding.[596] In 2,6-bis-aminomethyl phenols the phenolic hydrogen atom may participate in both the hydrogen bonded forms to an extent that depends on the nature of the ring substituent.[597,598]

Steric factors may influence the conformational equilibria of phenolic Mannich bases, as is demonstrated by the quantitative relationship observed between the molecular volume of alkyl substituents linked to the amino group and the thermodynamical parameters.[599,600]

Tautomerism and Other Equilibria

As mentioned before, racemization of optically active Mannich bases may take place due to 1, 3 rearrangement in cyclic derivatives or to the establishment of tautomeric

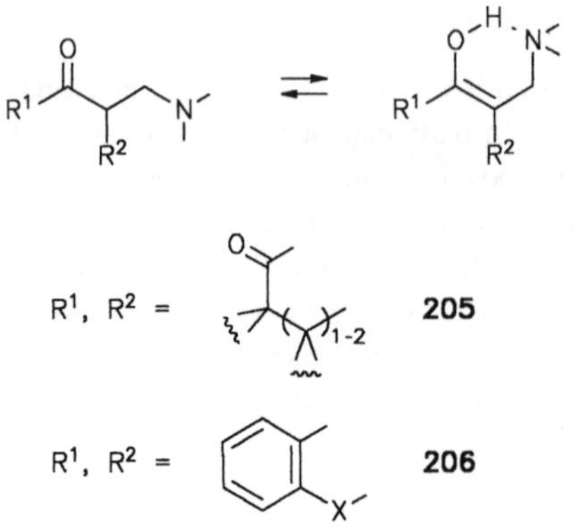

R^1, R^2 =   **205**

R^1, R^2 =   **206**

Fig. 75.   The tautomeric equilibrium in β-aminoketones.

keto-enolic equilibria (**201, 202**). The existence of enolic forms has been proved in Mannich derivatives of α-diketones (**205**) and heterocyclic ketones (**206**) (Fig. 75). In the case of compounds **205** only the enol form is present,[326,327] whereas in **206** the enolic form is predominant over the keto form, regardless of whether the Mannich base is free or is present as a hydrochloride salt.[173]

More complex equilibria (Fig. 76), also involving chemical modifications, are shown by the aminomethyl derivatives of α-arylidene-β-ketoesters (**207**), studied as anticancer drugs;[601,602] these have been found to be in equilibrium with cyclic quaternary ammonium salts, which in turn are characterized by the presence of keto-enol equilibrium.

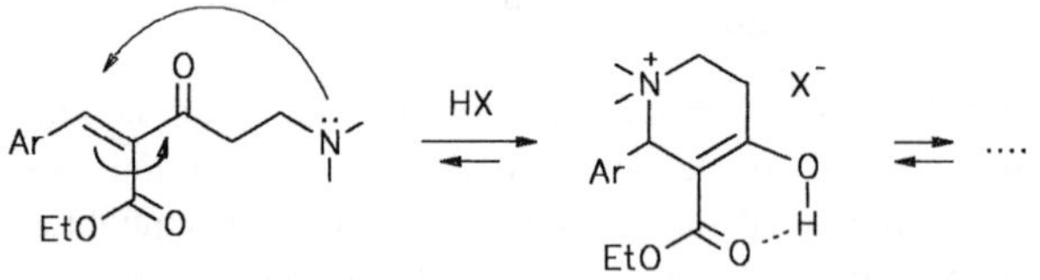

207

Fig. 76.   Equilibrium reactions involving cyclization in styryl Mannich ketobases.

In γ-piperidones **208** (Fig. 77), which have been thoroughly studied[603–608] using various techniques, including NMR and X-rays,[606–608] the enol form is not only particularly stable which allows its physical separation,[605,606] but it even leads to the isomerization reaction involving cleavage of the C—N bond (Chap. II, A.2).

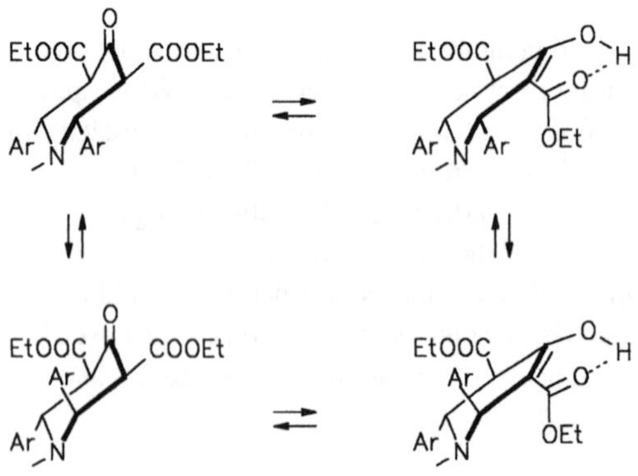

208

Fig. 77.   Equilibrium species afforded by γ-piperidone carboxyesters.

Breakage of the C—N bond is also observed in aminomethyl derivatives of benzotriazoles (**209**, Fig. 78)[280,310,609] and analogous systems,[432] where the preferred formation of the 2-aminomethyl derivative increases with decreasing polarity of the reaction medium. In this case, however, the breaking of the C—N bond produces deaminomethylation[609] (Chap. II, A.1), and the equilibrium appears affected, not only by the solvent's polarity, but also by the steric hindrance of the substituents.[310]

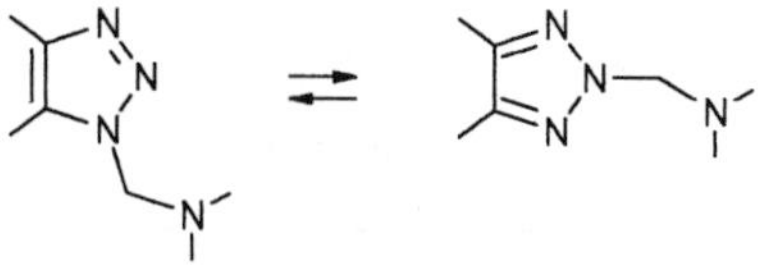

209

Fig. 78. Equilibrium species of aminomethyl-triazole derivatives.

D Survey of Substrates Subjected to Mannich Reaction

The vast and steadily increasing volume of research into Mannich reactions makes it impossible to exhaustively review here the papers published on the topic with the same thoroughness that characterized previous publications that dealt with the literature up to the 1970s.[4,6,7,10] Hence, we limit ourselves to a synthetical survey of the type of substrates employed in Mannich synthesis, with particular emphasis on the applications of the reaction. Updated references, however, enable the reader to search the literature to retrieve more comprehensive information on individual topics.

For the sake of clarity, substrates have been divided into two main categories, namely, compounds having the reactive hydrogen atom linked to carbon (affording C-aminomethylation) and those having it linked to a heteroatom X = N, P, O, S, etc., thus affording X-aminomethylation. As far as C-aminomethylation is concerned, substrates are further grouped according to the sp^3, sp^2, or sp hybridization of the reaction site (Tables 14, 15 to 17, and 18, respectively), whereas X-aminomethylations are listed in Tables 19, 20 (aliphatic and aromatic NH substrates, respectively), and 21 (remaining XH substrates with X $\neq$ N).

D.1 CH Substrates

C-Aminomethylation of compounds having the reactive center sp^3 hybridized (**210**, Table 14) requires the presence in the substrate molecule of an electron-acceptor or electron-attracting group (Sec. B.2) in order to favor the mobility of the hydrogen atom and its replacement by the aminomethyl moiety to give the Mannich base **211**. When this is insufficient for the reaction to occur, the presence of an additional activating group or the use of particularly reactive reagents (Sec. B.1) under more accurately selected reaction conditions may be necessary. Carboxy derivatives **210c**, for example, which are weakly reactive, may however produce decarboxylation when a further carbonyl group is linked to the substrate[251,617] (see Fig. 57). The presence of two RSO or RSO$_2$ groups **210j** is usually required in sulfoxide or sulfone substrates for obtaining Mannich derivatives.

Table 14

C-Aminomethylation Reactions on CH-Activated Aliphatic Substrates 210

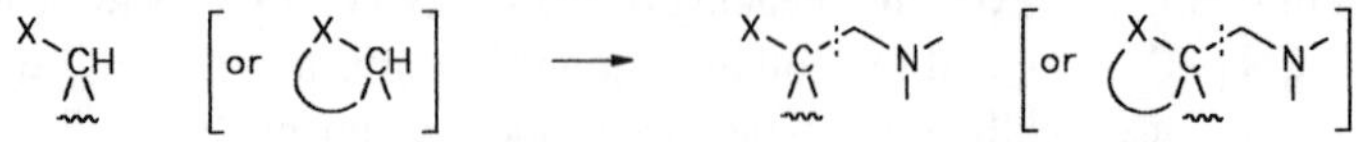

210   211

Substrate[a] 210	Mannich base 211	Reference
a Aliphatic Aldehydes	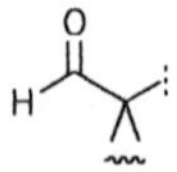	40,43,170,610
b Alkyl-Ketones (c)	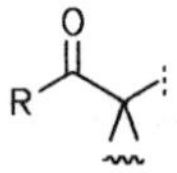	
Dialkyl-		44,186,611
Vinyl Alkyl-	(R = Alkyl, Ar,...)	354,602,612
Aryl Alkyl-		613-616
Heteroaryl Alkyl-		335
c Alkyl Carboxyl Deriv.s (c)	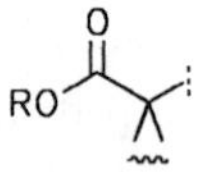	
Acids		617
Esters	(R = H, Alkyl)	49,177,618
Lactones		46,59
d Vinylogous Alkyl Carbonyl Deriv.s (c)	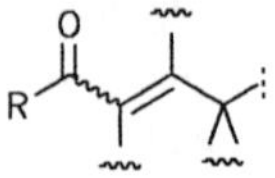	
Aldehydes		373
Ketones	(R = H, Alkyl,...)	41,361,619
Esters		620,621
e Nitro Toluenes		23,299
	(R = COOH, NO_2)	

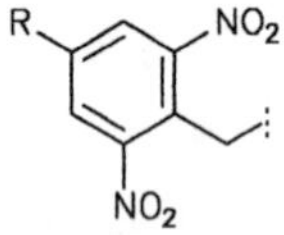

Alkenes that lack activating groups (**212a**, Table 15) or contain carbonyl **212b** and thiocarbonyl[488] groups, including push-pull olefines, and aza-alkenes such as oximes and hydrazones **212c,d**, yield unsaturated Mannich bases **213**. Concurrent side reactions may take place with these substrates, particularly with alkenes, whereas vinyl carbonyl substrates are frequently involved in chemo- or regioselective reactions (Secs. C.1 and 2, Table 11, and Ref. 270).

Table 14

(continued)

f	Schiff Bases, Hydrazones, Oximes	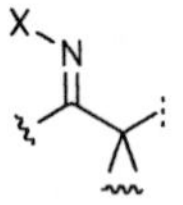	363,416
	(X = Alkyl, N, O)		
g	Alkyl N-heteroaryl Deriv.s (c)	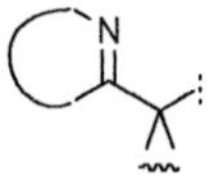	
	Five-membered rings		236,259,622
	Six-membered rings		235,623-625
h	Alkyl Nitriles	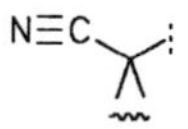	416,617
i	Nitro Alkanes	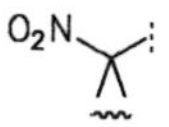	113,626-628
j	Alkyl Sulfones	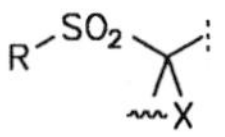	329,330,629,630
	(X = Activating group)		

[a] (c): cyclic substrates included

Homocyclic aromatic substrates **214a–f** (Table 16) include benzenoid derivatives **214a–c,** which are mostly represented by phenols and give aminomethylation in the ortho or para position (Sec. C.2). Mannich reactions on the corresponding phenolethers are much less frequent, and recent papers deal actually only with cyclizations (Sec. C.4).[251,270] Arylamines and sulfonamides are usually aminomethylated in the para position (**215b,c**); the sulfonamides, in particular, are unreactive when the H atom of NH—SO$_2$ is replaced by an alkyl group.

Table 15

C-Aminomethylation Reactions on Alkene and Aza-Alkene Substrates 212

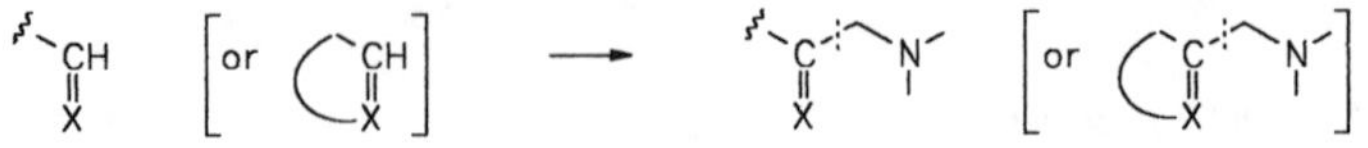

212   213

Substrate[a] 212	Mannich base 213	Reference
Alkenes		
a Alkenes (c)	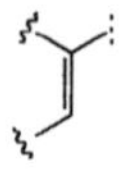	631 and Sec. C.5
b Vinyl Carbonyl Deriv.s (c)	(R = H, Alkyl, Ar, O)	
Cyclohexenones and Quinones	(Z = C)	96,359,632
Push-pull olefines	(Z = N)	393,478,633,634 and Sec. C.2
Pyrone Deriv.s, Hydroxycoumarins	(Z = O)	635-638
Azaalkenes	(R = Ar, NO₂,....)	
c Hydrazones	(Z = NH-Ar)	479,480
d Oximes	(Z = OH)	364,639

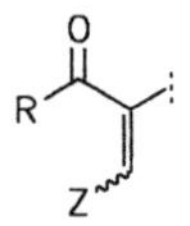

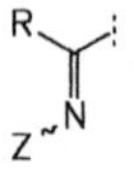

[a] (c): cyclic substrates included

Heterocyclic aromatic substrates (Table 17) constitute a very large and varied group of compounds, including five-membered **216a–c** as well as six-membered **216d, e** ring derivatives. The Mannich reaction on these substrates, in particular on **216a, c**, and **d**, may be chemo- and/or regioselective (Secs. C.1,2). The presence of tautomeric keto-enol equilibrium displaced towards the keto form, due to the nature of the substituent linked to the N atom, for example, as in the case of uracil Mannich bases **217e** and similar compounds,[669] is also observed.

Table 16

C-Aminomethylation Reactions on Aromatic Substrates 214

214 → **215**

Substrate **214**	Mannich base **215**	Reference
Benzenoid Deriv.s	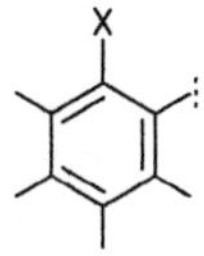	
a Phenol Deriv.s	(see structure) (o- and p-derivatives)	
Phenols	(X = OH)	408,640-642 and Sec. C.2
Phenolethers	(X = OAlkyl)	643 (see text)
b Arylamines	(X = NH$_2$, NHAlkyl,...)	415,644-646
c Sulfonamides	(X = NHSO$_2$R)	27
Nonbenzenoid Deriv.s	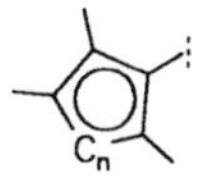	
d Azulenes	(n = 1 or 3)	647,648
e Tropolones		227,649
f Metallocenes		30,275,650

Substrates having *sp* hybridization of the reactive C atom (Table 18), such as alkynes, make it possible to widen considerably the synthetic potential of the Mannich reaction, as it may be applied to variously monosubstituted acetylene derivatives. Examples **218a–c**, selected from a large number of papers, are intended to stress the strong possibility of a wide range of combinations, among which are the insertion of unusual elements (Si, Ge, etc.) into the Mannich product and the interesting potential use of bis-Mannich bases **219c** as bifunctional monomers in macromolecular chemistry. Hydrogen cyanide **218d** is also important on account of its Mannich derivatives **219d**, which are the precursors of α-amino acids.

Table 17

C-Aminomethylation Reactions on Heterocyclic Aromatic Substrates 216

	Substrate 216	Mannich base 217	Reference
	Five-membered rings		
a	N-Heterocyclics[a]	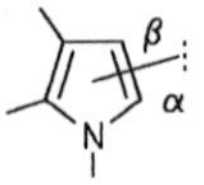	
	Pyrroles		211,375,651
	Indoles		418,421,652,653
	Indolizines		425,426,654
	Azaazulenones		655
b	O-, S-Heterocyclics	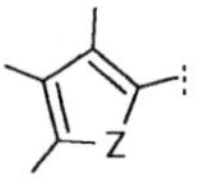	
	Furans and Oxaazulenones	(Z = O)	29,656-658
	Thiophenes	(Z = S)	196,349,427
c	Polyheteroatomic rings[a]	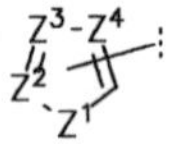 (Z^1 = N, S; Z^2-Z^4 = C, N)	
	Pyrazoles, Imidazoles, Thiazoles		431,495,659
	Tri-, Tetrazoles		660-662

D.2   XH Substrates

When amines **220a–c** (Table 19) are used as substrates in Mannich reaction, they are formally distinguished from the amines employed as reagents by their lower basicity. Acyclic and cyclic amides **220d–g** are also largely used as substrates. Among the latter, the isatins **220f**, their isomers such as phthalimide **220g** and its well-known sulfonyl analog, saccharin,[715] are worth mentioning. Some NH-substrates are studied mostly in relation to Mannich reactions leading to ring formation, as in the case of hydrazines **220b** and hydrazides **220e**. These compounds may also give more complex reactions (Sec. C.3 and Refs. 496 and 716), the same being true of imines **220c**, which are usually guanyl derivatives easily cyclizable through the vicinal amino group.

Table 17

(continued)

Six-membered rings

d	Pyridines and Quinolines (N-oxides included)[a]		
	3-Hydroxypyridines	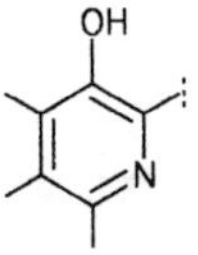	433,434,663
	Hydroxyquinolines and Naphthyridines	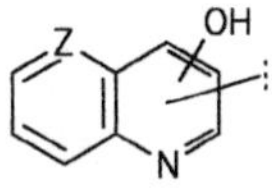 (Z = C, N)	435,493,664
e	Polyheteroatomic rings[a]		
	5-Hydroxypyrimidines and N-oxides	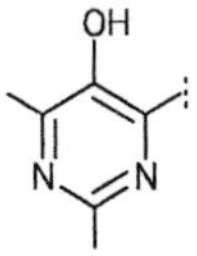	436,437,665
	Uracils	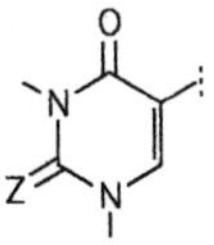 (Z = O, S)	666-668

[a] For the reaction site see Sec. C.2

Aromatic N-heterocyclic substrates (Table 20) are divided between five-membered derivatives **222a**, having the N—H reactive center not involved in tautomeric equilibria, and five- or six-membered derivatives **222b**, in which the reactive N—H group is the aminic form participating in a tautomeric imine-amine equilibrium (Fig. 79) (see also **70**, Table 8):

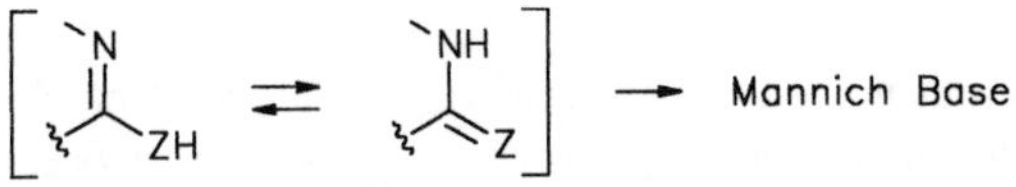

Fig. 79. The tautomeric equilibrium in N-heteroaromatic derivatives.

Table 18

C-Aminomethylation Reactions on Acetylenic Substrates and on Hydrogen Cyanide 218

$$X\equiv CH \longrightarrow X\overset{\text{---}}{=}\overset{\cdot\cdot}{C}\sim N\sim$$
$$(X = C, N)$$

| 218 | | 219 |

Substrate **218**	Mannich base **219**	Reference
Alkynes		
a　Ethynyl Deriv.s	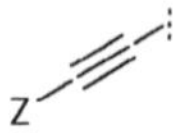 (Z = Ar, R₃Si, R₃Ge) $(Z = Ar, R_3Si, R_3Ge)$	24,670,671
b　2-Propynyl Deriv.s	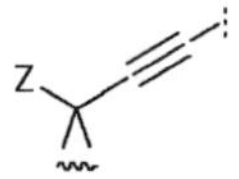	
Propargyl alcohol Deriv.s	$(Z = OH, OR, OSiR_3,...)$	672-675
Propargyl-amine, -amide Deriv.s	$(Z = NR_2, NRAcyl)$	676-679
Propargyl thio Deriv.s	$(Z = SR)$	680,681
c　Bis-Ethynyl Deriv.s	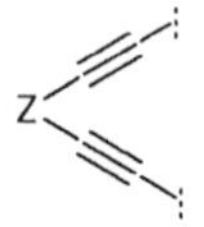 (Z = Phenylene, Bis-methyleneoxy,...)	682-685
Hydrogen Cyanide		
d　HCN	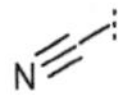	686-688

It is finally worth noting the synthesis of pyrrole N-Mannich bases,[717] which is actually a quite special type of intramolecular rearrangement occurring on the N-trimethylammonium salt of pyrrole.

Table 19

N-Aminomethylation Reactions on Aminic and Amidic Substrates 220

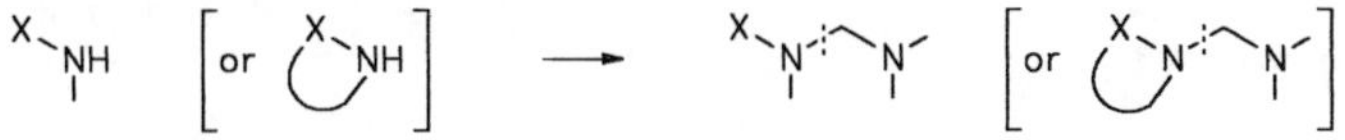

$$\text{X}_{\diagdown\text{NH}} \quad \left[\text{or} \quad \diagup\text{X}_{\diagdown\text{NH}} \right] \quad \longrightarrow \quad \text{X}_{\diagdown\text{N}\diagup\diagdown\text{N}\diagdown} \quad \left[\text{or} \quad \diagup\text{X}_{\diagdown\text{N}\diagup\diagdown\text{N}\diagdown} \right]$$

| | 220 | 221 | |

	Substrate[a] **220**	Mannich base **221**	Reference
	NH-Amines		
a	Ammonia, Prim. and Sec. Amines (c)	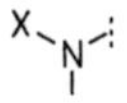 $\text{X}_{\diagdown\text{N}}$ (X = H, Alkyl, Ar)	212,689,690
b	Hydrazines and Hydroxylamines (c)	(X = NR$_2$, OR)	202,416,461
c	Imines	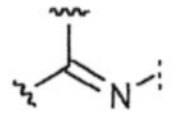	249,483
	NH-Amides		
d	Acyclic Amides	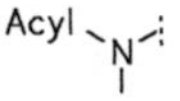 Acyl$_{\diagdown\text{N}}$	
	Carboxy Amides	(Acyl = RCO, R$_2$NCO)	230,691,692
	Carbothioamides	(Acyl = RCS)	693,694
	N-, S-, P-Acid Amides	(Acyl = NO$_2$, RSO$_x$,...)	71,258,416, 695-698
e	Hydrazides and Hydroxamic acid Deriv.s	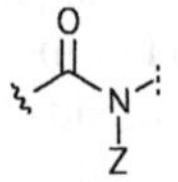 (Z = NR$_2$, OR)	699-701
f	Cyclic Amides	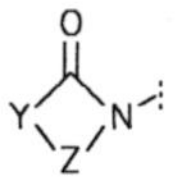	
	Lactams	(Y-Z = C$_{3-5}$)	702,703

Table 19		
(continued)		
Cyclic Urea and Thiourea Deriv.s	(Y-Z = N-C$_{2-3}$)	371,704-706
Isatin and Pyrazolone Deriv.s	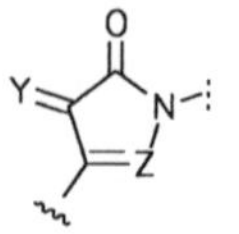 (Y = N, O; Z = C, N)	707-709
g Cyclic Imides	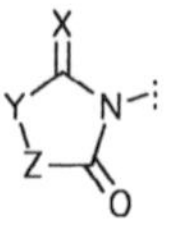 (X = O, N, S)	
Succinic, Glutamic,... Deriv.s	(Y-Z = C$_{2-3}$)	112,113,710
Phthalimide and Aza-Deriv.s	(Y-Z = e.g., 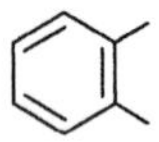)	711-713
Hydantoin Deriv.s	(Y-Z 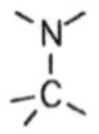 =)	367,714

^a (c): cyclic substrates included

Substrates **224**, employed in X-aminomethylation (X ≠ N, Table 21), bear heteroatoms belonging to groups V and VI of the periodic table. Apart from oxygen, all these elements are not in their maximum oxidation state, thus making it possible to obtain aminomethyl derivatives directly bound to the heteroatom. Many of these derivatives, particularly P- and S-Mannich bases **225a, d**, find practical application as complexing agents, antioxidants, etc. (Chap. V).

Table 20

N-Aminomethylaton Reactions on N-Heterocyclic Aromatic Substrates 222

222   **223**

Substrate **222** [a]	Mannich base **223**	Reference
Five-membered rings		
a Pyrroles and Aza-Deriv.s	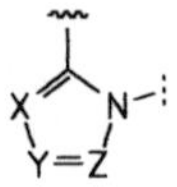 (X, Y, Z = C, N)	
Pyrroles, Indoles		365,717 and Sec. C.1
Pyrazoles		251,252
Imidazoles		222,277,718,719
Tri-, Tetrazoles		66,280,719,720
b Substrates reacting as tautomeric forms	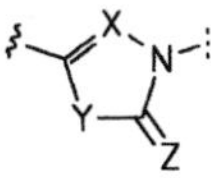 (Z = N, O, S)	
Oxazoles, Isoxazoles	(X = C; Y = O)	721,722
Thiazoles	(X = C; Y = S)	723-725
Triazoles, Oxa- and Thiadiazoles	(X = N; Y = N, O, S)	726-728
Six-membered rings		
c Substrates reacting as tautomeric forms	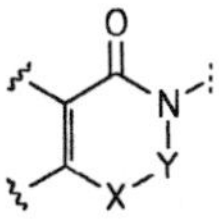	729,730
Maleic and Phthalic Hydrazides, Barbituric acid Deriv.s, Uracil Deriv.s	(X, Y = C, N)	
Cyanuric acid Deriv.s		731

[a] Fused ring derivatives included

Table 21

X-Aminomethylation Reactions on XH-Substrates 224

$$R-XH \quad \longrightarrow \quad R-X\overset{\cdot\cdot}{\diagup}\diagdown N\diagup$$
$$(X = P, As, O, S, Se)$$
$$\mathbf{224} \qquad\qquad\qquad \mathbf{225}$$

Substrate **224**	Mannich base **225**	Reference
a PH-Substrates		
Phosphines, Silyloxy-phosphines	$R_2P\overset{\cdot\cdot}{\diagup}$ (R = Alkyl, Ar, Me$_3$SiO)	34,589,732,733
Phosphine oxides	Ph$_2$PO$\overset{\cdot\cdot}{\diagup}$	734
Phosphorous acid and esters	RO$\diagdown$PO$\overset{\cdot\cdot}{\diagup}$ $\underset{OR}{\mid}$ (R = H, Alkyl)	735-738
b AsH-Substrates		
Arsines	$R_2As\overset{\cdot\cdot}{\diagup}$ (R = Alkyl, Ph)	203
c OH-Substrates		
Alcohols	AlkylO$\overset{\cdot\cdot}{\diagup}$	739-742
Picric acid, oximes,...		33,363,455
d SH-Substrates		
Hydrogen sulfide, Alkyl thiols	RS$\overset{\cdot\cdot}{\diagup}$ (R = H, Alkyl)	214,253,743,744
Aryl thiols	ArS$\overset{\cdot\cdot}{\diagup}$	745-748
Thiocarboxy acids	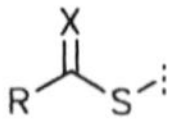 $R\overset{\overset{\displaystyle X}{\|}}{\diagdown}S\overset{\cdot\cdot}{\diagup}$ (X = O, S)	749,750
Sodium bisulfite, Sulfinic acids	RSO$_2\overset{\cdot\cdot}{\diagup}$ (R = ONa, Alkyl, Ar)	215,268,751
e SeH-Substrates		
Selenophenol	PhSe$\overset{\cdot\cdot}{\diagup}$	752

Chapter II

Reactions of Mannich Bases

The relevance of Mannich bases in organic chemistry has been pointed out in Chap. I, which stresses the versatility of the reaction (Fig. 3, Chap. I), with particular reference to the possibility of synthesizing a large number of derivatives starting from the most varied substrates.

A further, equally relevant contribution of Mannich bases to organic chemistry is their use as intermediates in the synthesis of other compounds, as indicated in Fig. 80. This shows the variety of reactions that Mannich bases may provide, such as the formation of unsaturated derivatives, amino group replacement, cyclization, etc. Other reactions (reduction and oxidation, acylation, etc.), involving substrate or amine moiety, are also reported in the literature. The polymerization reaction, not included in the figure, is treated in Chapter III.

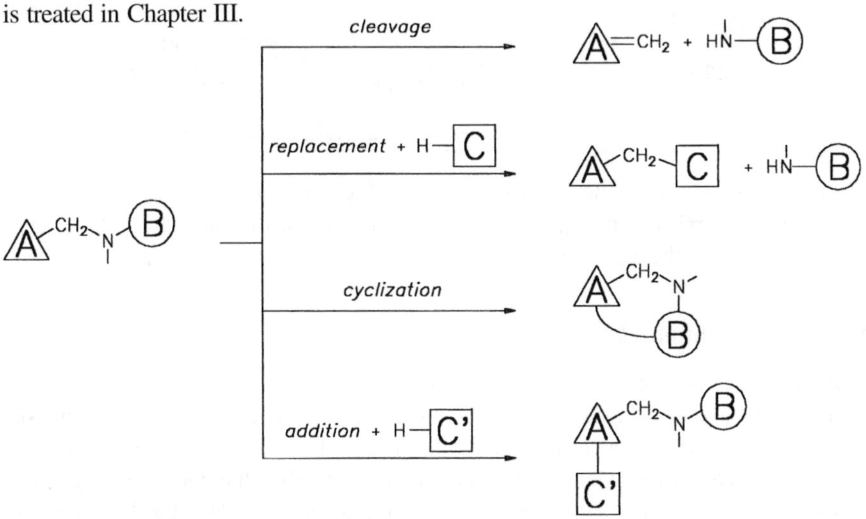

Fig. 80.   Main reactions given by Mannich bases.

A particularly important and typical reaction of Mannich bases is their cleavage, producing deaminomethylation or deamination. The reaction is strictly connected with the stability of these compounds and may usefully serve for the synthesis of derivatives and for other practical purposes.

A Cleavage

Under suitable, even relatively mild conditions, Mannich bases may undergo bond cleavage at the methylene bridge linking substrate and amine moiety. The reaction

(Fig. 81) may occur on either the substrate (**a**) or the amine (**b**) side, thus affording deaminomethylation or deamination, respectively. Although the possibility of simultaneous reaction exists, one process usually prevails over the other,[753,754] depending on the chemical nature of the base. In general, C-Mannich bases, particularly the ketonic ones, are more stable toward deaminomethylation, the reverse occurring with X-Mannich bases. There are, however, important exceptions to this behavior.

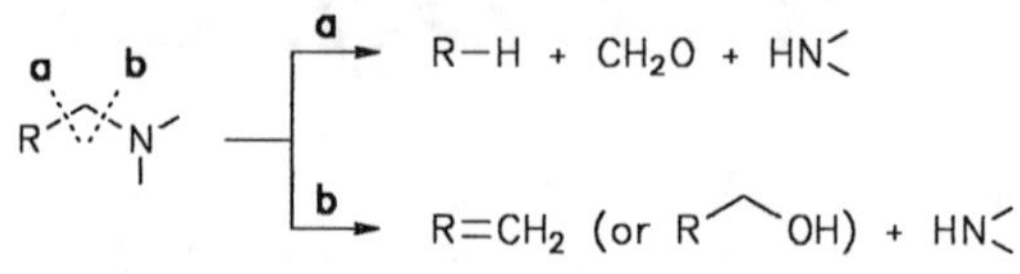

Fig. 81.   Cleavage reactions of Mannich bases.

The reaction conditions (pH, medium) may also affect the predominance of one type of cleavage, as reported for some oxadiazole derivatives (**226**, Fig. 82), which by deamination give a methylol derivative (**227**) instead of the unsaturated product.[755]

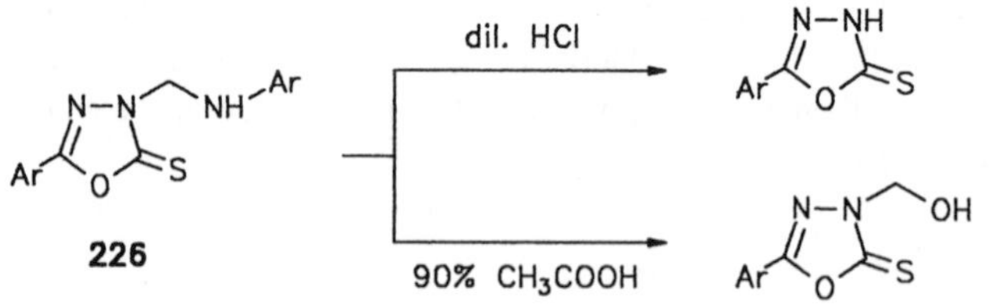

Fig. 82.   Deaminomethylation *vs.* deamination in oxadiazole derivatives.

Positive (*trans*-aminomethylation, synthesis of unsaturated derivatives, etc.) as well as negative (instability of the base, by-products in the course of Mannich synthesis, etc.) consequences may therefore be expected by the cleavage of Mannich bases.

A.1   Deaminomethylation

Deaminomethylation of Mannich bases (path **a** in Fig. 81), which is actually a retro-Mannich reaction, as it reproduces the same reactants employed in the direct synthesis, has been observed in practically every class of Mannich derivative (Table 22 and Figs. 186, 189 in Chap. V). C-Mannich bases are more stable than the corresponding derivatives, having the aminomethyl group linked to a heteroatom. This has been directly found, for example, in barbituric acid derivatives,[377] although the C-Mannich bases deriving from particularly activated substrates, such as nitroalkanes, also exhibit a remarkable tendency to decompose.[130]

Among cyclic O-Mannich bases, the dihydrooxazines deriving from phenols and primary amines (Chap. I, C.4), are useful starting materials for the clean production, by hydrolysis, of phenolic Mannich bases having a secondary amino group.[258,760] Hydrolytic deaminomethylation has been tried under acidic as well as alkaline conditions, with the former proving more effective. The acidic treatment requires only a few minutes at room temperature compared with neutral or alkaline hydrolysis, which occurs at higher temperatures and/or takes much longer.[377,630,759] The mechanism of hydrolysis has been thoroughly studied[130,761] on both C- and N- (amidic) Mannich bases, and the results

Table 22

Mannich Bases Submitted to Deaminomethylation Reaction

Mannich base		Reaction conditions[a]	Reference
C-Mannich bases			
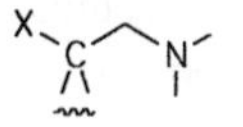	X = R—CO	B	754
	N-Heteroaryl	A	625
	NO_2	A,B	130
	RSO_2	A,B	630
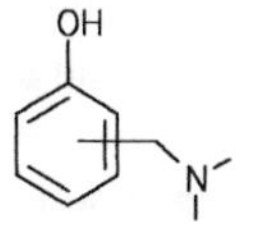	*ortho*-substituted	B	756
	para-substituted	-	189
N-Mannich bases			
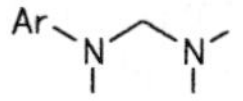	Ar = Cytosine, Guanine,...	-	757
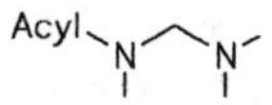	Acyl = R—CO	A,B	630,700,703
	R—SO_2	A	758
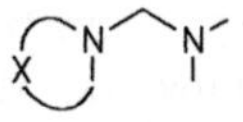	N-Heterocyclic, from Thia-, Oxadiazole, Barbituric Deriv.s	A,B	377,379,728,755
O-Mannich bases			
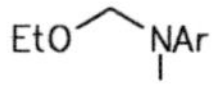 EtO⌒NAr	Ar = Adenosine, Cytidine,...	A,B	759
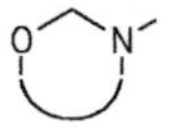		A	518,760

[a] A: acidic conditions; B: alkaline conditions

(Fig. 83) allow us to conclude that the reaction is actually a deaminomethylation of the free base, with cleavage of the C—X bond as the rate-determining step.

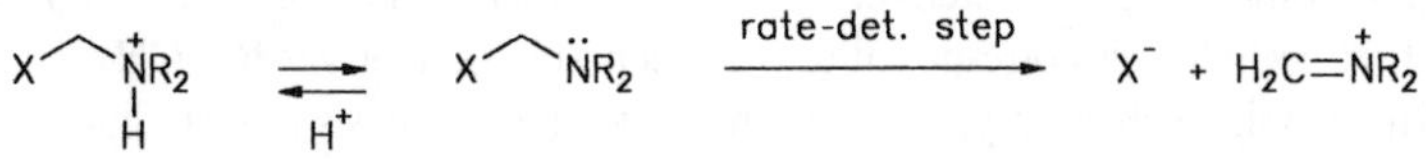

$$X\frown \overset{+}{N}R_2 \underset{H^+}{\overset{H}{\rightleftharpoons}} X\frown \overset{..}{N}R_2 \xrightarrow{\text{rate-det. step}} X^- + H_2C=\overset{+}{N}R_2$$

228 X = O_2N—CR'_2—
229 X = Acyl—NR'—

Fig. 83.   The mechanism of Mannich base deaminomethylation.

Steric hindrance as well as basicity of the amine moiety favor the cleavage of Mannich bases. In particular, deaminomethylation of **229** is enhanced by its capability to delocalize the negative charge on the amidic nitrogen.[15,761] A similar charge delocalization on carbonyl groups could be at the basis of the same reaction observed on aza-adamantanones.[762] On the other hand, studies of aminoalkyl sulfides in an aqueous acidic medium suggest the occurrence of an analogous cleavage directly given by the protonated Mannich base, the reaction being considered as an example of the S_{N1} mechanism afforded by the S-alkyl bond of a sulfide.[763]

The ionic products of deaminomethylation usually interact with the water molecules of the medium. The methyleneimmonium ion then evolves to aldehyde and amine or reacts with another substrate that may be present, thus affording transaminomethylation.

Deaminomethylation of Mannich bases may also occur in the presence of electrophilic species, capable of replacing the aminomethyl group linked to the substrate moiety, such as, for example, diazonium salts[764] acting on Mannich bases of phenols and indoles, or alcohols[765] acting on cyclic O-Mannich bases. The deaminomethylation reaction is also present in several equilibria afforded by aminomethyl derivatives of benzotriazole (see **209**, Chap. I, C.6).

Besides highlighting the reversible nature of the Mannich reaction, deaminomethylation implies as a consequence the reduced possibility of storing some types of Mannich bases, particularly when they are in the free base form, both neat and in solution. The acid salt derivatives are usually more stable, although hygroscopic to a greater or lesser extent; the hydrochlorides, however, are preferable to the carboxyacid salts, which appear in some cases to favor the decomposition of the Mannich base.

The formation of by-products[766] (Chap. I, B.3), as well as the racemization of intermediates in the synthesis of alkaloids[587] due to deaminomethylation, appear to constitute additional negative aspects of this type of reaction. However, deaminomethylation can be usefully exploited for intramolecular aminomethyl group transposition[283] and intermolecular transaminomethylation since the latter reaction (Fig. 84) involves Mannich bases as aminomethylating agents (Table 5, Chap. I, A.4). For instance, phthalimide or isatin N-Mannich bases (**230**, R = amido group) can aminomethylate alkyl ketones or sulfinic acids.[129,767]

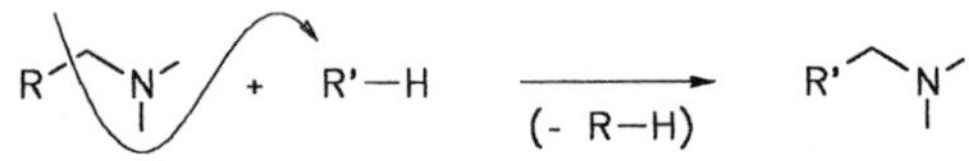

$$R \diagdown \diagup N \diagup + R'-H \xrightarrow[(- R-H)]{} R' \diagdown N \diagup$$

230

Fig. 84. The transaminomethylation reaction of Mannich bases.

Deaminomethylation, moreover, has some synthetic relevance in the conversion of Mannich bases into the corresponding methylene-bis-derivatives R—CH_2—R, as performed in tropolone chemistry,[768] and in the three-step synthesis of bispidinones starting from 3-oxoglutaric acid ester.[769] The readily occurring elimination of the aminomethyl group has been also applied as a protection method for the sulfonamide group in the synthesis of drugs.[758] The remarkable importance of deaminomethylation in the pharmacological field[770,771] is to be stressed, as several aminomethyl products are potentially useful prodrug candidates in the formation of NH-acidic compounds (amides, urea derivatives, etc.) and of primary and secondary amines. It is also noteworthy that deaminomethylation is related to the antitumoral activity of Mannich bases.[754,772,773]

A.2 Deamination

Except in rare cases, the deamination reaction (Fig. 85) is peculiar to C-Mannich bases and involves the elimination of a primary or secondary amine from **231** or of a tertiary amine from the quaternary ammonium salt **232**. The presence of a hydrogen atom in the β position with respect to the amino group is not essential, as other groups, such as the carboxyl group of β-amino acids **233** may also be eliminated in conjunction with the amine.

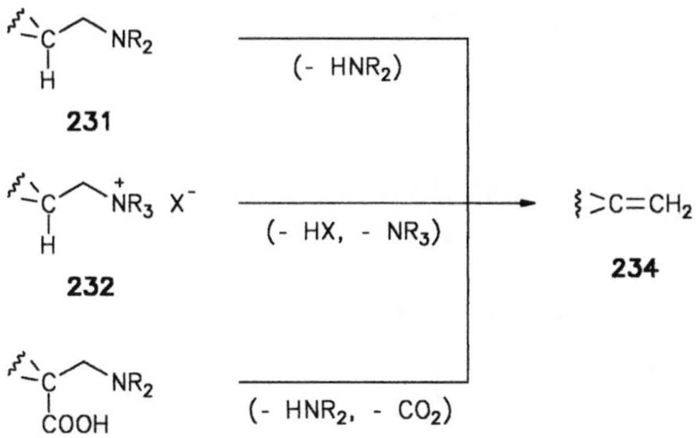

Fig. 85.   Deamination reactions given by different Mannich bases.

It is interesting to note that the production of the unsaturated derivative (**234**) and amine by deamination may be applied, for example, to the preparation of vinyl compounds for use as monomers in polymerization reactions[774] and also to industrial formulations requiring alkaline catalysis.[775]

Nearly all types of C-Mannich bases deriving from activated alkyl substrates (Table 23), including some heterocyclic bases,[776] have been submitted to deamination, the most widely studied being carbonyl and carboxyl derivatives (**234a** and **b**). Conjugated unsaturated systems starting from vinylogous ketobases and acetylenic Mannich bases (**234c** and **d**, respectively) are also obtained. Among vinylogous Mannich bases, natural products, including hormones, are frequently treated for deamination.

Several different deamination procedures are reported in Table 23, as the reaction may take place alternatively by modification of Mannich synthesis (A), by suitable treatment of a Mannich base (B), or by conversion of the base into the corresponding quaternary ammonium salt, which is more easily decomposable (C). Due probably to steric reasons, some particular amines, such as diethylamine, are more suited to the deamination process.[283,779,787]

Unsaturated products **234**, particularly α,β-unsaturated carbonyl derivatives, are sometimes hardly obtainable due to additional side reactions readily taking place in the presence of nucleophilic species, as discussed in Sec. B (following) or to dimerization and polymerization, or even isomerization, as partially illustrated in Figs. 30 and 31 (Chap. I, B.3).

The formation of dimers has been observed mainly in cyclic ketobases originating spiro derivatives of type **235** by deamination. The ketobases are usually six-membered

Table 23

Unsaturated Products Given by Deamination Reaction of Mannich Bases

Unsaturated product 234	Method [a]	Reference
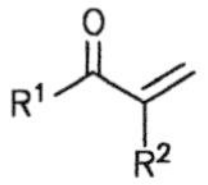 a $R^1 = H$; $R^2 = Alkyl$	A,B	777,778
$R^1 = Alkyl$, Vinyl, Ar; $R^2 = H$, Alkyl, Ar	A / B / C	779,780 / 781,782 / 333,783
R^1, $R^2 = ring$	A / B / C	283,784 / 45,785 / 42
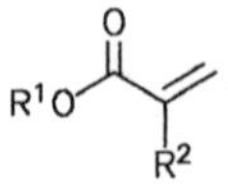 b $R^1 = R^2 = Me$ or R^1, $R^2 = ring$	A / B / C	786,787 / 262 / 39,59
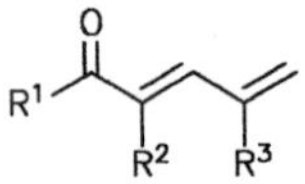 c $R^1 = H$; R^2, $R^3 = Alkyl$ or R^1-$R^3 = Hormone$ or Pinenone skeleton	A / B / C	788 / 361,373 / 41
 d $R = Alkyl$, Alkynyl	C	789
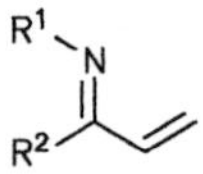 e $R^1 = OH$; $R^2 = Ar$ or R^1, $R^2 = ring$	A,C	237,790,791
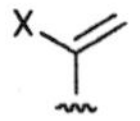 f $X = NO_2$, RSO_{1-2}	B	329,626,792

[a] A: reaction occurring during Mannich synthesis. B: deamination of the free Mannich base or the acid ammonium salt (hydrochloride, etc.). C: deamination of the tetraalkylammonium salt of the Mannich base (iodomethylate, etc.)

homocyclic ketones, including steroid derivatives[21,793,794] and sulfur-containing heterocycles.[795,796] In some cases polymeric derivatives are also formed,[783,793,797,798] with the value of the dimer-to-polymer ratio in the product being strictly related to the ring size of the ketobase.[797] Thus, polymeric products largely prevail over the dimeric ones when Mannich bases other than six-membered ones are involved in deamination.

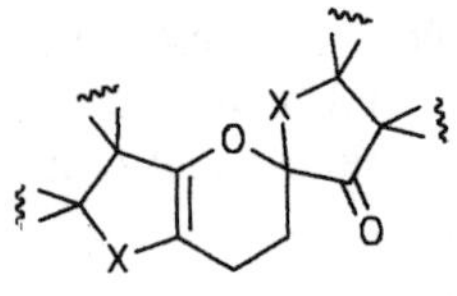

235

$X = -C_{1-3}-$; $-C-S-$

Isomerization may also modify the structure of the unsaturated product of deamination. In particular, vinyl derivatives afford more stable aromatic compounds, as reported in the case of the deamination of bis-dimethylaminomethyl cyclohexanone **236** (Fig. 86), which produces the dimethyl phenol **237** through a series of tautomeric equilibria.[799] The formation of methyl-hydroxy thiophene **238** is explained similarly.[800]

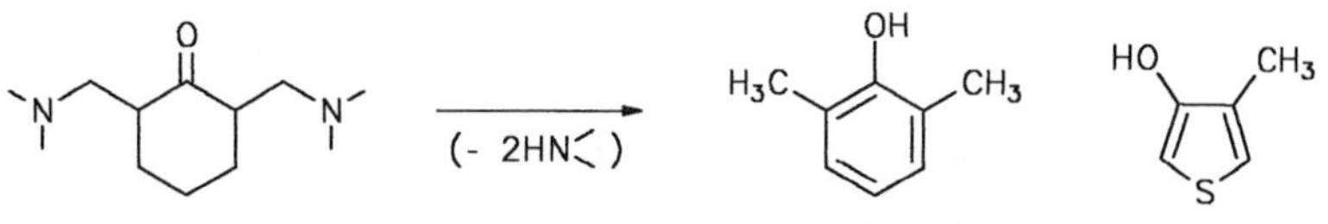

Fig. 86. Deamination of cyclohexanone Mannich bases leading to a phenol derivative.

Acetylenic Mannich bases, upon deamination, tend to produce isomerized derivatives such as allenes[801] or compounds of type **239**, bearing the alkyne moiety as an end group.[789] These compounds are occasionally found in the presence of vinyl acetylenes of type **234d** (R = Alkyl—C≡C—, Table 23).

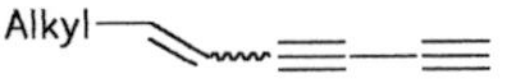

The mechanism of deamination has been frequently investigated.[258] Besides ketobases, phenol and indole derivatives have received particular

attention; the reaction is examined at different pH values of the medium. In some cases, the behavior of the corresponding tetraalkylammonium salts has also been studied. In addition, some interesting results are found in studies of the amino group replacement, as deamination represents the first step of this last reaction.

Deamination of β-aminoketones[264,802,803] involves a pre-equilibrium **a** (Fig. 87), which depends on the pH of the reaction, followed by the elimination of the amine (equilibrium **b**).

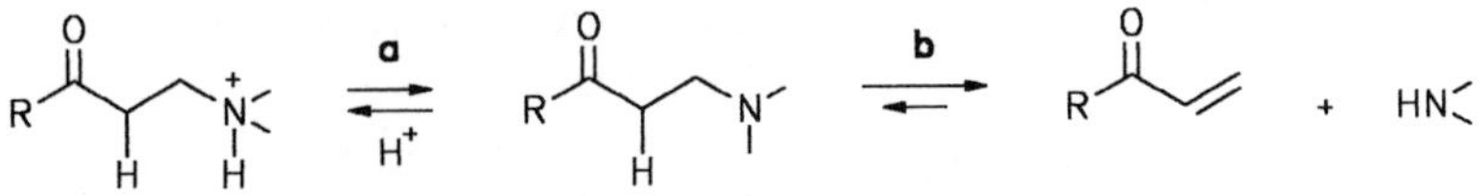

Fig. 87. The mechanism of β-aminoketone deamination.

Removal of the amine moiety in piperazine derivatives in acidic medium[803–805] is thought to occur from the enolic form of the ammonium salt, at least at very low pH values. However, step **b** has not been interpreted entirely from the mechanistic point of view, as far as the participation of ammonium/enolate structures is concerned. Several cyclic transition states have been envisaged, some of which involve the presence of a hydroxy group or the enolic form of the β-aminoketone. In particular, the participation of OH groups belonging to other molecules, such as carboxyacids present in the reaction medium, cannot be excluded.[258] By contrast, a nucleophilic attack by an acid anion on the carbonyl C atom has been claimed to be the rate-determining step for deamination in the reaction between β-aminoketones and triethyl phosphite in dimethylformamide.[806]

Deamination has also been recognized as the rate-determining step in some replacement reactions carried out on phenolic and indolic Mannich bases.[583,807,808] In phenolic derivatives, in particular (Fig. 88), a pre-equilibrium takes place similar to that described above for β-aminoketones, and is followed by the elimination of amine via an E_{1cB}-type mechanism.[807]

Similarly, the slow elimination of the amino group, occurring simultaneous with or subsequent to that of the NH proton on the indole ring, is suggested for indole derivatives.[583]

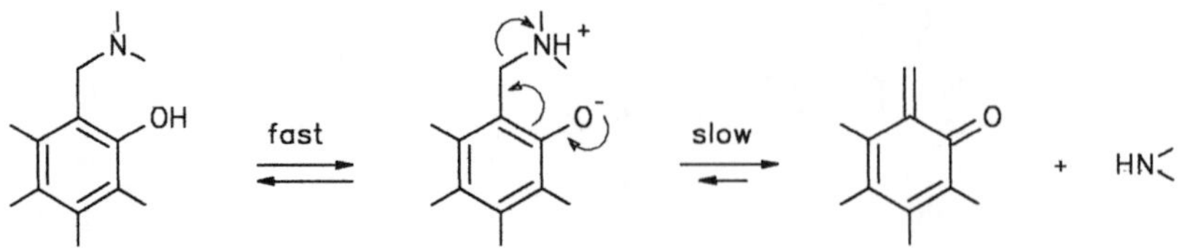

Fig. 88.   The mechanism of ortho-phenolic Mannich base deamination.

The base-catalyzed deamination of β-aminoketone methiodides has been claimed to follow an E_2 mechanism involving the partial carbanionic character of the intermediate **240**,[809,810] whereas the deamination of the corresponding indole derivatives would occur through the fast elimination of the indole NH proton in **241**, followed by the slow evolution of trialkylamine.[811]

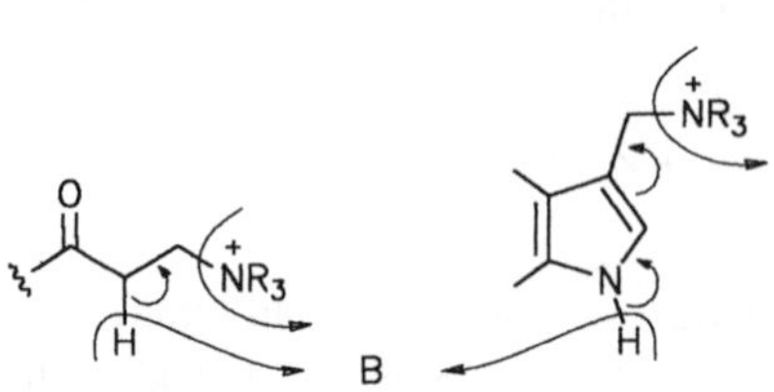

240   **241**

Fig. 89.   The C—N bond cleavage mechanism in deamination reactions.

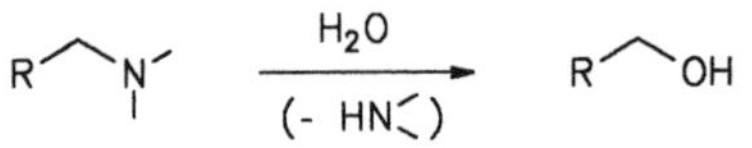

242   **243**

R = Dialkylphenol, oxadiazine, SO₃Na

Fig. 90.   Hydrolytic deamination reactions.

In several deamination reactions, formation of the hydroxymethyl derivatives **243** (Fig. 90) is frequently reported instead of the expected unsaturated product. Thus, deamination of para-aminomethyl phenols (**242**, R = *p*-hydroxyphenyl) produces an intermediate methylenequinone, which then evolves to hydroxymethyl phenol by reaction with water,[812] whereas deamination of aminomethane sulfonates (**242**, R = SO_3Na) proceeds through an S_{N2} attack by water on the zwitterionic form of the Mannich base to yield the hydroxylated derivative.[751]

Although characterized by a quite complex mechanism, the deamination of acetylenic Mannich bases by lithium-alkyl[813] yields interesting substituted acroleins R—CH=CH—CHO, that is, the isomeric form of the corresponding acetylenic hydroxymethyl derivatives R—C≡C—CH₂OH, as a consequence of alkyne metallation followed by hydrolysis.

Deaminodecarboxylation (Fig. 91) has been observed in Mannich bases **244** having the β-amino acid structure. The reaction, which may be accompanied by side reactions,[258] takes place regardless of the fact that the carboxy group is esterified, and even in derivatives possessing the formyl or the oxalyl group.[814,815]

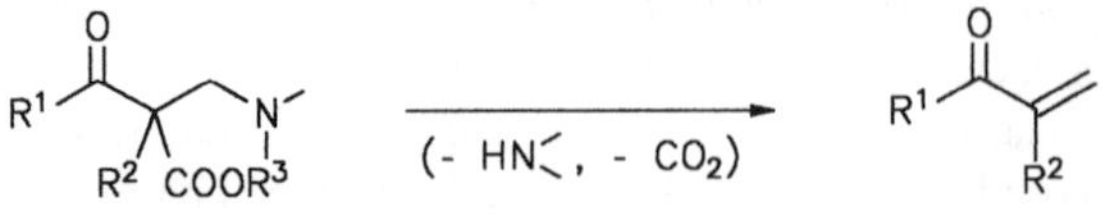

244   **245**

R¹ = Alkyl, OAlkyl; R² ≠ H; R³ = H, Alkyl
R¹, R² or R¹, R³ = ring

Fig. 91.   Deaminomethylation with decarboxylation of carboxylated Mannich bases.

The Mannich bases investigated are acetoacetic and malonic acid derivatives, γ-carboxyglutamic acid included.[258,816-819] Compounds having the lactone structure give preferably decarboxylation of the ester groups not involved in the lactone ring.[820]

In a manner similar to that indicated in Table 23, the unsaturated products **245** can be directly obtained from Mannich synthesis,[817,818] or by allowing the prepared base to react,[819] or by converting the Mannich base into the corresponding, more reactive iodomethyl derivative.[816,820]

Studies carried out on the decomposition of several β-amino acids lead to the conclusion that deaminodecarboxylation involves the fragmentation of the zwitterion **246** and is favored by pH values near the isoelectric point of the amino acid.[258,522]

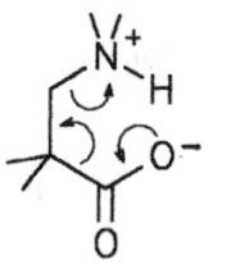

246

B Amino Group Replacement

The amino group of Mannich bases **247** can be replaced by various X residues by reaction with nucleophilic HX reactants to give the X-alkylated products **248** (Fig. 92):

This reaction, which also may be formally regarded as an X-alkylation by the R—CH$_2$— moiety of the Mannich base, is of considerable synthetic utility.[258,270,416]

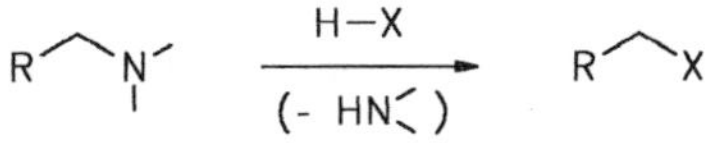

$$R\diagdown\diagup N\diagdown \xrightarrow[\text{(- HN$\langle$)}]{H-X} R\diagdown\diagup X$$

247 **248**

Fig. 92. The amino group replacement in Mannich bases.

Replacing the amino group with a hydrogen atom (i.e., H-alkylation), which actually corresponds to the hydrogenolysis of the CH$_2$—N bond of the Mannich base, is treated in Sec. C, which deals with the reduction reaction. Replacement of the amino group may also be carried out with a number of HX reagents that can give C-alkylation (compounds having an acidic C—H group), N-alkylation (amines, amides, and NH-heterocyclic compounds), S-alkylation (thiols and sulfinic acids), O- and P-alkylation, etc. This reactive behavior is one of the most interesting features of Mannich bases, as it enables us to apply the versatility of the replacement reaction to many different synthetic objectives, including the production of cyclic derivatives (Sec. E), which are often synthesized through replacement of the amino group.

The compounds suitable to X-alkylation are usually of the same nature as the substrates employed in Mannich synthesis, due to the analogous nucleophilic character and reactivity required by the reaction. As a consequence, the same general considerations concerning Mannich reactions made in the preceding chapter are valid here.

B.1 Replacement by CH Derivatives (C-Alkylation)

When Mannich bases are allowed to react with compounds containing nucleophilic CH moieties (**249**), such as activated alkyl derivatives, unsaturated and aromatic systems, HCN, etc. the amino group is replaced with the formation of the corresponding C-product and amine (Fig. 93):

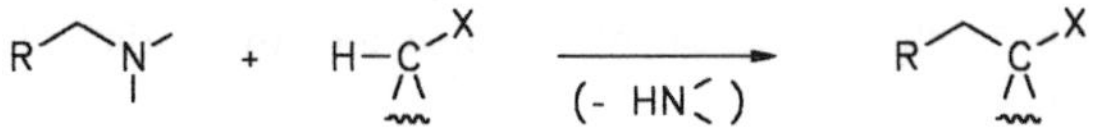

$$R\diagdown\diagup N\diagdown \ + \ H-C\diagdown^X \xrightarrow[\text{(- HN$\langle$)}]{} R\diagdown\diagup C\diagdown^X$$

249 **250**

Fig. 93. Amino group replacement by saturated CH derivatives affording C-alkylated products.

Table 24

C-Alkylated Products by Amino Group Replacement of Mannich Bases

C-Alkylated product **250**	Alkylating Mannich base[a] $R\diagup N\diagdown$ from:	Reference
From saturated alkyl derivatives		
R^1, R^2 = H, Alkyl, Ar[b] or R^1-R^2 = ring	Alkyl Ketones	821,822
	Phenols	823
	Chromones	824
	Thiols[c]	825
R^1 = COOAlkyl, CN,...; R^2 = Alkyl or R^1-R^2 = ring	Alkyl Ketones	826
	Phenols	827,828
	Ferrocene	826,829
	Theophylline	830
Analogous Amides, Nitriles, etc.	Alkyl Ketones	355
	Indoles	831
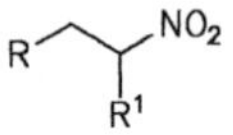 R^1 = H, Alkyl	Nitroalkanes	627
	Porphyrins	832
	Phenols	833
R^1 = COAr, SO$_2$Alkyl, CN,...	Alkyl Ketones	834
	Phenols	835,836
	Amides[d]	834
From unsaturated and aromatic derivatives		
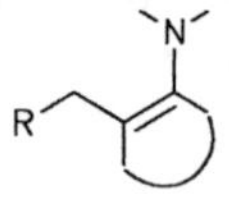	Alkyl Ketones	837,838
	Phenols	838
	Thiols[c]	839

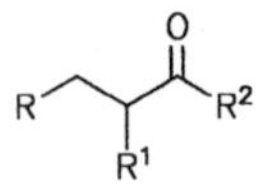

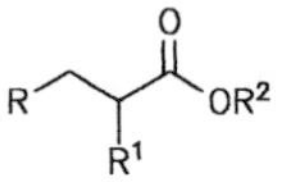

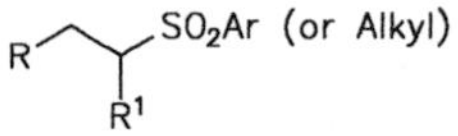

Thus many compounds are obtained (**250**, Table 24), which may be considered to be derived from C-alkylation of the nucleophilic reagent by the Mannich base. Alkyl ketonic and phenolic bases are mostly involved in this reaction, which exhibits many analogies with aminomethylation, particularly concerning chemo- and regioselectivity on aromatic and heteroaromatic derivatives. Ring activation by means of hydroxy and amino substituents in the alkylation of pyridine and pyrimidine derivatives is also required.

It is worth mentioning, moreover, the attractive possibility of the further conversion of certain products of the replacement reaction, such as nitriles, into the corresponding carboxylic acids.

Table 24

(continued)

Structure		
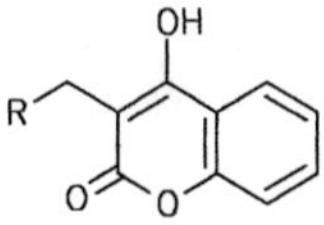 OH / R / O O (coumarin structure)	Alkyl Ketones	840,841
X (X = OH, NR$_2$) (o,p) R	Alkyl Ketones Thiols[c]	842 843
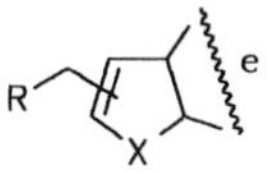 R (X = O, NH) e	Alkyl Ketones Phenols	844 845
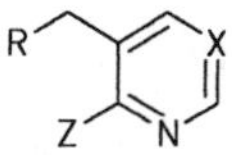 R X (X = CH, N) Z N (Z = OH, NR$_2$) Uracil, Barbituric acid deriv.s	Phenols, Indoles	846-848
From hydrogen cyanide R C≡N	Alkyl Ketones Phenols Ferrocenes Pyrroles, Indolizines	849 412,850 247 651,654

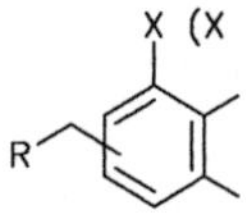

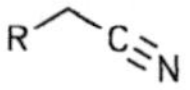

[a] C–Mannich bases, unless otherwise stated
[b] See also –SO$_2$Ar derivatives
[c] S–Mannich bases
[d] N–Mannich bases
[e] Fused ring derivatives included

The following two paths are conceivable, in principle, for the replacement reaction:

1. Elimination of the amino group of the Mannich base followed by addition of the CH-active reagent to the double bond of the vinyl derivative (Michael-type addition) so formed
2. Direct replacement of the amino group by the CH-active reagent (nucleophilic substitution)

Both the above mechanisms are proposed in the literature: with Mannich bases of nitroalkanes the substitution is clearly favored by the steric hindrance of the amine moiety, thus suggesting path 1,[851] whereas NMR studies on the reaction of β-amino-ketones with hydroxy coumarins do not reveal the presence of vinyl ketone intermediates.[841] Iodomethylated phenolic Mannich bases are also claimed to react according to path 2,[852] although the formation as by-products of dimers and methylene-bis-derivatives accounts for the participation of methylenequinone intermediates in the process.[828]

Intramolecular rearrangements involving the migration of the R—CH_2 group to the ortho and para positions of the aniline ring are observed in ferrocene/aniline[853] and thiophenol/aniline[854] Mannich bases, but this behavior is attributed to the presence of the cation R—CH_2^+ only in the case of the former reaction.

Finally, in the treatment of azulene Mannich base methiodide with zinc, the participation of a radical species of the type R—$CH_2 \cdot$ is proposed in order to explain the formation of the corresponding dimeric product.[648]

Reagents of type **251** (Fig. 94) are aldehydes or phosphoranes which, by reaction with ketonic and indolic Mannich bases, allow the production of β-diketones **252** or phosphoranes **253**, respectively. In the former case, the conversion of aldehyde to ketone is thus made possible,[855,856] and in the latter, the formation of unsaturated derivatives by subsequent reaction of the PPh_3 residue with carbonyl compounds is allowed.[857]

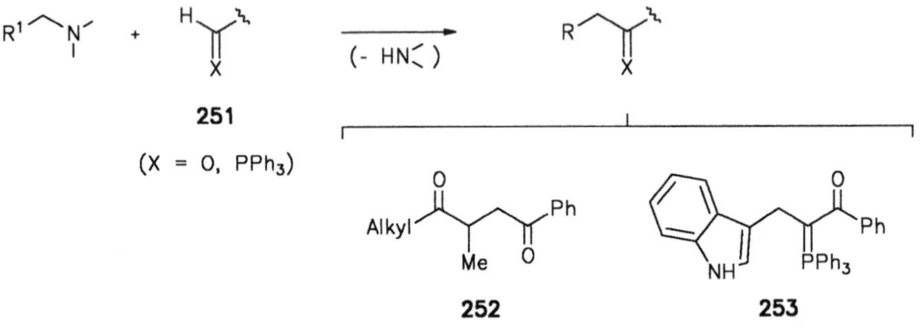

251

(X = O, PPh_3)

252   **253**

Fig. 94.   Amino group replacement by unsaturated CH derivatives affording C-alkylated products.

An anomalous behavior may be observed in **254** (Fig. 95), as the cleavage of the reagent molecule occurs in the course of the replacement reaction producing, simultaneously, C-alkylation of one fragment by the phenolic Mannich base and hydroxy group acylation by the other fragment.[836,858]

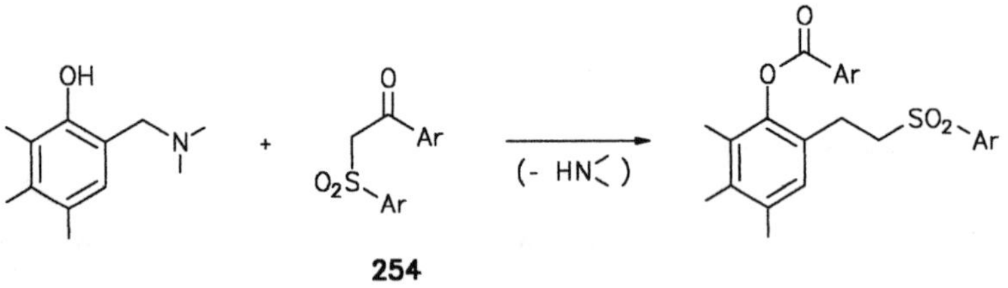

254

Fig. 95.   Reaction of phenolic Mannich bases with β-ketosulfones.

When two molecules of a Mannich base react together with the formation of a methylene-bis-derivative (**255**, Fig. 96), a particular type of C-alkylation takes place, simultaneously involving deaminomethylation and amine replacement, which could be defined as a self-condensation of the Mannich base.

This process is usually a side reaction in Mannich synthesis or in the reactions of Mannich bases, and it may predominate or even be the only reaction occurring.[846,859] When such a reaction is desired, it can be favored, in the case of phenolic or heterocyclic Mannich bases, by employing a solvent with a high boiling point.[360,637,846,859]

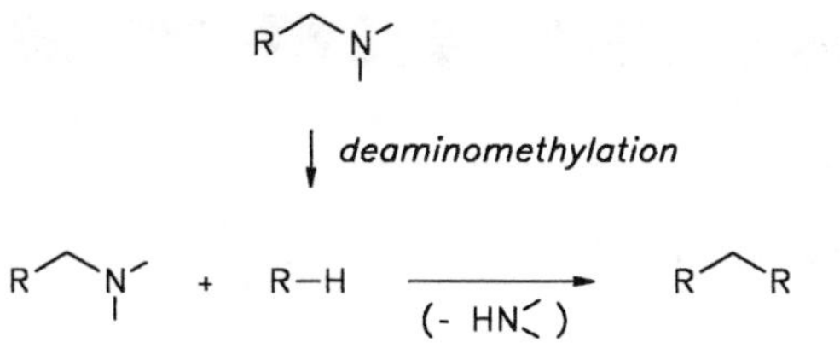

deaminomethylation

R = Phenolic or heterocyclic moiety

Fig. 96. The formation of methylene-bis-derivatives from Mannich bases.

B.2 Replacement by NH Derivatives (N-Alkylation)

Nucleophilic nitrogen-containing molecules may be employed in the replacement of Mannich base amino groups with formation of the derivatives **256** (Fig. 97), bearing various substituents on the nitrogen atom.

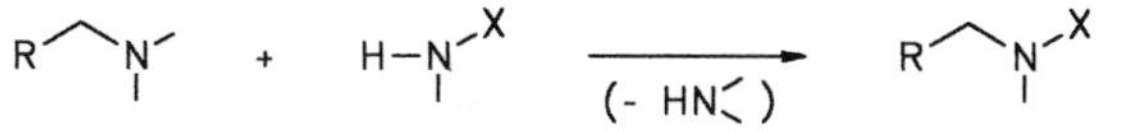

Fig. 97. The amino group replacement by NH derivatives affording N-alkylated products.

The reaction is successful with ammonia, primary and secondary amines, amides, and NH-heterocyclic compounds. A list of selected examples is reported in Table 25.

Any kind of C-Mannich base has been actually allowed to react with a variety of amines, including aminoalcohols[860] and amino acids.[861] Cyclic amines,[865,866] including optically active derivatives,[872] and amides[868,869] are also employed as are a number of aromatic heterocycles.

Although in many cases products **256** can be directly prepared by Mannich synthesis, the replacement reaction is particularly convenient (1) in the case of primary amines, as it yields a secondary amine derivative hardly obtainable through other synthetic routes, and (2) in the case of arylamines, as it makes it possible to avoid engaging an activated, reactive, aryl group in the direct Mannich reaction. Moreover, the method is frequently adopted in the synthesis of polymeric substances (Chap. III).

As with C-alkylation, the mechanism of amino group replacement can follow the elimination/addition or the nucleophilic replacement path. Both mechanisms are indeed mentioned in the literature and are occasionally claimed to occur concurrently. The elimination/addition path is suggested in the reaction of β-aminoketones with uracils, and a four-centered transition state is proposed for the same reaction with indole Mannich bases.[871]

A study of the reaction between 4-dimethylamino-butanone and N-deuterated arylamines[873] indicates the presence of both direct substitution (S_{N1} or S_{N2}) and elimination/addition; the nature of the arylamine is decisive in establishing the predominant mechanism.

Table 25

N-Alkylated Products by Amino Group Replacement of Mannich Bases

N-Alkylated product **256**		Alkylating Mannich base[a] 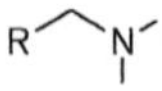 $R\frown N'$ from:	Reference
From amino derivatives			
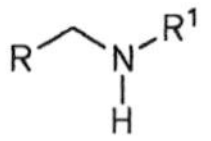 $R\frown N(H)-R^1$	R^1 = Alkyl	Alkyl Ketones	860,861
	Ar, Heteroaryl	Alkyl Ketones, Uracil	859,862
	NH_2	Phenols, Ferrocenes	863,864
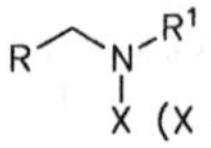 $R\frown N(X)-R^1$ (X ≠ H)	R^1 = Alkyl	Alkyl Ketones, Ferrocenes	865,866
	Ar	Ferrocenes	853
	OH (X = RCH_2)	Alkyl Ketones	867
From amides			
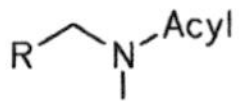 $R\frown N\text{-Acyl}$	Acyl = NO_2	Indoles	71
	COR'	Phenols	868
	Amides[b]		869
From NH-heteroaromatics			
$R\frown N\rangle$	Ring = Imidazole, Pyrazole,...	Alkyl Ketones, Phenols	842,870
(Heteroaromatic ring)	Uracil	Indoles	871

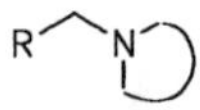

[a] C—Mannich bases, unless otherwise stated

[b] N—Mannich bases

The replacement reaction by polyfunctional nitrogen-containing derivatives, in particular, makes it possible to introduce multiple RCH_2 groups into the same molecule. Thus, the reaction of indole Mannich bases with ammonia yields bis- and tris-(3-indolyl) amines.[831] Likewise, primary amines,[493] hydrazine,[864] and hydroxylamine[867] can be bis-alkylated by several Mannich bases.

Some exchange reactions of cyclic ketobases, γ-piperidones,[874,875] and bispidinones[876] provide bis-alkylation with alkylamines involving interesting consequences determined by steric influence. For instance (Fig. 98), N-methyl-γ-piperidone methiodides **257** afford products **258** when the 2 position in **257** is unsubstituted; otherwise acyclic α,β-unsaturated β'-amino-ketones **259** are obtained.[874] However, if the R^2 group of the reacting amine is benzyl, that is, a group less bulky than cyclohexyl or *tert*-butyl, cyclic aminoketone is produced regardless of whether position 2 is substituted or not.

N-Alkylation may require careful selection of experimental conditions when several reactive moieties are present in the reactants. Thus, when a Mannich ketobase is allowed to react with hydroxylamine, it is preferable to use the iodomethylate of the base, since it is more easily decomposable, in order to reduce the extent of the concurrent oximation

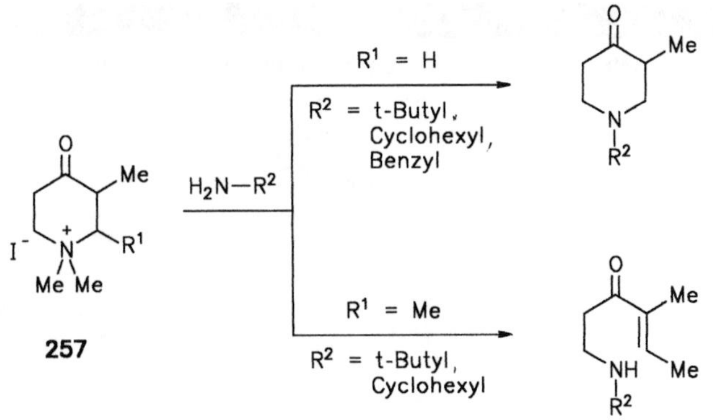

Fig. 98. Amino group replacement in γ-piperidone derivatives.

reaction.[867] In the reaction with aminoguanidine,[877] the expected exchange product **261** (Fig. 99) is formed at pH 5 to 6, whereas the hydrazone derivatives **260** or the N-heterocyclic compounds **262** are given under slightly different conditions.

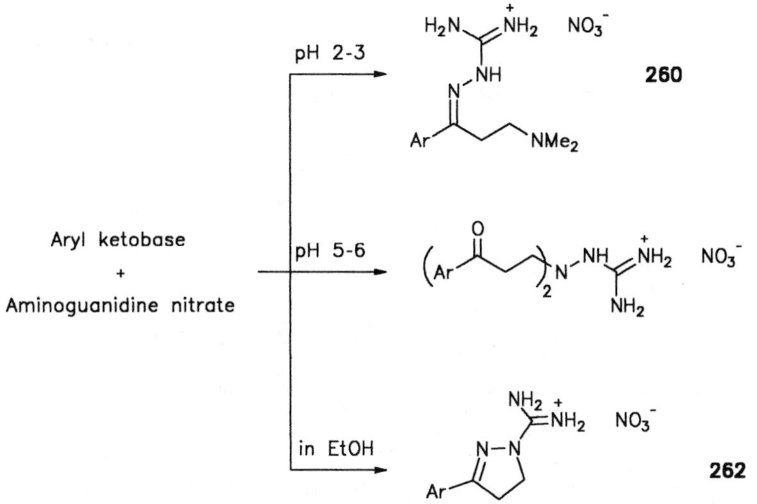

Fig. 99. Products derived from Mannich aryl ketobases and aminoguanidine under different reaction conditions.

B.3 Replacement by XH Derivatives (P-, As-, O- and S-Alkylation)

P- and As-Alkylation

Among the compounds containing elements of group V, in addition to those containing nitrogen, P and As derivatives may provide amino group replacement of Mannich bases. As with the requirements for the synthesis of X-Mannich bases, the oxidation state of the heteroatom involved in the replacement reaction is always lower than the maximum, so as to allow the formation of products **263** and **264** (Fig. 100).

Phosphines and arsines, in addition to phosphine oxides and alkyl phosphonates, may be used to give, respectively, compounds **263** and **264**, as detailed in Table 26.

Table 26

P- and As-Alkylated Products by Amino Group Replacement of Mannich Bases

X-Alkylated products 263 and 264	Alkylating Mannich base 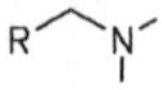 R⌒N⌐ from:	Reference
From phosphines		
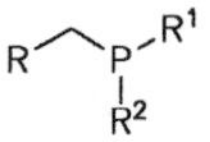 R⌒P(-R¹)(R²) R^1, R^2 = Alkyl, Ar	Alkyl Ketones	733,878
	Ferrocene	879
From arsines		
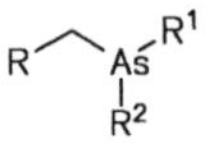 R⌒As(R¹)(R²) R^1, R^2 = Alkyl, Ar	Alkyl Ketones	878
From phosphinoxides and phosphites		
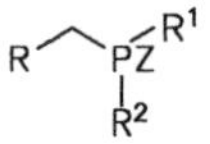 R⌒PZ(R¹)(R²) $Z = O; R^1, R^2 = Ph$	Alkyl Ketones, Ferrocenes	879,880
R^1, R^2 = OEt	Alkyl Ketones, Phenols	881,882
$Z = S; R^1, R^2 = OEt$	Phenols	883

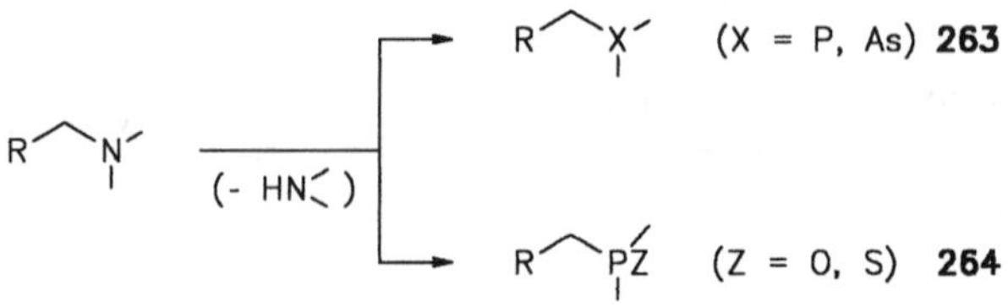

Fig. 100.　Amino group replacement by XH derivatives affording P- or As-alkylated products.

The Mannich bases subjected to the replacement reaction are usually derivatives of alkyl ketones, phenols, and ferrocene.

The reaction may also be carried out with tertiary phosphines (e.g., triethyl and triphenyl phosphine 265, Fig. 101) on either the free or the iodomethylated Mannich base, thus producing the corresponding phosphonium salt.[884] The phthalimidomethyl triphenyl phosphonium salt 266 is a pertinent example of this kind of derivative.

By-products can be formed in the synthesis of phosphines 263 with the iodome-thylates of ferrocenyl Mannich bases, as they tend also to give phosphonium salts 267 in amounts that depend on the duration of the reaction.[879] Furthermore, in the reaction with arsines, due to the reducing properties of such reagents, hydrogenolysis of the CH_2—N bond may occur with the formation of a propiophenone derivative instead of the expected product 263 (X = As).

Alkyl phosphonates 264 (Z = O, R^1R^2 = OEt, Table 26), analogously with the preparation of P-Mannich bases, can be obtained from trialkylphosphites as well as from

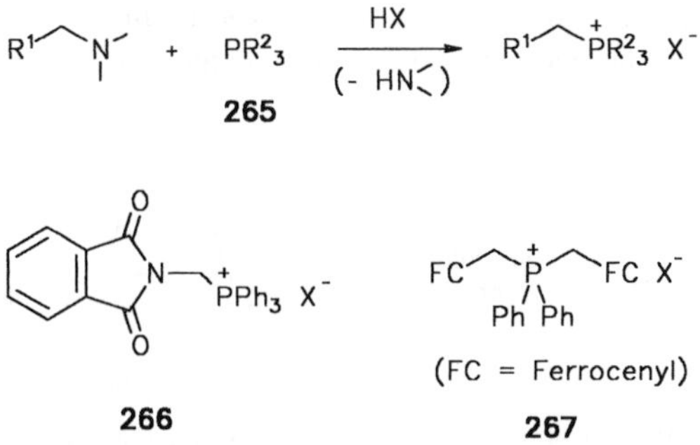

Fig. 101. Amino group replacement by tertiary phosphines.

phosphonic acid esters. The mechanism of the reaction, as studied for ketonic and phenolic Mannich bases,[258,806,881] is quite intricate.

O- and S-Alkylation

Among the compounds containing elements of group VI, oxygen and sulfur derivatives are commonly employed for the amino group replacement of Mannich bases. Figure 102 summarizes the substitution reactions given by oxygen derivatives.

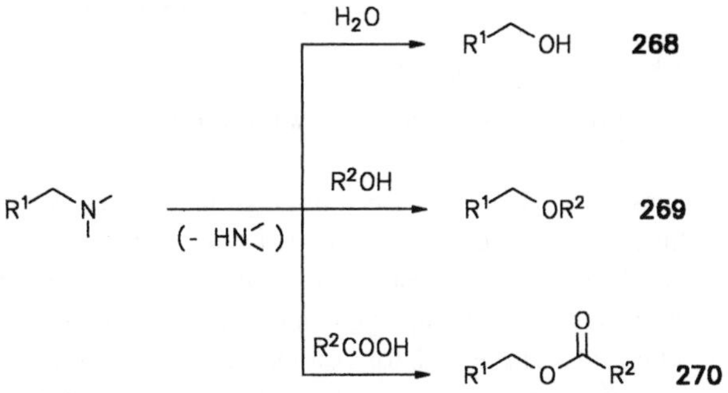

Fig. 102. The amino group replacement by OH derivatives affording O-alkylated products.

The use of water for the production of hydroxymethyl derivatives **268** is of little synthetical interest, except for a few cases,[885] as these compounds are better prepared by reaction of the Mannich substrate R^1—H with formaldehyde. It is in fact well known that methylol derivatives may be intermediate species of Mannich synthesis (Chap. I, B.2). They may be also formed as by-products,[879] in the course of the reaction, as a consequence of the hydrolitic deamination of a Mannich base (Fig. 18, Chap. I).

In the presence of a Mannich base, hydroxymethyl derivatives **268** may behave, in turn, as alcoholic reagents, thus affording amino group replacement with formation of the symmetrical ether **269** ($R^2 = R^1$—CH_2).[879] Indeed, alcohols and phenols allow the synthesis of ethers **269**, starting from phenolic Mannich bases[852] as well as from the methiodides of aminomethylferrocenes,[866,886] by reaction with alkoxide or phenoxide. Good results have been obtained with a thioporphirine Mannich base and zinc acetate.[832]

Esters **270** (Fig. 102) can be synthesized from various Mannich bases such as those derived from tropolones[768] or purines,[887] which are converted into acetoxymethyl derivatives by treatment with acetic acid or anhydride. In the case of ortho-phenolic substrates (Fig. 103), the hydroxy group may undergo reaction to give the diacetoxy compounds **271**.[888,889]

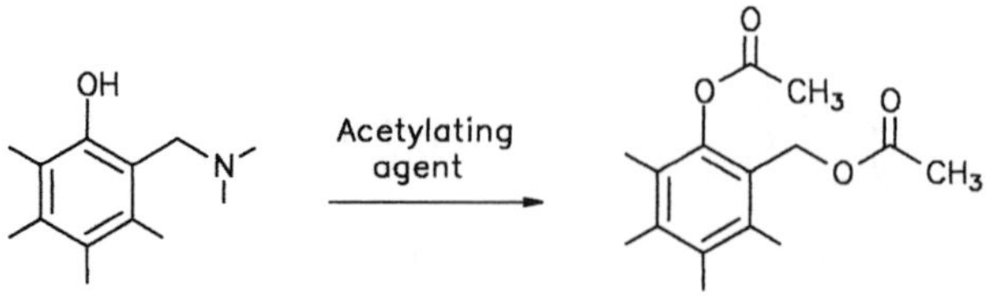

Fig. 103. Reaction of phenolic Mannich bases with acetylating reagents.

Finally, studies of alkylene bis-amines **272** (Fig. 104) indicate that the replacement reaction depends on the kind of reagent employed, as anhydrides[890] afford 1-aminoalkyl carboxylates **273**, whereas acyl halides give 1-amino-1-haloalkanes **274** in good yield.[891]

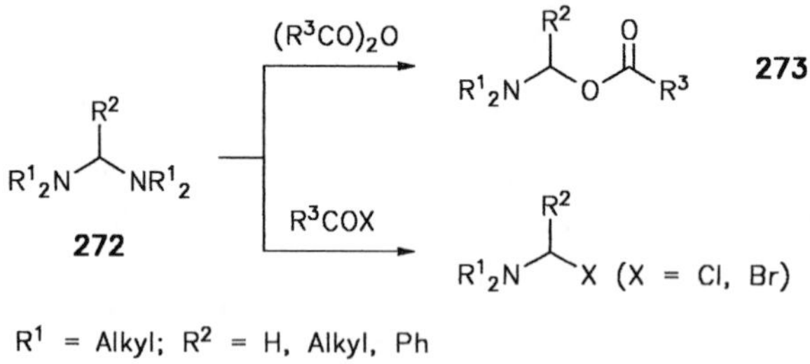

R^1 = Alkyl; R^2 = H, Alkyl, Ph

Fig. 104. Reaction of alkylene-bis-amines with acylating reagents.

The replacement of a Mannich base amino group by SH derivatives leads to thioethers (**275**), thiocarbamic acid esters (**276**), and sulfones (**277**) (Fig. 105 and Table 27). The reaction has been widely investigated for its synthetical relevance, due to the wide range of reagents that may be employed and to the possibility of further reactions of the products so obtained, as observed, for example, in the synthesis of diazoalkanes.[892] It is also used in biological studies with cysteine scavengers[893] and in several different applications, which are described in the appropriate chapters.

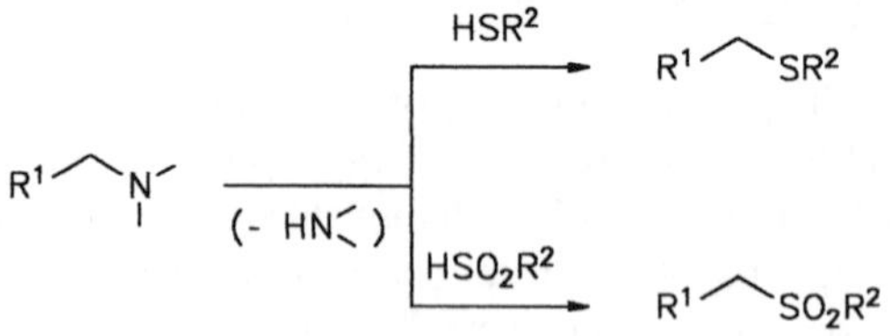

Fig. 105. Amino group replacement by SH derivatives affording S-alkylated products.

A very wide range of Mannich bases can be subjected to replacement reaction with R^2—SH derivatives, including ketobases producing optically active γ-ketosulfides [**275**, R^1 = PhCO-CH(Me)], when the reaction is carried out in the presence of catalytic amounts of chiral amine of the cinchonidine type.[902]

The reaction mechanism, as investigated for S-alkylation with ketonic,[264,809] naphtholic,[807] and indolic[811] Mannich bases (including the corresponding quaternary salts), involves amino group elimination followed by fast addition of the thiol to the vinyl

Table 27

S-Alkylated Products by Amino Group Replacement of Mannich Bases

S-Alkylated product		Alkylating Mannich base[a] 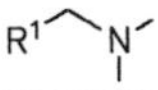 R^1 N from:	Reference
From thiols or dithiocarbamic acids			
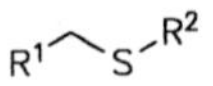 R^1 S R^2	**275**, R^2 = Alkyl, Ar	Alkyl Ketones	387,894
		Phenols	895,896
		Azaazulanone	655
		Uridine deriv.s	897
		Arylthiols[b]	898
	276, R^2 = 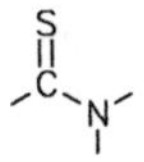	Alkyl Ketones	899,900
		Ferrocene	863
From sulfinic acids			
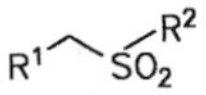 R^1 SO_2 R^2	**277**, R^2 = Alkyl, Ar	Alkyl Ketones, Alkyl Quinolines, Phenols	629,901
		Amides[c]	215,892

[a] C–Mannich bases, unless otherwise stated
[b] S–Mannich bases
[c] N–Mannich bases

intermediate (vinyl ketone, quinone methide, or imine) so formed. A more detailed description of the elimination step is reported in Sec. A.2. Amidic Mannich bases behave similarly[11] only when quaternary ammonium salts are used, with the formation of intermediate ArCO—N=CH$_2$. Amidic free bases, by contrast, react at neutral or alkaline pH, with the simultaneous production of the deaminomethylated derivative in consistent amounts.

Reagents other than R—SH derivatives are also employed. Thioethers **275** are successfully obtained by using thioesters in the presence of a phenolic catalyst,[903] and dithiocarbamic esters **276** (Fig. 106) are prepared by inserting CS$_2$ (or COS) between the methylene and the amino group of various Mannich bases.[904–906] The CS$_2$ insertion has been achieved on 2,6-dialkylphenols, in the course of Mannich synthesis,[828] presumably as a consequence of electrophilic CS$_2$ attack on the N atom with formation of the adduct **278**, followed by intramolecular rearrangement:[904]

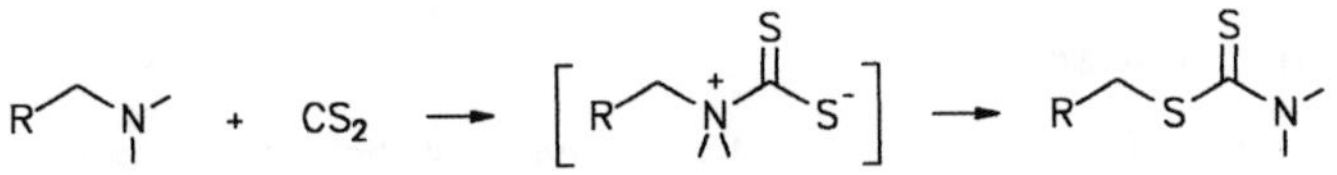

Fig. 106. Synthesis of thiocarbamic esters from Mannich bases.

With para-aminomethyl phenols, the participation of an intermediate para-methylenequinone is suggested.[905]

Formaldehyde sulfoxylate has been employed[907] in the synthesis of symmetrical dialkyl sulfones **279** (Fig. 107), through a reaction of general application carried out on the Mannich base as hydrochloride or iodomethylate on heating in a suitable solvent.

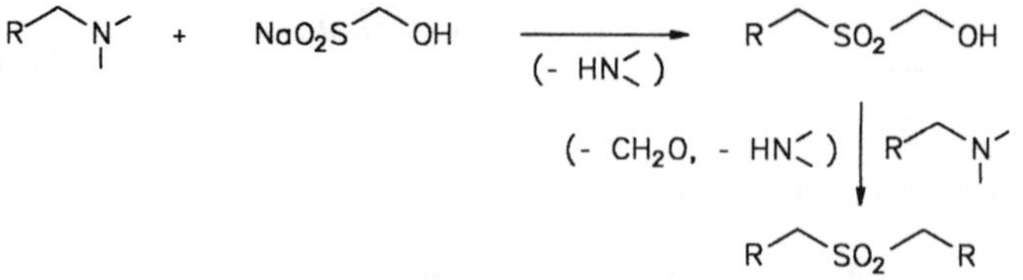

279

Fig. 107.   Synthesis of dialkylsulfones involving Mannich bases.

Polyfunctional Mannich bases may give differing reactions with thiols. In particular, acetylene derivatives of type **280** are scarcely suitable for S replacement of the amino group. The ester group reacts quantitatively with thiol,[908] and unsaturated ketobases such as the styryl ketones **281** give rise to three different products, deriving respectively from amino group replacement, addition to double bond, and contemporary replacement and addition.[894,909,910] Reaction yield and type of product are affected by the nature of both amino group and thiol, by the solvent, and by the reaction duration. The addition product prevails when the duration of the reaction is short, whereas the replacement derivative predominates after more prolonged periods.

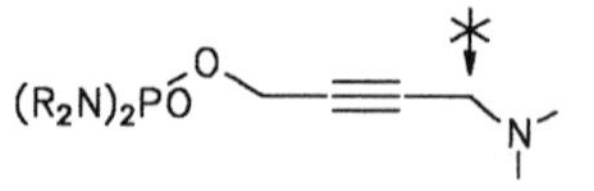

280 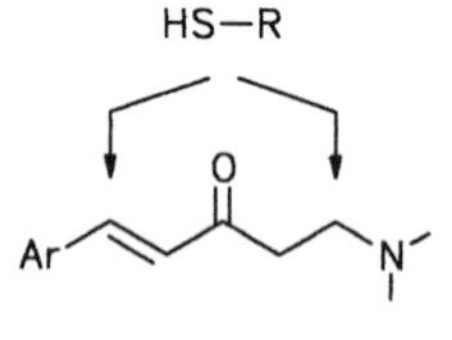   **281**

Aryl sulfinic acids behave similarly to thiols toward unsaturated ketobases. The addition to double bond is preferred, particularly under acidic conditions, to amino group replacement.[911]

C Reduction

Reducing agents have two main effects on Mannich bases, namely, hydrogenolysis of σ bonds and reduction of unsaturated functional groups.

C.1   Hydrogenolysis

This important reaction (Fig. 108) involves cleavage of the CH_2—N bond of Mannich base, and in most cases it is used as a method for generating a CH_3 group or, more rarely, an alkyl group.[155] It may be formally considered the replacement of an amino group by hydrogen.

Table 28
Hydrogenolysis of CH$_2$—N Bond of Mannich Bases

Mannich base from:	Hydrogenation agent[a] (Reference)
Alkyl Ketones	M/A (326), MH (912), H/C (913)
Phenols	MH (642), H/C (914,915)
Phenols[b]	MH (916)
Heterocyclics	MH (912), H/C (243)
Heterocyclics[b]	MH (443), H/C (655)
Heptamethylcorrine[b]	H/C (176)

[a] M/A: Metal and acid. MH: Metal hydride. H/C: Hydrogen and catalyst

[b] Mannich base as quaternary ammonium salt

As reported in Table 28, hydrogenolysis is performed only on C-Mannich bases, usually as quaternary ammonium salts, employing hydrogen or metal hydride.

$$R\!-\!N\!\!<\quad \xrightarrow[(-\ HN\!\!<)]{H_2}\quad R\!-\!CH_3$$

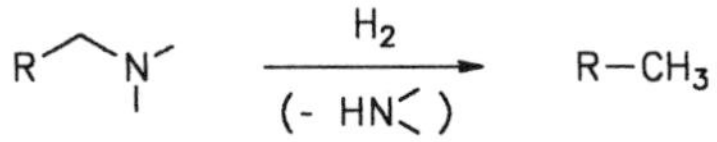

Fig. 108. The hydrogenolysis of the aminomethyl group.

Besides the traditional reagents (hydrogen and catalyst, zinc and acid), some reducing agents have been more recently introduced, including borohydride and tri-butyltin hydride. Sodium amalgam has also proved its efficacy in this type of reaction.[319,917] Diphenyl arsine can provide hydrogenolysis of the C—N bond, as reported in connection with attempts at replacing amino groups by As in β-amino-propiophenones[878] (Fig. 100, Sec. B.3). In one case the methyl derivative was obtained simply by heating the Mannich base in triethylamine, in the absence of any reducing agent.[918]

Nevertheless, in many cases, hydrogenolysis does not occur readily; however, the ammonium salt of the Mannich base appears more suitable for a successful reaction. Thus, several ortho-phenolic Mannich bases are hydrogenated as hydrochlorides, in order to avoid the presence of the intramolecular hydrogen bond between OH and N, which is responsible for the weak reactivity of the free Mannich base.[919] The ferrocenyl Mannich base **282**, containing both tertiary and quaternary amino groups, gives 2-di-methylaminomethyl-1-(2-methylferrocen-1-yl) ferrocene (**283**) upon reaction of the more reactive quaternary ammonium moiety.[917]

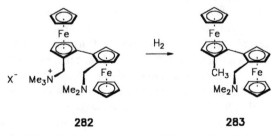

282   **283**

Fig. 109. Hydrogenolysis reaction of ferrocene Mannich bases.

The hydrogenolysis product of the Mannich base is also obtained through the formation of a suitable intermediate, in some cases not isolated or purified, which is subsequently subjected to hydrogenation. This may be an unsaturated derivative ($R=CH_2$) produced by deamination of Mannich base,[785,913,920] or the $R—CH_2—X$ ($X = O$-Acyl[921] or S-Alkyl[922]) product deriving from amino group replacement.

Debenzylation of a Mannich base by hydrogenolysis is widely adopted.[236,498,738] It has been used, for example, in the synthesis[738] of aminomethylphosphonic acid $(HO)_2PO—CH_2NH_2$.

Other hydrogenolytic reactions, involving the $R—CH_2$ bond between the substrate and the methylene group of the Mannich base, are carried out on the Mannich bases of succinimide,[923] on some thiazolidine derivatives[503] and on cyclic β-aminoketones.[551]

Borohydride reduction of compounds **284** (Fig. 110) makes it possible to obtain N-methyl arylamines, but is limited to NH Mannich bases, as the N-methyl derivatives undergo only carbonyl reduction.

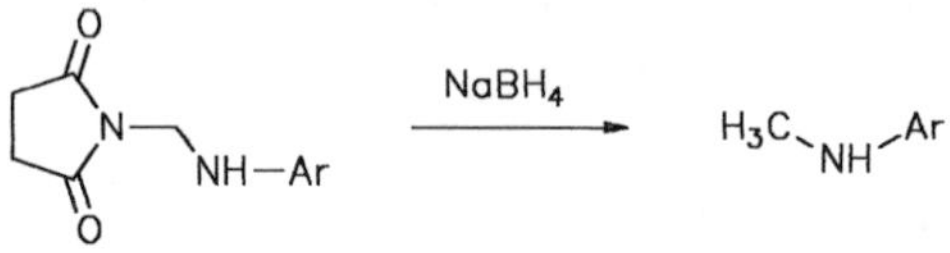

284

Fig. 110.   Hydrogenolysis reaction of succinimide Mannich bases.

Double bond migration may occur when metal hydrides are allowed to react with acetylenic iodomethylated Mannich bases, with the formation of allene **285** (Fig. 111), instead of the expected methyl-acetylene derivative.[924]

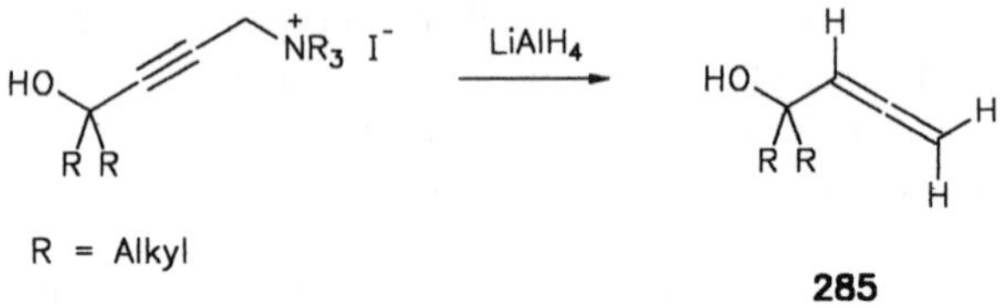

R = Alkyl

285

Fig. 111.   Hydrogenolysis reaction of acetylenic Mannich bases.

C.2   Hydrogenation of Unsaturated Groups

The reduction of unsaturated moieties present in Mannich bases is a fruitful topic, due not only to the large number of studies reported in the literature, but also to relevant stereochemical implications connected with the prochiral nature of most of the unsaturated groups that can be subjected to this reaction. Thus, the examples range from the hydrogenation of acetylenic derivatives[682,925,926] and the quinoline ring[624] to the reduction of nitro derivatives[23,927] and pyridine and pyridazine N-oxides.[434,440] However, β-aminoketones **286** (Fig. 112) are by far the most employed, as the reaction provides secondary aminoalcohols (**287**) of prominent pharmacological interest.[258,616,928,929] The exhaustive carbonyl reduction to methylene by catalytic hydrogenation has also been performed.[929]

Aminoalcohols **287** may be useful intermediates in the synthesis of drugs, as they give by dehydration the corresponding allylamino derivatives (R^1—CH=CH—CH$_2$—N<).[616,930]

R = mainly Ar

286

287

Fig. 112. The reduction reaction of β-aminoketones.

Cases of Mannich bases containing different reducible groups are reported quite frequently in the literature; the most common are ketobases bearing vinyl groups, such as styryl ketobases (Ar—CH=CH—CO—CH$_2$CH$_2$—N<)[931,932] and vinylogous Mannich bases (R^1—CO—CR^2=CR^3—CH$_2$CH$_2$—N<),[362] as well as acetylenic bases possessing carbonyl groups.[933] In all these cases the chemoselectivity of the reaction is determined by the type of reducing agent; usually, hydrogen and catalyst is selected for hydrogenating double or triple carbon-carbon bonds[362,933] and metal hydride for reducing carbonyl groups.[931,933]

When the unsaturated moiety is prochiral, the reaction stereoselectivity may deeply affect the nature of the product. In the reduction of carbonyl Mannich bases containing a chiral center, the formation of a diastereomeric mixture of aminoalcohols (Fig. 113) is to be expected. The relative amounts of isomers are determined by steric hindrance of the asymmetric center as well as by the nature of the reducing agent. On varying the reaction conditions, it is often possible to affect the diastereomeric ratio of the aminoalcohols produced, so as to obtain the predominance of either isomer. This can be particularly relevant in the synthesis of pharmacologically active substances.

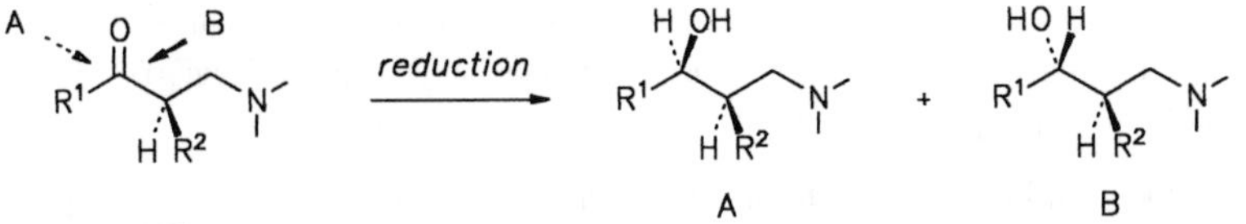

288

reduction

289

A

B

290

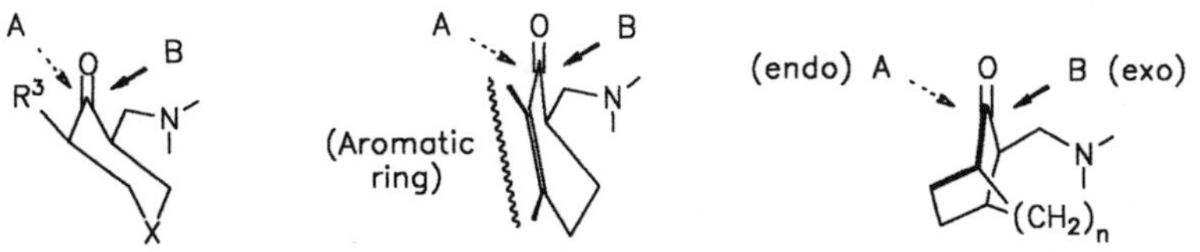

291

(endo) A

B (exo)

292

R^1, R^2 = Alkyl, Ar; R^3 = H, Ph; X = CH$_2$, O, S; n = 1-2

Fig. 113. The carbonyl group reduction of α-chiral β-aminoketones (only one enantiomer of the racemic pair is represented).

Although this topic is covered by a large amount of literature, most of which has been exhaustively assembled in a review paper on the stereoselective synthesis of diastereomeric aminoalcohols,[591] the stereochemical aspects of this type of reduction are frequently only partially outlined with a generic indication of the presence of diastereomers among the reaction products. However, in many cases, diastereomeric ratios as well as relative, or even absolute, configurations of the aminoalcohols produced are accurately investigated.[591] Mannich bases having the chiral center in the α position with respect to the prochiral carbonyl group are the most frequently studied, although the presence of the asymmetric induction center in other sites of the molecule, including the amino group, is also examined.[231,934] The stereochemical results[270,591] of carbonyl reduction in Mannich bases can be summarized as follows:

1. The reduction of acyclic β-amino ketones **288** with metal hydride or hydrogen and catalyst generally gives products derived from predominant attack from the **A** side of the carbonyl plane (Fig. 113); the **B** attack prevails slightly only with selected reagents or when R^1 has low steric hindrance (methyl, styryl). Quite special reaction conditions, for example, the use of cyanoborohydride in tetrahydrofuran, are needed for obtaining a consistent predominance of isomer **289B.**

2. Homocyclic ketones **290** (X = CH$_2$) and **291** predominantly undergo a type **B** attack, although the reduction with sodium borohydride increases the amount of the minor **A** isomer. Attack from the **A** side prevails slightly only in the 6-phenyl-substituted Mannich base **290** (X = CH$_2$, R = Ph).[386]

3. With other homocyclic derivatives, such as aminomethyl norbornanones **292** ($n = 1$), the prevailing attack also takes place from the **B** (*exo*) direction. However, the nature of the reducing agent as well as the type of amino group (particularly its steric hindrance) may affect the reaction course and reduce selectivity. In Mannich bases of bicyclo[2.2.2]octane-2-one (**292**, $n = 2$) the stereoselectivity is inverted, and the **A** aminoalcohol is predominantly produced.

4. The heterocyclic Mannich bases of oxa- and thia-cyclohexanone (**290**, X = O, S) obey the general rule of predominant **B** attack in hydride reductions. The synthesis of **A** diastereomer requires the use of aluminum isopropoxide (Meerwein-Pondorf), which always favors the opposite direction of attack with respect to metal hydride.

When the chiral center of ketonic Mannich bases is located in a different position from the α, a coherent behavior toward reduction becomes much more difficult to observe, and the discussion has to be limited to specific cases. Actually, a slight change in a few variables of the system under examination is sufficient to obtain an inversion of the predominant attack.[591]

For instance, the sodium borohydride reduction of conformers **203** (Chap. I, C.6) is affected by the reaction solvent. In conformer **203b**, in fact, the hydride attack occurs from the side of the oxygenated ring, regardless of the solvent employed, as is expected on the basis of lower steric hindrance, whereas with **203a**, the predominant attack by hydride occurs from the side of the N ring in water/dioxane and from the opposite side in methanol.[592]

The stereochemical results of Mannich ketobases are not readily interpretable and different opinions are expressed on this point (see Ref. 591). A cyclic model of the transition state is often suggested in order to explain the stereochemical pathway of the reduction in a homogeneous series of compounds affording a steadily prevailing isomer. It involves the formation of a cyclic chelate, provided by the metal linking together the carbonyl oxygen and the amine nitrogen atoms of the base, followed by attack of the reducing species from the less hindered side of the ring, that is, the opposite side with respect to the α substituent. Alternatively, an open-chain model would be responsible for the results. According to other authors, different conformers of the same cyclic model would play a role related to their relative abundance in determining the diastereomeric ratio of the products.

The reduction of acetylenic Mannich bases exhibits several interesting stereochemical features. The triple bond can be partially hydrogenated with di-isobutyl aluminum hydride, thus giving E-allylamines **293**. This reagent provides very good stereoselectivity, the mechanism of which has been investigated.[935]

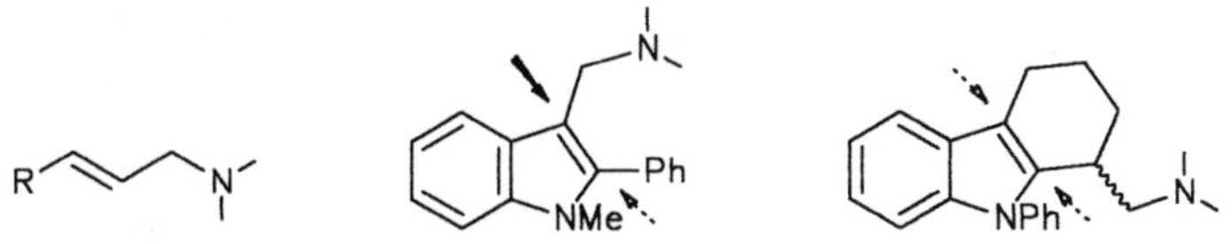

293   **294**   **295**

Stereoselective hydrogenation of the indole ring of various Mannich bases is also reported. In Mannich base **294** the two entering hydrogen atoms are linked in the *trans* configuration,[936] whereas in **295** the adjacent rings are predominantly *cis*-fused to an extent related to the type of reducing agent.[937]

D Addition of Organometals

Mannich bases deriving from alkylketones (β-aminoketones), phenols, alkynes, thiols, etc., are used as substrates in the addition of organometals. A remarkably different behavior, related to the nature of the Mannich base, is observed, and a dependence on the type of metal in the reagent has also been noticed in several experiments. Most of the reactions are aimed at inducing modifications to the Mannich base without affecting the aminomethyl group; however, in some cases, cleavage of the methylene moiety is purposely pursued.

D.1 Reactions Not Involving Cleavage of the Mannich Base

The most frequently performed reaction with organometals concerns the well-known addition to the carbonyl group of ketobases. This reaction (Fig. 114) exhibits many affinities with reduction, as it affords an aminoalcohol (**296**), in this case tertiary, and involves analogous stereochemical features when chiral ketobases are employed. Again, the aminoalcohols produced can be further converted, by dehydration, into allylamines (R^1R^2C=CH—CH$_2$—N<), which are useful in pharmaceutical chemistry.[938–940]

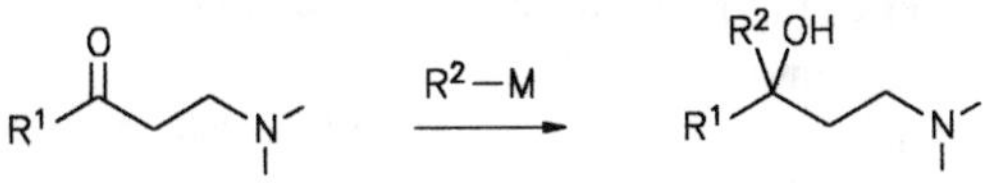

286   **296**

Fig. 114. The addition reaction of organometals to β-aminoketones.

A great variety of organometal reagents may be involved in the reaction, depending on the group that has to be introduced into the ketobase. As evidenced in Table 29, rather unusual functionalities are also employed for synthetical purposes. Thus, acetylenic moieties, nitrogen-containing molecules of different basicity (ranging from trialkylamines to pyridines), precursors of acid such as carboxyesters, are used for reaction with the Mannich base, in addition to the common alkyl, aryl, and heteroaryl groups.

Table 29		
Tertiary Aminoalcohols 296 from β-Aminoketones and Organometals R²M		
R² group in **296**	M group	Reference
Alkyl, Ar	MgX, Li	258,270,929,934,938
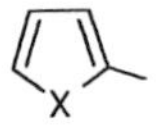 (furyl/thienyl structure) (X = O, S)	Li	941
R−C≡C−, HC≡C─\	Li, MgX	942,943
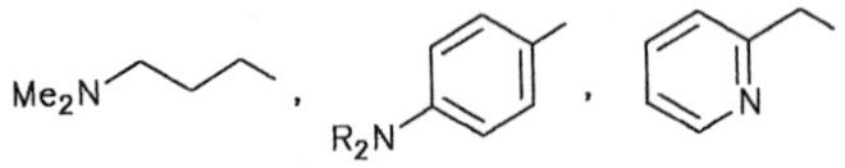 Me₂N─\\\\\ , (aryl-R₂N structure), (pyridyl structure)	MgX, Li	223,944,945
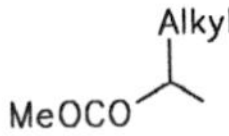 MeOCO (Alkyl substituted)	ZnBr	946,947
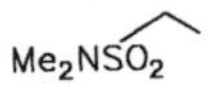 Me₂NSO₂─\\	Li	948

Organomagnesium reagents are mainly employed, with organolithiums being mostly used for introducing alkynyl groups, besides the phenyl and thienyl group. The Reformatsky reagent (zinc and α-halo-carboxyester) makes it possible to perform the synthesis of carboxyesters (see below). The mechanism of this reaction has been accurately interpreted on the basis of a cyclic transition state possessing an O-metal bond.[947,949]

In contrast to the reduction reaction, organometal addition to Mannich bases proceeds less cleanly, and unexpected reactions[950] as well as appreciable amounts of by-product[591,947] may be observed. Indeed, the enolization of the ketone carbonyl group,[951] which consumes a certain amount of reagent, thus causing incomplete reaction; the reduction to secondary alcohol, which is typical of alkylmetals having a hydrogen atom on the C-β position; and the Mannich base decomposition are all factors contributing to reduction of the reaction yield. In the Reformatsky reaction the quaternization of amino group also has to be taken into account.[947]

Ketobases possessing a chiral center give rise to diastereomeric mixtures of aminoalcohols upon organometal addition (Fig. 115). The same occurs when the asymmetric induction center is located on the reagent (see below).

This reaction is highly stereoselective,[447] usually to a larger extent than the corresponding reduction. The following are the main conclusions on the stereochemistry of organometal addition to α-asymmetric β-aminoketones that can be drawn from the great wealth of results reported in the literature:[11,591]

1. Acyclic β-aminoketones (**297**) are alkylated from the less hindered side (**A**) of the carbonyl plane, similarly to hydrogenation, but with higher stereoselectivity. The proposed stereochemical model involves a cyclic transition state in which the magnesium

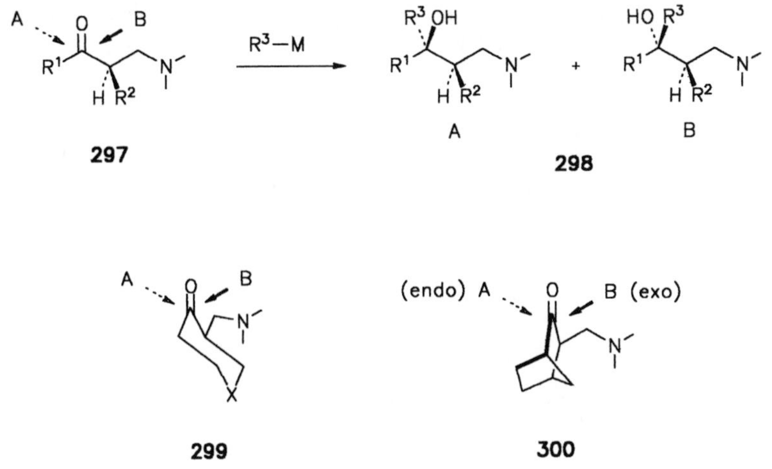

297

298

299

300

R¹, R², X: as in Fig. 113

Fig. 115. The addition of organometals to α-chiral β-aminoketones (only one enantiomer of the racemic pair is represented).

atom is linked to both the carbonyl oxygen and the amine nitrogen of the ketobase, thus
originating two conformational isomers leading, respectively, to either of the diastereomeric products.

2. Opposite stereochemical results with respect to hydrogenation are obtained with cyclic
β-aminoketones of type **299**, which include the Mannich bases of cyclohexanone (X =
CH_2) and several oxa- and thia-derivatives (X = O, S). In this case, the preferred
direction of attack is usually **A**, regardless of the substituent's position, type of amino
group, and homo- or heterocyclic ring, the only exceptions being those of the lithium-
acetylene reagent with dimethylaminomethyl cyclohexanone derivatives.

3. As far as the reactions on Mannich bases of norbornanone **300** are concerned, the
predominant attack is again from the **B** (*exo*) side, producing the *endo*-aminoalcohol,
although with rather low stereoselectivity.

When the asymmetric center is located in a different site from the C-α position, a
much more detailed interpretation of the results is required (see Ref. 591 for specialized
literature).

Interesting results are provided by reactions performed with the chiral organometal
reagent deriving from the α-bromo ester **301** and zinc (Reformatsky reaction) (Fig. 116),
which produces the diastereomeric hydroxy-aminoesters **303** from the achiral β-ami-
nopropiophenones **302**.

The prevalence of diastereomer **303B** is interpreted on the basis of a competition
between cyclic and open-chain transition states.[949]

Like the carbonyl derivatives discussed above, acetylenic Mannich bases **304** (Fig.
117) undergo addition to the unsaturated moiety by the organomagnesium reagent.[952]
Besides the stereoselectivity, however, the regioselectivity of the reaction also has to be
taken into account, as the allylamines produced may derive from attack by the organometal R² residue on either of the carbon atoms engaged in the triple bond. Usually,
allylamines **305**, produced by *anti* addition of the reagent and attack by R² on the

unsaturated carbon atom farthest from the amino group, are predominant as compared with the isomeric products **306** and **307**. Allene derivatives may be also formed in small amounts.

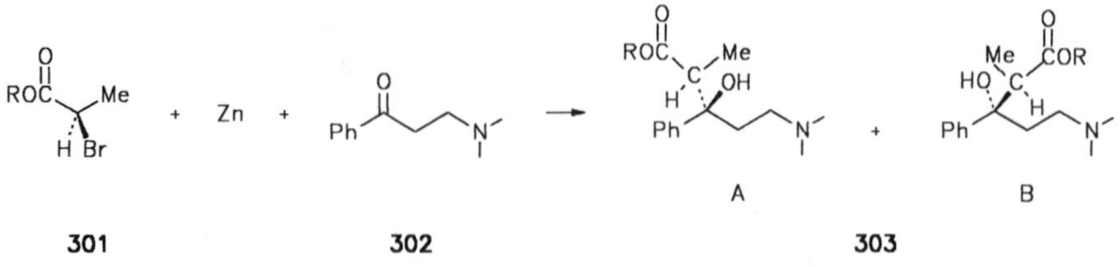

R = Me, Et

Fig. 116. Reformatsky reaction on β-aminoketones with chiral reagents (only one enantiomer of the racemic pair is represented).

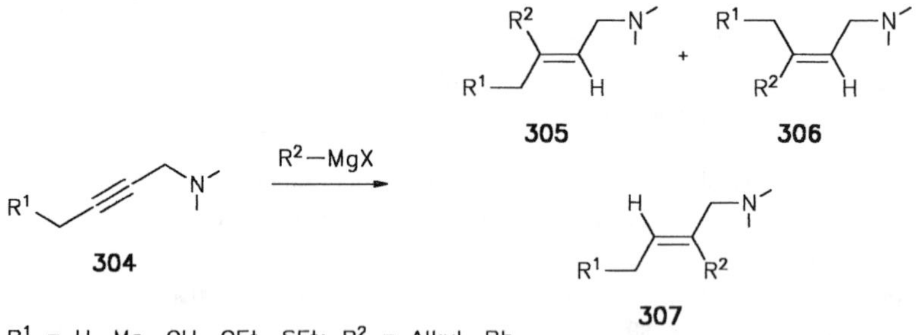

R^1 = H, Me, OH, OEt, SEt; R^2 = Alkyl, Ph

Fig. 117. The addition of organometals to acetylenic Mannich bases.

Allylamines **307** are the main products when the reaction is carried out with the allylmagnesium reagent, and relevant amounts of allenes are formed from **304** when R^1 = Alkyl—X—CH_2— (X = O, S). The production of allenes is in fact observed in the reaction of Mannich bases **308** (Fig. 118) with Grignard reagents, due to elimination of the methoxy group. Studies of the mechanism[953] indicate that reaction conditions and steric factors are determining in establishing the balance between allenes **309** and **310**, obtained by Grignard addition and reduction, respectively.

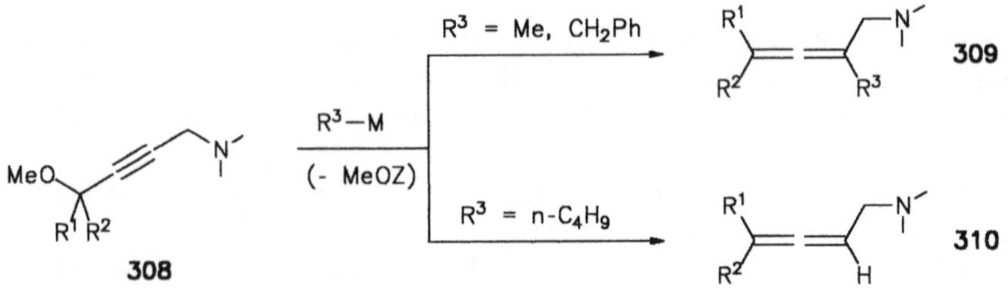

R^1, R^2 = H, C_{1-4} Alkyl

Fig. 118. Synthesis of allenes by organometal reaction on α-methoxy acetylenic Mannich bases.

Lithium derivatives of type **311** (route **a** in Fig. 119) can be obtained from Mannich bases that have suitably reactive moieties toward organometals, by reaction with alkyllithium. Phenolic Mannich bases exhibit such behavior,[954] although different products can also be formed (see below). Ferrocenyl Mannich bases also undergo metallation, favored to some extent by the aminomethyl group, as the first lithium atom is linked to the vicinal carbon atom belonging to the ring, whereas the second is attached to the other unsubstituted ring.

Metallation with lithium is used to introduce into the molecule functional groups such as halogens, carbinol moieties (**312**), and even isotopes.[954,955]

In the reaction of butyllithium or lithium di-isopropylamine with the Mannich bases derived from hydrogen cyanide,[956] phosphine oxides, and phosphorous esters,[957,958] as well as from phenols,[642] the metal atom is prevalently bound to the CH_2—N moiety (**313** in Fig. 119, route **b**). This intermediate is then allowed to react with halides, epoxides, and other alkylating reagents in order to link an alkyl group to methylene. Under proper conditions, aldehydes, ketones and enamines can be prepared by this method.

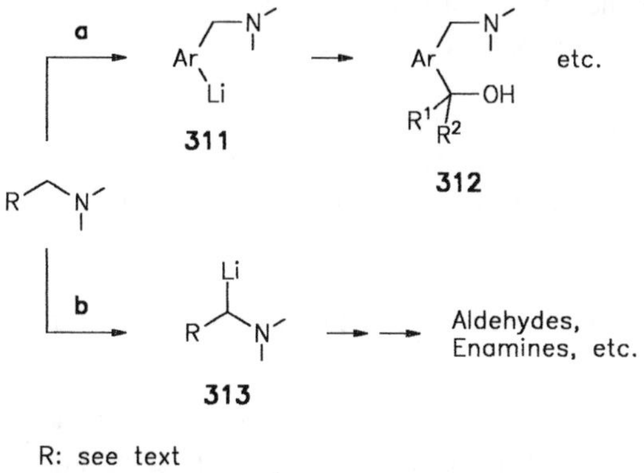

R: see text

Fig. 119. Lithiation reaction of Mannich bases.

D.2 Reactions with Cleavage of the Mannich Base

A few examples are reported in the literature of organometal addition to Mannich bases causing bond cleavage at the methylene group on either the substrate or the amino group side (**a** and **b,** respectively, in Fig. 120). In the former case, aminomethyl phenyl sulfides (**314**, R = PhS), in the presence of aliphatic or aromatic Grignard reagents, permit the preparation of alkylamines (**315**) with an alkyl chain of C_{n+1} atoms starting from alkyl halides having a C_n chain length.[254,959,960] The reaction can be also applied to the synthesis of benzylamines as well as of allylamines, when suitable unsaturated organometals are employed.[961]

By contrast, some phenolic Mannich bases (**314**, R = phenolic residue), heated in the presence of organomagnesium in toluene, afford amine elimination with formation of the R^1-substituted product (**316**).[642] Analogous behavior is observed with organoboranes.[962] By this method, the otherwise unfeasible replacement of an amine by alkyl or aryl groups is thus made possible.

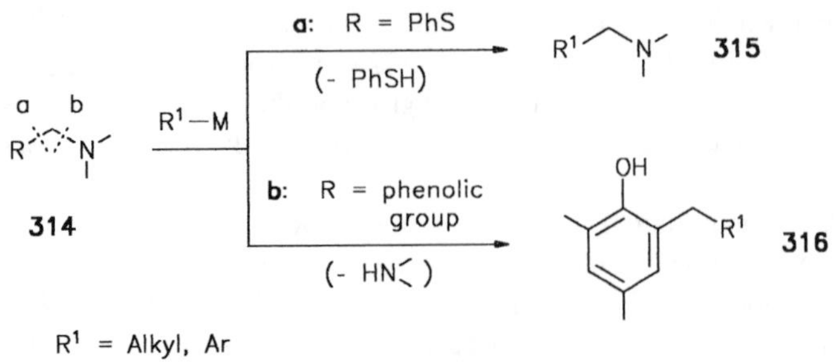

Fig. 120.   Addition of organometals to thiolic or phenolic Mannich bases.

E Cyclization

The multiplicity of the various reactions, drawn from the whole range of organic chemistry, that enable us to produce cyclic derivatives from Mannich bases, calls for the methodical assembly of the great wealth of experimental results reported in the literature. One possible method is to classify the reactions according to the manner in which the Mannich base is inserted into the final product, thus enabling us to distinguish between reactions that do not affect the original molecular skeleton of the base and reactions that cause cleavage of the base and consequent amino group elimination.

E.1   Cyclization Not Involving Amino Group Elimination

The reactions belonging to this group include ring formation involving only the original substrate moiety of the Mannich base (**A**) or both the substrate and amine moieties (**B**), with the production of heterocyclic derivatives containing the amine nitrogen atom.

Only two types of Mannich bases, namely, ketonic and acetylenic, are actually employed in the synthesis of cyclic derivatives of group **A**. The former afford ring products, as depicted in Fig. 121, either (path **a**) by photochemically induced cyclization producing interesting amino-cyclopropanols **317**,[963,964] or (path **b**) by the common methods applied to the synthesis of N-heteroaromatic (pyrroles, pyridines, and rings containing multiple nitrogen atoms)[333,965–967] as well as aliphatic cycles.[968] Such methods usually involve both the carbonyl group and the substituent located in the α position.

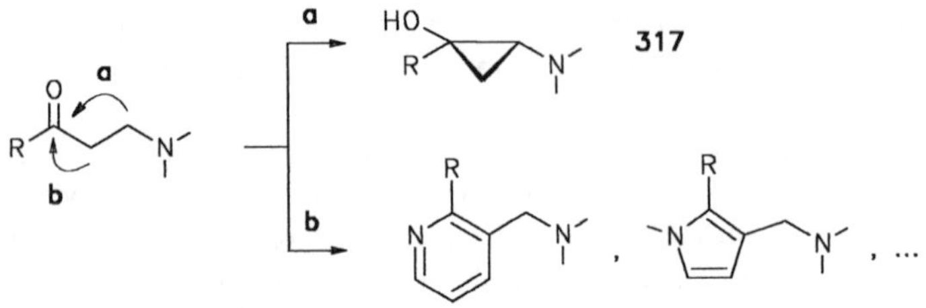

Fig. 121.   Cyclization reaction of β-aminoketones not involving amine elimination.

Examples of synthesis following path **b** are given by the formation of benzopyrazine derivative starting from N-aminomethyl-isatine[969] and by the preparation of pyiridazine derivative **318** (Fig. 122) from aminomethyl-γ-ketocarboxyacid and hydrazine. In this last case, however, reductive deamination takes place concomitantly with formation of the methyl derivative instead of the aminomethyl derivative.[614]

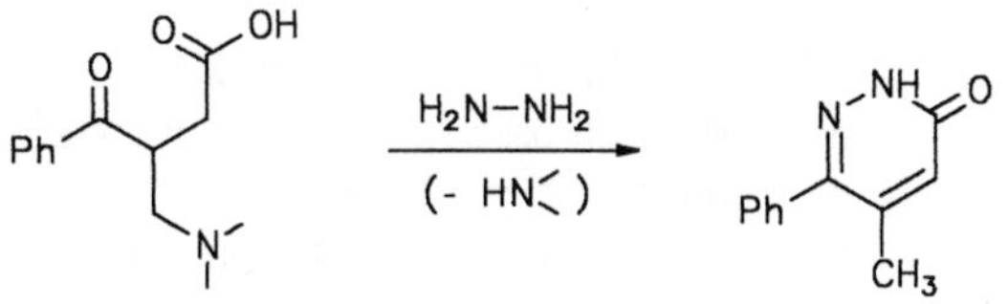

Fig. 122.   Cyclization reaction of γ-ketocarboxyacid Mannich bases.

Acetylenic Mannich bases are also used for producing homo- as well as heterocyclic derivatives (Fig. 123). For instance, the cyclopentene ring is formed by cyclization of the bis-Mannich base **319**, prepared from heptane-1,6-di-yne with butyllithium.[970] However, the reaction is strongly affected by the steric hindrance of the amino group, as it occurs almost instantaneously with the dimethylamino compound, whereas the diethylamino- and pyrrolidino derivatives do not react.

As far as heterocyclic derivatives are concerned, oxazolidinones **320** are synthesized from hydroxyacetylenic Mannich bases and phenyl isocyanate.[971]

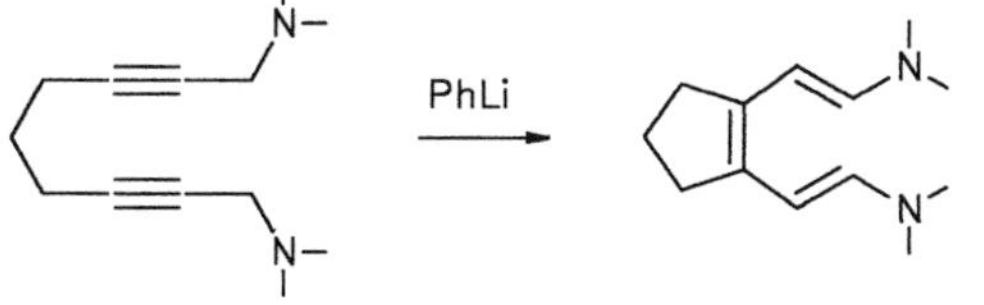

319

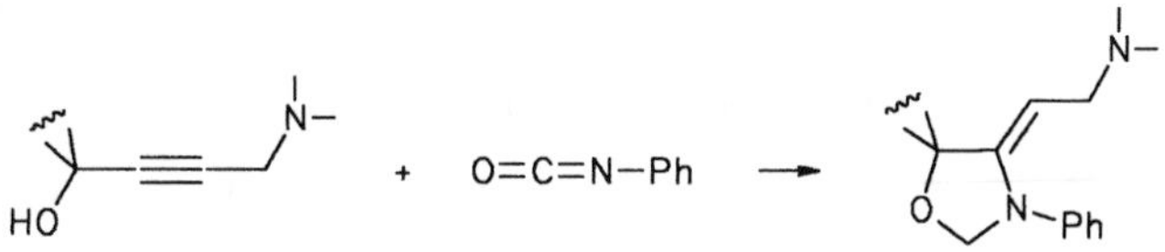

320

Fig. 123.   Cyclization reaction of acetylenic Mannich bases.

Heterocyclic compounds originated by the involvement of the amine moiety of the base (group **B**) can be prepared from several different classes of Mannich derivatives; β-aminoketones in particular, are largely employed. Thus, 4-hydroxypiperidines **322** are obtained[972–974] from bis-ketones **321** (Fig. 124), prepared by Mannich synthesis with primary alkylamine. Both the mechanism[973] and the stereochemistry[972] of the reaction have been investigated; it has been demonstrated that cyclization proceeds stereospecifically, and the resulting product has the configuration depicted in **322**.

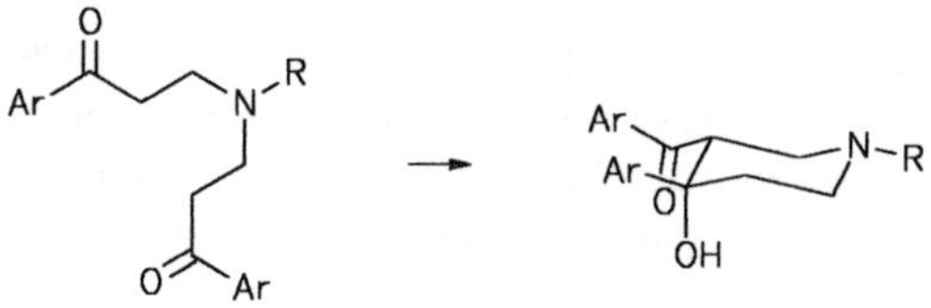

321

322

Fig. 124.   Synthesis of 4-hydroxy-piperidine derivatives.

Intramolecular addition of the secondary amine moiety to the styrene double bond in styryl ketobases Ph—CH=CH—CO—CH$_2$CH$_2$—NH—Ar also produces piperidino derivatives.[975]

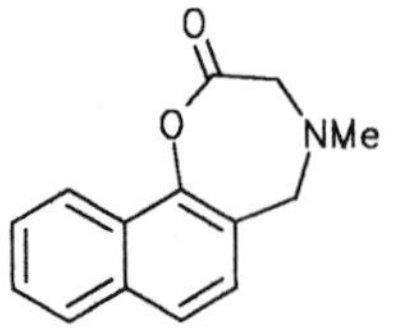

Interesting cyclic derivatives are obtained from phenolic Mannich bases having the amino acid moiety; thus, the seven-membered heterocycle **323** is formed from the α-amino acidic residue and the phenolic hydroxy group by reaction with thionyl chloride.[976]

Phenolic Mannich bases of anthranilic acid (**324**, Fig. 125) behave differently in the cyclization reaction with anhydride, depending on the type of anhydride and reaction conditions, with the formation of dibenzo-oxazocinones **325**, quino- (**326**) or benzoxazino-benzoxazinone (**327**) derivatives.[977]

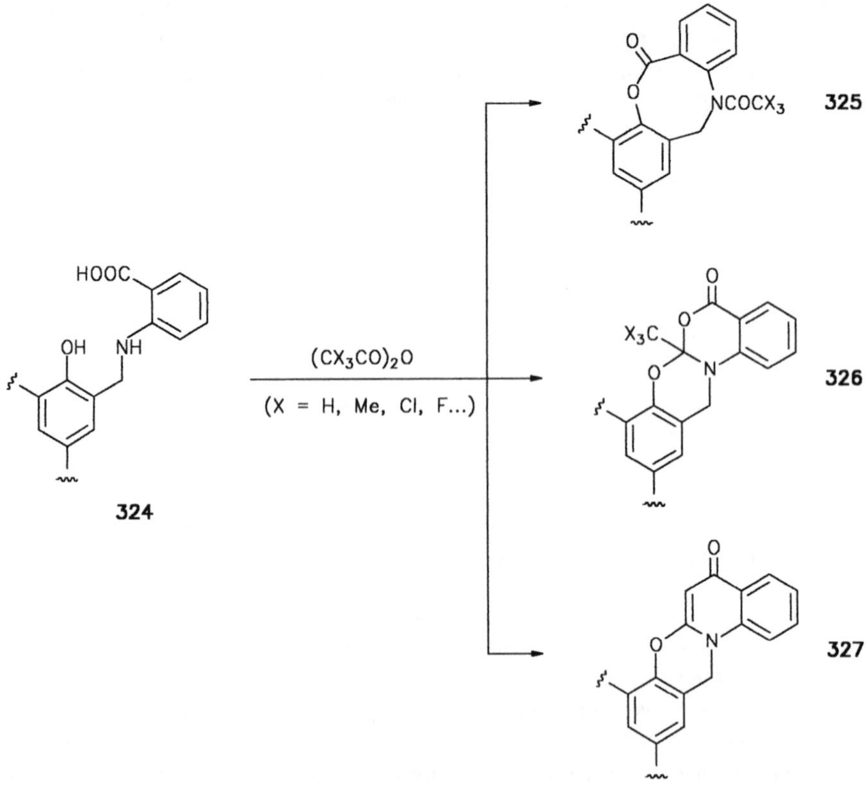

Fig. 125.   Cyclization reactions of phenolic Mannich bases of anthranilic acid.

Oxadiazole derivatives (sydnonimines) **328** and oxazoles **329** (Fig. 126) can be synthesized from Mannich bases of hydrogen cyanide possessing a secondary amino group, as ring closure may take place by condensation with nitrous acid[978–980] or anhydride,[240] respectively. Analogous syntheses of spiroidantoins are reported.[981]

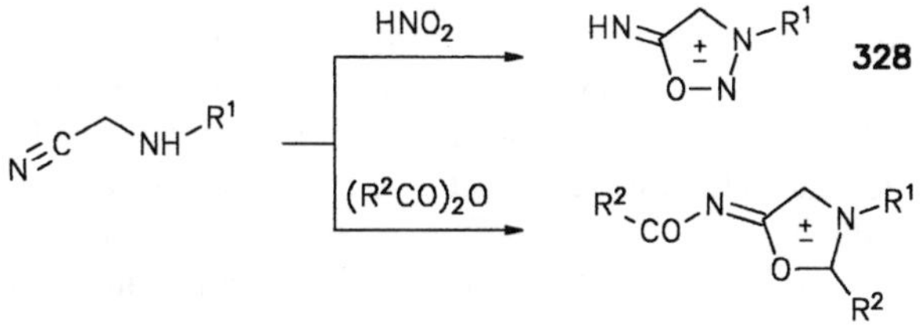

R^1 = Alkyl, Amine, OH
R^2 = Alkyl

Fig. 126. Synthesis of oxadiazoles and oxazoles.

Tertiary Mannich bases having a halogen atom in position 4 with respect to the amino group (Fig. 127) can give cyclic products by intramolecular N-alkylation, with formation of the heterocyclic quaternary salts **330**, as shown by the N-Mannich bases of chloroacetohydrazide[202] and by some brominated O-Mannich bases.[982]

Both the mechanism as well as the stereochemical course of the photochemically induced cyclization of a large series of phthalimide Mannich bases (**331**, Fig. 128) have been accurately investigated.[983–985] It has been found that the main reaction is bond formation between the carbonyl

R ≠ H; X = O, NR'; Z = Halogen

Fig. 127. Cyclization reaction of tertiary Mannich bases containing a halogen atom.

and the C-α atom of the substituent linked to the amino group. This produces the imidazolidine derivative **332**, although in some cases the formation of complex mixtures of products is observed. The phthalimide Mannich base **331** can be used in the synthesis of perhydrooxadiazoles[986] and perhydrobenzodiazepines[987] as well.

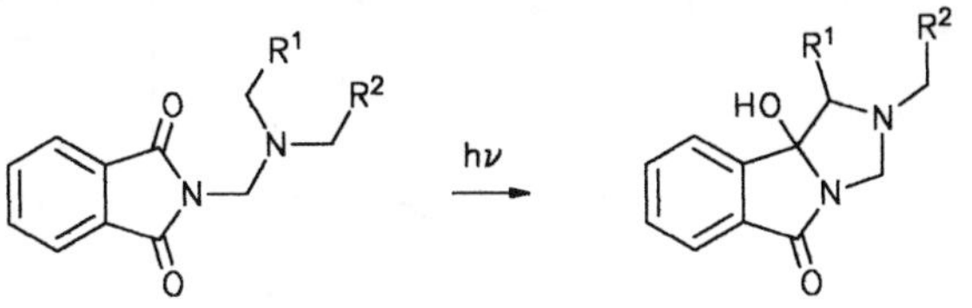

331   **332**

Fig. 128. Photochemical cyclization of dialkylamino Mannich bases of phthalimide.

Further details concerning the synthesis of individual cyclic derivatives of Mannich bases are found in specific surveys.[258,416]

E.2 Cyclization with Amino Group Elimination

Ring closure with concomitant amino group elimination is actually a replacement reaction, as depicted below, strictly connected with the replacement reaction described in Sec. B. Similarly, the cyclization mechanism may or may not involve the formation of a vinyl intermediate (vinylketone, methylenequinone, etc.), originated by the deamination of the initial Mannich base, which then undergoes addition.

Mostly β-aminoketones, along with phenolic Mannich bases, are usually employed in the synthesis of homo- and heterocyclic derivatives. As far as β-aminoketones are concerned, in particular, homocyclic compounds (Fig. 129) are prepared by the Diels-Alder reaction of the Mannich base, acting actually as dienophile, with a conjugated diene.[766,988,989] Interestingly, styryl ketobases[990] give styryl cycloalkenyl ketones (**333**, R = Ar—CH=CH—) by the same type of reaction, without any involvement of the styryl group.

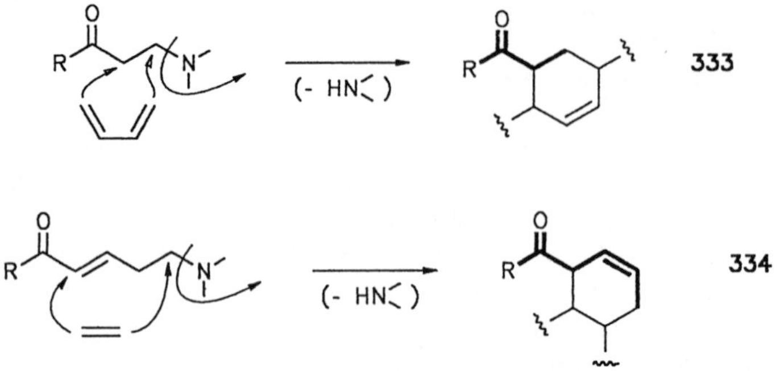

Fig. 129. Synthesis of acylated homocyclic compounds from Mannich ketobases (bold lines represent the original skeleton of Mannich base).

Inversely, vinylogous Mannich bases give the adducts **334** by reaction with dienophiles such as maleic anhydride or with a vinyl group of their deamination product, thus behaving as dienes.[362]

Homocyclic derivatives can also be obtained (Fig. 130) by amino group replacement of the Mannich base by nucleophilic reagents, such as mono- and dialkyl ketones.[822,991–994] The same β-aminoketone may behave as a nucleophilic reagent in self-condensation reactions.[991] Arylketobases (**335**, Ar in the place of RCH₂, path **a**) have also been studied.[992]

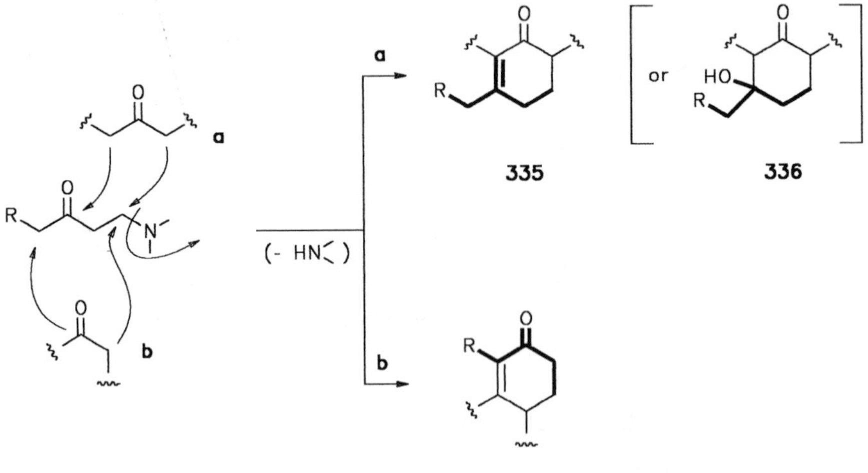

Fig. 130. Synthesis of homocyclic ketones from Mannich ketobases (bold lines represent the original skeleton of the Mannich base).

The resulting cyclic products **335** and **337** are usually unsaturated ketones, except in the case that the hydroxy group arising from aldol condensation is not eliminated (e.g., **336**), that double bond migration occurs,[994] and further consecutive reactions, such as decarboxylation, etc., take place.[822]

Intramolecular amino group replacement by aromatic heterocycle may lead[343] to interesting ring-fused cycloketone derivatives of type **338** (Fig. 131).

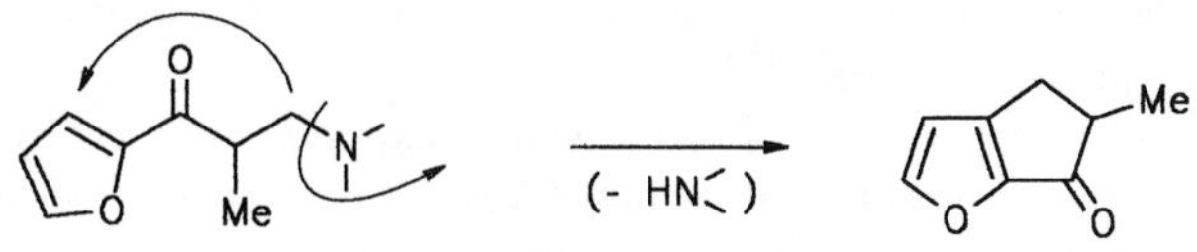

Fig. 131. Intramolecular amino group replacement leading to cycloketone derivative.

Among heterocyclic derivatives obtained from ketonic Mannich bases, nitrogen-containing rings are by far the most prevalent, and the products involved are quite numerous and varied.[258] As a consequence, only selected synthetic methods are mentioned here.

Pyrazolines and oxazolines (**339**, Fig. 132; see also **262** in Sec. B.2) are prepared[258,344,995] by the well-known reaction with hydrazines and hydroxylamines, respectively. Investigations carried out with isotopes allow the exclusion of an elimination-addition mechanism for the reaction, at least in the case of the oxazoline synthesis.[996]

The preparation of pyridines and quinolines has been even more thoroughly investigated and may occur by two different routes, as depicted in Fig. 132.

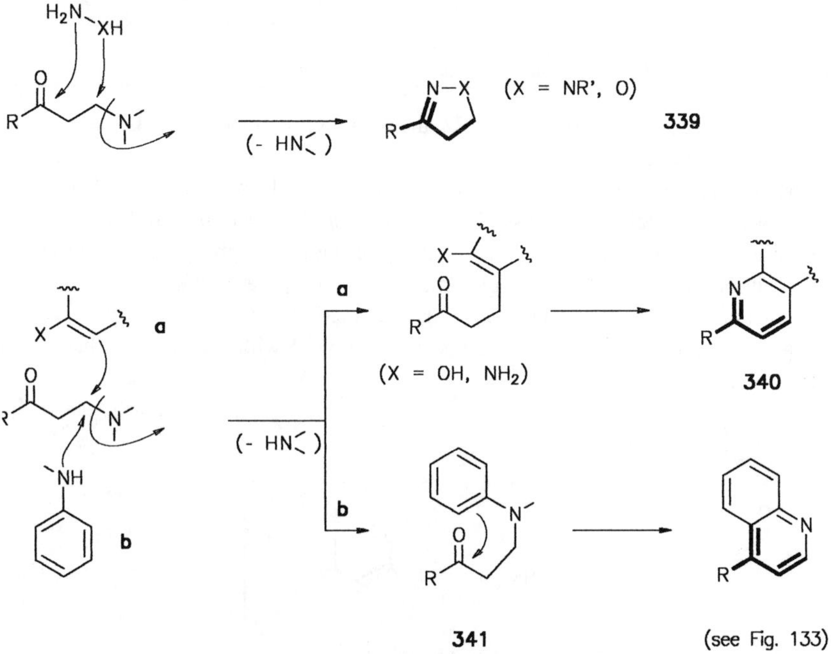

Fig. 132. Synthesis of N-heterocyclic compounds from Mannich ketobases (bold lines represent the original skeleton of the Mannich base).

Path **a** can be considered as a C-alkylation by the Mannich base of enamine or enol derivatives; the latter acts in the presence, for example, of hydroxylamine, followed by intramolecular condensation leading to ring closure and formation of the aromatic nucleus **340**.[997–999] By contrast, path **b** involves amino group replacement by arylamine (N-alkylation by the Mannich base) producing the β-arylaminoketone **341**, directly obtainable also by Mannich synthesis.[1000–1004]

The intramolecular condensation of **341** would be expected to give dihydroquinoline **342** (Fig. 133), but a dismutation reaction takes place, affording the tetrahydroquinoline **343** along with approximately equal amounts of the aromatic derivative **344**, as is frequently observed[1000,1001] when the reaction is performed in the absence of an oxidizing agent. In polyphosphoric acid medium the proposed mechanism involves migration of the hydride ion from position 2 of **342** toward the cationic site of a protonated dihydroquinoline molecule.[1001]

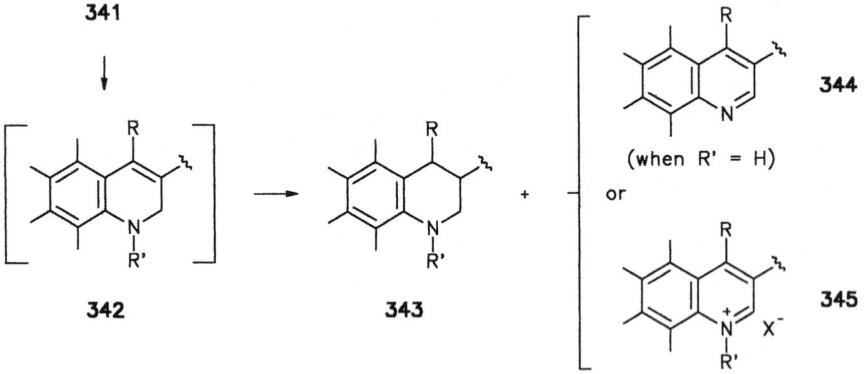

Fig. 133.   Formation of quinolines and quinolinium salts from β-arylamino ketones.

When R′ is different from H in **342**, the resulting product[1002–1004] is the quaternary quinolinium salt **345**.

Mannich ketobases may also be useful materials for the synthesis of O-heterocycles (Fig. 134), as they behave as masked heterodienes in the reaction with dienophiles. The α,β-unsaturated ketone formed by elimination of the amino group can in fact undergo cycloaddition by the dienophile to give the dihydro-4H-pyrans **346**. In the event that an HX molecule is eliminated, the pyran ring is formed.[766,793,847,988] The above reaction is also quite favored by dienes, and it can be competitive[766] with the diene synthesis shown in Fig. 129.

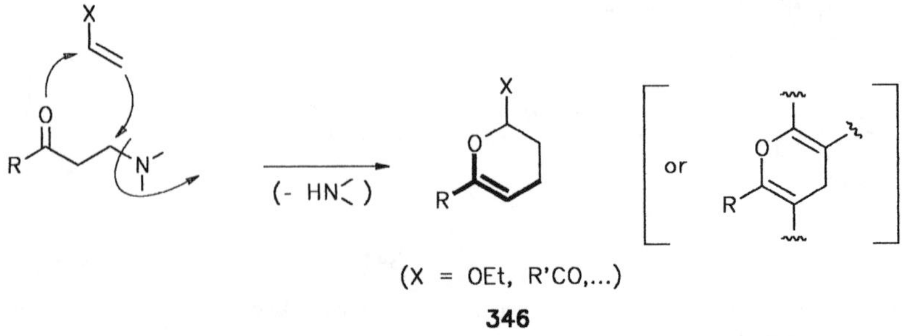

(X = OEt, R'CO,...)

346

Fig. 134.   Synthesis of pyran derivatives from β-aminoketones (bold lines represent the original skeleton of the Mannich base).

The dimerization of vinylketones formed by deamination,[793] already mentioned in Chap. I, B.3, can also be included in the scheme of Fig. 134, as it gives products **346** (X = RCO).

Finally, acyl-epoxides **347** may be produced in appreciable quantity by deamination of the iodomethyl derivative of Mannich ketobase in the presence of hydrogen peroxide,[1005] and cyclic sulfones **348** are yielded by reaction of sodium hydroxymethane sulfinate, $HOCH_2SO_2Na$, with the Mannich base of 1,5-diphenyl-pentanedione.[1006]

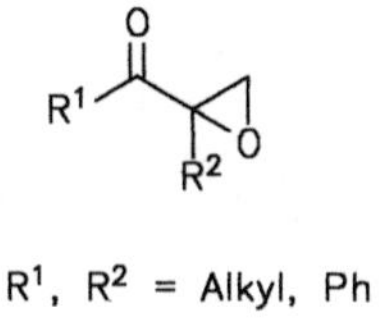

R^1, R^2 = Alkyl, Ph

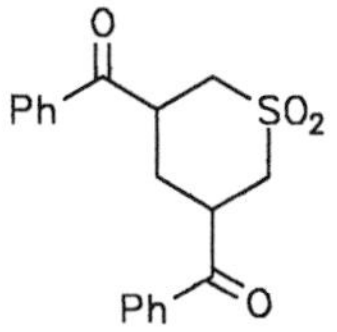

Phenolic Mannich bases afford cyclic derivatives as the result of deamination, O-heterocycles being prepared to a larger extent than N-heterocycles, although an interesting example of homocycle synthesis (Fig. 135) from para-aminomethylated phenolic Mannich bases is worth mentioning.[1007]

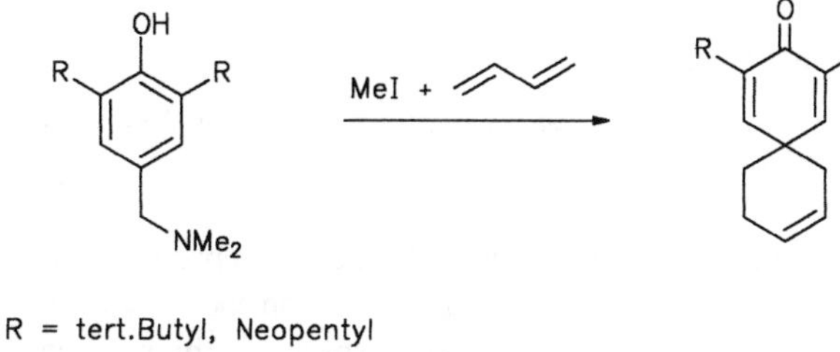

R = tert.Butyl, Neopentyl

Fig. 135. Spiro derivatives from para-phenolic Mannich bases.

Spiroketones **349** can be obtained in good yield even using phenol and butadiene in the presence of catalytic amounts of any amine capable of giving a Mannich reaction.

Nitrogen-containing heterocycles are obtained (Fig. 136) in a way similar to that described above for the synthesis of pyridines from Mannich ketobases and arylamines. Thus, phenolic bases[1008] as well as analogous derivatives (e.g., hydroxy-quinolones)[859] give dihydrophenazines **350**, which can be readily dehydrogenated by the usual oxidizing agents to yield the corresponding aromatic derivatives **351**. The presence in the reaction mixture of reducible species may also lead directly to derivatives **351** as the predominant product. Indeed, the reduced species α-methyl-β-naphthol has been isolated in one case.[1008] The methylene-bis-derivative R—CH₂—R may turn out to be the predominant product in the reaction with aminopyridines.[859]

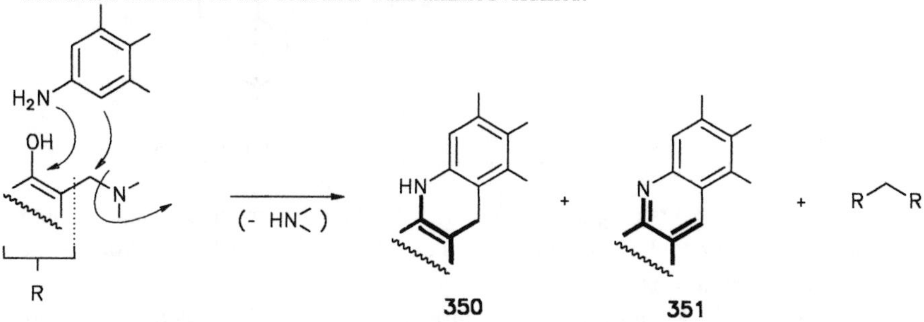

Fig. 136. Synthesis of benzoquinoline derivatives from phenolic Mannich bases (bold lines represent the original skeleton of the Mannich base).

Five-membered O-heterocyclic derivatives of the type dihydrobenzofuran (**352** in Fig. 137) are formed by reaction with diazomethane[1009] or analogous reagents.[827,1010,1011] It is to be stressed, in particular, that the deamination of a Mannich base may be favored by the formation of amine N-oxide,[1010] instead of the corresponding quaternary ammonium salt, as is usually required.

Triethylphosphite[1012] and analogous phosphorus-containing reagents[1013] make it possible to prepare oxa-phospha-indans with the structure **353**.

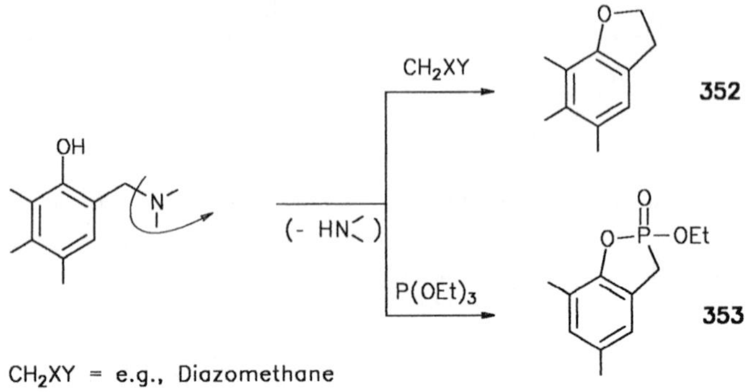

CH$_2$XY = e.g., Diazomethane

Fig. 137.   Synthesis of oxa- or oxa-phospha-indan derivatives from phenolic Mannich bases.

The majority of O-heterocyclic derivatives obtained from phenolic Mannich bases, however, are six-membered rings. In particular, 4-H-chromene (**354** in Fig. 138) and 3,4-dihydrocoumarin (**355**) derivatives are given, respectively, by reaction with alkyl-ketones having two hydrogen atoms in the α position to the carbonyl group[823,1014] and by reaction with esters[823,826,827] or the analogous Mannich bases of coumarin.[857]

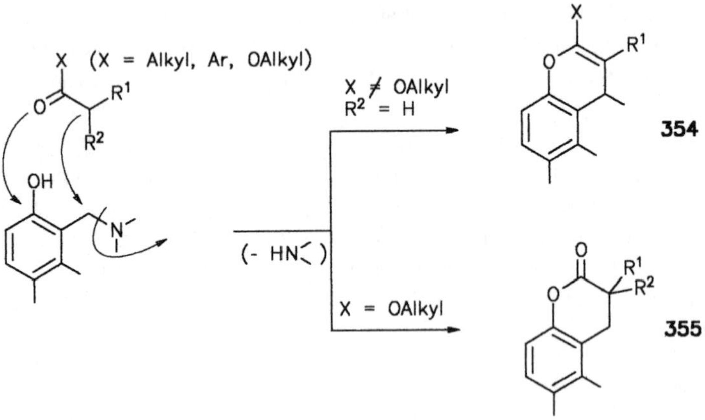

Fig. 138.   Synthesis of six-membered O-heterocycles from phenolic Mannich bases.

Usually, the reagents employed in the above synthesis contain activating groups (R^1 and/or R^2 = carbonyl, etc.); however, concurrent reactions such as hydrolysis and decarboxylation are also observed.

Finally, phenolic Mannich bases may provide heterodiene formation at the ortho-methylenequinone moiety originated by deamination (Fig. 139; see also Fig. 134); this produces derivatives **356**, usually benzopyrans, dihydrobenzopyrans, or less frequently, dihydrobenzoxazines (X = N). These last compounds are given by cyclic imine[1020] or potassium thiocyanate.[1021] A large series of acylbenzopyrans is obtained from alkynyl ketones, whereas benzopyrans are also formed starting from enamines.

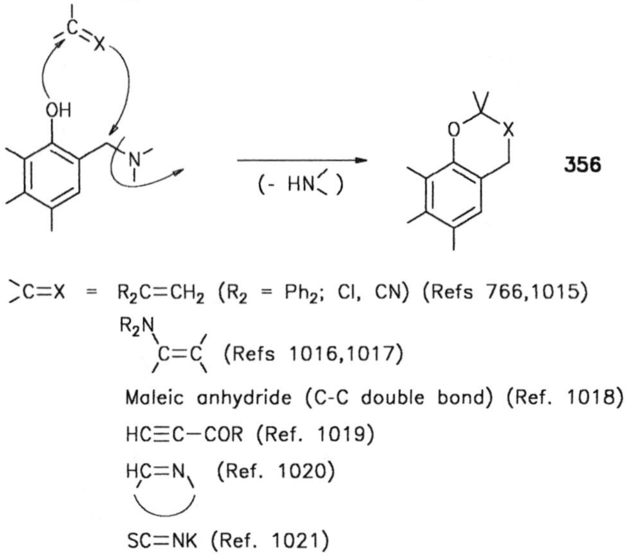
OH
(- HN<)
356
>C=X = R2C=CH2 (R2 = Ph2; Cl, CN) (Refs 766,1015)
R2N
C=C (Refs 1016,1017)
Maleic anhydride (C-C double bond) (Ref. 1018)
HC≡C—COR (Ref. 1019)
HC=N (Ref. 1020)
SC=NK (Ref. 1021)

Fig. 139. Synthesis of benzopyrans and benzoxazines from phenolic Mannich bases.

In addition to the ketonic and phenolic bases, Mannich bases of ferrocene and pyrrole provide further notable examples of ring-forming synthesis. Ferrocenes (Fig. 140) may in fact undergo two types of cyclization leading to the dihydrofuran ring (**357**) or to a bridged product (**358**), deriving, respectively, from amine replacement involving the vicinal diphenyl carbinol group[1022] or from replacement of both the trialkylammonium groups linked to the cyclopentadiene rings.[1023]

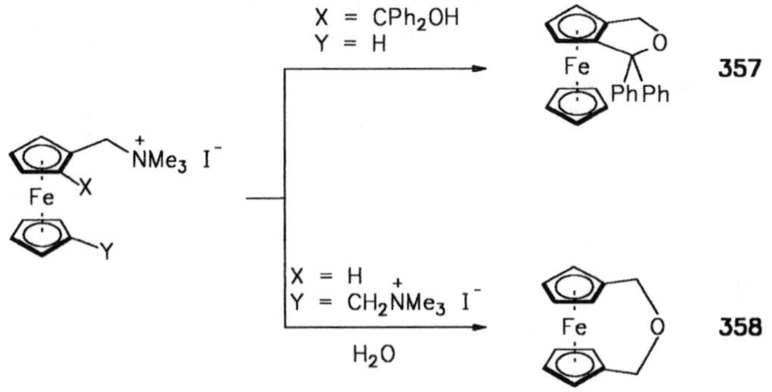
X = CPh2OH
Y = H
Fe
O
PhPh
357
NMe3 I-
X
Fe
Y
X = H
Y = CH2NMe3 I-
H2O
Fe
O
358

Fig. 140. Cyclization reactions on ferrocene Mannich bases.

Pyrrole Mannich bases are intermediates in the synthesis of porphirines **360** (Fig. 141) starting from alkylene-bis-pyrroles **359**. Although the aminomethyl derivative is not isolated, it may be allowed to react with different pyrroles, thus producing macrocycles having different substituents.[191]

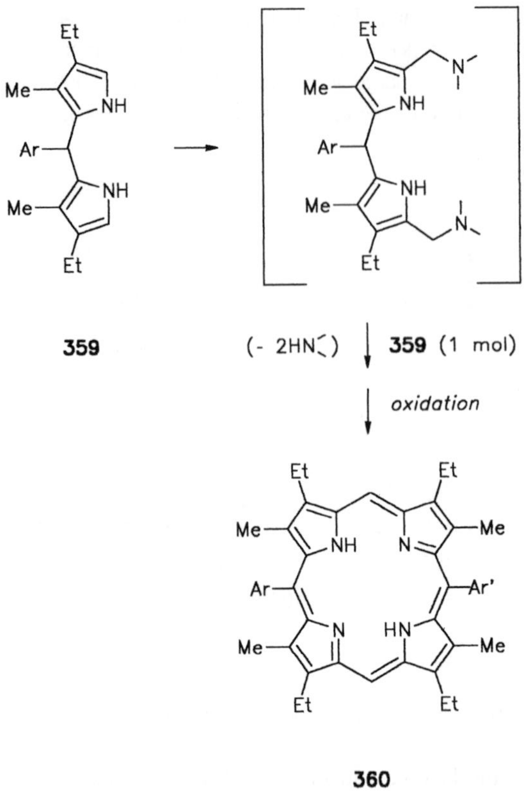

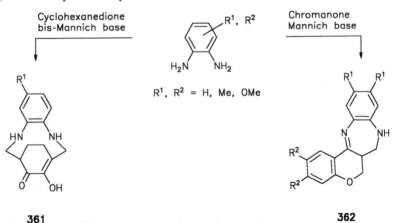

360

Fig. 141.   Synthesis of porphirines involving pyrrole Mannich bases.

Macrocyclic derivatives of type **361** (Fig. 142) can be prepared from bis-Mannich bases of diketones and *o*-phenylenediamines, as the replacement of both the amino groups of the base takes place instead of the expected cyclization involving the carbonyl groups.[1024] Seven-membered ring compounds having cancerostatic properties, such as **362**, are similarly derived by condensation of the Mannich bases of chromanone.[1025]

361   **362**

Fig. 142.   Cyclization reactions involving Mannich bases and *o*-phenylenediamines.

When the *o*-amino nitroso derivative **363** is subjected to reaction with various N-indolic or phenolic Mannich bases, closure of the imidazole ring by the methylene moiety occurs, thus making it possible to obtain the interesting purine derivatives **364**.[1026]

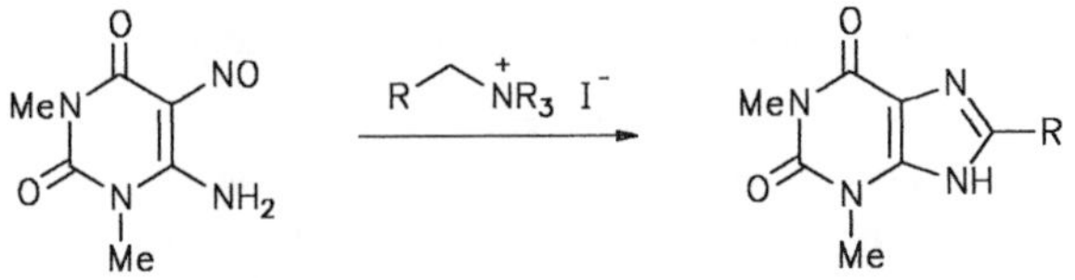

363   **364**

Fig. 143.   Formation of imidazole ring by Mannich bases.

F Miscellaneous Reactions

Less-common reactions of Mannich bases, not included in the preceding sections, are reported and summarized here (Table 30), along with brief comments on the type of reaction and functional groups involved.

Acylation—The mobile hydrogen atoms of phenolic hydroxy groups and secondary amino groups of Mannich bases are replaced mainly by the acyl, usually acetyl, group. Amino group replacement by O-COCH$_3$ can also take place.[1028] Isocyanate and thioisocyanate[1027,1029] may be used as acylation agents.

Nitrosation and nitration—The NH amino group of secondary amidic Mannich bases, upon reaction with nitrous acid, gives the N-nitroso derivative, which may then decompose in an alkaline medium to give diazoalkanes.[1031] The nitration of Mannich bases of nitroalkanes has been studied. Nitrolysis of C—N bonds of the base is frequently observed.[452,1030]

N-Alkylation—Secondary Mannich bases can give the corresponding tertiary derivatives by treatment with particular alkylation agents, such as epoxides (affording β-aminoalcohols) and acrylic derivatives.[1029] Tertiary Mannich bases, mostly, are submitted to N-alkylation in order to produce stable quaternary ammonium salts to be subsequently subjected to deamination (Sec. A.2). However, different quaternary ammonium by-products can be readily given by the reaction.[1033] For instance, a base-catalyzed rearrangement is afforded by allyl ammonium salts **365** (Fig. 144), obtained by N-alkylation of acetylenic Mannich bases with allyl halides.[1034] In the presence of sodium hydride, the compounds **365** yield a wide range of 3-amino-5-hexen-1-yne derivatives **366**.

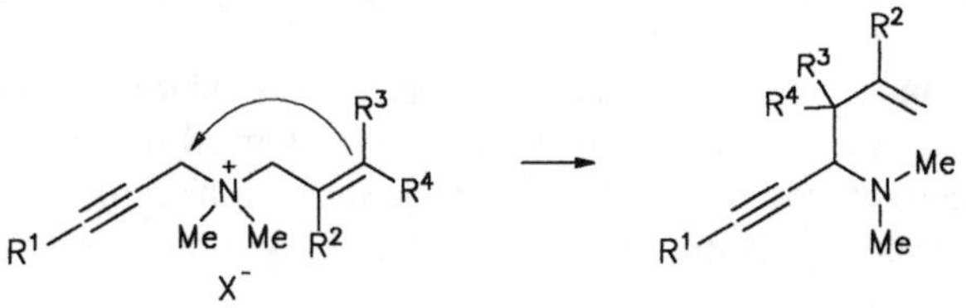

365   **366**

R^1-R^4 = H, Me, Ph

Fig. 144.   Rearrangement of acetylenic Mannich bases.

Table 30
Miscellaneous Reactions on Mannich Bases Reported in the Literature

Reaction type	Mannich base[a] derived from:	Notes	Reference
Acylation	Phenols, Hydroxy coumarins	O-Acylation	914,1027,1028
	Pyrrole, Imines [b]	N-Acylation	249,320,1029
	Amides, Nitroalkanes	N-Nitrosation, -Nitration	452,1030,1031
Alkylation	Phenols, Alkynes,... (see text)	N-Alkylation on amino group	1029,1032- -1034
Azocoupling	See Dyes, Chap. V, A.2		
Complex formation	See Complexing agents, Chap. V, A.2		
Decarboxylation	Acyl and Cyanoacetic acid derivatives (γ-Piperidones included)	Carboxy group elimination	473,617,1035, (see text)
Halogenation	Alkyl Ketones	Halogenation at the α or α' position of carbonyl	320,1036,1037
	Alkynes, Allyl Alcohols[c]	Addition to the unsaturated bond	982,1038
	Various substrates	OH Replacement	634,926
Hydrazone and Oxime form.	Alkyl Ketones, Isatin[b]	Hydrazone form.	877,969,1039
	Alkyl Ketones	Oxime form.	932,1040,1041
Hydrolysis and Hydration	Alkynes, Phenols	Ester and Amide hydrolysis	679,1042,1043
	Alkynes	Water addition to unsaturated bond	933,1044
Oxidation	Phenols mainly		see text
Photochemical	Alkyl Ketones mainly		see text

[a] C–Mannich bases, unless otherwise stated
[b] N–Mannich bases
[c] O–Mannich bases

Azocoupling and complex formation—These reactions, strictly connected with various applications, are discussed in Chap. V, which deals with the relationships between the nature and properties of Mannich bases.

Decarboxylation—Carboxyacids and -esters, previously subjected to acid hydrolysis, undergo decarboxylation. The reaction has been performed on azabicycloketones in refluxing aqueous hydrogen chloride. The replacement of the carboxy group by deuterium is also possible.[1035] Further examples of decarboxylation are reported in Sec. A.2 and in Chap. I, C.5.

Halogenation—The introduction of halogens into Mannich bases is usually carried out according to the well-known synthetic methods of organic chemistry (Table 30). In particular, the replacement of hydrogen atoms in the α or α' position of β-aminoketones is performed with the aim of controlling the regioselectivity of the reaction[1037] and the replacement of hydroxy groups is obtained, for example, by converting die-thanolamino Mannich bases into the corresponding dichloroethyl derivatives which have improved antitumoral activity.[926] The quaternization of heterocyclic amino groups by bromine,[161] the synthesis of chloroamines from secondary Mannich bases[686] and the reaction of iodine with ketobases in order to investigate kinetically the enolization re-action[1045] are also worth mentioning. Halogen-containing reagents, such as cyanogen bromide or acetyl chloride, afford amino group replacement by a halogen when allowed to react with Mannich bases of alkynes[674] or nitroamides,[1046] respectively (see also Fig. 104).

Formation of hydrazones and oximes—Only carbonyl compounds (β-aminoketones and several heterocyclic derivatives) undergo condensation with hydrazines and hy-droxylamines (Table 30). Hydrazo derivatives are mostly prepared from β-aminopro-piophenones or by reaction at position 3 of the N-Mannich base of isatin.[969] Hydrazine may also be involved in amino group replacement (see, e.g., Fig. 99), with the che-moselectivity of the reaction controlled by pH and the nature of the solvent. The same occurs in the reaction with hydroxylamine, where the structure of the reagent molecule appears to play a significant role.[867] Oximes may react further with suitable halogenation reagents to give the corresponding *gem*-chloro-nitroso derivatives[1047] or to undergo cy-clization.[1048] The Beckmann rearrangement (Fig. 145) has been investigated[790] on oximes deriving from Mannich bases of acetophenone, which afford the 3-aminopro-panoic acid anilides **367**.

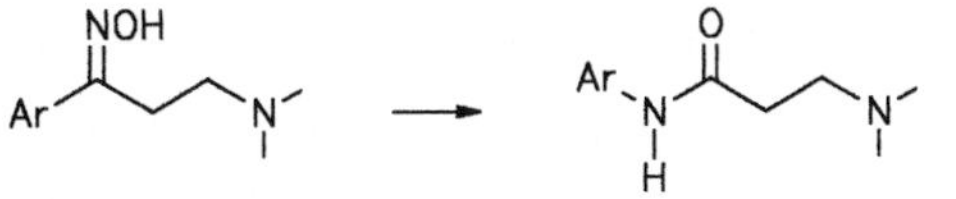

Fig. 145.   Beckmann rearrangement of ketoxime Mannich bases.

Hydrolysis and hydration—Hydrolysis of functional groups present in Mannich bases is best carried out under acidic conditions in order to improve the stability of the base. Hydrazine has been used in the hydrolysis of phthalimide derivatives to obtain products having a primary amino group[1049] (see also Fig. 15, Chap. I). The triple bond of acet-ylenic Mannich bases has been subjected to the addition of water in the presence of mercury salts with the aim of producing β-aminoketones (Table 30).

Oxidation—Several Mannich bases, particularly the phenolic derivatives, have been frequently treated with oxidizing agents.

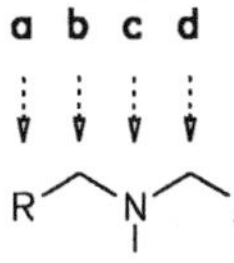

The reaction may involve different reactive centers (**368, a–d**), depending on the reagent employed, as follows:

a. The aromatic ring of phenolic bases is oxidized by N-oxides with the forma-tion of aminomethyl-para-quinones[1050] or the corresponding deaminomethylated derivatives.[1051]

b. Chromic anhydride and mercuric oxide react at the methylene moiety with production of aldehyde. Hydroxybenzaldehydes,[1052,1053] ferrocenyl aldehydes,[1054] and N,N-dialkyl-formamides[1055] are thus synthesized from phenolic and ferrocenyl Mannich bases, re-spectively, as well as from methylene-bis-amines. The methylene group can also undergo thionation, as is observed in the case of hexahydrotriazines, which give[1056] the thiourea derivatives **369** on treatment with sulfur followed by ring opening.

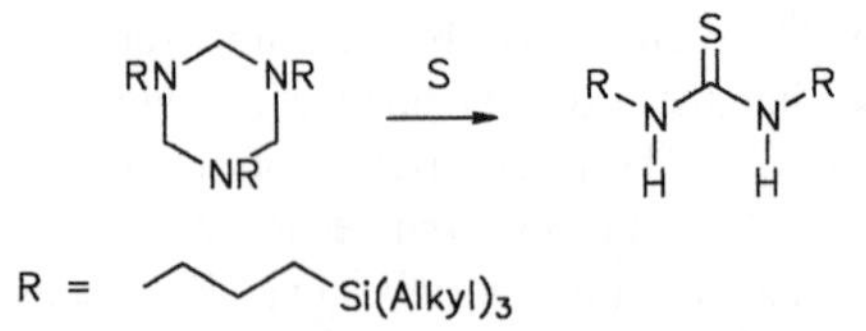

Fig. 146.   Thionation reaction of hexahydrotriazines.

c. The amine N atom is oxidized[1057,1058] by hydrogen peroxide or peracids with the for-
    mation in good yield of the corresponding N-oxide. Interestingly, the N-oxides of mor-
    pholino Mannich bases **370** (Fig. 147), on thermolysis, undergo rearrangement leading
    to allene derivatives.[1010]

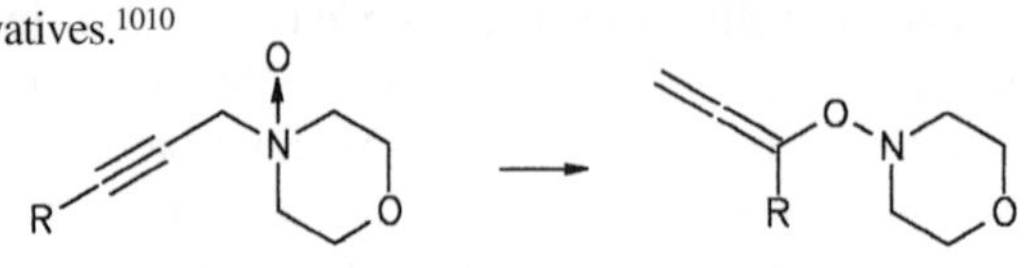

370

R = H, Me

Fig. 147.   Rearrangement of the N-oxide of acetylenic
             Mannich bases.

d. Disodium mercuric ethylenediamine tetraacetic acid oxidizes the methylene group in
    position 2 of cyclic amines, such as piperidine, to give the corresponding lactam.[156,1059]

Photochemical reactions—Mannich ketobases, mainly, are submitted to photochemical
reactions leading to ring formation (see Figs. 121 and 128). Irradiation in methanol of
several aminoaryl ketones, however, gives the imines **371** (Fig. 148) and deaminoal-
kylated compounds are also formed as byproducts, due to fragmentation
reactions.[1060,1061]

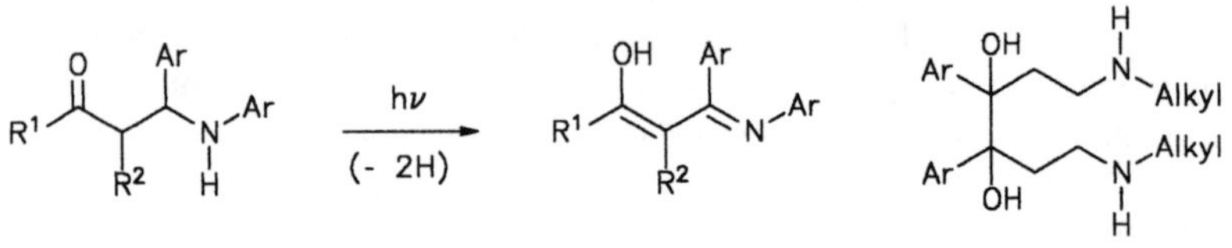

R^1, R^2 = Ar or cyclic   **371**   **372**

Fig. 148.   Photochemical reactions of β-aminoketones.

By contrast, irradiation in benzene of β-aminoketones having a secondary amino
group produces photopinacolization[82] with formation of the diols **372**.
Finally, the Mannich bases derived from cycloalkanones undergo photolysis in
aqueous medium (Fig. 149) to give ω-aminocarboxylic acids **373**.[1062]

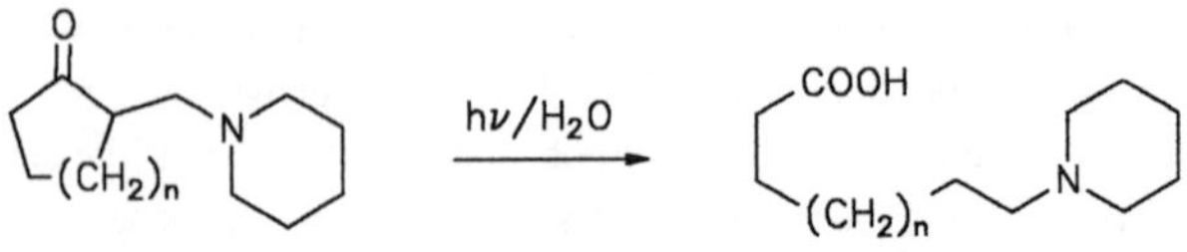

n = 1, 3, 4   **373**

Fig. 149.   Photolysis of cycloketone Mannich bases.

■ References for Introduction and Chapters I and II

1. Neidlein, R., Historisches zu der nach Carl Mannich benannten 'Mannich-reaktion', *Dsch. Apoth. Ztg.,* 117, 1215, 1977.
2. Friedrich, C. and Dallman, C., Carl Mannich und die Pharmazie, *Pharm. Ztg.,* 136, 9, 1991.
3. Böhme, H., Carl Mannich, 1877–1947, *Chem. Ber.,* 88, 1, 1955.
4. Blicke, F. F., The Mannich reaction, in *Organic Reactions,* Vol. 1, 4th ed., John Wiley & Sons, London, 1947, 303.
5. Hellmann, H. and Opitz, G., Aminomethylierung—Eine Studie zur Aufklärung und Einordung der Mannich-reaktion, *Angew. Chem.,* 68, 265, 1956.
6. Reichert, B., *Die Mannich Reaktion,* Springer-Verlag, Berlin, 1959.
7. Hellmann, H. and Opitz, G., *α-Aminoalkylierung,* Verlag Chemie, Weinheim, 1960.
8. Miocque, M., Les réactions d'aminométhylation selon Mannich, *Ann. Pharm. Fr.,* 27, 381, 1969.
9. Thompson, B. B., The Mannich reaction. Mechanistic and technological considerations, *J. Pharm. Sci.,* 57, 715, 1968.
10. Tramontini, M., Advances in the chemistry of Mannich bases, *Synthesis,* 703, 1973.
11. Tramontini, M. and Angiolini, L., Further advances in the chemistry of Mannich bases, *Tetrahedron,* 46, 1791, 1990.
12. Further references on specific topics treated in papers in Polish, Russian, Japanese, and Chinese language are reported by the following *Chem. Abstr.:* 110, 172353, 1989; 111, 135219, 1989; 103, 122498, 1985; 92, 214372, 1980; 103, 195859, 1985; 97, 24174, 1982; 115, 115577, 1991; 92, 314372, 1980.
13. Varma, R. S., The application of Mannich reaction in the field of drug research, *Labdev B,* 12, 126, 1974; *Chem. Abstr.,* 85, 45637, 1976.
14. Sweeney, T. R. and Pick, R. O., 4-Aminoquinolines and Mannich bases, *Handbo. Exp. Pharmacol.,* 68, 363, 1984.
15. Bundgaard, H., Formation of prodrugs of amines, amides, ureides and imides, *Methods Enzymol.,* 112, 347, 1985.
16. Riera de Narváez, A. J. and Ferreira, E. I., *Quim. Nova Rio de Janeiro,* 38, 1985; *Chem. Abstr.,* 107, 198116, 1987.
17. Zalai, A., Effect of detergent-dispersant engine oil additives, in 5th. Int. Koll. *Addit. Schmierst. Arbeitsflussigkeiten,* Budapest, 1986, 2, 9/4/1.

18. Fedtke, M., Acceleration mechanism in curing reactions involving model systems, *Makrom. Kem. Makrom. Symp.,* 7, 153, 1987.
19. Tramontini, M., Angiolini, L., and Ghedini, N., Mannich bases in polymer chemistry, *Polymer,* 29, 771, 1988.
20. Andrisano, R., Baroncini, L., and Tramontini, M., Ricerche sulla reattività delle basi di Mannich. III. Su alcuni coloranti reattivi contenenti il gruppo β-ammino-etil-chetonico, *Chim. Ind.,* 47, 173, 1965.
21. Mauli, R., Ringold, H. J., and Djerassi, C., Steroids. CXLV. 2-Methylandrostane derivatives. Demonstration of boat form in the bromination of 2-α-methyl-androstan-17β-ol-3-one, *J. Am. Chem. Soc.,* 82, 5494, 1960.
22. Phillips, S. D. and Castle, R. N., Quino[1,2-c]quinazolines. I. Synthesis of quino[1,2-c]quinazolinium derivatives as analogues of antitumor benzo[c] phenanthridine alkaloids, *J. Heterocycl. Chem.,* 1489, 1980.
23. Vinogradova, E. V., Gorbacheva, L. I., Terent'ev, A. P., and Krylov, V. D., Synthesis based on 3,5-dinitrotoluic acid, *Zh. Org. Khim.,* 6, 362, 1970; *Chem. Abstr.,* 72, 110966, 1970.
24. Voronov, M. G., Mirskov, R. G., Kuznetsov, A. L., and Proidakov, A. G., Trialkylaminopropargylgermanes, *Izv. Akad. Nauk SSSR,* 1452, 1978; *Chem. Abstr.,* 89, 109830, 1978.
25. Noll, B., Weinelt, F., Weinelt, H., Hauptmann, S., Mann, G., Erhardt, D., and Mertens, W., East German Patent DD 287,490, 1991; *Chem. Abstr.,* 115, 49075, 1991.
26. Dubina, V. L., Zakatov, V. V., and Stakhovskaya, V. O., Effect of an arylsulfonamide group on substitution in an aromatic nucleus. I. Aminomethylation of sulfonic acid arylamides, *Zh. Org. Khim.,* 22, 601, 1986; *Chem. Abstr.,* 106, 84085, 1987.
27. Lis, R. and Marisca, A. J., Methanesulfonanilides and the Mannich reactions, *J. Org. Chem.,* 52, 4377, 1987.
28. Bernardi, L. and Temperilli, A., Ergoline derivatives. Note XI. 2-Aminomethyl- and 2-methylergolines, *Chim. Ind.,* 54, 998, 1972.
29. Yang, P. W., Liu, R. F., and Lin, L. C., Syntheses of 3-aminoalkyl-1-oxaazulan-2-ones, *Hua Hsueh,* 74, 1974; *Chem. Abstr.,* 85, 32731, 1976.
30. Rausch, M. D. and Genetti, R. A., Organometallic π complexes. XXII. The chemistry of π-cyclopentadienyltetraphenylcyclobutadienecobalt and related compounds, *J. Org. Chem.,* 35, 3888, 1970.

31. Lapenko, V. L., Potapova, L. B., Slivkin, A. I., and Vasil'eva, E. V., Synthesis and some analogs of D-galacturonic acid amide and hydrazide, *Izv. Vyssh. Uchebn. Zaved. Khim. Khim. Tekhnol.*, 30, 38, 1987; *Chem. Abstr.*, 108, 38260, 1988.

32. Huettenrauch, R. and Keiner, I., Dissoziationskonstanten von Mannich-Basen der Tetracyclin-Antibiotika, *Arch. Pharm.*, 301, 97, 1968.

33. Georgescu, M. A. and Leonte, M. V., α-Aminoalkylation. V. Aminomethylation of nitrophenols, *Bull. Univ. Galati*, 2, 69, 1979.

34. Prishchenko, A. A., Livantsov, M. V., Pisarnitskii, D. A., Shagi-Mukhametova, N. M., and Petrosian, V. S., Amino- and amidomethylation of bis(trimethylsiloxy)phosphine, *Zh. Obshch. Khim.*, 60, 699, 1990; *Chem. Abstr.*, 113, 78533, 1990.

35. Dutra, G. A., U.S. Patent, 4,083,898, 1978; *Chem. Abstr.*, 89, 109957, 1978.

36. Redmore, D., Chemistry of phosphorous acid: new routes to phosphonic acids and phosphate esters, *J. Org. Chem.*, 43, 992 and 996, 1978.

37. Otal Olivan, J. V. and Perez Esteban, L. E., Spanish Patent ES 538,013, 1985; *Chem. Abstr.*, 106, 19042, 1987.

38. Paraskewas, S. M., Aminomethylierung der bis-[pyridin-3-carboxamid]-PdCl₂-Komplexes, *Synthesis*, 47, 1978.

39. Danishefsky, S., Kitahara, T., McKee, R., and Schuda, P. F., Reaction of silyl enol ethers and lactone enolates with dimethyl(methylene)ammonium iodide. The bis-α-methylenation of pre-vernolepin and pre-vernomenin, *J. Am. Chem. Soc.*, 98, 6715, 1976; see also *J. Am. Chem. Soc.*, 99, 6066, 1977.

40. Holy, N. L., Fowler, R., Burnett, E., and Lorenz, R., The Mannich reaction. II. Derivatization of aldehydes and ketones using dimethyl(methylene)ammonium salts, *Tetrahedron*, 35, 613, 1979.

41. Danishefsky, S., Prisbylla, M., and Lipisko, B., Regioselective Mannich reactions via trimethylsilyl enol ethers, *Tetrahedron Lett.*, 21, 805, 1980.

42. Danishefsky, S., Kahn, M., and Silvestri, M., An anomalous Mannich reaction of a trimethylsilyl enol ether, *Tetrahedron Lett.*, 23, 1419, 1982.

43. Miyano, S., Hokari, H. and Hashimoto, H., Carbon-carbon bond formation by the use of chloroiodomethane as a C₁ unit. III, *Bull. Chem. Soc. Jpn.*, 55, 534, 1982.

44. Hosomi, A., Iijima, S., and Sakurai, H., A novel aminomethylation of silyl enol ethers with aminomethyl ethers catalyzed by iodotrimethylsilane or trimethylsilyl trifluoromethane sulfonate, *Tetrahedron Lett.*, 23, 547, 1982.

45. Lajunen, M. and Krieger, H., Versuche zur Darstellung von α-Methylen-3(10)carenen, *Rapp. Univ. Oulu. Ser. Chem.*, 19, 1985.

46. Oida, T., Tanimoto, S., Ikehira, H., and Okano, M., Reaction of O-silylated enolates of carboxylic esters and of lactones with aminomethylethers, *Bull. Chem. Soc. Jpn.*, 56, 645, 1983.

47. Renaud, R. N., Stephens, C. J. and Brochu, G., Electrochemical oxidation of 4-substituted N,N-dimethylaniline in the presence of silyl enol ether. Effect of the substituent on the formation of Mannich base. II, *Can. J. Chem.*, 62, 565, 1984; see also *Can. J. Chem.*, 61, 1379, 1983.

48. Rochin, C., Babot, O., Dunoguès, J., and Duboudin, F., A new convenient synthesis of dialkyl(methylene)ammonium chloride, *Synthesis*, 228 and 667, 1986.

49. Ikeda, K., Achiwa, K., and Sekiya, M., Trifluoromethanesulfonic acid-promoted reaction of hexahydro-1,3,5-triazines, *Chem. Pharm. Bull. Tokyo*, 34, 1579, 1986.

50. Pilli, R. A. and Russowsky, D., Secondary Mannich bases via trimethylsilyl trifluoromethane sulphonate promoted addition of silyl enol ethers to Shiff bases, *J. Chem. Soc. Chem. Commun.*, 1053, 1987.

51. Ikeda, K., Achiwa, K., and Sekiya, M., Trifluoromethanesulfonic acid-promoted reaction of hexahydro-1,3,5-triazines with ketene silyl acetals. Convenient synthesis of alkyl β-aminocarboxylates, *Tetrahedron Lett.*, 24, 913, 1983.

52. Okano, K., Morimoto, T., and Sekiya, M., N,N-bis(trimethylsilyl)methoxymethylamine as a convenient synthetic equivalent for —CH₂NH₂: primary aminomethylation of esters, *J. Chem. Soc. Chem. Commun.*, 883, 1984.

53. Fairhurst, R. A., Heaney, H., Papageorgiu, G., Wilkins, R. F., and Eyley, S. C., Mannich reaction of oxazolidines, *Tetrahedron Lett.*, 30, 1433, 1989.

54. Kusnetsov, A. L., Mirskov, R. G., Voronkov, M. G., and Rakhlin, V. I., 2,2-Diorganyl-1,3-dioxa-6-aza-2-sylacyclooctanes in aminomethylation reaction, *Zh. Obshch. Khim.*, 59, 483, 1989; *Chem. Abstr.*, 112, 198476, 1990.

55. Overman, L. E. and Goldstein, S. W., Enantioselective total synthesis of allopumiliotoxin A alkaloids 267A and 339B, *J. Am. Chem. Soc.*, 106, 5360, 1984.

56. Kunz, H. and Schanzenbach, D., Carbohydrates as chiral templates: stereoselective synthesis of β-aminoacids, *Angew. Chem. Int. Ed. Engl.*, 28, 1068, 1989.

57. McClure, N. L., Dai, G. Y., and Mosher, H. S., exo,endo-3-(Dimethylaminomethyl)-d-camphor: d-Camphor Mannich products, *J. Org. Chem.*, 53, 2617, 1988.

58. Kunz, H. and Pfrengle, W., Carbohydrates as chiral templates: stereoselective tandem Mannich-Michael reactions for synthesis of piperidine alkaloids *Angew. Chem. Int. Ed. Engl.*, 28, 1067, 1989.

59. Roberts, J. L., Borromeo, P. S., and Poulter, C. D., Addition of Eschenmoser salt to enolates of ketones, esters and lactones, *Tetrahedron Lett.*, 1621, 1977.

60. Seebach, D., Betschart, C., and Schiess, M., Diastereoselektive Synthese neuartiger Mannich Basen mittels Titanderivaten, *Helv. Chim. Acta*, 67, 1593, 1984.

61. Katritzky, A. R. and Harris, P. A. Benzotriazole-assisted synthesis of novel Mannich bases from ketones and diverse aldehydes, *Tetrahedron*, 46, 987, 1990.

62. Nolen, E. G., Allocco, A., Vitarius, J., and McSorley, K., Mannich reaction of carbonyl compounds via boron enolates and N,N,N',N'-tetramethyldiaminomethane, *J. Chem. Soc. Chem. Commun.*, 1532, 1990.

63. Hooz, J., Oudenes, J., Roberts, J. L., and Benderly, A., A new regiospecific synthesis of enol-boranes of methyl ketones, *J. Org. Chem.*, 52, 1347, 1987.

64. Nolen, E. G., Allocco, A., Broody, M., and Zuppa, A., Diastereoselectivity in the synthesis of Mannich bases, *Tetrahedron Lett.*, 32, 73, 1991.

65. Hooz, J. and Bridson, J. N. A method for the regiospecific synthesis of Mannich bases. Reaction of enolborinates with dimethyl (methylene) ammonium iodide, *J. Am. Chem. Soc.*, 95, 6220, 1973.

66. Kalinin, A. V., Apasov, E. T. Ioffe, S. L. Kozyukov, V. P. and Kozyukov, Vi. P., Nonaqueous aminomethylation of silylated polynitrogen heterocycles, *Izv. Akad. Nauk. SSSR Ser. Khim.*, 1447, 1985; *Chem. Abstr.*, 104, 168415, 1986.

67. Orlova, N. A., Belavin, L. Y., Sergeev, Y. N., Shipov, A. G., and Bankov, Y. I. Trimethylsilyl derivatives of N-methylol compounds as N-amido- and N-aminomethylating agents, *Zh. Obshch. Khim.*, 54, 717, 1984; *Chem. Abstr.*, 101, 110668, 1984.

68. Kalinin, A. V., Apasov, E. T., Ioffe, S. L., Kozyukov, V. P., and Kozyukov, Vi. P., Nonaqueous condensation of silylated C- and N-nitrocompounds with [[(trimethylsilyl)oxy]-methyl]dialkylamines, *Izv. Akad. Nauk. SSSR Ser. Khim.*, 2635, 1985; *Chem. Abstr.*, 105, 191186, 1986.

69. Böhme, H. and Ziegler, F., Zur Aminomethylierung von 1-Cyan-isochroman und 1-Cyan-isothiochroman, *Arch. Pharm.*, 307, 287, 1974.

70. Fridman, A. L., Zalesov, V. S., and Mukhamershin, F. M., Formation of γ,γ-dinitro-δ-valerolactames, *Zh. Org. Khim.*, 10, 2053, 1974; *Chem. Abstr.*, 82, 72754, 1975.

71. Unterhalt, B. and Thamer, D., Nitramine. IV. Aminomethyl-nitramine, *Synthesis*, 676, 1973.

72. Zinner, G., Kliegel, W., Hitze, M., and Vollrath, R., Aminomethylierung von N-Hydroxyphthalimid. II, *Liebigs Ann. Chem.*, 745, 207, 1971.

73. Johnson, P. Y., Silver, R. B., and Davis, M. M., The Mannich reaction. -6-Alkoxytetrahydro-5,5-dimethyl-1,3-oxazines, *J. Org. Chem.*, 38, 3753, 1973.

74. Bredereck, K., Banzhaf, L., and Koch, E., Untersuchungen zur Methylierung, Hydroxymethylierung und Aminomethylierung von Anthrachinonen, *Chem. Ber.*, 105, 1062, 1972.

75. Ivanov, B. E., Ageeva A. B., Krokhina, S. S. Valitova, L. A., Chichkanova, T. V., and Gaidai, V. T., Phosphonomethylation of p-substituted anilines, *Izv. Akad. Nauk.. SSSR Ser. Khim.*, 178, 1982; *Chem. Abstr.*, 96, 162834, 1982.

76. Gaudette, R. R., Ohlson, J. L., and Scanlon, P. M., U.S. Patent 4,338,460, 1982; *Chem. Abstr.*, 97, 198563, 1982.

77. Mitsubishi Gas Chem. Co., Inc., Japanese Kokai Tokkyo Koho JP 82 80,346, 1982; *Chem. Abstr.*, 97, 198560, 1982.

78. Dronov, V. I. and Nikitin, Y. E., Thioalkylation reaction, *Usp. Khim.*, 54, 941, 1985; *Chem. Abstr.*, 103, 214237, 1985.

79. Petersen, H. A. Cross-linking with formalde-hyde-containing reactants, in *Handbook of Fiber Science and Technology, Vol 2: Chemical Processing of Fibers and Fabrics: Functional Finishes, Part A*, Lewin, M. and Sello, S. B., Eds., Dekker, New York, 1983; see, in particular, paragraphs 3, 5, and 6.

80. Mazur, S. G., Morgun, T. M., and Starinchikova, A. F., Synthesis of weakly basic anion-exchange resins by amidomethylation, *Plast. Massy*, 15, 1987; *Chem Abstr.*, 107, 176953, 1987.

81. In this context, the chapter on "Vinylogous Aminoalkylation" of Ref. 7, p. 216, is worthy of interest.

82. Roth, H. J., Abdul-Baki, A., and Schraut, T., Versuche zur photochemischen Synthese ephedrinähnlicher Verbindungen. I. Photopinakolisierung von 3-Amino-propiophenonen, *Arch. Pharm.*, 309, 2, 1976.

83. Magarian, R. A. and Sorenson, W. G., Adamantanamine derivatives. Antimicrobial activities of certain Mannich bases, *J. Med. Chem.*, 19, 186, 1976.

84. Möhrle, H. and Tröster, K., Über Mannich Basen. XIX. Mannich-reaktionen mit 1-Naphthol und primären Aminen, *Arch. Pharm.*, 315, 619, 1982.

85. Möhrle, H., Scharf, U. Rühmann, E., and Schmid, R., Über Hexahydrotriazine. IV. Addition N-monosubstituirter Amide an Azomethine, *Arch. Pharm.*, 316, 222, 1983.

86. Gafurov, R. G., Korepin, A. G. Sogomomyan, E. M., Salakhova, A. N., and Eremenko, L. T., 3-Fluoro-3,3-dinitro-1-aminopropane in the Mannich reaction, *Izv. Akad. Nauk*, 1876, 1972; *Chem. Abstr.*, 77, 151388, 1972.

87. Bhargava, P. N. and Sharma, S. C., Some possible antihystaminics and antispasmodics. II. Synthesis of Mannich bases, *Bull. Chem. Soc. Jpn.*, 38, 912, 1965.

88. Novikov, S. S., Ducinskaya, A. A., Hakarov, N. V., and Khmel'nitskyi, L. T., Dinitroamines in the Mannich condensation. I. Condensation of ethylenedinitroamine with aliphatic diamines, *Izv. Akad. Nauk SSSR Ser. Khim.*, 1833, 1967; *Chem. Abstr.*, 68, 29687, 1968.

89. Ungnade, H. E. and Kissinger, L. W., β,β-Dinitroalkylamines and -nitroamines and related compounds, *J. Org. Chem.*, 30, 354, 1965.

90. Harrel, W. B., Mannich bases from 1,2-diphenylindolizine: ephedrine and methamphetamine as amine components, *J. Pharm. Sci.*, 59, 275, 1970.

91. Thiele, K., Belgian Patent 630, 296, 1963; *Chem. Abstr.*, 61, 1800, 1964.

92. Bobbitt, J. M., Kulkarni, C. L., Dutta, C. P., Kofod, H., and Ng Chiong, K., Synthesis of indoles and carbolines via aminoacetaldehyde acetals, *J. Org. Chem.*, 43, 3541, 1978.

93. Bruening, H., Darling, C. M., Magarian, R. A., and Nobles, W. L., Use of N-methyl-tetrahydrofurfurylamine in the Mannich reaction, *J. Pharm. Sci.*, 54, 1537, 1965.

94. Soc Anon. Dabeer, French Patent Appl. 2,354,993, 1978; *Chem. Abstr.*, 89, 215057, 1978.

95. Murase, I., Synthesis of N,N'-bis(2-hydroxybenzyl)ethylenediamine-N,N'-diacetic acid, *Mem. Fac. Sci. Kyushu Univ.*, 8, 25, 1972; *Chem. Abstr.*, 76, 140123, 1972.

96. Vaughan, W. R., Habib, M. S., McElhinney, R. S., Takahashi, N., and Waters, J. A., Synthesis of potential anticancer agents. IX. Lawsone derivatives containing an alkylating function, *J. Org. Chem.*, 26, 2392, 1961.

97. Magarian, R. A. and Nobles, W. L., Potential antiinfective agents. I. Quinoline, phenolic and β-aminoketone derivatives, *J. Pharm. Sci.*, 56, 987, 1967.

98. Varma, R. S., and Nobles, W. L., Synthesis and antibacterial activity of certain 3-substituted benzoxazolinones, *J. Pharm. Sci.*, 57, 39, 1968.

99. Luts, H. A., and Nobles, W. L., Heptamethyleneimine in the Mannich reaction. I. Substituted β-aminoketones and γ-aminoalcohols, *J. Pharm. Sci.*, 54, 67, 1965.

100. Luts, H. A., Grattan, J. F., Yankelowitz, S., and Nobles, W. L., Octamethyleneimine in the Mannich reaction. I. Substituted β-aminoketones and substituted α-aminoalcohols, *J. Pharm. Sci.*, 56, 1114, 1967.

101. Potti, N. D. and Nobles, W. L., Uses of 3-azabicyclo[3.2.1]octane in the Mannich reaction, *J. Pharm. Sci.*, 57, 1487, 1968.

102. Angiolini, L., Costa Bizzarri, P., Scapini, G., and Tramontini, M., Stereochemistry of aminocarbonyl compounds. X. The influence of the amine moiety on 1–2, 1–5 and 1–6 asymmetric induction in Grignard additions and hydride reductions, *Tetrahedron*, 37, 2137, 1981.

103. Arya, V. P., Kaul, C. L., and Grewal, R. S., Synthesis and antitussive activity of 3-azabicyclo[3.2.2]nonane derivatives, *Arzneim. Forsch.*, 27, 1648, 1977.

104. Schneider, W. and Dechov, H. J., Mannichbasen des 2-Aza-bicyclo[2,2,2]octans, *Arch. Pharm.*, 299, 279, 1966.

105. Cavalla, J. F., Lockhart, I. M., Webb, N. E., Winder, C. V., Welford, M., and Wong, A., Analgetics based on the pyrrolidine ring. V, *J. Med. Chem.*, 13, 794, 1970.

106. Sandoz AG, Basel, 1,3,4,9b-Tetrahydro-2H-indeno[1,2-c]pyridines, *Angew. Chem. Int. Ed. Engl.,* 11, 857, 1972.

107. Tandon, M., Tandon, P., Barthwal, J. P., Bhalla, T. N., and Bhargava, K. P., Antiinflammatory and antiproteolytic activities of newer indolyl isoquinolines, *Arzneim. Forsch.,* 32, 1233, 1982.

108. Ivanov, I. C., Troanska, G. R., Christova, K. S., Dantchev, D. K., Sulay, P. B., and Waltchanova, R. P., Derivate des 2-Amino-1,2,3,4-tetrahydronaphthalins. III. Synthese einiger N'-substituierter N-(trans-3-Hydroxy-1,2,3,4-tetrahydro-2-naphthyl)-piperazine, *Arch. Pharm.,* 310, 925, 1977.

109. Cignarella, G., Occelli, E., Maffii, G., and Testa, E., Bicyclic homologs of piperazine, synthesis of pharmacologically active 8-methyl-3,8-diazabicyclo[3.2.1]octanes, *J. Med. Chem.,* 6, 29, 1963.

110. Khullar, K. K., and Chatten, L. G., Synthesis of some new Mannich bases using 2,6-dimethylmorpholine as amine moiety, *J. Pharm. Sci.,* 56, 328, 1967.

111. Stärk, H. C., Siemer, H., and Doppstadt, A., British Patent 852,727, 1960; *Chem. Abstr.,* 55, 11443, 1961.

112. Iordanova, K., Danchev, D., Gakhniyan-Mircheva, R., and Mikhailova, R., Derivatives of 5,5a,9a,10-tetrahydronaphtho[2,3-b]morpholine. Mannich bases of some physiologically active compounds, *Farmatsiya Sofia,* 26, 1, 1976; *Chem. Abstr.,* 86, 83621, 1977.

113. Bogatsky, A. V., Lukyanenko, N. G., Pastushok, V. N., and Kostyanovsky, R. G., Macroheterocycles. XIV. A convenient synthesis of azacrown ether derivatives via aminomethylation, *Synthesis,* 992, 1983.

114. Roscoe, C. W., Phillips, J. W., and Gillchriest, W. C., New compounds: organoboron derivatives of tetracyclines, *J. Pharm. Sci.,* 66, 1505, 1977.

115. Csuk, R., Hönig, H. Weidmann, H., and Zimmerman, H. K., Aminoalkoholesther von Hydroxyboranen. X. Tetracyclin-bor-Mannich-basen als potentielle Antitumorwirkstoffe, *Arch. Pharm.,* 317, 336, 1984.

116. Lucka-Sobstel, B. and Zejc, A., Mannich base derivatives of pharmacologically active compounds. XIV. 3-Methyl- and cis- and trans-2,5-dimethylpiperazine in the reaction of aminoalkylation of alkyl- and arylsuccinimides, *Diss. Pharm. Pharmacol.,* 23, 135, 1971; *Chem. Abstr.,* 75, 35937, 1971.

117. Varma, R. S. and Nobles, W. L., Application of 1,3-di(4-piperidyl)propane in the Mannich reaction. Synthesis of β-aminoketones, *J. Med. Chem.,* 11, 195, 1968.

118. Tsuchida, E. and Hasegawa, E., Polyamine oligomers obtained from Mannich polymerization, *J. Polym. Sci. Polym. Lett. Ed.,* 14, 103, 1976.

119. Luk'yanenko, N. G., Kostyanovskii, R. G., Pastushok, V. N., and Bogatskii, A. V., Macroheterocycles. 24. Synthesis of new derivatives of diaza-18-crown-6, *Khim. Geterotsikl. Soedin.,* 413, 1986; *Chem. Abstr.,* 106, 50175, 1987.

120. Short, J. H. and Ours, C. W., Use of aminoacids in the Mannich reaction, *J. Heterocycl. Chem.,* 12, 869, 1975.

121. Griengl, H. and Bleikolm, A., Siebengliedrige Ringe durch 1,5-Cycloadditionen. II. Octahydrofurano[3,2- f]-1,4-oxazepine und Octahydro-6H-pyrano[3,2-f]-1,4-oxazepine, *Liebigs Ann. Chem.,* 1783, 1976; see also *Liebigs Ann. Chem.,* 1792, 1976.

122. Tyka, R., Novel synthesis of α-aminophosphonic acids, *Tetrahedron Lett.,* 677, 1970.

123. Walter, M., U.S. Patent 2,795,613, 1957; *Chem. Abstr.,* 52, 437, 1958.

124. Becker, H. G. O. and Fanghaenel, E., Versuche zur Darstellung von Mannich Basen mit reversibel blockierten Stickstoffatom, *J. Prakt. Chem.,* 26, 58, 1964.

125. Wako Pure Chem. Ind., Ltd., Japanese Kokai Tokkyo Koho JP, 60,104,050, 1985; *Chem. Abstr.,* 103, 195826, 1985. See also Bestmann, H. J. and Wölfel, G., *Angew. Chem. Int. Ed. Engl.,* 23, 53, 1984.

126. Muminov, A., Yudin, L. G., Zinchenko, E. Y., Romanova, N. N., and Kost, A. N., Synthesis of (aminomethyl)indoles, *Khim. Geterotsikl. Soedin.,* 1218, 1985; *Chem. Abstr.,* 104, 129741, 1986.

127. Muhi-Eldeen, Z., Shubber, A., Musa, N., and Khayat, A., Synthesis and biological evaluation of L-5-[N-(4-tert-amino-2-butynyl)carbamoyl]-2-pyrrolidones and their racemates, *Eur. J. Med. Chem.,* 17, 49, 1982.

128. Choudhuri, M. D. and Heindee, N. D., U.S. Patent 4,950,770, 1990; *Chem. Abstr.,* 114, 42576, 1991.

129. Nobles, W. L. and Potti, N. D., Mechanism of the Mannich reaction, *J. Pharm. Sci.,* 57, 1097, 1968.

130. Belikov, V. M., Belokon, Y. N., Dolgaya, M. M., and Martinkova, N. S., The kinetics and mechanism of Mannich base dissociations in aqueous buffers, *Tetrahedron,* 26, 1199, 1970.

131. Roth, H. J. and Mühlenbruch, M., Bildung-stendenz symmetrischer bis-Mannich Basen des Piperazins, *Arch. Pharm.*, 303, 156, 1970.

132. Shakhgel'diev, M. A., Babaeva, G. B., Nabiev, O. G., and Kostyanovskii, R. G., 1,3-Bis(alkoxymethyl)imidazolidines, *Khim. Geterotsikl. Soedin.*, 712, 1987; *Chem. Abstr.*, 108, 55804, 1988.

133. Aversa, M. C. and Giannetto, P., 1,2-Di-hydro-1-(2-hydroxybenzoyl)-3,1-benzoxazin-4-ones as stable intermediates in the synthesis of Mannich products from phenols and an-thranilic acid, *J. Chem. Res.*, 200, 1984.

134. Hinman, R. L., Ellefson, R. D., and Camp-bell, R. D., Alkylhydrazines in the Mannich reaction: a convenient synthesis of Δ^3-pyra-zolines, *J. Am. Chem. Soc.*, 82, 3988, 1960.

135. Ito, K. and Sekiya, M., Preparation of N-(al-kylthiomethyl) derivatives of hydroxyla-mines, *Chem. Pharm. Bull. Tokyo*, 27, 1691, 1979.

136. Mozolis, V. and Jokubaityte, S., Benzotria-zole and thiourea in the Mannich reaction, *Liet. TSR Mokslu Akad. Darb. Ser. B*, 129, 1970; *Chem. Abstr.*, 73, 77152, 1970.

137. Diem, H. and Matthias, G., Amino resins, in *Ullmann's Encyclopedia of Industrial Chemistry*, Vol. A2, 5th ed., Gerhartz, W., Ed., VCH, Weinheim, 1985, 115; see, in partic-ular, pp. 127–130.

138. Kumlin, K. and Simonson, R., Urea-formal-dehyde resins, *Angew. Makromol. Chem.*, 93, 27, 1981.

139. Speckamp, W. N. and Hiemstra, H., Intra-molecular reactions of N-acyliminium inter-mediates, *Tetrahedron*, 41, 4367, 1985.

140. Tsuruta, S., Historical note on the chemistry of synthetic resins. XXXVII. An introduction to ammonia resols, *Netsu Kokasey Juski*, 10, 25, 1989; *Chem. Abstr.*, 111, 135219, 1989.

141. Reuss, G., Disteldorf, W., Grundler, O., and Hilt, A., Formaldehyde, in *Ullmann's Ency-clopedia of Industrial Chemistry*, Vol. A11, 5th ed., Gerhartz, W., Ed., VCH, Weinheim, 1988, 619.

142. Petersen, H. and Petri, N., Formaldehyde: General situation, analytical methods and use in finishing of textiles, *Melliand Textilber.*, 66, 217, 1985; see also *Melliand Textilber.*, 66, 285 and 363, 1985.

143. Miyano, S., Mori, A., Hokari, H., Ohta, K., and Hasimoto, H., Carbon-carbon bond for-mation by the use of chloroiodomethane as a C_1 unit. IV, *Bull. Chem. Soc. Jpn.*, 55, 1331, 1982.

144. Matsumoto, K., Synthesis under high pres-sure: Mannich reaction of ketones and esters with dichloromethane and secondary amines, *Angew. Chem. Int. Ed. Engl.*, 21, 922, 1982.

145. Schönenberger, H., Petter, A., and Külhling, V., Azomethine und Oxazolidine des Norfenefrins. Versuche zur Entwicklung von Verbindungen mit langanhaltender blutruck-steigernder Wirkung, *Arch. Pharm.*, 308, 717, 1975.

146. Ivanova, Z. M., Kim, T. V., Suvalova, E. A., Boldeslkul, I. E., and Gololobov, Y. G., Aminophosphonates. I. 1-(Alkylamino)alkyl-phosphonates, *Zh. Obshch. Khim.*, 46, 236, 1976; *Chem. Abstr.*, 85, 21540, 1976.

147. van Tilborg, W. J. M., Dooyewaard, G., Steinberg, H., and de Boer, T. J., The chem-istry of small ring compounds. XVII. Man-nich type reaction of cyclopropanone, *Tetrahedron Lett.*, 1677, 1972.

148. Roth, H. J. and Schumann, E., Mannichbasen als Zwischenstufen bei Chinolinsynthesen, *Arch. Pharm.*, 303, 268, 1970.

149. Baliah, V. and Usha, R., 7-Aza-3-thia- and 3,7-diazabicyclo[3.3.1]nonanes, *Indian J. Chem.*, 10, 319, 1972.

150. Azerbaev, I. N., Omarov, T. T., and Al'mu-khanova, K. A., Stereospecificity of the Man-nich reaction, *Kratk. Tezisy-Vses. Sovesh. Probl. Mekh. Geterolit. Reakts.*, 150, 1974; *Chem. Abstr.*, 85, 77503, 1976.

151. Quast, H., Müller, B., Peters, E. M., Peters, K., and von Schnering, H. G., Synthese und Struktur diastereomerer 2,4,6,8-Tetraaryl-3-azabicyclo[3.3.1]nonane. Stereochemie der Mannich Reaktion, *Chem. Ber.*, 116, 424, 1983.

152. Miyano S. and Abe, N., Synthesis of 3,3-dimethyl-2,3,4,5,10,11-hexahydro-11-phenyl-1H-dibenzo[b,e][1,4]diazepin-1-one, a new tricyclic system, *Chem. Pharm. Bull. Tokyo*, 20, 1588, 1972.

153. Schreiber, J., Vermuth, C. G., and Meyer, A., Cétoacides fonctionnalisés. I. Aminoalcoyla-tions avec l'acide glyoxylique, *Bull. Soc. Chim. Fr.*, 625, 1973.

154. Couquelet, J., Couquelet, J., Boyer, J. B., Cluzel, R., and Tronke, P., De quelques acides α-amino-β-aroyl-propioniques à effet antibactérien, *Chim. Ther.*, 4, 127, 1969.

155. Leston, G., U.S. Patents 4,475,001 and 4,480,140, 1984; *Chem. Abstr.*, 102, 5926 and 5929, 1985.

156. Moehrle, H. and Miller, C., Konformative Einflüsse bei Dehydrierung von Phenol-Man-nichbasen, *Monatsh. Chem.*, 105, 1151, 1974.

157. Mattioda, G. and Christidis, Y., Glyoxilic acid, in *Ullmann's Encyclopedia of Industrial Chemistry*, Vol. A12, 5th ed., Gerhartz, W., Ed., VCH, Weinheim, 1989, 495.

158. Julia, M., Bagot, J., and Siffert, O., Sur une nouvelle voie d'accès aux tryptamines, *Bull. Soc. Chim. Fr.*, 1424, 1973.

159. Haerter, H. P. and Schindler, O., Seven membered heterocycles. XXIV. Synthesis of 1,4-diazepino[1,2-a]indoles, *Chimia*, 32, 362, 1977.

160. Demerson, C. A., Philipp, A. H., Humber, L. G., Kraml, M. J., Charest, M. P., Tom, H., and Vàvra, I., Pyrrolo[4,3,2-de]isoquinolines with central nervous system and hypertensive activities, *J. Med. Chem.*, 17, 1140, 1974.

161. Douglass, J. E. and Dial, R., A facile synthesis of 2-aryl-1,3-dimethyl-1,4,5,6-tetrahydropyrimidinium salts, *Synthesis*, 654, 1975.

162. Parish, H. A. and Gilliom, R. D., Syntheses and diuretic activity of 1,2-dihydro-2-(3-pyridyl)-3H-pyrido[2,3-d]pyrimidin-4-one and related compounds, *J. Med. Chem.*, 25, 98, 1982.

163. Katritzky, A. R., Borowiecka, J., and Fan, W. Q., A novel synthesis of α-aminoaldehydes from glyoxal monoacetal, *Synthesis*, 1173, 1990.

164. Eiden, F., Wiedemann, H., and Schnabel, K., Über die Ringaufspaltung von 1,2-Dihydro-3,1-benzoxazinen, *Arch. Pharm.*, 308, 130, 1975.

165. Lukszo, J. and Tyka, R., New protective groups in the synthesis of 1-amino-alkane-phosphonic acids and esters, *Synthesis*, 239, 1977.

166. Bethge, H., Drauz, K., Kleemann, A., Martens, J., and Weigel, J., German Patent DE 3,202,295, 1983; *Chem. Abstr.*, 99, 105700, 1983.

167. Ducher, D., Couquelet, J., Cluzel, R., and Couquelet, J., α-Amino-γ-céto-esters à activité antibactérienne, *Chim. Ther.*, 8, 552, 1973.

168. Bosch, J., Rubiralta, M., and Moral, M., Mannich cyclization involving the α-position of ketals; Synthesis of 2-aryl-4-piperidones and 2-aryl-3-acyl-pyrrolidines, *Heterocycles*, 19, 473, 1982.

169. Roth, H. J. and Assadi, F., Herstellung von Mannich Basen des α-Tetralons und Indanons durch Anlagerung an Diarylimine, *Arch. Pharm.*, 303, 29, 1970.

170. Kinast, G. and Tietze, L. F., A new variant of the Mannich reaction, *Angew, Chem. Int. Ed. Engl.*, 15, 239, 1976.

171. Jasor, Y. Gaudry, M., Luche, M. J and Marquet, A., Synthèse régiosélective de bases de Mannich de cétones dissymétriques, *Tetrahedron*, 33, 295, 1977.

172. Gaudry, M., Jasor, Y., and Trung, B. K., Regioselective Mannich condensation with dimethyl(methylene)ammonium CF₃COO⁻; 1-Dimethylamino-4-methyl-3-pentanone, *Org. Synth.*, 59, 153, 1980.

173. Schaefer, M., Weber, J., and Faller, P., Synthèse de bases de Mannich très instables. Aminométhylation d'étérocyclanones par les chlorures de dialcoyl-N,N-méthylèneimonium, *Bull. Soc. Chim. Fr.*, 241, 1978.

174. Bryson, T. A., Bonitz, G. H., Reichel, C. J., and Dardis, R. E., Preformed Mannich bases: a facile preparation of dimethyl(methylene)ammonium iodide, *J. Org. Chem.*, 45, 524, 1980.

175. Böhme, H., and Sickmüller, A., Zur Acylspaltung von Formaldehyd-N,O-acetalen, *Chem. Ber.*, 110, 208, 1977.

176. Schreiber, J., Maag, H., Hashimoto, N., and Eschenmoser, A. E., Dimethyl(methylene)ammonium iodide, *Angew. Chem. Int. Ed. Engl.*, 10, 330, 1971.

177. Bidan, G. and Genies, M. Utilisation en synthèse du cation iminium généré *in situ* par oxydation électrochimique d'amines tertiaires, *Tetrahedron*, 37, 2297, 1981.

178. Niyazymbetov, M. E. and Petrosyan, V. A., Electrochemical variation of the Mannich reaction, *Izv. Akad. Nauk SSSR Ser. Khim.*, 1676, 1984; *Chem. Abstr.*, 102, 78053, 1985.

179. Ahond, A., Cavé, A., Kan-Fan, C., and Potier, P., Modification de la réaction de Polonovsky: préparation d'un nouveau réactif de Mannich, *Bull. Soc. Chim. Fr.*, 2704, 1970.

180. Volz, H., and Kilz, H. H., Methyleniminmonium Salze durch Hydrid-abstraction aus tertiären Aminen, *Liebigs Ann. Chem.*, 752, 86, 1971.

181. Stenlake, J. B., Urwin, J., and Waigh, R. D., Biodegradable neuromuscular blocking agents. II. Quaternary ketones, *Eur. J. Med. Chem.*, 14, 85, 1979.

182. Böhme, H. and Hilp, M., Über α-halogenierte Amine. XXIV. Darstellung und Eigenschaften von Fluoromethyl-dialkylaminen, *Chem. Ber.*, 103, 104, 1970.

183. Krieger, H., Lumme, H., Petäjä, T., Talvitie, A., and Vainio, U. M., Aminomethylierung von 2,3-Dimethy-2-buten, *Rep. Ser. Chem. Univ. Oulu.*, 18, 1984.

184. Möhrle, H. and Schaltenbrand, R., Konkurrierende Reaktion der Phenol- und 1,3-Dicarbonyl-Funktion unter Mannich Bedingungen, *Pharmazie,* 40, 307, 1985; see also *Pharmazie,* 40, 697 and 767, 1985.

185. Dimmock, J. R., Raghavan, S. K., Logan, B. M., and Bigam, G. E., Antileukemic evaluation of some Mannich bases derived from 2-arylidene-1,3-diketones, *Eur. J. Med. Chem.,* 18, 248, 1983.

186. Moilanen, M. and Manninen, K., Aminomethylation of bicyclo[2.2.1]heptene-2,5-dione, *Acta Chem. Scand.,* 44, 857, 1990.

187. Krieger, H., Lajunen, M., and Myllyla, M., Reduktion von 3-N,N-Dimethylaminomethyl-bicyclo[2.2.2]octan-2-on, *Finn. Chem. Lett.,* 25, 1984.

188. Möhrle, H. and Tröster, K., Über Mannich basen. XVII. Amino- und Amidoalkylierung von 2-Methylnaphthol bzw. 1-Naphthol, *Arch. Pharm.,* 314, 690, 1981; see also *Arch. Pharm.,* 315, 222, 1982.

189. Möhrle, H. and Tröster, K., Über Mannich Basen. 18 Mitt. Mannich Reaktionen mit 1-Naphthol und sekundären Aminen, *Arch. Pharm.,* 315, 397, 1982.

190. Pochini, A., Puglia, G., and Ungaro, R., Selective synthesis of phenolic Mannich bases under solid-liquid phases transfer conditions, *Synthesis,* 906, 1983.

191. Hombrecher, H. K. and Horter, G., Synthese von 5,15-diarylsubstituierten Porphyrinen. über Aminomethylierung von bis(4-Ethyl-3-methyl-2-pyrryl)phenylmethanen, *Liebigs Ann. Chem.,* 219, 1991.

192. Gall, M., Kamdar, B. V., Lipton, M. F., Chidester, C. G., and DuChamp, D. J., Mannich reactions of heterocycles with dimethyl(methylene)ammonium chloride: A high yield, one-step conversion of estazolam to adinazolam, *J. Heterocycl. Chem.,* 25, 1649, 1988.

193. Kozikowski, A. P. and Ishida, H., Use of N,N-dimethyl(methylene)ammonium chloride in the functionalization of indoles, *Heterocycles,* 14, 55, 1980.

194. Sliva, H. and Blondeau, D., Synthesis and NMR study of Mannich bases of 8-acetoxyindolizines, *Heterocycles,* 16, 2159, 1981.

195. Suh, J. J. and Hong, Y. H. Dialkylaminomethylation of 1,4-dihydropyridine, *Yakhak Hoechi,* 33, 3280, 1989; *Chem. Abstr.,* 114, 42504, 1991.

196. Dowle, M. D., Hayes, R., Judd, D. B., and Williams, C. N., A convenient synthesis of Mannich bases of thiophene and substituted thiophenes, *Synthesis,* 73, 1983.

197. Risch, N. and Esser, A. Iminiumtetrachloroaluminate als leistungsfähige Mannich-Reagenzien. α-Aminoalkylierung von Enaminen, *Z. Naturforsch. Teil B,* 44B, 208, 1989.

198. Möhrle, H. and Arz, P., Reaktionsverhalten substituierter Enhydrazine unter Mannich-bedingungen, *Arch. Pharm.,* 319, 303, 1986.

199. Möhrle, H. and Novak, H. J., Aminomethylierung von 3-Amino-inden-1-onen, *Z. Naturforsch. Teil B,* 37B, 669, 1982.

200. Möhrle, H. and Herbrüggen, G. S., Reaktionsverhalten 4-aminosubstituierter Naphthochinone unter Mannich Bedingungen, *Arch. Pharm.,* 323, 433, 1990.

201. Böhme, H. and Stammberger, W., Notiz über bis-Dialkylaminomethyl-cyanamide, *Chem. Ber.,* 104, 3354, 1971.

202. Böhme, H. and Martin, F., Zur Aminomethylierung von (2-Chloräthyl)hydrazinen und Chloressigsäure-hydrazinen, *Chem. Ber.,* 106, 3540, 1973.

203. Kellner, K., Seidel, B., and Tzschach, A., Organoarsen-verbindungen. XXXIII. Synthese und Reaktionsverhalten der α-Aminomethylphosphine und -arsine, *J. Organomet. Chem.,* 149, 167, 1978.

204. Böhme, H. and Viehe, H. G., *Iminium Salts in Organic Chemistry,* John Wiley & Sons, New York, 1979, parts 1 and 2.

205. Damour, D., Pornet, J., Randrianoelina, B., and Miginiac, L., Synthèse régiosélective d'amines α-alléniques par aminométhylation-désilylation de propargyltriméthyl silanes, *J. Organomet. Chem.,* 396, 289, 1990.

206. Grieco, P. A. and Bahsas, A., Reactions of allylstannanes with *in situ* generated immonium salts in protic solvent: a facile aminomethane destannylation process. *J. Org. Chem.,* 52, 1378, 1987.

207. Reich, H. J., Schroeder, M. C., and Reich, I. L., Organoselenium chemistry. Preparation of allyl and homoallyl amines by aminomethylation of phenylselenosubstituted allyltin reagents, *Isr. J. Chem.,* 24, 157, 1984; *Chem. Abstr.,* 101, 229473, 1984.

208. Cooper, M. S. and Heaney, H., Mannich reactions of aryl trialkyl stannates via preformed iminium salts, *Tetrahedron Lett.,* 27, 5011, 1986.

209. Larsen, S. D. and Grieco, P. A., Aza Diels-Alder reactions in aqueous solution: Cyclocondensation of dienes with siruple iminium salts generated under Mannich conditions, *J. Am. Chem. Soc.,* 107, 1768, 1985.

210. Heaney, H., Papageorgiou, G., and Wilkins, R. F., Mannich reactions of nucleophilic aromatic compounds involving animals and α-aminoethers activated by chlorosilano derivatives, *J. Chem. Soc. Chem. Commun.*, 1161, 1988.

211. Eyley S. C., Heaney, H., Papageorgiou, G., and Wilkins, R. F., Mannich reaction of π-excessive heterocycles using bis-(dialkylaminomethanes) and alkoxy dialkylaminomethanes activated with acetylchloride or sulphur dioxide, *Tetrahedron Lett.*, 29, 2997, 1988.

212. Nabiev, O. G., Shakhgel'diev, M. A., Chervin, I. I, and Kostyanovskii, R. G., Aminomethylation of methyl-2-aziridine carboxylate, *Izv. Akad, Naukk SSSR Ser. Khim.*, 716, 1985; *Chem. Abstr.*, 103, 53883, 1985.

213. Fernandez, J. E. and Calderazzo, J. M., Synthesis of methanediamines and Mannich bases of 2-naphthol, *J. Chem. Eng. Data*, 10, 402, 1965.

214. Reynolds, D. D. and Cossar, B. C., 1,3,5-Trisubstituted hexahydrotriazines as Mannich reagents. I–III, *J. Heterocycl. Chem.*, 8, 597, 1971; see also pp. 601 and 611.

215. Messinger, P. and Gompertz, J., Notiz zur Synthese von α-Amino- und α-Amidosulfonen, *Arch. Pharm.*, 307, 653, 1974.

216. Xiu-Juan, X. and Guang-Xu, C., Mannich reaction of arylamines, *Acta Chim. Sinica*, 40, 463, 1982; *Chem. Abstr.*, 97, 162500, 1982.

217. Griengl, H., Prischl, G., and Bleikolm, A., Siebengliedrige Ringe durch 1,5-Cycloadditionen. V. Perhydro-1,4-diazepine, *Liebigs Ann. Chem.*, 400, 1979; see also *Liebigs Ann. Chem.*, 1573, 1980.

218. Iwai, I. and Yura, Y., Acetylenic compounds. XXXIII. New synthetic method for aminoacetylenic compounds, *Chem. Pharm. Bull. Tokyo*, 11, 1049, 1963.

219. Fernandez, J. E., Powell, C., and Fowler, J. S., Reaction of 2-naphthol with N-piperidinomethyl and N-morpholinomethyl alkyl ethers, *J. Chem. Eng. Data*, 8, 600, 1963.

220. Epsztein, R. and Le Goff, N., Syntèse de bases de Mannich acétyleniques cycliques, *Tetrahedron Lett.*, 26, 3203, 1985.

221. Griengl, H., Bleikolm, A., Grubbauer, W., and Söllradl, H., Siebengliedrige Ringe durch 1,5-Cycloadditionen. IV. Umsetzung mit 1,3-Oxazolidinen in Gegenwart von Protonensäuren, *Liebigs Ann. Chem.*, 392, 1979.

222. Griengl, H., Hayden, W., Kalchauer, W., and Wanek, E., Aminoalkyl Derivate von 6- Chlor- und 6-Methylthio-purin und von Uracil, *Arch. Pharm.*, 317, 193, 1984.

223. Ulbrich, H., Priewe, H., and Schröder, E., Substituierte Perhydro-pyrido[1,2-a][1,4]diazepine und Perhydro-pyrido[2,1-c][1,4]oxazepine, *Eur. J. Med. Chem.*, 11, 343, 1976.

224. Engel, J., British Patent Appl. GB 2,087,397, 1982; *Chem. Abstr.*, 97, 127237, 1982.

225. Xiu-Juan, X. and Guang-Xü, C., Exchange reaction between acetophenone and benzimidazole N-Mannich base, *Acta Chim. Sinica*, 40, 362, 1982.

226. Danchenko, M. N. and Gololobov, Y. G., Dithiobutyric acid esters in the Mannich reaction: synthesis of O (S) esters of 3-dialkylamino-2,2-dimethylpropane dithioic acid, *Zh. Org. Khim.*, 19, 717, 1983; *Chem. Abstr.*, 99, 87620, 1983.

227. Pauson, P. L., Kelly, P. B., and Porter, R. J., Mannich condensation of tropolones, *J. Chem. Soc. C*, 1323, 1970.

228. Overman, L. E. and Wild, H., Preparation of functionalized hydroindol-3-ols via tandem aza-Cope rearrangement-Mannich cyclizations. Total synthesis of (±) 6α-epipretazetine and related alkaloids, *Tetrahedron Lett.*, 30, 647, 1989.

229. Overman, L. E., Mendelson, L. T., and Jacobsen, E. J., Applications of cationic aza-Cope rearrangements for alkaloid synthesis. Stereoselective preparation of cis-3a-aryloctahydroindoles and a new short route to amaryllidacea alkaloids, *J. Am. Chem. Soc.*, 105, 6629, 1983.

230. Kovalenko, A. L., Serov, Y. V., and Tselinskii, I. V., Aminoalkylation of lower aliphatic carboxamides by N-methylene-tert-butyl-amine, *Zh. Org. Khim.*, 26, 1240, 1990; *Chem. Abstr.*, 113, 211357, 1990.

231. Claxton, G. P., Grisar, J. M., Roberts, E. M., and Fleming, R. W., Synthesis of α-[p-(fluoren-9-ylidenemethyl)phenyl]-2-piperidineethanol, an inhibitor of platelet aggregation, *J. Med. Chem.*, 15, 500, 1972.

232. Werner, W. and Mühlstädt, M., Aminomethylierung von Alkyl-aryl-ketonen, *Liebigs Ann. Chem.*, 693, 197, 1966.

233. Littell, R., Greenblatt, E. N., and Allen, G. R., Jr., Mannich bases of 2,3-dihydro-4(1H)-carbazolones as potential psychotropic agents, *J. Med. Chem.*, 15, 875, 1972.

234. Roth, H. J. and Ergenzinger, K., Synthese polyfunktioneller Heterocyclen durch Aminoalkylierung von Nitroalkanen, *Arch. Pharm.*, 311, 492, 1978.

235. Kamogawa, H., Kubota, K., and Nanasawa, M., Syntheses of aromatic Mannich bases involving nitrogen heterocycles by means of amine-exchange reaction, *Bull. Chem. Soc. Jpn.*, 51, 1571, 1978.

236. Cattanach, C. J., Cohen, A., and Heath-Brown, B., Studies in the indole series. VII. Indolo[1,7-a,b][1]benzoazepines and related compounds, *J. Chem. Soc. Perkin Trans. 1*, 1041, 1973.

237. Kagan, E. S. and Ardashev, B. I., Synthesis of vinyl derivatives of quinoline by Mannich reaction, *Khim. Geterotsikl. Soedin.*, 701, 1967; *Chem. Abstr.*, 68, 11023, 1968.

238. Manninen, K. J., The reaction of 2-phenyl bicyclo[2.2.1]hept-2-ene with formaldehyde and methylammonium chlorides in acetic acid, *Acta Chem. Scand.*, B32, 691, 1978.

239. Miocque, M. and Vierfond, J. M., Aminométhylation sélective dans la série des N-propargylanilines, *Ann. Pharm. Fr.*, 31, 721, 1973.

240. Götz M. and Zeile, K., Mesoionic 3-amino-5-imino oxazoline derivatives, *Tetrahedron*, 26, 3185, 1970.

241. Oelschläger, H., Ewert, M., and Götze, G., C-Mannich-basen des 5-Hydroxybenzotriazols, *Arch. Pharm.*, 307, 622, 1974; see also *Arch. Pharm.*, 308, 550, 1975.

242. Hung, J. and Werbel, L. M., Camoform analogs as potential agents against mefloquine resistant malaria, *Eur. J. Med. Chem.*, 18, 61, 1983.

243. Seela, F. and Lüpke, H., Mannich-Reaktion am 2-Amino-3,7-dihydropyrrolo[2,3-d]pyrimidin-4-on, dem Chromophor des Ribonucleosids Q, *Chem. Ber.*, 110, 1462, 1977.

244. Abushanab, E., Lee, D. Y., and Goodman, L., Imidazo[1,5-a]pyrazines. IV. Aromatic substitution reactions, *J. Org. Chem.*, 40, 3373, 1975.

245. Meister, C. and Scharf, H. D., Synthese von 3(2H)-Furanonen und 3-Methoxyfuranen, *Synthesis*, 737, 1981.

246. Knox, G. R., Morrison, I. G., and Pauson, P. L., Ferrocene derivatives. XVI. The aminomethylation of methylthio- and bis-methylthio-ferrocene, *J. Chem. Soc. C*, 1842, 1847 and 1851, 1967.

247. Hisatome, M., Ichida, S., Yamaguchi, T., and Yamakawa, K., Organometallic compounds, XXXI. N,N-Dimethylaminomethylation of biferrocene and conversion of the products into several biferrocene derivatives, *J. Organomet. Chem.*, 217, 221, 1981.

248. Watase, Y., Terao, Y., and Sekiya, M., Synthesis of N-(alkylaminomethyl)amides, *Chem. Pharm. Bull. Tokyo*, 21, 2775, 1973.

249. Svetkin, Y. V., Vasil'eva, S. A. and Pronina, V. M., Aminomethylation of some 4-thiazolidinones, *Khim. Gerotsikl. Soedin.*, 365, 1974; *Chem. Abstr.*, 81, 13421, 1974.

250. Csuk, R., Haas, J., Hönig, H., and Weidann, H., Aminoalkoholester von Hydroxyboranen. IX. Salicylamid-bor-Mannich Basen als potentielle Antitumorwirkstoffe, *Monatsh. Chem.*, 112, 879, 1981.

251. Peglion, G. L., Grenier, J., Pastor, R., and Cambon, A., Réactivité des F-alkylpyrazoles. II. Régiospécificité des réactions de Michaél et de Mannich, *Bull. Soc. Chim. Fr.*, 181, 1982.

252. Dumont, J. M., Yanes, T., Paris, J., and Petavy, A. F., Synthèse et étude de l'activité antihelminthique d'indazoles et nitro-indazoles, *Eur. J. Med. Chem.*, 18, 469, 1983.

253. Terao, Y., Matsunaga, K., and Sekiya, M., Synthesis of α-thio-, α-sulfinyl- and α-sulfonyl-substituted nitrosoamines, *Chem. Pharm. Bull. Tokyo*, 25, 2964, 1977.

254. Pollak, I. E., Trifunac, A. D., and Grillot, G. F., The reaction of Grignard reagents with N,N-dialkyl-N-(phenylthiomethyl)amines, *J. Org. Chem.*, 32, 272, 1967.

255. Messinger, P. and Myer, H., Darstellung von N-Imidomethyl-N-sulfonylmethyl-anilinderivaten, *Arch. Pharm.*, 312, 1007, 1979.

256. Grim, S. O., and Matienzo, L. J., The synthesis and characterization of some novel polydentate phosphorous-nitrogen ligands, *Tetrahedron Lett.*, 2951, 1973.

257. Moedritzer, K. and Irani, R., The direct synthesis of aminomethylphosphonic acids. Mannich-type reactions with orthophosphorous acid, *J. Org. Chem.*, 31, 1603, 1966.

258. See also the survey in Ref. 10.

259. Hester, J. B., New synthesis of 8-chlor-1-[2-(dimethylamino)ethyl]-6-phenyl-4H-s-triazolo[4,3-a][1,4]-benzodiazepine, which has antidepressant properties, *J. Org. Chem.*, 44, 4165, 1979.

260. Jacobsen, E. J., Levin, J., and Overman, L. E., Scope and mechanism of tandem cationic aza-Cope rearrangement-Mannich cyclization reactions, *J. Am. Chem. Soc.*, 110, 4329, 1988.

261. Heimes, A., Brienne, M. J., Jacques, J., Johnston, D. B. R., and Windholz, T. B., Some reactions of 5-α-A-nor-androstan-3-ones, in *Proc. 2nd Int. Congr. Hormonal Steroids*, Milan, 1966.

262. Chhabra, B. R., Bolte, M. L., and Crow, W. D., Monoalkylidenes of Meldrum's acid, *Aust. J. Chem.,* 37, 1795, 1984.

263. Eichberger, G., Griengl, H., and Paar, W., Zur Reaktion von 1,3-Oxazolidinen mit Dicarbonsäureanhydriden, *Monatsh. Chem.,* 117, 545, 1986.

264. Andrisano, R., Angeloni, A. S., De Maria, P., and Tramontini, M., Reactivity of Mannich bases. X. The mechanism of the reaction between β-aminoketones and thiophenols, *J. Chem. Soc. C,* 2307, 1967.

265. Volkov, V. S., Ivanova, G. A., Poverennyi, A. H., and Sverdlov, E. D., Aminoacids as catalysts in the binding reaction of formaldehyde with adenine residue of polyadenylic acid, *Bioorg. Khim.,* 13, 805, 1987; *Chem. Abstr.,* 107, 97053, 1987.

266. Grakauskas, V. and Baum, K., Mannich reaction of 2-fluoro-2,2-dinitroethanol, *J. Org. Chem.,* 36, 2599, 1971.

267. Dixneuf, P. and Dabard, R., III. Synthèse et propriété d'amines dérivées du ferrrocéne. Action du N,N,N',N'-tetraméthyldiaminométhane sur des dérivés ammoniométhylferrocènes, ferrocénylcarbinoil et ferrocényl-1 éthyléniques, *Bull. Soc. Chim. Fr.,* 2838, 1972.

268. Messinger, P. and Greve, H., Einige Reaktionen mit aktivierten Sulfinsäuren, *Arch. Pharm.,* 311, 827, 1978.

269. Märkl, G., Jin, G. Yu, and Schoerner, C., Chiral aminomethylphosphines and aminomethyldiphosphines, *Tetrahedron Lett.,* 1409, 1980.

270. See also the survey in Ref. 11.

271. Capasso, R., Randazzo, G., and Pecci, L., Reaction of γ-carboxyglutamic acid with aldehydes, *Can. J. Chem.,* 61, 2657, 1983.

272. Rigo, B., Fossaert, E., de Quilacq, J., and Kolocouris, N., Étude dans la série des pyrrolidinones: réaction de Mannich du glutamate de diéthyle, *J. Heterocycl. Chem.,* 21, 1381, 1984.

273. Altosaar, H., Christjanson, P., and Siimer, K., Amidomethylation reactions. XII. Study of phenylamine reactions with 2-pyrrolidone and formaldehyde, *Tr. Tallin Politekh. Inst.,* 459, 41, 1978; *Chem. Abstr.,* 92, 40897, 1980.

274. Wurziger, H., Hydroxymethylierung, *Kontakte,* 36, 1988.

275. Beckwith, A. L. J. and Vickery, G. G., Displacement of the OH group from ferrocenylmethanol by amines, *J. Chem. Soc. Perkin Trans, 1,* 1818, 1975.

276. Kamienski, B., The di(2-alkyl-2-nitroalkyl)-methylamines from reactions of nitroalkanes with formaldehyde and methylamine, *Tetrahedron,* 30, 2777, 1974.

277. Varma, R. S. and Kapoor, A., Potential biologically active agents. XXV. 1-Arylaminomethyl-5- and -6-chlorobenzimidazoles, *Eur. J. Med. Chem.,* 15, 536, 1980.

278. Hahn, W. E. and Bartnik, R., Isonitroso ketones. I. Aminomethylation of α-isonitrosoalkyl isopropyl, ethyl, and methyl ketones, *Rocz. Chem.,* 47, 2089, 1973; *Chem. Abstr.,* 80, 95183, 1974; see also *Rocz. Chem.,* 48, 475, 1974; *Chem. Abstr.,* 81, 63448, 1974.

279. He, D., Xing, Q., Jin, S., and Lu, M., Kinetic study of the Mannich reaction of 3-thiolocoumarin with diphenylamine, *Wuli Huaxue, Xuebao,* 6, 699, 1990; *Chem. Abstr.,* 114, 142389, 1991.

280. Katrizky, A. R., Rachwal, S., and Rachwal, B., The chemistry of benzotriazole. III. The aminoalkylation of benzotriazole, *J. Chem. Soc. Perkin Trans. 1,* 799, 1987.

281. Khan, M. N., Aqueous degradation of N-(hydroxymethyl)phthalimide in the presence of specific and general bases. Kinetic assessment of N-hydroxymethyl derivatives of nitrogen heterocyclics as possible prodrugs, *J. Pharm. Biomed. Anal.,* 7, 685, 1989.

282. Berger, J. G. and Schoen, K., Reaction of pyrrole ketones with formaldehyde. Formation of N-pyrrole methanols, *J. Heterocycl. Chem.,* 9, 419, 1972.

283. Miller, R. B. and Smith, B. F., Synthesis of α-methylene carbonyl compounds: unexpected product from the Mannich reaction of 2-carbomethoxy cyclohexanone, *Synth. Commun.,* 3, 129, 1973.

284. Abrams, W. R. and Kallen, R. G., Equilibria and kinetics of N-hydroxymethylamine formation from aromatic exocyclic amines and formaldehyde, *J. Am. Chem. Soc.,* 98, 7777, 1976; see also *J. Am. Chem. Soc.,* 94, 576 and 4731, 1972.

285. Kallen, R. G., The mechanism of reactions involving intermediates Schiff bases. Formation of thiazolidine from L-cysteine and formaldehyde, *J. Am. Chem. Soc.,* 93, 6236, 1971.

286. McDonald, C. J. and Beaver, R. H., The Mannich reaction of polyacrylamide, *Macromolecules,* 12, 203, 1979.

287. Chapuis, G., Gauvreau, A., Klaebe, A., Lattes, A., and Perie, J. J., Condensation de diamines-1,2 sur les composés carbonylés. Synthèses d'imidazolidines: méchanisme de la réaction, *Bull. Soc. Chim. Fr.,* 977, 1973.

288. Bourguignon, J. J. and Wermuth, C. G., Lactone chemistry: synthesis of β-substituted, γ-functionalized butanolides and butenolides and succinaldehydic acids from glyoxylic acid, *J. Org. Chem.,* 46, 4889, 1981.

289. Masul, M., Fujita, K., and Ohmori, H., Direct observation of a Mannich intermediate in solution, *J. Chem. Soc. Chem. Commun.,* 182, 1970.

290. Hilton, A. and Leussing, D. L., Reaction of ethylenediamine and C, C, C′,C′-tetramethylethylenediamine with glyoxylate in the presence and absence of zinc(II) or nickel(II), *J. Am. Chem. Soc.,* 93, 6831, 1971.

291. Lambert, J. B. and Majchrzak, M. W., Ring-chain tautomerism in 1,3-diaza and 1,3-oxaza heterocycles, *J. Am. Chem. Soc.,* 102, 3588, 1980.

292. Alva Astudillo, M. E., Chokotho, N. C. J., Jarvis, T. C., Johnson, C. D., Lewis, C. C., and McDonnell, P. D., Hydroxy Schiff base-oxazolidine tautomerism. Apparent breakdown of Baldwin's rules. *Tetrahedron,* 41, 5919, 1985.

293. Krasnov, V. L., Matyukov, E. V., and Bodrikov, I. V., Dynamic system of states of 1-anilino-1-arylmethanesulfonates as a universal reagent for aminoalkylation of carbon acids, *Izv. Vyssh. Uchebn. Zaved. Khim. Khim. Tekhnol.,* 30, 38, 1987; *Chem. Abstr.,* 108, 131170, 1988.

294. Möhrle, H. and Scharf, U., Über Hexahydrotriazine, III. Mannich Reagenzien aus, 1,3,5-Trialkylhexahydro-1,3,5-triazinen, *Arch. Pharm.,* 313, 435, 1980.

295. Paris, J., Couquelet, J., and Tronche, P., N-Aminométhylation en série thiazolique: Application à quelques amino-2-benzothiazoles, *Bull. Soc. Chim. Fr.,* 672, 1973.

296. Mornet, R. and Gouin, L., Aminométhylation de dérivés propargyliques. Méthode générale applicable avec différents types d'amines secondaires, *Bull. Soc. Chim. Fr.,* 206, 1974.

297. Matsumoto, K., Hasimoto, S., Otani, S., Amita, F., and Osugi, J., Mannich reaction under high pressure. Dimethylaminomethylation of ketones with bis(dimethylamino)methane under mild conditions, *Synth. Commun.,* 14, 585, 1984.

298. Tychopulos, V. and Tyman, J. H. P., Enhancement of the rate of Mannich reactions in aqueous media, *Synth. Commun.,* 16, 1401, 1986.

299. Fernandez, J. E., Mones, J. D., Schwartz, M. L., and Wulff, R. E., A study of the Mannich reaction. IX. Reaction of 2,4,6-trinitro-

toluene with methylene-bis-piperidine, *J. Chem. Soc. B,* 506, 1969.

300. Mironov, G. S. and Farberov, M. I., Mechanism of the Mannich reaction, *Uch. Zap. Yarosl. Tekhnol. Inst.,* 11, 127, 1969; *Chem. Abstr.,* 74, 22305, 1971.

301. Jagannadham, V., Sethuram, B., and Rao, T. N., A kinetic study of the Mannich reaction, *Indian J. Chem. B,* 17B, 598, 1979.

302. Curulli, A. and Sleiter, G., Electrophilic heteroaromatic substitutions. VIII. Studies on the mechanism of the α-side-chain aminomethylation and hydrogen/deuterium isotope exchange reactions of α-methylpyrroles, *J. Org. Chem.,* 50, 4925, 1985.

303. Raines, S. and Kovacs, C. A., A study on the condensation of pyrroles, formaldehyde and primary amine hydrochlorides, *J. Heterocycl. Chem.,* 7, 223, 1970.

304. Poirier, M. A., Curran, N. M., McErlane, K. M., and Lovering, E. G., Impurities in drugs. III. Trihexyphenidyl, *J. Pharm. Sci.,* 68, 1124, 1979.

305. Gaidai, V. I., Valitova, L. A., Krokina, S. S., and Ivanov, B. E., Phosphonomethylation of o-aminophenol and o-anisidine, *Izv. Akad. Nauk SSSR Ser. Khim.,* 611, 1979; *Chem. Abstr.,* 91, 20604, 1979.

306. Tome, D., Naulet, N., and Martin, G. J., NMR Studies of the reaction of formaldehyde with the aminogroup of alanine and lysine versus pH, *J. Chim. Phys. Chim. Biol.,* 79, 361, 1982.

307. Trézl, L., Rusznák, I., Náray-Szabo, G., Szarvas, T., Csiba, A., and Ludanyi, A., Essential differences in spontaneous reactions of L-lysine and L-arginine with formaldehyde, *Period. Polytech. Chem. Eng. Budapest,* 32, 251, 1988; *Chem. Abstr.,* 112, 99170, 1990.

308. Pettit, G. R. and Settepani, J. A., Condensation of formaldehyde with N-bis(haloethyl)amines: study of the products, *Chem. Ind.,* 1805, 1964.

309. Böhme and Ort, H., Über Chloromethyl-bis(β-chloräthyl)-amin und Versuche zur Darstellung der Aminale von β-Chloräthyl- und β-Acetoxyäthyl-aminen. 16. Über α-Halogenierte Amine, *Arch. Pharm.,* 300, 148, 1967.

310. Katritzky, A. R., Rachwal, S., and Wu, J., Substituent effects in the bis(benzotriazolylmethylation) of aromatic amines, *Can. J. Chem.,* 68, 446, 1990.

311. Becker, H. G. O., Ecknig, W., Fanghaenel, E., and Rommel, S., Simple synthesis of secondary Mannich bases, *Wiss. Z. Tech.,* 11, 38, 1968; *Chem. Abstr.,* 71, 60938, 1969.

312. Salakhutdinov, N. F., Krysin, A. P., and Koptyug, V. A., Selective functionalization of aromatic compounds. I. O-Aminomethylation of phenol and o-tert-butylphenol, *Zh. Org. Khim.*, 26, 775, 1990; *Chem. Abstr.*, 113, 151939, 1990.

313. Müller, W., Flügel, R., and Stein, C., Basisch Di- und Tri-hydroxy-antrachinon-derivate as Modell-verbindungen der Anthracyclin-antibiotika, *Leibigs Ann. Chem.*, 754, 15, 1971.

314. Ramsh, S. M., Khebrova, E. S., and Shamina, L. P., Aminomethylation of pemoline, *Khim. Geterotsikl. Soedin.*, 1670, 1990; *Chem. Abstr.*, 114, 247251, 1991.

315. Litkei, G., Patonay, T., and Kardos, J., Synthesis of Mannich compounds from flavanones, *Org. Prep. Proced. Int.*, 22, 47, 1990.

316. Afsah, R. M., Hammouda, M., and Khalifa, M. M., A study on the Mannich reaction with 1,3-indandione, *Z. Naturforsch. Teil B*, 45B, 1065, 1990.

317. Möhrle, H. and Herbke, J., Aminomethylierung von Enaminonen des Dimedons, *Pharmazie*, 35, 288, 1980.

318. Möhrle, H, and Arz, P., Aminoalkylierung von Enhydrazinonen, *Z. Naturforsch. Teil B*, 42B, 1035, 1987.

319. Mndzhoyan, O. L. and Gevorkyan, G. A., Derivatives of aminoketones. IX. Products of the Mannich reaction using methyl ether ketones, *Arm. Khim. Zh.*, 26, 220, 1973; *Chem. Abstr.*, 79, 65719, 1973.

320. Metha, M. D., British Patent 1,199,731, 1970; *Chem. Abstr.*, 73, 120344, 1970.

321. Schake, D., Substituierte und Kondensierte Heterocyclische Spyrocyclohexadienone, *Dissertation (Doktor grade)*, Univ. Düsseldorf, 1990.

322. Zayed, S. M. A. D., Sidky, M. M., and Emran, A., Mannich reaction of 5-hydroxybenzo[b]thiophene, *J. Prakt. Chem.*, 315, 244, 1973.

323. Delia, T. J. and Sami, S. M., Synthesis of 6-benzyl- and 6-phenethyl-2,4-diamino-5,6,7,8-tetrahydropyrimidino[4,5-d]pyrimidines, *J. Heterocycl. Chem.*, 18, 929, 1981.

324. Bazanova, I. N., Neplyuev, V. M., and Loziuskii, M. O., 1-(Phenylsulphonyl)-1-cyanoethene, *Zh. Org. Khim.*, 21, 1577, 1985; *Chem. Abstr.*, 104, 206855, 1986.

325. Dhawan, B. and Redmore, D., Phosphonomethylation of 2-aminoethanethiol and thiazolidine. An unexpected product, *J. Heterocycl. Chem.*, 25, 1273, 1988.

326. Ohashi, M., Takahashi, T., Inoue, S., and Sato, K., The Mannich reaction of alicyclic α-diketones. A novel synthesis of 2-hydroxy-3-methyl-2-cyclohexen-1-one, *Bull. Chem. Soc. Jpn.*, 48, 1892, 1975.

327. Greenhill, J. V., Ingle, P. H. B., and Ramli, M., Mannich reactions on 1,2-diketones, *J. Chem. Soc. Perkin Trans. 1*, 1667, 1972.

328. Amer, F. A. K., Afsah, E. S., and Etman, H., Condensation of 2-acetyl-1,3-indandione with amines and diazonium salts, *Z. Naturforsch. Teil B*, 34B, 867, 1979.

329. Stetter, H. and Steinbeck, K., Kondensationen mit Methylendisulfonen, *Liebigs Ann. Chem.*, 1315, 1974.

330. Böhme, H. and Clement, B., Alkylierung und Aminomethylierung von α-Sulfinylsulfonen, *Tetrahedron Lett.*, 1737, 1979.

331. Ruppert, J., Eder, U., Sauer, G., Haffer, G., and Wiechert, R., German Offen. 2,251,976, 1974; *Chem. Abstr.*, 81, 25087, 1974.

332. Akopyan, Z. G. and Tatevosyan, G. T., Aminoketone derivatives. 2-Substituted-5-oxo-7-aminoenanthic acids and some indole derivatives obtained from them, *Arm. Khim. Zh.*, 29, 1039, 1976; *Chem. Abstr.*, 87, 22946, 1977.

333. Chumak, A. D., Pavel, G. V., and Tilichenko, M. N., Reaction of 1,5-diketones. VIII. Aminomethylation of 1,3-diphenyl-1(α,α-dimethyltetrahydro-γ-pyron-β-yl)-3-propanone, *Khim. Geterotsikl. Soedin.*, 738, 1973; *Chem. Abstr.*, 79, 105104, 1973.

334. Okuda, T. and Matsumoto, U., Alkylation with Mannich bases. V. Mannich reaction of acyl-1-naphthols, *Yakugaku Zasshi*, 79, 1140, 1959; *Chem. Abstr.*, 54, 3452, 1960.

335. Natova, L., Mondeshka, D., and Zhelyazkov, L., Synthesis and structure of some Mannich bases containing five-membered heterocycles, *God. Vissh. Khimikotekhnol. Inst. Sofia*, 24, 47, 1978; *Chem. Abstr.*, 95, 220040, 1981; see also *God. Vissh. Khimikotekhnol Inst. Sofia* 24, 57, 1978.

336. Kuliev, A. M., Guseinov, M. S., Sardarova, S. A., and Iskenderova, T. Y., Alkylation by Mannich bases. Synthesis and study of morpholino and piperazino derivatives of hydroxypropiophenones, *Z. Org. Khim.*, 8, 1301, 1972; *Chem. Abstr.*, 77, 101527, 1972.

337. Gautier, J. A., Miocque, M., and Quan, D. Q., Mannich reaction on p-hydroxyacetophenone, *Compt. Rend.*, 258, 3731, 1964.

338. Brandes, R and Roth H. J., Die 3 Isomeren Morpholin-Mannich Basen des Phenols, *Arch. Pharm.*, 300, 1005, 1967.

339. Hagen, H. E., Frahm, A. W., Brandes, R., and Roth, H. J., Darstellung und Acetolyse des 2,4-Dihydroxy-3-morpholinomethyl-acetophenons. 11 Mitt. Acetolyse von Mannich Basen, *Arch. Pharm.*, 303, 988, 1970.

340. Gautier, J. A., Micocque, M., Mascrier-Demagny, L., Raynaud, G., Thomas, J., Gouret, C., and Pourrias, B., Cétones arylaliphatiques à fonction acétylénique. V. Phénylalcynylcétones parasubstituées: préparation, réactivité, triage pharmacologique, *Chim. Ther.,* 8, 14, 1973.

341. Meshkauskaite, I. and Urbonaite, L., Alkylation of aromatic amines by Mannich bases, *Tr. Pyatoi Nauchn. Tekhn. Konf. Stud.,* 39, 1961; *Chem. Abstr.,* 58, 6794, 1963.

342. Szmuszkovicz, J., 3-Acylindole Mannich bases and their transformation products, *J. Am. Chem. Soc.,* 82, 1180, 1960.

343. Sam, J. and Mozingo, J. R., Preparation of some Mannich bases of bicyclic furans, *J. Pharm. Sci.,* 58, 1030, 1969.

344. Suciu, D., Preparation of some Mannich bases from 2-amino- and 2-allylamino-4-methyl-5-acetylthiazole, *J. Prakt. Chem.,* 313, 193, 1971.

345. Pan, Y. G. and Hochman, L. L., European Patent Appl. EP 373,668, 1990; *Chem. Abstr.,* 113, 211559, 1990.

346. Hartung, H., Cavagna, F., Martin, W., and Dürckheimer, W., Zur struktur aminomethylierter Tetracycline, *Arch. Pharm.,* 307, 651, 1974.

347. Gottstein, W. J., Minor, W. F., and Cheney, L. C., Carboxamido derivatives of the tetracyclines, *J. Am. Chem. Soc.,* 81, 1198, 1959.

348. Azerbaev, I. N., Dzhailauov, S. D., and Bosyakov, Y. G., Allyl esters of N-substituted aminomethyl phosphonic acids, *Izv. Akad. Nauk Kaz. SSR Ser. Khim.,* 28, 51, 1978; *Chem. Abstr.,* 89, 215498, 1978. See also *Izv. Akad. Nauk. Kaz. SSR Ser. Khim.,* 28, 57, 1978.

349. Shvedov, V. I., Kharizomenova, I. A., Medvedeva, N. V., and Grinev, A. N., Functional derivatives of thiophene. XIV. Alkylaminomethylation of α-acylaminothiophene and synthesis of 1,2,3,4-tetrahydro[2,3-d]pyrimidine, *Khim. Geterotsikl. Soedin.,* 918, 1975; *Chem. Abstr.,* 83, 193212, 1975.

350. Dhaneshwar, S. R., Khadikar, P. V., Katiyar, J. C., Dhawan, B. N., and Chaturvedi, S. C., Synthesis and anthelmintic activity of some Mannich bases of mebendazole, *Indian J. Pharm. Sci.,* 52, 261, 1990; *Chem. Abstr.,* 115, 114422, 1991.

351. Saldabols, N., Zeligman, L. L., and Ritevskaya, L. A., Mannich bases and iodomethylates of 6-(2-furyl)imidazo[2,1-b]thiazole and its derivatives, *Khim. Geterotsikl. Soedin.,* 1208, 1975; *Chem. Abstr.,* 84, 30960, 1976.

352. Stavrovskaya, V. I. and Drusviatskaya, S. K., Acid amides in the Mannich reaction. II. Aminomethylation of salicyl- and 5-chlorosalicylamides, *Probl. Poluch. Poluprod. Prom. Org. Sin. Akad. Nauk SSSR,* 164, 1967; *Chem. Abstr.,* 68, 12820, 1968.

353. Dimmock, J. R., Nyathi, C. B., and Smith, P. J., Syntheses and bioactivities of 1-(hydroxyphenyl)-1-nonen-3-ones and related ethers and esters, *J. Pharm. Sci.,* 68, 1216, 1979; see also *J. Pharm. Sci.,* 67, 1543, 1978.

354. Edwards, M. L., Ritter, H. W., Stemerick, D. M., and Stewart, K. T., Mannich bases of 4-phenyl-3-buten-2-one. A new class of antiherpes agents, *J. Med. Chem.,* 26, 431, 1983.

355. Fišnerova, L., Kakác B., and Nemecek, O., New 4-substituted 1,2-diphenyl-3,5-dioxopyrazolidines, *Collect. Czech. Chem. Commun.,* 39, 624, 1974.

356. Möhrle, H. and Reinhardt, H. W., Vinyloge Harnstoffe als Substrat für die Mannich Reaktion, *Chem. Zg.,* 107, 370, 1983.

357. Möhrle, H. and Reinhardt, H. W., Aminomethylierung von Pusch-pull-olefinen, *Arch. Pharm.,* 315, 716, 1982.

358. Möhrle, H. and Reinhardt, H. W., β-Aminoenone als Komponenten in der Mannich Reaktion, *Arch. Pharm.,* 317, 156, 1984; see also p. 1017 and *Arch. Pharm.,* 314, 767, 1981.

359. Torii, S., Tanaka, H., and Takao, H., Preparation of Mannich bases from 6-methoxy-2H-pyran-3(6H)-one and its epoxide, *Bull. Chem. Soc. Jpn.,* 50, 2823, 1977.

360. Tamura, Y., Chen, L. C., Fujita, M., and Kita, Y., Synthesis of 1-substituted 3-anilino-4-diethylaminomethyl-5-oxo-3,4-dehydropiperidines, *J. Heterocycl. Chem.,* 17, 1, 1980.

361. Krieger, H., Kojo, A., and Oikarinen, A., Aminomethylation of 2-pinen-4-one, *Finn. Chem. Lett.,* 185, 1978; *Chem. Abstr.,* 90, 23283, 1979.

362. Roth, H. J. and Langer, G., Vinyloge Aminomethylierung des Isophorons, *Arch. Pharm.,* 301, 695, 1968; see also p. 707.

363. Unterhalt, B. and Koeler, H., Styrylalkyloxime. XV. Sauerstoff Mannich Basen, *Synthesis,* 265, 1977.

364. Hahn, W. E. and Cebulska, Z., Aminomethylation of oximes of azaarenecarboxaldehydes, *Pol. J. Chem.,* 60, 305, 1986; *Chem. Abstr.,* 108, 21684, 1988.

365. Wolinsky, J. and Sundeen, J. E., The reaction of 3-propylindole with aldehydes. Preparation of 2(α-aminoalkyl)indoles, *Tetrahedron,* 26, 5427, 1970.

366. Rida, S. M., Farghaly, A. M., and Ashour, F. A., Mannich bases of theophylline, *Pharmazie,* 34, 214, 1979.

367. Kwon, K. S., Aminomethylation of 5-phenylidantoin, *Yakhak Hoe Chi*, 26, 111, 1982; *Chem. Abstr.*, 97, 162890, 1982.

368. Ramsh, S. M., Zheltonog, N. G., and Khrabrova, E. S., Geometric isomerism of 2'-(piperidinomethyl)-5-arylidenecreatinines, *Khim. Geterotsikl. Soedin.*, 1401, 1989; *Chem. Abstr.*, 113, 23762, 1990.

369. Böhme, H. and Hotzel, H. H., Über N-Hydroxymethyl- und N-Dialkylaminomethylthiocarbonsäureamide, *Arch. Pharm.*, 300, 241, 1967.

370. Halasa, A. F. and Smith, G. E. P., Study of the Michael and Mannich reactions with benzotriazole-2-thiol, *J. Org. Chem.*, 36, 636, 1971.

371. Sawlewicz, J., Wisterowicz, K., and Vogel, S., Synthesis of Mannich bases from 2-imidazolidinethione derivatives, *Acta Pol. Pharm.*, 32, 435, 1975; *Chem. Abstr.*, 85, 21214, 1976.

372. Kadyrov, A., Saidaliev, Z. G., and Abduvaliev, A. A., Condensation reaction of the aniline series, *Tr. Tashk. Politekh. Inst.*, 119, 19, 1974; *Chem. Abstr.*, 84, 121720, 1976.

373. Mironov, G. S., Farberov, M. I., and Bespalova, I. I., Synthesis of carbonyl monomers based on the Mannich reaction. V. Synthesis of pentadienals, *Zh. Obshch. Khim.*, 34, 1642, 1964; *Chem. Abstr.*, 61, 5505, 1964.

374. Bodendorf, K. and Kloss, P., Reaktion mit vinilogen Methyl-aryl-ketonen am Beispiel des Propenyl-anisyl-ketons, *Liebigs Ann. Chem.*, 677, 95, 1964.

375. Berger, J. G., Teller, S. R., and Pachter, I. J., Cycloalkylpyrrolones via decarboxylative ring closure of pyrrole-3-alkanoic acids and derivatives, *J. Org. Chem.*, 35, 3122, 1970.

376. Bansal, P. C., Pitman, I. H., Tam, J. N. S., Mertes, M., and Kaminski, J. J., N-Hydroxymethyl derivatives of nitrogen heterocycles as possible prodrugs. I. N-Hydroxymethylation of uracils, *J. Pharm. Sci.*, 70, 850, 1981.

377. Šladowska, H., 1,3-Dicyclohexyl-5-alkyl-5-aminomethylbarbituric acids as potential antiinflammatory agents, *Farmaco Ed. Sci.*, 32, 866, 1977; see also *Farmaco Ed. Sci.*, 34, 979, 1979.

378. Werner, W., Zschiesche, W., Güttner, J., and Heinecke, H., Struktur-wirkungs-beziehungen bei Mannich Basen mit und ohne Stikstofflostgruppen und einigen Vergleisverbindungen als potentielle Immunsuppressiva, *Pharmazie*, 31, 282, 1976.

379. Karmouta, M. G., Lafont, O., Combet, Farnoux, G., Miocque, M., Rigothier, M. C., Louchon, B., and Gayral, P., Chemical study and antiparasitic activity of iminobarbituric derivatives, *Eur. J. Med. Chem.*, 15, 341, 1980.

380. Musabekov, Y. Y., Moskvin, A. F., Yablonskii, O. P., Voronenkov, V. V., and Mironov, G. S., Mixture composition and structure of α,β-unsaturated ketones prepared by the Mannich reaction, *Zh. Org. Khim.*, 8, 2288, 1972; *Chem. Abstr.*, 79, 4936, 1973.

381. Guang-Xu, C., Xiu-Juan, X., and Lijun, L., Mannich reaction with arylamines as amine components, *Chem. J. Chinese Univ.*, 3, 83, 1982.

382. Thiele, K. and Posselt, K., U.S. Patent 3,733,340, 1973; *Chem. Abstr.*, 79, 32182, 1973.

383. Dimmock, J. R., Qureshi, A. M., Noble, L. M., Smith, P. J., and Baker, M. A., Comparison of antimicrobial activity of nuclear substituted aromatic esters of 5-dimethylamino-1-phenyl-3-pentanol and 3-dimethylamino-1-phenyl-1-propanol with related cyclic analogs, *J. Pharm. Sci.*, 65, 38, 1976.

384. Buchbauer, G., Mannichbases des 1-(3,3-Dimethyl-2-exo-norbornyl)-ethanons und ein neues Akineton-analogon, *Monatsh. Chem.*, 108, 21, 1977.

385. Hahn, W. E., Sokolowska, A., Szalecki, W., and Siekierska, M., Ethanoanthracenes. IV. Synthesis and transformations of derivatives of 11-(3-amino-1-oxopropyl)-9,10-dihydroanthracene, *Pol. J. Chem.*, 54, 349, 1980; *Chem. Abstr.*, 93, 220503, 1980.

386. Dimmock, J. R. and Turner, W. A., Synthesis and stereochemistry of 2-dimethylaminomethyl-6-phenyl cyclohexanols and related esters with antimicrobial activity, *Can. J. Pharm. Sci.*, 9, 33, 1974.

387. Kuliev, A. M., Gasanov, F. I., Mamedov, F. N., and Kuliev, A. G., Synthesis of aminomethylthiomethyl derivatives of cyclopentanone and -hexanone and conversion of the resulting substances, *Zh. Org. Khim.*, 13, 1193, 1977; *Chem. Abstr.*, 87, 101999, 1977.

388. Viterbo, R., Mastursi, M., and Perri, G. C., German Offen. 2,241,578, 1974; *Chem. Abstr.*, 80, 108064, 1974.

389. Descotes, G. and Laurent, S., Application of the Mannich reaction to 3-alkylcyclohexanones, *Compt. Rend.*, 68, 8358, 1968.

390. Golovin, E. T., Glukhov, B. M., Botsman, L. S., and Burdeleva, T. V., Mannich reactions in six-membered heterocyclic γ-ketones. VIII. *Khim. Geterotsikl. Soedin.*, 903, 1975; *Chem. Abstr.*, 84, 4791, 1976.

391. Sawa, Y. Kato, T., Hattori, T., and Kawai, K., Japanese Kokai Tokkyo Koho 76 56,434, 1976; *Chem. Abstr.,* 85, 176952, 1976.

392. Golovin, E. T., Glukhov, B. M., Yastrebov, V. V., and Unkovskii, B. V., Mannich reactions in six-membered heterocyclic γ-ketones. VII. *Zh. Org. Khim.,* 9, 840, 1973; *Chem. Abstr.,* 79, 53149, 1973.

393. Krishna, K. R. S., Rao, M., and Devi, Y. U., A study of the Mannich reaction in the isoxazole series. *Proc. Indian Acad. Sci. A,* 84, 79, 1976; *Chem. Abstr.,* 86, 16584, 1977.

394. Möhrle, H. and Miller, C. H. R., Aminomethylierung verschieden substituierter Phenole mit secundären Aminen, *Pharmazie,* 33, 500, 1978.

395. Hodgkin, J. H. and Allen, R. J., Cyclopolymerization. XIII. Cyclopolymerization of diallylaminomethylphenols, *J. Macromol. Sci. Chem.,* A11, 937, 1977.

396. Abdullaev, G. K., Agamalieva, E. A., and Abasova, N. A., Aminomethylation of halophenols, *Azerb. Khim. Zh.,* 59, 1972; *Chem. Abstr.,* 79, 52918, 1973.

397. Monti, S. A. and Johnson, W. O., Position selective Mannich reaction of some 5- and 6-hydroxyindols, *Tetrahedron,* 26, 3685, 1970.

398. Kuriakose, A. P., Mannich reaction and chloromethylation of some dihydroxynaphthalenes, *Indian J. Chem.,* 13, 1149, 1975.

399. Mehta, R. H., Synthesis of Mannich bases from coumarin derivatives and screening for their biological activity, *J. Indian Chem. Soc.,* 60, 201, 1983.

400. Tyukavkina, N. A., Kalabin, G. A., Kononova, V. V., and Kushnarev, D. F., Structure of products of the aminoalkylation of 5- and 7-hydroxyflavones, *Khim. Geterotsikl. Soedin.,* 609, 1978; *Chem. Abstr.,* 89, 129356, 1978.

401. Bell, M. R., Oesterlin, R., Beyler, A. L., Harding, H. R., and Potts, G. O., Isomeric Mannich bases derived from ethyl 5-hydroxy-2-methylindole-3-carboxylate, *J. Med. Chem.,* 10, 264, 1967.

402. Kuckländer, H., Mannich-bases von 6-Hydroxy-indol-derivaten, *Arch. Pharm.,* 311, 966, 1978.

403. Troxler, F., Borman, G., and Seemann, F., Synthesen von Mannich Basen von Hydroxyindolen, *Helv. Chim. Acta,* 51, 1203, 1968.

404. Monti, S. A., Johnson, W. O., and White, D. H., The Mannich reaction of hydroxyindoles, *Tetrahedron Lett.,* 4459, 1966.

405. Burkhalter, J. H. and Leib, R. I., Amino- and chloromethylation of 8-quinolinol. Mecha-

nism of preponderant ortho substitution in phenols under Mannich conditions, *J. Org. Chem.,* 26, 4078, 1961.

406. Fernandez, J. E. and Ferree, W. I., Mannich reaction of 2-naphthol with secondary amines, *Q. J. Fl. Acad. Sci.,* 29, 13, 1966; *Chem. Abstr.,* 67, 21247, 1967.

407. Gevorkyan, G. A., Gabrielyan, S. A., Apoyan, N. A., Chilingaryan, D. G., and Mndzhoyan, O. L., Ethyl-5-(aminomethyl)-salicylates and biological properties of their salts, *Khim. Farm. Zh.,* 14, 128, 1980; Chem. Abstr., 94, 65281, 1981.

408. Arold, H., German Offen. 2,601,782, 1977; *Chem. Abstr.,* 87, 201067, 1977.

409. Sucharda-Sobczyk, A. and Ritter, S., New Mannich bases, *Pol. J. Chem.,* 52, 1555, 1978; *Chem Abstr.,* 90, 138389, 1979.

410. Hansell, D. P., Mannich reaction of 2,5-xylenol, morpholine and formaldehyde, *Liebigs Ann. Chem.,* 54, 1978.

411. Sun, M., Jin, Y., Huang, L., Zhou, X., Chen, A., and He, Q., Mannich reaction of substituted phenols, *Huaxue Xuebao,* 43, 306, 1985; *Chem. Abstr.,* 103, 160165, 1985.

412. Short, J. H., Dunnigan, D. A., and Ours, C. W., Synthesis of phenethylamines from phenylacetonitriles obtained by alkylation of cyanide ion with Mannich bases from phenols and other benzylamines, *Tetrahedron,* 29, 1931, 1973.

413. Postovskii, I. Y., Novikova, A. P., Chechulina, L. A., and Lyubomudrova, L. N., Antioxidant and antiirradiation activity of Mannich bases in a series of pyrogallols and hydroxyhydroquinones, *Tr. Inst. Khim. Ural.,* 37, 24, 1978; *Chem. Abstr.,* 92, 128829, 1980.

414. Miocque, M. and Vierfond, J. M., Application de la réaction de Mannich aux arylamines. I. Aminométhylation des dialcoyl anilines, *Bull. Soc. Chim. Fr.,* 1896, 1970; see also p. 1901.

415. Grinev, A. N. Lomanova, E. V., Alekseeva, L. M., Turchin, K. F., and Sheinker, Y. N., Synthesis and aminomethylation of derivatives of pyrazino[3,2,1-jk]carbazole and diazepino[3,2,1-jk]carbazole derivatives, *Khim. Geterotsikl. Soedin.,* 1660, 1983; *Chem. Abstr.,* 100, 209758, 1984.

416. See also survey in Ref. 7.

417. See also survey in Ref. 6.

418. Acheson, R. M., Littlewood, D. M, and Rosenberg, H. E., Synthesis of 1-methoxyindoles, *J. Chem. Soc. Chem. Commun.,* 671, 1974.

419. Flaugh, M. E., Crowell, T. A., Clemens, J. A., and Sawyer, B. D., Synthesis and evaluation of the antiovulatory activity of a variety of melatonin analogues, *J. Med. Chem.*, 22, 63, 1979.

420. Kotlyarevskii, I. L., Pavlyukina, L. A., and Al't, Y. L., 3-Aminomethyl substituted 5-aminopropynyl indoles, *Izv. Akad. Nauk*, 2380, 1979; *Chem. Abstr.*, 92, 76226, 1980.

421. Samsoniya, S. A., Chikvaidze, I. S., Kereselidze, D. A., and Suvorov, N. N., Bis indoles. II. Electrophilic substitution in the bis(5-indolyl) sulfone series and data on reactivity indexes of some bisindoles, *Khim. Geterosikl. Soedin.*, 1653, 1982; *Chem. Abstr.*, 98, 125033, 1983.

422. Yakhontov, L. N. and Lapan, E. I., Azaindole derivatives. XLI. Synthesis of 3-substituted 5-azaindoles, *Khim. Geterotrsikl. Soedin.*, 1528, 1972; *Chem. Abstr.*, 78, 43322, 1973.

423. Benghiat, E. and Crooks, P. A., Reaction of 2-acetamido-3,7-dihydropyrrolo[2,3-d]-pyrimidin-4-one with dimethylamine and formaldehyde. Formation of two isomeric Mannich bases, *J. Heterocycl. Chem.*, 20, 1023, 1983.

424. Akimoto, H., Imamiya, E., Hitaka, T., Nomura, H., and Nishimura, S., Synthesis of queuine, base of naturally occurring hypermodified nucleoside (queuosine) and its analogues, *J. Chem. Soc. Perkin Trans. 1*, 1637, 1988.

425. Kuo, H. S., Tsai, S. L., and Tung, Y. C., Mannich bases from 2,7-dimethylpyrrolo[1,2-a]quinoline: synthesis and biological activities, *T'ai-wan Yao Hsueh Tsa Chih*, 32, 79, 1981; *Chem. Abstr.*, 96, 68780, 1982.

426. Guilford, J. and Harrel, W. B., Mannich bases derived from 1-phenyl-2-(p-chlorophenyl)-indolizine, *Tex. J. Sci.*, 36, 33, 1986; *Chem. Abstr.*, 106, 49945, 1987.

427. Barker, J. M., Huddleston, P. R., and Wood, M. L., Mannich reactions of methoxythiophenes, *Synth. Commun.*, 5, 59, 1975; *Chem. Abstr.*, 83, 9661, 1975.

428. Stocker, F. B., Kurtz, J. L., Gilman, B. L., and Forsyth, D. A., The Mannich reaction of imidazoles, *J. Org. Chem.*, 35, 883, 1970.

429. Parthasarathy, P. C., Desai, H. K., and Saindane, M. T., Nitroimidazoles. XVIII. Mannich reaction of azomycin-2-nitro-4,5-bisaminoalkyl imidazoles, *Indian. J. Chem.*, 22B, 157, 1983.

430. Prescott, B., Caldes, G., Piggott, W. R., and James, W. D., Some new derivatives of 4,4'-diaminodiphenyl sulfone as potential antimicrobial agents. III. Mannich bases and related compounds, *Antimicrob. Agents Chemother.*, 419, 1966; *Chem. Abstr.*, 68, 3805, 1968.

431. Dorn, H. and Zubek, A. Potentielle Cytostatica. XX. Bis-(2-chlor-äthyl)-aminomethyl-pyrazol-derivate, *J. Prakt. Chem.*, 313, 211, 1971.

432. Leonova, T. S. and Yashunskii, V. G., Some reactions of 4-aminopyrazolo[3,4-d]pyrimidines, *Khim. Geterotsikl. Soedin.*, 982, 1982; *Chem. Abstr.*, 97, 162932, 1982.

433. Smirnov, L. D., Zhuravlev, V. S., Lozina, V. P., Zaitsev, B. E., and Dyumaev, K. M., Reactions of 5-benzyl-3-hydroxypyridine, *Izv. Akad. Nauk SSSR Ser. Khim.*, 2801, 1973; *Chem. Abstr.*, 80, 95679, 1974.

434. Dyumaev, K. M. and Lokhov, R. E., Aminomethylation in a series of 3-hydroxypyridine N-oxides, *Zh. Org. Khim.*, 416, 1972; *Chem. Abstr.*, 76, 126728, 1972.

435. Gol'tsova, L. V., Lezina, V. P., Kuz'min, V. I., and Smirnov, L. D., Study of electrophilic reactions of 3-hydroxyquinoline-1-oxide, *Izv. Adad. Nauk SSSR Ser. Khim.*, 1660, 1984, *Chem. Abstr.*, 102, 6156, 1985.

436. Gashev, S. B. and Smirnov, L. D., Electrophilic reactions of 5-hydroxypyrimidine, *Khim. Geterotsikl. Soedin.*, 393, 1982; *Chem. Abstr.*, 97, 38908, 1982.

437. Gashev, S. B., Sedova, V. F., Smirnov, L. D., and Mamaev, V. P., Study of some electrophilic reactions of 4-phenyl-5-hydroxypyrimidine and its 1-oxide, *Khim. Geterotsikl. Soedin.*, 1257, 1983; *Chem. Abstr.*, 100, 51543, 1984.

438. O'Brien, D. E., Springer, R. H., and Cheng, C. C., A new Mannich reaction of pyrimidines, *J. Heterocycl. Chem.*, 3, 115, 1966.

439. Kamiya, S., Okusa, G., Osada, M., Kumagai, M., Nakamura, A., and Koshinuma, K., C-Alkylaminomethylation of pyridazinol N-oxides. IV. The Mannich reaction of 3- and 5-pyridazinol 1-oxides using primary amines, *Chem. Pharm. Bull. Tokyo*, 16, 939, 1968.

440. Kamiya, S. and Okusa, G., Syntheses of 3- and 4-alkylaminomethylpyridazines, *Chem. Pharm. Bull. Tokyo*, 21, 1510, 1973; see also *Chem. Pharm. Bull. Tokyo*, 23, 923, 1975.

441. Burckhalter, J. H., Seiwald, R. J., and Scarborough, H. C., The amino- and chloromethylation of uracil, *J. Am. Chem. Soc.*, 82, 991, 1960.

442. Elderfield, R. C. and Wood, J. R., Application of the Mannich reaction with β,β-dichlorodiethylamine to derivatives of uracil, *J. Org. Chem.*, 26, 3042, 1961.

443. Salisbury, S. A. and Brown, D. M., Electrophilic substitution in dyhydrouracils, *J. Chem. Soc. Chem. Commun.*, 656, 1979.

444. Xu, J., Guo, R., Zhen, F., Lu, R., Huang, J., and Wang, Y., Studies on nucleic acids. VI. Synthesis of ribothymidine-3'-phosphate, *Huaxue Xuebao*, 39, 681, 1981; *Chem. Abstr.*, 97, 127997, 1982.

445. Lyapova, M. and Kurtev, B., On the interaction of desoxybenzoin with azomethine compounds, *C. R. Acad. Bulgar. Sci.*, 21, 447, 1968; see also p. 905.

446. Thompson, M. D., Smith, G. S., and Berlin, K. D., Synthesis of 3-selena-7-azabicyclo[3.3.1]nonanes and certain derivatives, *Org. Prep. Proced. Int.*, 18, 329, 1986; *Chem. Abstr.*, 106, 176358, 1987.

447. Llama, E. F. and Trigo, G. G., Synthesis and structural study of new derivatives of 6,8-diaryl-3-thia-7-azabicyclo[3.3.1]nonane systems, *Heterocycles*, 24, 719, 1986.

448. Hine, J., Li, W. S., and Zeigler, J. P., Stereoselective bifunctional catalysis of dedeuteration of cyclopentanone-2,2,5,5-d$_4$ by (1R,2S,3R,4R)-3-dimethylaminomethyl-1,7,7-trimethyl-2-norbornanamine, *J. Am. Chem. Soc.*, 102, 4403, 1980.

449. Sabie, R., Fillion, H., Pinatel, H., and Fenet, B., Synthesis and proton NMR stereochemical study of 7,8,9-substituted-7,8-dihydro-4H,9H-furo[2',3',4',4,4a,5]-napth[2,1-a][1,3]oxazin-4-ones, *J. Heterocycl. Chem.*, 27, 1893, 1990.

450. Urbanski, T., Primary nitro compounds as a source of some heterocyclic systems, *Synthesis*, 613, 1974.

451. Yi, L. and Xu, X., Mannich reaction with arylamine as amine component. VIII. Mannich reaction of arylamine with 2,4-pentanedione or 3-methyl-2,4-pentanedione, *Beijing Shifan Daxue Xuebao*, 54, 1990; *Chem. Abstr.*, 114, 122251, 1991.

452. Farminer, A. F. and Webb, G. A., Nitration and nitrosation reactions of 7-nitro-1,3,5-triazaadamantane and derivatives, *J. Chem. Soc. Perkin Trans. 1*, 940, 1976.

453. Daigle, D. J., Pepperman, A. B., and Vail, S. L., Synthesis of a monophosphorous analog of hexamethylenetetramine, *J. Heterocycl. Chem.*, 11, 407, 1974.

454. Curtze, J. and Thomas, K., Reaktionen von halogenierten Thiobenzamiden mit Formaldehyd und aliphatischen Aminen, *Liebigs Ann. Chem.*, 2318, 1975.

455. Katritzky, A. R., Baker, V. J., Brito-Palma, F. M. S., Sullivan, J. M., and Finzel, R. B., The conformational analysis of saturated heterocycles. XCII. Conformational equilibria of 1,2-dioxa-4,5-diazacyclohexanes, *J. Chem. Soc. Perkin Trans. 2*, 1133, 1979.

456. Afsah, E. M., Hammouda, M., and Abou-Elzahab, M. M., A study of the double Mannich reaction with 1,3-diphenylacetone, *Monatsh. Chem.*, 115, 581, 1984.

457. House, H. O., Wickham, P. P., and Müller, H. C., The synthesis of certain azabicyclic ketones, *J. Am. Chem. Soc.*, 84, 3139, 1962.

458. Afsah, E. M., Metwally, M. A., and Khalifa, M. M., Synthesis of 2,5-bis(pyridinomethyl)-piperidine and 1,5-bis(aminomethyl)-3-azabicyclo-[3.2.1]-octanones, *Monatsh. Chem.*, 115, 303, 1984.

459. Haller, R., Zur Kenntnis substituierter 3,7-Diaza- und 3-Oxa-7-aza-bicyclo-[3.3.1]-nonanone, *Arzneim. Forsch.*, 15, 1327, 1965.

460. Settimj, G., Landi Vittory, R., Gatta, F., Sarti, N., and Chiavarelli, S., Sintesi nella serie dell'1,5-difenil-bispidin-9-one e -9-olo. XI, *Gazz. Chim. Ital.*, 96, 604, 1966.

461. Nabiev, O. G., Shakhgel'diev, M. A., Chervin, I. L., and Kostyanovskii, R. G., Geminal systems. XXVII. Asymmetric nitrogen. XLIII. Aminomethylation of diaziridines by methylenediamines, alkoxymethyl- and bis(alkoxymethyl)amines, *Dokl. Akad. Nauk SSSR*, 284, 872, 1985; *Chem. Abstr.*, 105, 133781, 1986.

462. Rochling, H., Sachse, B., and Gattner, H., British Patent 1,580,485, 1980; *Chem. Abstr.*, 95, 7289, 1981.

463. Petersen, H., Syntheses of cyclic ureas by α-ureidoalkylation, *Synthesis*, 243, 1973.

464. Unterhalt, B., Seebach, E., and Thamer, D., Substituirte 1,2,4,6-Thiatriazacyclohexan-1,1-dioxide, *Arch. Pharm.*, 311, 47, 1978.

465. Thewalt, U. and Bugg, C. E., Röntgenographische Charakterisierung des Reaktions Produktes von bis(S-Amino-dithionitrito)-nickel(II) mit Ammoniak, Formaldehyd und Methanol, *Chem. Ber.*, 105, 1614, 1972.

466. Howes, P. D., Payne, J. J., and Pianka, M., Synthesis of 5,6-dihydro-4H-1,3,5-dithiazines, 2,3-dihydro-6-thioxo-6H-1,3-thiazine and 6-amino-1,3-dithiins, *J. Chem. Soc. Perkin Trans. 1*, 1038, 1986.

467. Issleib, K., Leissring, E., and Riemer, M., Aminomethylation of o-phenylenediphosphines, *Z. Chem.*, 25, 172, 1985; *Chem. Abstr.*, 105, 24339, 1986.

468. Chiavarelli, S., Toffler, F., Mazzeo, P., and Gramiccioni, L., Sintesi nella serie dell'1,5-difenilbispidin-9-one. Nota XIV, *Farmaco Ed. Sci.*, 23, 360, 1968; *Chem. Abstr.*, 69, 52115, 1968.

469. Risch, N., Dreifache Mannich-cyclisierung. Aufbau von drei benachbarten quartären sp^3-Kohlenstoffzentren in einem Reaktionsschritt, *Z. Naturforsch. Teil B,* 41B, 787, 1986.

470. Sokolova, V. A., Boldyrev, M. D., Gidaspov, B. V., and Timoteeva, T. N., Reduction of 2,4,6-trinitrotoluene and -xylene by sodium borohydride, *Zh. Org. Khim.,* 8, 1243, 1972; *Chem. Abstr.,* 77, 101363, 1972.

471. Sargeson, A. M., Caged metal ions, *Chem. Br.,* 15, 23, 1979.

472. Afsah, E. M., Sarhan, A. A., and Ibraham, M. R., Mannich reaction with 5-phenylcyclohexane-1,3-dione, *J. Prakt. Chem.,* 326, 683, 1984.

473. Hahn, W. E. and Korzeniewski, C., Application of the Mannich reaction for the synthesis of heterocyclic systems. VI. Bicyclo[4.3.1]decane derivatives and analogs, *Rocz. Chem.,* 40, 37, 1966; *Chem. Abstr.,* 65, 3829, 1966.

474. Dold, O., N-Substituierte 2-Carbamoyl-6-cyan-piperidine und 2-Carbamoyl-5-cyanpyrrolidine, *Chem. Ber.,* 96, 2052, 1963.

475. Arya, V. P., Synthesis of new heterocycles. XXIII. Syntheses of 3-azabicyclo[3.3.1]nonane derivatives and related heterocycles, *Indian J. Chem. B,* 14B, 982, 1976.

476. Bobowski, G. and Yates, P., The Mannich-reaction. 1-Alkyl-3,3-diphenyl-4-piperidiones, 1,6′-dialkyl-3′,4′,5′,6′,7′,8′-hexahydro-5,5,8′,8′-tetraphenylspiro[piperidine-3,-2′-[2H]pyrano[3,2-c]pyridin]-4-ones and their derivatives, *J. Org. Chem.,* 50, 1900, 1985.

477. Bahagji, E. H., Tronche, P., Couquelet, J., Harraga, S., Panouse-Perrin, J., and Rubat, C., Studies on immunostimulating derivatives. Synthesis of some pyrrolo[1,2-c]pyrimidines, *Chem. Pharm. Bull. Tokyo,* 39, 2126, 1991.

478. Hammouda, M., Hamama, W. S., and Afsah, E. M., The Mannich reaction with 1-phenylamino-3-indenone, *Z. Naturforsch. Teil B,* 42B, 94, 1987.

479. Hahn, W. E. and Cebulska, Z., Aminomethylation of phenylhydrazones of derivatives of 2-formylquinoxalines, *Pol. J. Chem.,* 53, 779, 1979; *Chem. Abstr.,* 91, 157688, 1979.

480. Hahn, W. E. and Zawadzka, H., Application of Mannich reaction in the synthesis of heterocyclic systems. V., *Rocz. Chem.,* 38, 557, 1964; *Chem. Abstr.,* 61, 10685, 1964.

481. Dychenko, A. I. and Pupko, L. S., Mannich reaction in a series of nitroformaldehyde arylhydrazones, *Ukr. Khim. Zh.,* 40, 1220, 1974, *Chem. Abstr.,* 82, 57648, 1975.

482. Solov'eva-Yavits, S. Y., Ramsh, S. M., and Ginak, A. I., Study of the reactivity and tautomerism of azolidines. XL. 2-Iminothiazolidin-4-one in a Mannich reaction, *Khim. Geterotsikl. Soedin.,* 477, 1981; *Chem. Abstr.,* 95, 80900, 1981; see also *Chem. Abstr.,* 93, 204518, 1980.

483. Solov'eva, S. Y., Ramsh, S. M., and Ginak, A. I., Study of the reactivity and tautomerism of azolidines. XLI. Anomalous products of aminomethylation of 2-iminothiazolidin-4-one by aqueous formaldehyde and primary amines, *Khim. Geterotsikl. Soedin.,* 1204, 1983; *Chem. Abstr.,* 100, 68262, 1984.

484. Georgescu, M. A., Zugravescu, I., and Leonte, M. V., α-Aminoalkylation. IV. Application of Mannich reaction for synthesis of substituted dihydro-1,3-benzoxazines and their derivatives, *Bull. Univ. Galati,* 2, 59, 1979; *Chem. Abstr.,* 95, 7180, 1981.

485. Georgescu, M. A. and Leonte, A. I., α-Aminoalkylation. VI. Application of the Mannich reaction to the synthesis of purpurogallin derivatives, *Bull. Univ. Galati,* 2, 75, 1979; *Chem. Abstr.,* 95, 61715, 1981.

486. Kristián, P. and Bernát, J., Neue Darstellungsmethode für 3,5-disubstituierte 2-Thiontetrahydro-1,3,5-thiadiazinen, *Tetrahedron Lett.,* 679, 1968.

487. Geigy, J. R., A.-G., French M. 6521, 1969; *Chem. Abstr.,* 74, 88072, 1971.

488. Möhrle, H. and Bangert, R., Aminomethylierung von cyclischen Enaminothionen, *Pharmazie,* 45, 37, 1990.

489. Rehse, U., Mannich Bases aus β-Dicarbonylverbindungen, *Arch. Pharm.,* 308, 881, 1975.

490. Abbiss, T. P., Soloway, A. H., and Mark, V. H., Ammonoacetals. I. A new class of potential antineoplastic compounds, *J. Med. Chem.,* 7, 644, 1964.

491. Billman, J. H. and Khan, M. S., Hexahydropyrimidines. IX. Synthesis of 2-substituted 1,3-bis[2-methyl-4-[N,N-bis(2-chloroethyl)-amino]benzyl]hexahydropyrimidines as transport molecules for tumors inhibition, *J. Pharm. Sci.,* 57, 1817, 1968.

492. Galík, V. and Landa, S., Über stickstoffhaltige Adamantanverbindungen. II. Synthese von 1,3-Diazaadamantan, *Collect. Czech. Chem. Commun.,* 38, 1101, 1973.

493. Abbasi, M., Nasr, M., Zoorob, H. H., and Michael, J. M., Reactivity of 4-hydroxy-2-methyl-7,8,9,10-tetrahydrobenzo[h]quinoline towards base-catalyzed cyclization, Mannich and Turpin reactions, *J. Heterocycl. Chem.,* 15, 649, 1978.

494. Baker, V. J., Katritzky, A. R., Majoral, J. P., Martin, A. R., and Sullivan, J. M., The conformational analysis of saturated heterocycles. LXXVII. Rationalisation of the equilibria of tetraalkylhexahydro-1,2,4,5-tetrazines, *J. Am. Chem. Soc.,* 98, 5748, 1976.

495. Masui, M., Suda, K.,Yamauchi, M., and Yoshida, N., Novel Mannich reaction product from 2,4(5)-diisopropylimidazole, *Chem. Pharm. Bull. Tokyo,* 21, 1387, 1973.

496. Hammerum, S., The chemistry of hexahydro-1,2,4,5-tetrazines. IV. The reactions of alkylhydrazines with formaldehyde, *Acta Chem. Scand.,* 27, 779, 1973.

497. Swan, G. A., Studies of the reaction of benzoyl peroxide with N,N-disubstituted aromatic amines and related compounds. V, *J. Chem. Soc. C,* 2880, 1971.

498. Maier, L., Preparation and properties of N-(hydroxy-carbonylmethyl)aminomethyl alkyl- and arylphosphinic acids and derivatives, *ACS Symp. Ser.,* 171 (Phosphorus Chem.), 251, 1981; *Chem. Abstr.,* 96, 85657, 1982.

499. Benkovic, S. J., Benkovic, P. A., and Chrzanowski, R., Studies on models of tetrahydrofolic acid. II. Additional observations on the mechanism for condensation of formaldehyde with tetrahydroquinoxaline analogs, *J. Am. Chem. Soc.,* 92, 523, 1970.

500. Eiter, K., Hebenbrock, K. F., and Kabbe, H. J., Neue offenkettige und cyclische α-Nitrosaminoalkyl-äter, *Liebigs Ann. Chem.,* 765, 55, 1972.

501. Kellner, K., Hanke, W., and Tzschach, A., Reaction of aminoacids with formaldehyde and secondary phosphines, *Z. Chem.,* 24, 193, 1984; *Chem. Abstr.,* 101, 192416, 1984.

502. Barkworth, P. M. R. and Crabb, T. A., Compounds with bridgehead nitrogen. XLIII. The reaction between trans-2-aminocycloalkanethiols and formaldehyde, *J. Chem. Soc. Perkin Trans. 1,* 2777, 1982.

503. Melchiorre, C., Giardinà, D., and Angeli, P., Reductive opening of the thiazolidine ring. Selective N-methylation of cysteamines, *J. Heterocycl. Chem.,* 1215, 1980.

504. Dean, R. T. and Rapoport, H., Stereospecific cyclizations of iminium salts from α-aminoacid decarboxylation. Synthesis of 8- and 13-methylberbines, *J. Org. Chem.,* 43, 4183, 1978.

505. Mathison, I. W., Solomons, W. E., and Jones, R. H., Synthesis of cyclopentano-1,2,3,4-tetrahydroisoquinolines. Novel heterocyclic systems, *J. Org. Chem.,* 39, 2852, 1974.

506. Kametani, T., Matsumoto, H., Satoh, Y., Nemoto, H., and Fukumoto, K., Studies on the synthesis of heterocyclic compounds. DCLXXIX. A stereoselective total synthesis of (±)-ophiocarpine, *J. Chem. Soc. Perkin Trans. 1,* 376, 1977.

507. Berney, D. and Jauner, T., 1-Aralkylated tetrahydro-2-benzazepines. II. Synthesis from 3-(3,4-dimethoxyphenyl)-propylamine, *Helv. Chim. Acta,* 59, 623, 1976.

508. Bruice, T. C. and Lombardo, A., Catalytic reactions involving azomethines. XI. The kinetics of condensation of histamine with 3-hydroxypyridine-4-aldehyde. An intramolecular Mannich reaction, *J. Am. Chem. Soc.,* 91, 3009, 1969.

509. Katritzky, A. R. and Patel, R. C., The conformational analysis of saturated heterocycles. XC. Syntheses and conformational analysis of 1,2,4-trimethyl- and 1,2,3,4-tetramethyl-1,2,4-triazacyclohexanes, *J. Chem. Soc. Perkin Trans. 2,* 984, 1979.

510. Riddel, F. G. and Turner, E. S., Conformational equilibria and nitrogen inversion in tetrahydro-1,2,5-oxadiazines, *Tetrahedron,* 35, 1311, 1979.

511. Gatta, F., Landi Vittory, R., Nunez Barrios, G., and Tomassetti, M., Synthesis of 7,8-dimethoxy-1,4-benzodiazepines, *Ann. 1st Sup. Sanità,* 7, 533, 1971; *Chem. Abstr.,* 77, 61967, 1972.

512. Barrows, T. H., Farina, P. R., Chrzanowski, R. L., Benkovic, P. A., and Benkovic, S. J., Studies on models for tetrahydrofolic acid. VII. Reactions and mechanism of tetrahydroquinoxaline derivatives at the formaldehyde level of oxidation, *J. Am. Chem. Soc.,* 98, 3678, 1976.

513. Hiramitsu, T. and Maki, Y., A Mannich-type cyclisation to thiazepines. Synthesis of pyrimido[5,4-f]benzo[b]-1,4-thiazepines, a new tricyclic system, *Synthesis,* 177, 1977.

514. Calcagni, A., Rossi, D., and Lucente, G., α-Hydroxymethylation of Schiff bases derived from α-amino acid esters, *Synthesis,* 445, 1981.

515. Baker, V. J., Ferguson, I. J., Katritzky, A. R., Patel, R., and Rahimi-Rastgoo, S., The conformational analysis of saturated heterocycles. LXXXIV. Conformational consequences of internal β-heteroatoms, *J. Chem. Soc. Perkin Trans. 2,* 377, 1978.

516. Geue, B. J., Snow, M. R., Springborg, J., Herlt, A. J., Sargeson, A. M., and Taylor, D., Condensation of formaldehyde with chelated glycine and ethylenediamine: a new macrocycle synthesis, *J. Chem. Soc. Chem. Commun.*, 285, 1976.

517. Harsányi, K., Kiss, P., and Kórbónits, D., Mannich-type reactions of an isobasic isoquinoline, *J. Heterocycl. Chem.*, 10, 435, 1973.

518. Ishiwata, S. and Shiokawa, Y., Benzimidazoles and related compounds. III. Intramolecular Mannich reaction of 2-alkylaminomethylbenzimidazoles, *Chem. Pharm. Bull. Tokyo*, 18, 1245, 1970.

519. Raines, S., Chai, S. Y., and Palopoli, F. P., The preparation and some reactions of 9(disubstituted-amino)-9H-pyrrolo[1,2-α]indoles, *J. Org. Chem.*, 36, 3992, 1971.

520. Petersen, J. S., Toteberg-Kaulen, S., and Rapoport, H., Synthesis of (±)-ω-aza[x.y.1]bicycloalkanes by an intramolecular Mannich reaction, *J. Org. Chem.*, 49, 2948, 1984.

521. King, F. D., A facile synthesis of quinolizidines and indolizidines, *Tetrahedron Lett.*, 24, 3281, 1983.

522. Szántay, C. and Rohály, J., Über die Bildung ungesättigter Ketone in den Mannich-reaktionen substituierter Acetessigsäuren, *Chem. Ber.*, 96, 1788, 1963.

523. Nobles, W. L. and Thompson, B. B., Application of Mannich reaction to sulfones. I, *J. Pharm. Sci.*, 54, 576, 1965; see also p. 709.

524. Schauble, J. H. and Hertz, E., A reinvestigation of the Mannich reaction of 4-nitro- and 2,4-dinitro-phenylacetic acid, *J. Org. Chem.*, 35, 2529, 1970.

525. Adrian, G., *Bull. Soc. Chim. Fr.*, 4160, 1971.

526. Braun, W., Waechter, R., and Weissauer, H., German Patent 1,132,269, 1962; *Chem. Abstr.*, 59, 12954, 1963.

527. Yakhontov, L. N., Mrachkovaskaya, L. B., Ermakov, A. I., Turchin, K. F., and Vlasova, T. F., Mannich reaction of piperazine with malonic acid derivatives, *Zh. Org. Khim.*, 10, 868, 1974; *Chem. Abstr.*, 81, 13460, 1974.

528. Harhash, A. H., Mansour, A. K., Elnagdi, M. H., and Elmoghayer, M. R. H., Reaction of arylhydrazones of some α,β-diketoesters, *J. Prakt. Chem.*, 315, 235, 1973.

529. Hishmat, O. H., Zohair, M. M. Y., El-Ebrashi, N. M. A., and Soliman, F. M. A., Synthesis of some visnaginone and kellinone derivatives of possible biological activity, *Pharmazie*, 35, 682, 1980.

530. Hamman, A. G., Ali, A. S., and Youssif, N. M., Reactions with (arylmethylene)cycloalkanones. 6. Synthesis and deacetylation of 2-acetyl-5-aryl-2,3,6,7,8,9-hexahydro-5H-thiazolo[2,3-b]-quinazolin-3-ones, *Egypt. J. Chem.*, 26, 461, 1983; *Chem. Abstr.*, 101, 230459, 1984.

531. Bredereck, K., Metwally, S. A., Koch, E., and Weckmann, R., Antrachinone. VI. Hydroxymethylierung und Aminomethylierung von Antrachinonen, *Liebigs Ann. Chem.*, 972, 1975.

532. Baires, S. V., Ivanov, V. B., Krokhina, S. S., and Ivanov, B. E., Aminomethylation of dimethylsulfite, *Zh. Obshch. Khim.*, 57, 2387, 1987; *Chem. Abstr.*, 108, 204193, 1988.

533. Eiden, F. and Rehse, U., Aminoalkylierung von Chromonen, *Chem. Ber.*, 107, 1057, 1974.

534. Solov'eva, S. Y., Ramah, S. M., and Ginak, A. I., Study of the reactivity and tautomerism of azolidines. 42. 2-Iminothiazolidin-4-ones in a Mannich reaction with secondary amines, *Khim. Geterotsikl. Soedin.*, 1352, 1983; *Chem. Abstr.*, 100, 103230, 1984.

535. Dauchenko, M. N. and Golobov, Y. G., Esters of dithioisobutyric acid in a Mannich reaction. 3-Dialkylamino-2,2-dimethylpropanethioic O-acid esters, *Zh. Org. Khim.*, 17, 199, 1981; *Chem. Abstr.*, 94, 208321, 1981.

536. Petersen, H. J., 2-p-Chlorophenyloxazolin-5-ones in the Mannich reaction, *Tetrahedron Lett.*, 1557, 1969.

537. Yasuo, H., Synthesis and properties of S-aminomethylthiamine and N-hydroxymethylthiamine, *Chem. Pharm. Bull. Tokyo*, 24, 845, 1976.

538. Nitrokemia Ipartelepek, Belgian Patent 875,501, 1979; *Chem. Abstr.*, 92, 6690, 1980.

539. Prishchenko, A. A., Livantsov, M. V., and Petrosyan, V. S., Alkoxy- and dialkylaminomethylation of methylenebis[dichlorophosphine] and its derivatives, *Zh. Obshch. Khim.*, 60, 1420, 1990; *Chem. Abstr.*, 114, 6635, 1991.

540. Khananashvili, L. M., Kopylov, V. M., Shkol'nik, M. I., Zaitseva, M. G., Sraml, I., and Chvalovsky, V., Study of the reaction of hydroorganosilazanes with a tertiary nitrogen atom with vinylorganocyclosiloxanes, *Zh. Obshch. Khim.*, 50, 1565, 1980; *Chem. Abstr.*, 93, 220820, 1980.

541. Gololobov, Y. G., Malenko, D. M., and Respina, L. A., Synthesis of ylides and phosphazo compounds in the α-aminomethylphosphonate series, *Zh. Obshch. Khim.*, 50, 1206, 1980; *Chem. Abstr.*, 93, 186465, 1980.

542. Manninen, K. and Parhi, S., The Prins reaction of 2-phenylbicyclo[2.2.1]hept-2-ene, *Acta Chem. Scand.*, B35, 45, 1981.

543. Krieger, H. and Manninen, K., Auftreten eines Aldehyds und seiner Folgeprodukte bei der Aminomethylierung eines bicyclischen Alkens, *Tetrahedron Lett.*, 6483, 1966.

544. Cohen, T. and Onopchenko, A., Competing hydride transfer and ene reactions in the aminoalkylation of 1-alkenes with N,N-dimethylmethyleniminium ions. A literature correction, *J. Org. Chem.*, 48, 4531, 1983.

545. Yrjänheikki, E., The reaction of certain pinenes with dimethylamine and formaldehyde, *Acta Univ. Oulu.*, A103, 1980.

546. Krieger, H., Aristila, A. and Koskenniska, L., Aminomethylierung von Camphen, *Rep. Ser. Chem. Univ. Oulu.*, 10, 1983.

547. Södervall, M., The reaction of 2-methylene-norbornane with dimethylamine and formaldehyde, *Acta Univ. Oulu.*, A67, 1978.

548. Krieger, H., Alavuotunki, E., Keränen, H., Oraviita, P., and Peltonen, S., Aminomethylierung von Santen, *Rep. Ser. Chem. Univ. Oulu.*, 20, 1986.

549. Manninen, K. and Haapala, J., The reaction of 2-phenyl-2-norbornane with formaldehyde and dimethylamine, *Acta Chem. Scand.*, B28, 433, 1974; see also p. 603.

550. Aritomi, J. and Nishimura, H., Mannich reaction of dihydropyridine derivatives. II. Reaction with primary amines, *Chem. Pharm. Bull. Tokyo*, 29, 1193, 1981.

551. Berney, D. and Schuh, K., Heterocyclic spironaphthalenones. I. Synthesis and reactions of some spiro[(1H-naphthalenone)-1,3′-piperidines], *Helv. Chim. Acta*, 61, 1262, 1978.

552. Chaaban, I., Greenhill, J. V., and Akhtar, P., Enaminones in the Mannich reaction. II. Further investigations of internal Mannich reactions, *J. Chem. Soc. Perkin Trans. 1*, 1593, 1979.

553. Sohar, P., Lazar, J., and Bernath, G., Isolation and identification of byproducts of the aminomethylation of methylstyrene, *Kem. Kozl.*, 63, 181, 1985; *Chem. Abstr.*, 108, 21683, 1988.

554. Earley, W. G., Jacobsen, E. J., Meier, G. P., Oh, T., and Overman, L. E., Synthesis studies directed towards gelsemine. A new synthesis of highly functionalized cis-hydroisoquinolines, *Tetrahedron Lett.*, 29, 3781, 1988.

555. Shishido, K., Hiroia, K., Fukumoto, K., and Kametani, T., An efficient and highly regioselective intramolecular Mannich-type reaction: a construction of AEF ring system of aconitine-type diterpene alkaloids, *Tetrahedron Lett.*, 27, 1167, 1986.

556. Guyot, B., Pornet, J., and Miginiac, L., Synthèse de N-alkyl-tetrahydro-2,3,6,7-azépines et de N-alkyl-hexahydro-1,2,3,4,7,8-azocynes, *J. Organomet. Chem.*, 386, 19, 1990.

557. Schmidle, C. J. and Mansfield, R. C., U.S. Patent 2,807,613, 1957; *Chem. Abstr.*, 52, 8210, 1958.

558. Kruglikova, R. I. and Kundryutskova, L. A., Preparation and cyclization of enzyme aminoalcohols, *Zh. Org. Khim.*, 9, 2477, 1973; *Chem. Abstr.*, 80, 82534, 1974.

559. Simirskaya, N. I., Nguyen, C. H., Mavrov, M. V., and Serebryakov, E. P., Anomalous aminomethylation of Z-1-hydroxy-3-methyl-2-penten-4-yne, *Izv. Akad. Nauk SSSR Ser. Khim.*, 1198, 1987; *Chem. Abstr.*, 108, 55791, 1988.

560. Kotlyarevskii, I. L., Myasnikova, R. N., and Bardamova, M. I., Aminoacetylene derivatives of benzofuran, *Izv. Akad. Nauk SSSR*, 202, 1971; *Chem. Abstr.*, 75, 5594, 1971.

561. Overman, L. E., Kakimoto, M., Okazaki, M. E., and Meier, G. P., Carbon-carbon bond formation under mild conditions via tandem cationic aza-Cope rearrangement-Mannich reactions. A convenient synthesis of polysubstituted pyrrolidines, *J. Am. Chem. Soc.*, 105, 6622, 1983.

562. Doedens, R. J., Meier, G. P., and Overman, L. E., Transition state geometry of [3.3]-sigmatropic rearrangements of iminium ions, *J. Org. Chem.*, 53, 685, 1988.

563. Natarajan, S. N., Pai, B. R., Rajaraman, R., Swaminathan, C. S., Nagarajan, K., Sudarsanam, V., Rogers, D., and Quick, A., Studies in protoberberine alkaloids. XII. Novel transformation of some 1-(2-bromo-α-methyl-benzyl)-1,2,3,4-tetrahydroisoquinolines to isoquinobenzazepines during Mannich reaction, *Tetrahedron Lett.*, 3573, 1975.

564. Möhrle, H., and Schnädelbach, D., Die Mannich Reaktion mit primären Aminen, Formaldehyd und Isobutyraldehyd als CH-acider Komponente, *Arch. Pharm.*, 308, 352, 1975.

565. Taha, A. M., Omar, N. M., Bayomi, S. M., Ammar, E. M., and Afifi, A., Synthesis and screening of cyclic benzylhydrazine congeners as antidepressants, *J. Pharm. Sci.*, 63, 395, 1974.

566. Cascaval, A., 2-Hydroxyketones. XII. A simple synthesis of 3,3,6,8-tetrasubstituted 4-chromanones, *Synthesis*, 579, 1983.

567. Hikoya, H., Yasuhiro, Y., Seiko, Y., Yuriko, Y., Kenichi, T., and Kazuo, N., N-Sulfomethylation of guanine, adenine and cytosine with formaldehyde bisulfite, *Nucleic Acids Res.,* 10, 6281, 1982; *Chem. Abstr.,* 98, 89805, 1983.

568. O'Brien, G., Patterson, J. M., and Meadow, J. R., Amino acid derivatives of kojic acid, *J. Org. Chem.,* 27, 1711, 1962.

569. Veber, D. F., Milkowski, J. D., Varga, S. L., Denkewalter, R. G., and Hirschmann, R., Acetamidomethyl. A novel thiol protecting group for cysteine, *J. Am. Chem. Soc.,* 94, 5456, 1972.

570. Astik, J. K. and Thaker, K. A., Mannich reaction of optically active (+)-ethyl-2-methylbutylacetoacetate, *J. Inst. Chem. Calcutta,* 47, 68, 1975; *Chem. Abstr.,* 83, 163594, 1975.

571. Hussain, S. A., Sarfaraz, T. B., Sultana, N., and Murtaza, N., Studies on intramolecular Mannich reaction of (S)-2-(α-hydroxyethyl)benzimidazole, *Heterocycles,* 31, 1245, 1990.

572. Bhatt, B. D. and Thaker, K. A., Preparation of optically active aminomethyl-thioethers by the Mannich reaction, *J. Inst. Chem. Calcutta,* 47, 29, 1975; *Chem. Abstr.,* 83, 113591, 1975.

573. Von Thiele, K., Schimassek, U., and von Schlichtegroll, A., Neue herzwirksame β-Aminoketone, *Arzneim. Forsch.,* 16, 1064, 1966.

574. Chaftez, L. and Chen, T. M., Enhancement of optical rotation of levodopa by cyclization, *J. Pharm. Sci.,* 63, 807, 1974.

575. Brush, J. R., Magee, R. J., O'Connor, M. J., Teo, S. B., Geue, R. J., and Snow, M. R., Nature of the copper(II) complexes formed in the reaction of formaldehyde with bis [(S)serinato]copper(II), *J. Am. Chem. Soc.,* 95, 2034, 1973.

576. Angiolini, L., Costa Bizzarri, P., and Tramontini, M., Stereochemistry of Mannich bases. II. Stereospecific synthesis and absolute configuration of diastereomeric 1-phenyl-1,2-dimethyl-3-dimethylaminopropan-1-ols, *Tetrahedron,* 25, 4211, 1969.

577. Andrisano, R., Angeloni, A. S., Gottarelli, G., Marzocchi, S., Samorì B. and Scapini, G., Approach to the conformational analysis of Mannich bases, *J. Org. Chem.,* 41, 2913, 1976.

578. Newman, P., *Optical Resolution Procedures for Chemical Compounds,* Vol. 1: *Amines and Related Compounds,* Optical Resolution Information Center, New York, 1979, 237, 263, 356, and 394.

579. Pohland, A., Peters, L. R., and Sullivan, H. R., Analgesics. Stereoselective synthesis of α (+)- and α(−)-4-dimethylamino-1,2-diphenyl-3-methyl-2-propionoxybutane, *J. Org. Chem.,* 28, 2483, 1963.

580. Angeloni A. S., Gottarelli, G., and Tramontini, M., Stereochemistry of Mannich bases. I. Absolute configuration of some α-methyl-β-aminoketones, *Tetrahedron,* 25, 4147, 1969.

581. Newman, P., *Optical Resolution Procedures for Chemical Compounds.* Vol. 1: *Amines and Related Compounds,* Optical Resolution Information Center, New York, 1979, 356 (Ref. 621) and 385 (Ref. 679).

582. MacConaill, R. J. and Scott, F. L., The first synthesis of optically active Δ^2-isoxazolines, *Tetrahedron Lett.,* 2993, 1970.

583. Albright, J. D. and Snyder, H. R., Reactions of optically active indole Mannich bases, *J. Am. Chem. Soc.,* 81, 2239, 1959.

584. Cannata, V., Samorì, B. and Tramontini, M., Stereochemistry of Mannich bases. VI. Absolute configuration of β-piperidinobutyrophenone and of some 1-phenyl-3-dialkylaminobutanes, *Tetrahedron,* 27, 5247, 1971.

585. Guiles, J. W. and Meyers, A. I., Asymmetric synthesis of benoquinolizidines: a formal synthesis of (−)-emetine, *J. Org. Chem.,* 56, 6873, 1991.

586. Meyers, A. I., Miller, D. B., and White, F. H., Chiral and achiral formamidine in synthesis. The first asymmetric route to (−)-yohimbone and an efficient total synthesis of (±)-yohimbone, *J. Am. Chem. Soc.,* 110, 4778, 1988.

587. Feldman, P. L. and Rapoport, H., Synthesis of optically pure Δ^4-tetrahydroquinolinic acids and hexahydroindolo[2,3-a]quinolines from L-aspartic acid. Racemization on the route to vindoline, *J. Org. Chem.,* 51, 3882, 1986.

588. Harada, K., Okawara, T., and Matsumoto, K., Sterically controlled syntheses of optically active organic compounds. XIX. Asymmetric syntheses of aminoacids by the Strecker reaction, *Bull. Chem. Soc. Jpn.,* 46, 1865, 1973.

589. Kellner, K., Tzschach, A., Nagy-Magos, Z., and Marko, L., Optically active N-phosphinomethylated α-aminoacids: synthesis and application as ligands in asymmetric hydrogenation with Rh complexes, *J. Organomet. Chem.,* 193, 307, 1980.

590. Casy, A. F. and Myers, J. L., Mechanism of a Mannich base exchange reaction, *J. Chem. Soc.,* 4639, 1964.

591. Tramontini, M., Stereoselective synthesis of diastereomeric aminoalcohols from chiral aminocarbonyl compounds by reduction or by addition of organometallic reagents, *Synthesis*, 605, 1982.

592. Haller, R. and Ashauer, U., Konformere 3-Oxa-7-azabicyclo[3.3.1]nonanone und ihre Reduktionsprodukte, *Arch. Pharm.*, 318, 700, 1985.

593. Salimov, M. A., Kosynkina, L. S., Bilalov, S. B., Rustamova, S. N., and Ibragimov, N. Y., Infrared spectra of Mannich bases synthesized from p-methoxyphenol, *Azerb. Khim. Zh.*, 23, 1970; *Chem. Abstr.*, 76, 13305, 1972.

594. Rumynskáya, I. G., Study of intramolecular hydrogen bonding in phenolic Mannich bases, Deposited doc., 1981, VINITI 4213–81, 234; *Chem. Abstr.*, 98, 34145, 1983.

595. Rospenk, M. and Sobczyk, L., Proton NMR studies of proton transfer in ortho-Mannich bases, *Magn. Res. Chem.*, 27, 445, 1989; *Chem. Abstr.*, 111, 193989, 1989.

596. Teimel'baum, A. B., Kudryavtseva, L. A., Bel'skii, V. E., and Ivanov, B. E., Basicity of o-aminomethylphenols in non-aqueous media, *Izv. Akad. Nauk SSSR Ser. Khim.*, 2253, 1980; *Chem. Abstr.*, 94, 30007, 1981.

597. Sucharda-Sobczyk, A. and Sobczyk, L., Tautomerism of 6-nitro derivatives of ortho Mannich bases, *J. Chem. Res. Synop.*, 208, 1985; *Chem. Abstr.*, 103, 177892, 1985.

598. Brycki, B., Maciejewska, H., Brzezinski, B., and Zundel, G., Preparation and NMR characterization of hydrogen bonding in 4-substituted 2- and 2,6-bis(diethylaminomethyl)phenols, *J. Mol. Struct.*, 246, 61, 1991; *Chem. Abstr.*, 115, 135859, 1991.

599. Bel'skii, V. E., Kudryavtseva, L. A., Shishkina, N. A., Zyablikova, T. A., Il'yasov, A. V., and Ivanov, B. E., Study of intramolecular hydrogen bonding in o-(aminomethyl)-phenols, *Izv. Akad. Nauk SSSR Ser. Khim.*, 331, 1977; *Chem. Abstr.*, 87, 4879, 1977.

600. Rospenk, M., The influence of steric effects on proton-transfer equilibrium in intramolecular hydrogen bonds, *J. Mol. Struct.*, 221, 109, 1990; *Chem. Abstr.*, 113, 96913, 1990.

601. Dimmock, J. R., Erciyas, E., Bigam, G. E., Kirkpatrik, D. L., and Duke, M. M., Cytotoxicity of Mannich bases of α-arylidene-β-ketoesters and related compounds against EMT6 mammary carcinoma cells, *Drug Des. Delivery*, 7, 51, 1990; *Chem. Abstr.*, 115, 84809, 1991.

602. Dimmock, J. R., Raghavan, S. K., and Bigam, G. E., Evaluation of Mannich bases of 2-arylidene-1,3-diketones versus murine P388 leukemia, *Eur. J. Med. Chem.*, 23, 111, 1988.

603. Merz, K. W., Müller, E., and Haller, R., Zur Desmotropie bei Piperidondicarbonsäure Estern, *Chem. Ber.*, 98, 2317, 1965.

604. Mistryukov, E. A. and Smirnova, G. N., Conformational analysis of 4-ketopiperidines. The conformational energy of C_2 and C_3 methyl substituent of N-methylpiperidones-4, *Tetrahedron*, 27, 375, 1971.

605. Chiavarelli, S., Gramiccioni, L., Toffler, F., and Valsecchi, G. P., Alcune osservazioni sulla sintesi degli 1,5-dicarbossibispidin-9-oni. Nota I, *Gazz. Chim. Ital.*, 97, 1231, 1967.

606. Caujolle, R., Castera, P., and Lattes, A., Stéréochimie des formes énoliques de pipéridones précurseurs de bispidones, *Bull. Soc. Chim. Fr.*, 52, 1983.

607. Holzgrabe, U., Piening, B., Hesse, K. F., Höltje, H. D., and Worch, M., Stereochemistry of 2,6-dipyridine substituted N-benzyl-4-piperidone mono- and dicarboxylates and of the corresponding reduction products, *Z. Naturforsch. Teil B*, 44B, 565, 1989.

608. Holzgrabe, U., Fridrichsen, W., and Hesse, K. F., Keto-enol tautomerism and configurational isomerism of 2,6-disubstituted 4-piperidone-3,5-dicarboxylates, *Z. Naturforsch. Teil B*, 46B, 1237, 1991.

609. Smith, J. R. L. and Sadd, J. S., Isomerism of 1- and 2-(N,N-disubstituted aminomethyl) benzotriazoles, *J. Chem. Soc. Perkin Trans. 1*, 1181, 1975.

610. Bernhagen, W., Falk, W., Springer, H., Weber, J., Wiebus, E., and Kneip, K., European Patent Appl. EP 46,288, 1982; *Chem. Abstr.*, 96, 180791, 1982.

611. Gabriel, J., Indruch, M., and Mayer, J., Czechoslovakian Patent 185,393 1980; *Chem. Abstr.*, 95, 80989, 1981.

612. Dimmock, J. R. and Taylor, W. G., Synthesis and physical properties of substituted 4-dimethylaminomethyl-1-phenyl-1-nonen-3-ones possessing antitumor properties, *J. Pharm. Sci.*, 63, 69, 1974.

613. Pathak, V. N. and Singh, R. P., Studies in fluorinated Mannich bases. 2. Synthesis and biological activity of some new 3-alkylaminopropiophenones, *Pharmazie*, 35, 434, 1980.

614. Eirin, A., Santana, L., Raviña, E., Fernandez, F., Sanchez-Abarca, E., and Calleja, J., Synthesis of 3-aminomethyl-1-tetralones as potential neuroleptic agents, *Eur. J. Med. Chem.*, 13, 533, 1978.

615. Welch, W. M., Harbert, C. A., Sarges, R., Stratten, W. P., and Weissman, A., Analgesic and tranquilizing activity of 5,8-disubstituted-1-tetralone Mannich bases, *J. Med. Chem.*, 20, 699, 1977.

616. Rudinger-Adler, E. and Büchi, J., Synthese einiger Benzoylphenyl-derivate mit lokalanästetischer Wirkung, *Arzneim. Forsch.*, 29, 1326; see also *Arzheim. Forsch.*, 29, 591, 1979.

617. Möhrle, H. and Dörnbrack, S., Mannich bases. 5. Aminomethylation of phenylcyanoacetic acid, *Pharmazie*, 27, 799, 1972.

618. Böhme, H., Lauer, R., and Matusch, R., Zur Aminomethylierung von 1-Alkoxycarbonylethan- und 1-Cyanoethan-phosphonsäuredialkylestern, *Arch. Pharm.*, 312, 49, 1979.

619. Aritomi, J., Ueda, S., and Nishimura, H., Mannich reaction of 1,4-dihydroquinoline derivatives, *Chem. Pharm. Bull. Tokyo*, 29, 3721, 1981.

620. Avetisyan, A. A., Kagramanyan, A. A., and Melikyan, G. S., Studies of unsaturated lactones. Aminomethylation and cyanoethylation reactions in γ- and δ-lactones, *Arm. Khim. Zh.*, 42, 633, 1989; *Chem. Abstr.*, 113, 6097, 1990.

621. Aritomi, J., Shozo, U., and Nishimura, H., Mannich reaction of dihydropyridine derivatives. I. Reaction with secondary amines, *Chem. Pharm. Bull. Tokyo*, 28, 3163, 1980 and Jpn. Kokai Tokkyo Koho 80 47,656 1980; *Chem. Abstr.*, 94, 15570, 1981.

622. Curulli, A., Giardi, M. T., and Sleiter, G., Electrophilic heteroaromatic substitutions. VI. The reaction of some 3,4,5-tri- and 1,3,4,5-tetrasubstituted 2-methylpyrroles with Mannich reagents, *Gazz. Chim. Ital.*, 113, 115, 1983.

623. Hahn, W. E. and Koziolkiewicz, W., Cycloparaffins condensed with heterocyclic rings. XVII. Aminomethylation of 1,2,3,4,7,8,9,10-octahydrophenantridine, *Lodz. Tow. Nauk. Pr. Wydz. 3*, 15, 71, 1970; *Chem. Abstr.*, 75, 98419, 1971.

624. Shchukina, M. N., Vasil'eva, V. F., and Zagrutdinova, R. A., Substituted 1,2,3,4-tetrahydroquinolines. III. 2-β(dialkylaminoethyl) derivatives, *Khim. Farm. Zh.*, 9, 24, 1975; *Chem. Abstr.*, 82, 170631, 1975.

625. Yamanaka, H., Ogawa, S., and Konno, S., Studies on pyrimidine derivatives. XVIII. Reaction of active methyl groups on pyrimidine N-oxides, *Chem. Pharm. Bull. Tokyo*, 28, 1526, 1980.

626. Muehlstaedt, M. and Schulze, B., Nitrovinylverbindungen. I. Dinitroalkadiene als thermische Abbauprodukte von Mannich Basen dihydrochloriden, *J. Prakt. Chem.*, 313, 205, 1971; see also *J. Prakt. Chem.*, 313, 745, 1971.

627. Lagrenée, M., Synthesis of ketones and β-diketones by variation of Nef-reaction, *C. R. Acad. Sci. C*, 284, 153, 1977.

628. Belayev, V. F. and Grushevich, V. I., Mannich bases based on unsaturated nitro-ketones, *Vestsi Akad. Navuk BSSR*, 137, 1977; *Chem. Abstr.*, 88, 169719, 1978.

629. Messinger, P. and Gompertz, J., Sulfone als chemische transportformen germicid wirkender Stoffe. 8 Mitt. Sulfonylderivate von Chinaldin-, Pyrrol- und Phenol-Mannich Basen, *Arch. Pharm.*, 310, 249, 1977.

630. Zayed, S. M. A. D. and Farghaly, M., Aminomethylierung von (Arylthio)- und (Arylsulfonyl)-acethydroxamsäuren, *Liebigs Ann. Chem.*, 195, 1973.

631. Grier, N., Synthesis and antimicrobial evaluation of quaternary salts of 4-phenyl-1,2,3,6-tetrahydropyridine and 3,6-dimethyl-6-phenyl-tetrahydro2H-1,3-oxazine, *J. Pharm. Sci.*, 68, 407, 1979.

632. Fleischhacker, W. and Koehl, M., Basische Tetrahydrodibenzofuran-derivate mit einem Abstand von drei C-Atomen zwischen dem Stickstoff und dem quartären C-Atom, *Monatsh. Chem.*, 109, 1099, 1978.

633. Möhrle, H. and Herbke, J., Aminomethylierung N-monosubstituierter vinyloger Formamide, *Arch. Pharm.*, 312, 641, 1979.

634. Nasr, M., Nabih, I., and Burckhalter, J. H., Synthesis of pyrimido[5,4-c]quinolines and related quinolines as potential antimalarials, *J. Med. Chem.*, 21, 295, 1978.

635. Eiden, F. and Herdeis, C., Über die Mannichreaktion von 4-Pyron, *Arch. Pharm.*, 309, 764, 1976.

636. Romussi, G. and Ciarallo, G., Mannich bases from methoxychromones, *Farmaco Ed. Sci.*, 32, 635, 1977; *Chem. Abstr.*, 88, 50594, 1978.

637. Ermili, A., Roma, G., Mazzei, M., Ambrosini, A., and Passerini, N., Ricerche chimiche e farmacologiche su derivati del 1H-nafto[2,1-b]pirano, *Farmaco Ed. Sci.*, 29, 237, 1974; *Chem. Abstr.*, 80, 133188, 1974. See also *Farmaco Ed. Sci.*, 29, 247, 1974.

638. Trkovnik, M., Bobarevic, B., Kekic, M., and Krtalic N. Z., Synthese neueartiger Mannich Bases des 4-Hydroxy-cumarins, *Z. Naturforsch. Teil B*, 28B, 373, 1973.

639. Sidky, M. M. Soliman, F. M., and El-Kateb, A. A., Mannich bases of glyoxylanilide 2-oximes and their effect on photosynthetic electron transport, *Z. Naturforsch. Teil B,* 27B, 797, 1972.

640. Stokker, G. E., Deana, A. A., deSolms, S. J., Smith, R. L., Cragoe, E. J., Baer, J. E., Russo, H. F., and Watson, L. S., 2-(Aminomethyl-phenols, a new class of saluretic agents. IV. Effects of oxygen and/or nitrogen substitution, *J. Med. Chem.,* 25, 735, 1982.

641. Ng, G. P. and Dawson, C. R., Synthesis of compounds structurally related to poison ivy vrushiol. 7. 4-, 5-, and 6-(Piperidinomethyl)-3-n-pentadecylcatechols, *J. Org. Chem.,* 43, 3205, 1978.

642. Katritzky, A. R., Lan, X., and Lam, J. N., o(α-Benzotriazolylalkyl)phenols: versatile intermediates for the synthesis of substituted phenols, *Chem. Ber.,* 124, 1809, 1991.

643. Fritsch, P., Über ein neues Verfahren zur Darstellung von p-Alkyloxybenylanilin und dessen Homologen, *Liebigs Ann. Chem.,* 315, 138, 1901.

644. Shoffner, J. P., U.S. Patent 4,157,343, 1979; *Chem. Abstr.,* 91, 107788, 1979.

645. Lehuédé, J., Vierfond, J. M., and Miocque, M., Aminométhylation de l'homoacridane, de l'iminodibenzyle et de l'iminostylbène, *Bull. Soc. Chim. Fr.,* 185, 1981.

646. Vierfond, J. M., Lehuédé, J., and Miocque, M., Aminométhylation de la phénothiazine et de la phénoxazine. Étude chimique et pharmacologique, *Eur. J. Med. Chem.,* 18, 35, 1983.

647. Morrill, T. C., Opitz, R., Replogle, L. L., Katsumoto, K., Schroeder, W., and Hess, B. A., Correspondence between theoretically predicted and experimentally observed sites of electrophilic substitution on a fused tricyclic heteroaromatic (azulene) system, *Tetrahedron Lett.,* 2077, 1975.

648. Fujimura, M., Nakazawa, T., and Murata, I., A novel coupling reaction of 1-azulylmethyl-trimethylammonium iodide: synthesis of 1,2-bis(1-azulyl)ethane and [2.2.2.2](1.3)-azulenophane, *Tetrahedron Lett.,* 825, 1979.

649. Ogura, K., The Mannich bases of troponoid and its application. V. On the Mannich bases of 3-isopropyltropolone (α-thujoplicin), *Bull. Chem. Soc. Jpn,* 34, 839, 1961.

650. Kalennikov, E. A. and Dashevskaya, R. I., Aminomethylation of alkyl derivatives of ferrocene, *Khim. Khim. Tekhnol. Minsk,* 2, 116, 1977; *Chem. Abstr.,* 88, 105525, 1978.

651. Gillet, C., Dehoux, E., Kestens, J., Roba, J., and Lambelin, G., Dérivés d'acides pyrrole-acétiques à activité antiinflammatoire et analgésique, *Eur. J. Med. Chem.,* 11, 173, 1976.

652. Muramatsu, M., Ishizumi, K., and Katsube, Y., Japanese Kokai Tokkyo Koho 78 87,352, 1978; *Chem. Abstr.,* 90, 22807, 1979.

653. Martinez, S. J. and Joule, J. A., Synthesis of 5-hydroxy-2,6-dimethyl-6H-pirido[4, 3-b]-carbazole, *J. Chem. Soc. Chem. Commun.,* 818, 1976.

654. Rosseels, G., Inion, H., Matteazzi, J. R., Peiren, M., Prost, M., Descamps, M., Tornay, C., Colot, M., and Charlier, R., Indolizinyl pyridyl cétones. Synthèse et recherche d'effets antiinflammatoire et antalgique, *Eur. J. Med. Chem.,* 10, 579, 1975.

655. Sato, A., Nozoe, S., Toda, T., Seto, S., and Nozoe, T., Application of the Mannich reaction to 1-azaazulan-2-one. Syntheses of seven-membered analogues of tryptophan and related compounds, *Bull. Chem. Soc. Jpn.,* 46, 3530, 1973.

656. Sun, C. and Bai, D., Synthesis of some compounds related to tanshinquinone, *Yaoxue Xuebao,* 20, 39, 1985; *Chem. Abstr.,* 103, 196286, 1985.

657. Oleinik, A. F., Dozorova, E. N., Solov'eva, N. P., Polukhina, L. M., Filitis, L. N., Poliakova, O. N., and Pershin, G. N., Synthesis and tuberculostatic activity of 3- and 5-substituted 2-aryl-furans, *Khim. Farm. Zh.,* 17, 928, 1983; *Chem. Abstr.,* 99, 194736, 1983.

658. Koei Chemical Co., Ltd., Japanese Kokai Tokkyo Koho JP 60 25989, 1985; *Chem. Abstr.,* 103, 6209, 1985.

659. Singh, J. M., Saxena, G. P., and Tripathi, B. N., Chemistry of azole derivatives. VI. Potential antispasmodic and antihistaminic Mannich bases, *Indian J. Appl. Chem.,* 32, 194, 1969; *Chem. Abstr.,* 75, 20261, 1971.

660. Gall, M., Hester, J. B., Rudzik, A. D., and Lahti, R. A., Synthesis and pharmacology of novel ansiolytic agents derived from 2[(di-alkylamino)methyl-4H-triazol-4-yl]benzophenones and related heterocyclic benzophenones, *J. Med. Chem.,* 19, 1057, 1976.

661. Zinner, G., Moderhack, D., Hantelmann, O., and Bock, W., Hydroxylamine in der Vierkomponenten-kondensation nach Ugi, *Chem. Ber.,* 107, 2947, 1974.

662. Karavai, V. P. and Gaponik, P. N., Aminomethylation of 1-substituted tetrazoles, *Khim. Geterotsikl. Soedin.,* 66, 1991; *Chem. Abstr.,* 115, 71484, 1991.

663. Smirnov, L. D., Stolyarova, L. G., Zhuravlev, V. S., Lezina, V. P., and Dyumaev, K. M., Study of electrophilic substitution of 5-hydroxypicolinic acid, *Izv. Akad. Nauk,* 1658, 1976; *Chem. Abstr.,* 85, 159837, 1976.

664. Zbarskii, A. E., Smirnov, L. D., Zakharov, V. F., Zvolinskii, V. P., and Dyumaev, K. M., Study of aminomethylation of 3-hydroxy-1,5-naphthyridine, *Izv. Akad. Nauk,* 2359, 1979; *Chem. Abstr.,* 92, 76362, 1980.

665. Gashev, S. B., Lezina, V. P., and Smirnov, L. D., Aminomethylation of substituted 5-hydroxy pyrimidines, *Khim. Geterotsikl. Soedin.,* 681, 1980; *Chem. Abstr.,* 93, 168219, 1980.

666. Delia, T. J., Scovill, J. P., and Munslow, W.D., Synthesis of 5-substituted aminomethyluracils via the Mannich reaction, *J. Med. Chem.,* 19, 344, 1976.

667. Fabrissin, S., de Nardo, M., Nisi, C., and Morasca, L., Synthesis and anticancer activity of 5-diethylaminomethyl derivatives and nitrogen mustards of uracil and 2-thiouracils, *J. Med. Chem.,* 19, 639, 1976.

668. Badman, G. T. and Reese, C. B., Reactions between methiodides of nucleoside Mannich bases and carbon nucleophiles, *J. Chem. Soc. Chem. Commun.,* 1732, 1987.

669. Tomisawa, H., Wang, C. H., and Kato, H., 1-Alkyl-2(1H)pyridone derivatives. XXI. Mannich reaction of 2-methyl-1(2H)-thioisoquinolone, *Yakugaku Zasshi,* 94, 124, 1974; *Chem. Abstr.,* 80, 133215, 1974.

670. Kwatra, M. M., Simon, D. Z., Salvador, R. L., and Cooper, P. D., Acetylenics. II. Synthesis and pharmacology of certain N,N-dialkyl-3-phenylpropyn-2-amines, *J. Med. Chem.,* 21, 253, 1978.

671. Yarosh, O. G., Komarov, N. V., and Shergina, N. I., Aminomethylation of ethynylsilanes, *Izv. Akad. Nauk SSSR Ser. Khim.,* 2818, 1969; *Chem. Abstr.,* 72, 79148, 1970.

672. Hoffmann, H., Graefje, H., Koernig, W., and Winderl, S., German Offen. 2,637,425, 1978; *Chem. Abstr.,* 88, 152020, 1978.

673. Karaev, S. F., Dzhafarov, D. S., and Askerov, M. E., Propargyl ether of α-phenylethyl alcohol and its derivatives, *Zh. Org. Khim.,* 16, 928, 1980; *Chem. Abstr.,* 93, 167775, 1980.

674. Bicking, J. B., Robb, C. M., Smith, R. I., and Cragoe, E. J., Jr., 11,12-Secoprostaglandins. I. Acylhydroxyalkanoic acids and related compounds, *J. Med. Chem.,* 20, 35, 1977.

675. Movsumzade, M. M., Guseinov, S. H., Guidoyanez, Q., and Karaev, S. F., Synthesis of methyl di(iso)alkylpropargyloxysilanes with isopropyl, isobutyl and pentyl substituents of silicon, *Dokl. Akad. Nauk Az. SSR,* 39, 41, 1983; *Chem. Abstr.,* 101, 23574, 1984.

676. Frehel, D., Meymes, A., Maffrand, J. P., Eloy, F., Aubert, D., and Rolland, F., Synthèse et propriétés pharmacologiques d'azétidinols-3, *Eur. J. Med. Chem.,* 12, 447, 1977.

677. Resul, B., Ringdahl, B., and Dahlbom, R., Compounds of potential pharmacological value. XXXV. β-Lactame analogues of oxotremorine, *Eur. J. Med. Chem.,* 16, 379, 1981.

678. Muhi-Eldeen, Z., Shubber, A., Musa, N., Muhammed, A., Khayat, A., and Gantous, H., Synthesis and biological evaluation of N-(4-γ-amino-2-butynyloxy)- and N-(4-γ-amino-2-butynyl)-phthalimides, *Eur. J. Med. Chem.,* 15, 85, 1980.

679. Corbel, B., Paugam, J. P., and Sturtz, G., Aminomethylation of N-propargylphosphoramides. Synthesis of unsymmetrical acetylenic diamides, *Can. J. Chem.,* 58, 2183, 1980.

680. Bouet, G., Mornet, R., and Gouin, L., L'addition des organomagnésiens aliphatiques saturés sur les triples liaisons disubstituées non conjuguées. IV. Influence d'un groupement fonctionnel thioéther en α de la triple liaison, *J. Organomet. Chem.,* 135, 151, 1977.

681. Muhi-Eldeen, Z., Al-Jawad, F., Eldin, S., Abdul-Kader, S., Gantous, H., and Garabet, M., Synthesis and biological evaluation of 2(4-tert-amino-2-butynyl)thio-5-aryl-1,3,4-thiadiazoles, *Eur. J. Med. Chem.,* 17, 479, 1982.

682. Andrievskaya, E. K., Kotlyarevskii, I. L., and Fischer, L. B., Mannich reactions with p-diethynylbenzene and its derivatives, *Khim. At-setilena,* 92, 1968; *Chem. Abstr.,* 70, 106076, 1969.

683. Simon, D. Z., Brookman, S., Beliveau, J., and Salvador, R. L., Synthetic acetylenic antifungal agents, *J. Pharm. Sci.,* 66, 431, 1977.

684. Eldeen, Z. M., Cosmo, A. N., Ghantous, H., and Khayat, A., Synthesis and biological evaluation of N,N′-di-(4-t-amino-2-butynyl) pyromellitic acid diimides, *Eur. J. Med. Chem.,* 16, 91, 1981.

685. Azerbaev, I. N., Bosyakov, Y. G., and Dzhailanov, S. D., Aminomethylation of the dipropargyl ester of an α-hydroxyphosphonic acid containing a tetrahydropyran ring, *Zh. Obshch. Khim.,* 45, 2391, 1975; *Chem. Abstr.,* 84, 59669, 1976.

686. Boyer, J. H. and Kooi, J., N-Alkylformimidoyl cyanides and isocyanides, *J. Am. Chem. Soc.*, 98, 1099, 1976.

687. Shen, C. Y., European Patent Appl. EP 102,935, 1984; *Chem. Abstr.*, 100, 209192, 1984.

688. Widder, R., Gousetis, C., Oftring, A., and Birnbach, S., German Offen. DE 3,814,291, 1989; *Chem. Abstr.*, 112, 216251, 1990.

689. Issleib, K. and Hannig, R., N-Alkyl-N-aminoalkyl-α-aminoalkanephosphonic acid esters, *Z. Chem.*, 16, 150, 1976; *Chem. Abstr.*, 85, 21555, 1976.

690. Yasuo, H. and Yoneda, N., Aminomethylation of thiamine disulfides and formation of cyclobismethylenethiamine from N-aminomethylated thiamine disulfide, *Chem. Pharm. Bull. Tokyo*, 24, 1128, 1976.

691. Bundgaard, H., Johansen, M., Stella, V., and Cortese, M., Prodrugs as drug delivery systems. XXI. Preparation, physicochemical properties and bioavailability of a novel water-soluble prodrug type for carbamazepine, *Int. J. Pharm.*, 10, 181, 1982; *Chem. Abstr.*, 97, 11708, 1982.

692. Ribalta, J. M., Artùs, J. J., Salvador, L., Roma, E., Vilageliu, J., Freixes, J., and Bruseghini, L., Synthesis and pharmaceutical evaluation of N-morpholinomethylurea derivatives with platelet antiaggregant activity, *Arzneim. Forsch.*, 31, 1782, 1981.

693. Böhme, H., Ahrens, K. H., and Hotzel, H. H., Eigenschaften und Umsetzungen von N-[α-Hydroxyalkyl]-thiocarbonsäureamiden, *Arch. Pharm.*, 307, 748, 1974.

694. Foks, H. and Manowska, W., Aminomethylation of pyridine- and pyrazine-carbothioamides. V. 6-Chloro- and 6-aminopyrazine-2-carbothioamides in the Mannich reaction, *Acta Pol. Pharm.*, 33, 55, 1976; *Chem. Abstr.*, 86, 72575, 1977.

695. Haake, M., Pothmann, R., Ahrens, K. H., and Fritschi, E., Canadian Patent CA 1,102,803, 1981; *Chem. Abstr.*, 96, 6376, 1982.

696. Wenschuh, E., Guenther, W., and Plewinski, K., Beitrage zum reaktiven Verhalten von Sulfinsäurederivaten. VI. Zur α-Aminomethylierung von Cyclohexylsulfinamiden, *J. Prakt. Chem.*, 319, 297, 1977.

697. Sieger, G. M., Barringer, W. C., and Krueger, J. E., Mannich derivatives of medicinals. II. Derivatives of some carbonic anhydrase inhibitors, *J. Med. Chem.*, 14, 458, 1971.

698. Issleib, K. and Oehme, H., Aminomethylation of phosphoro-, phosphono-, phosphinoamidoates and amidothioates, *Z. Anorg. Allg. Chem.*, 428, 16, 1977.

699. Zinner, G., Krause, T., and Dörschner, K., Ringschlussreaktionen von Semicarbaziden und Carbazinsäureestern mit Aldehyden, *Arch. Pharm.*, 310, 642, 1977.

700. Zayed, S. M. A. D. and Fahmy, A. M., Aminomethylierung und Reduktion der Aryloxyacet- und (Arylamino)acethydroxamsäuren, *Arch. Pharm.*, 303, 933, 1970.

701. Payard, M., Paris, J., Espinasse, M., Fousson, C., and Bastide, J., Dérivés N-aminométhylés de quelques acides hydroxamiques à visée anti-inflammatoire, *Eur. J. Med. Chem.*, 10, 125, 1975.

702. Singh, G. B. and Jetley, A., Studies on the aminoalkylation of some lactams as potential psyco pharmacological agents. I, *Indian Drugs*, 19, 1973.

703. Bolotov, V. V. and Drugovina, V. V., Synthesis and properties of N-aminomethyl derivatives of 2-indolinones, *Farm. Zh. Kiev*, 47, 1978; *Chem. Abstr.*, 89, 108918, 1978.

704. Sidzhakova, D., Danchev, D., Galabov, A. S., Velichkova, E., Karparov, A., and Chakova, N., Structure and properties of cyclic polymethylene ureas. III. Synthesis and biological activity of Mannich bases of tetrahydro-2-(1H)-pyrimidinone, *Arch. Pharm.*, 315, 509, 1982.

705. Harhas, A. H. and Amer, F. A. K., Some reactions on 4-arylhydrazono-3-methyl-2-pyrazolin-5-ones, *Mansoura Sci. Bull.*, 4, 223, 1976; *Chem. Abstr.*, 91, 193224, 1979.

706. Singh, G. B. and Jetly, A., Aminolakylation of some 4,5-disubstituted imidazolone derivatives as potentials antihystaminic and anticonvulsant drugs, *Indian Drugs Ann.*, 51, 1972.

707. Lucka-Sobstel, B., Zeic, A., Borysiewicz, J., and Potec, Z., In vitro antiviral activity of Mannich bases derived from 5-bromoisatin and its β-thiosemicarbazone, *Acta Pharm. Jugosl.*, 4, 95, 1974; *Chem. Abstr.*, 82, 25652, 1975.

708. Movrin, M. and Medic-Saric, M., Biologically active Mannich bases derived from isatin and 5-nitroisatin, *Eur. J. Med. Chem.*, 13, 309, 1978.

709. Joshi, K. C., Pathak, V. N., and Chand, P., Possible psychopharmacological agents. X. Synthesis of some fluorine-containing indole-2,3-dione derivatives, *J. Prakt. Chem.*, 322, 314, 1980.

710. Kutlu, H., N-Substituted derivatives of suc-cinimide, *Istanbul Univ. Eczacilik Fak. Mecm.*, 11, 1, 1975; *Chem. Abstr.*, 83, 193121, 1975.

711. Varma, R. S., Abnormal products in a con-densation reaction, *J. Indian Chem. Soc.*, 51, 647, 1974.

712. Sabirov, K., Alimov, E., and Iskandarov, S. I., Deposited doc., 1976, VINITI 310–76; *Chem. Abstr.*, 88, 74278, 1978.

713. Fickentscher, K. and Günter, E., Synthese von Pyrimidin-4,5-dicarbonsäureamid, *Arch. Pharm.*, 308, 118, 1975.

714. Shalaby, A. F. A., Daboun, H. A., and Aziz, M. A. A., Reactions with 2-thiohydantoin; synthesis of imidazothiazolone derivatives and Mannich bases of 5-arylidene-2-thiohy-dantoin, *Z. Naturforsch. Teil B*, 31B, 111, 1976.

715. Kutlu, H., Condensation products of sac-charin with some amines. V., *Istanbul Univ. Eczacilik Fak. Mecm.*, 5, 11, 1975; *Chem. Abstr.*, 84, 30945, 1976; see also *Chem. Abstr.*, 79, 53220, 1973.

716. Lamberton, J. A. and Nelson, E. R., The re-action of formaldehyde with mixtures of arylhydrazines and amines, *Aust. J. Chem.*, 29, 1853, 1976.

717. Zeltner, P. and Bernauer, K., 1-[(Dimethy-lamino)methyl]pyrrol aus Trimethyl(1-pyr-rolyl)-ammonium Ion, *Helv. Chim. Acta*, 66, 1860, 1983.

718. Steck, E. A. and Brundage, R. P., Benzimi-dazole derivatives, *Org. Prep. Proced. Int.*, 7, 6, 1975; *Chem. Abstr.*, 83, 131520, 1975.

719. Collino, F. and Volpe, S., Basi di Mannich di benzimidazoli, benzotriazoli e altri analoghi composti d'interesse farmacologico, *Boll. Chim. Farm.*, 121, 167, 1982; *Chem. Abstr.*, 97, 162889, 1982.

720. Rudnicka, W. and Sawlewicz, J., Aminome-thylation, cyanoethylation and acylation of 1,2,4-triazole-3-thione, *Acta Pol. Pharm.*, 33, 321, 1976; *Chem. Abstr.*, 87, 53222, 1977.

721. Varma, R. S., Potential biologically active agents. VII. Synthesis of 3-substituted-6-chloro-2-benzoxazolinones, *Pol. J. Phar-macol. Pharm.*, 26, 449, 1974; *Chem. Abstr.*, 82, 57591, 1975.

722. Domagalina, E., Kleinrok, Z., Bien, I., and Kolasa, K., Polish Patent PL 109, 387,1981; *Chem. Abstr.*, 96, 199665, 1982.

723. Zubenko, V. G., Synthesis of azolidine deriv-atives. VIII. Aminomethylation of azolidines with possible hypoglycemic effects, *Farm. Zh. Kiev*, 26, 11, 1971; *Chem. Abstr.*, 76, 126843, 1972.

724. Misra, R. S., Barthwal, J. P., Parmar, S. S., and Brumleve, S. J., Substituted benzoxazol-, benzothiazol-2-thiones. Interrelationship be-tween anticonvulsant activity and inhibition of nicotinamide adenine dinucleotide-de-pendent oxidations and monoamine oxidase, *J. Pharm. Sci.*, 63, 401, 1974.

725. Holbova, E. and Odlerova, Z., Benzothiazole compounds. XVI. Preparation and antimyco-bacterial activity of N,N'-bis[(2-thioxo-3-benzothiazolinyl)methyl]hydrazides, *Chem. Zvesti*, 34, 399, 1980; *Chem. Abstr.*, 94, 30613, 1981.

726. Nath, T. G. S. and Srinivasan, V. R., Propar-gylation and Mannich reactions on 4,5-disub-stituted 1,2,4-triazole-3-ones(thiones), *Indian J. Chem. B*, 15B, 603, 1977.

727. Mazzone, G. and Bonina, F., Sintesi e attività antimicotica di 3-metilamminoderivati di alcuni 3-mercapto-5-aril-1,3,4-ossadiazoli, *Farmaco Ed. Sci.*, 34, 390, 1979; *Chem. Abstr.*, 91, 39396, 1979.

728. Abdel-Megeid, F. M. E., Elkaschef, M. A. F., and Ghattas, A. A. G., Mannich reaction on some 1,3,4-thiadiazole derivatives, *Egypt. J. Chem.*, 20, 235, 1979; *Chem. Abstr.*, 93, 8094, 1980.

729. Strumillo, J., Synthesis of Mannich N-bases with expected pharmacological activity, *Acta Pol. Pharm.*, 32, 287, 1975; *Chem. Abstr.*, 84, 150580, 1976.

730. Werner, W. and Fritzsche, H., Potentielle Cy-tostatika durch Aminomethylierung NH-acider Hypnotika, *Arch. Pharm.*, 302, 188, 1969.

731. Dorgham, C., Richard, B., Richard, M. and Lenzi, M., Synthèse de certains dérivés mono- et dialkylés de l'acide cyanurique par réactions d'aminométhylation, *Bull. Soc. Chim. Fr.*, 414, 1991.

732. Kostyanovskii, R. G., El'natanov, Y. I., and Shikhaliev, S. M., Aminomethylation of phosphines by alkoxymethylamines and dia-minomethane, *Izv. Akad. Nauk*, 1590, 1979; *Chem. Abstr.*, 91, 140919, 1979.

733. Kellner, K. and Tzschach, A., Die Mannich Reaktion als Synthese Konzept in der Phos-phinchemie, *Z. Chem.*, 24, 365, 1984.

734. Broekhof, N. L. J. M., Van Elburg, P., and Van der Gen, A., Synthesis of α-aminosub-stituted diphenylphosphine, *Recueil*, 103, 312, 1984.

735. Mitsubishi Gas Ch. Co., Japanese Kokai Tokkyo Koho JP 82 75,990, 1982; *Chem. Abstr.*, 97, 145013, 1982.

736. Tyka, R., Hagele, G. and Peters, J., (N-Alkyl-N-phosphonomethylamino)methane phosphinic acids, *Phosphorus Sulfur,* 34, 31, 1987; *Chem. Abstr.,* 109, 110520, 1988.

737. Ivanova, Z. H., Kim, T. V., Suvalova, E. A., Boldeskol, I. E., and Gololobov, Y. G., Aminophosphonates. I. 1(Alkylamino)alkylphosphonates, *Zh. Obshch. Khim.,* 46, 236, 1976; *Chem. Abstr.,* 85, 21540, 1976.

738. Szczepaniak, W. and Kuczynski, K., Polish Patent 107,626, 1980; *Chem. Abstr.,* 95, 43349, 1981.

739. Gasanova, M. M., Babakhanov, R. A., Arabov, A. K., and Akhundov, A. A., Synthesis and study of bis(dimethylbenzyl)-aminoethoxyaminomethanes, *Azerb. Khim. Zh.,* 70, 1980; *Chem. Abstr.,* 95, 80336, 1981.

740. Karaev, S. F., Kuliev, R. M., Guseinov, S. O., Askerov, M. E., and Movsumzade, M. M., Synthesis and reactions of dimethylbenzylsilylpropargyl alcohol, *Zh. Obshch. Khim.,* 52, 1160, 1982; *Chem. Abstr.,* 97, 92385, 1982.

741. Iovu, M., Mahara, P., Cristea, A., and Angheliki, M., New cycloalkyl aminomethyl ethers with analgesic activity, *Rev. Chim. Bucharest,* 38, 14, 1987; *Chem. Abstr.,* 107, 133928, 1987.

742. Mustafaev, S. A., Gasanov, V. S., Dzhafarov, I. A., and Alekperov, R. K., Synthesis and study of some reactions of alkylthioalkanols, *Izv. Vyssh. Uchebn. Zaved. Khim. Khim. Tekhnol.,* 33, 100, 1990; *Chem. Abstr.,* 114, 23374, 1991.

743. Lerman, B. M., Umanskaya, L. I., Galin, F. Z., and Tolstikov, G. A., Preparation of 1-thiomethylenamino derivative of 2,4,6,8-tetrathiaadamantane and adamantane, *Khim. Geterotsikl. Soedin.,* 658, 1976; *Chem. Abstr.,* 85, 46536, 1976.

744. Evana, S., European Patent Appl. EP 286,595, 1988; *Chem. Abstr.,* 110, 94520, 1989.

745. Weatherbee, C., Nieft, J. W., and Munie, G. C., Mannich condensation of thiophenols. Reaction of p-halothiophenols with primary aliphatic amines and formaldehyde, *Trans. Ill. State Acad. Sci.,* 70, 91, 1977; *Chem. Abstr.,* 89, 163165, 1978.

746. Kuliev, A. M., Farzaliev, V. M., Allakhverdiev, M. A., and Mamedov, C. I., Synthesis of some S-substituted derivatives of 4,6-di-tert-butyl-2-mercaptophenol, *Zh. Obshch. Khim.,* 52, 2122, 1982; *Chem. Abstr.,* 97, 215664, 1982.

747. Mamedov, F. N., Movsum-Zade, M., Zeinalov, A. M., and Ibad-Zade, A., Synthesis of aminomethyl derivatives of 1-hydroxy-2-thionaphthol, *Azerb. Khim. Zh.,* 66, 1978; *Chem. Abstr.,* 89, 215099, 1978.

748. Orudzheva, I. M., Efendiev, T. E., and Aliev, S. M., Synthesis of some derivatives of 2-mercaptopyridine, *Zh. Org. Khim.,* 17, 410, 1981; *Chem. Abstr.,* 94, 192081, 1981.

749. Lysenko, M. N., Condensation of primary amines with paraformaldehyde and thiocarboxylic acids, *Zh. Org. Khim.,* 10, 2049, 1974; *Chem. Abstr.,* 82, 43246, 1975.

750. Korotkov, A. A., Kutbiddinov, K., Saidaliev, Z. G., and Safaev, A., Deposited doc., 1982, VINITI 3121; *Chem. Abstr.,* 98, 215272, 1983.

751. Neumann, M. G. and De Groote, R. A. M. C., Reaction of sodium hydroxymethanesulfonate with substituted anilines, *J. Pharm. Sci.,* 67, 1283, 1978.

752. Pollak, I. E. and Grillot, G. F., Aminomethylation of selenophenol, *J. Org. Chem.,* 31, 3514, 1966.

753. Angeloni, A. S. and Tramontini, M., Contributo alla conoscenza delle reazioni di decomposizione di alcune basi di Mannich in mezzo acquoso, *Ann. Chim. Rome,* 54, 745, 1964; *Chem. Abstr.,* 62, 1541, 1965.

754. Dimmock, J. R., Patil, S. A., Leek, D. M., Warrington, R. C., and Fang, W. D., Evaluation of acrylophenones and related bis-Mannich bases against murine P 388 leukemia, *Eur. J. Med. Chem.,* 22, 545, 1987.

755. Ram, V. J. and Pandey, H. N., Mannich bases derived from 5-(2-chlorophenyl)-3H-1,3,4-oxadiazol-2-thione. IV, *J. Med. Chem.,* 12, 537, 1977.

756. Monti, S. A. and Castillo, G. D., The reverse Mannich reaction of some 5-hydroxyindoles, *J. Org. Chem.,* 35, 3764, 1970.

757. Sloan, K. B. and Siver, K. G., The aminomethylation of adenine, cytosine and guanine, *Tetrahedron,* 40, 3997, 1984.

758. Mueller, M. and Otto, H. H., Properties and reactions of substituted 1,2-thiazetidine 1,1-dioxides: Synthesis of N-aminoalkylated 1,2-thiazetidine 1,1-dioxides, *Arch. Pharm.,* 322, 515, 1989.

759. Bridson, P. K., Jiricny, J., Kemal, O., and Reese, C. B., Reactions between ribonucleoside derivatives and formaldehyde in ethanol solution, *J. Chem. Soc. Chem. Commun.,* 208, 1980.

760. Mathai, K. P., Mannich reaction on biphenols, *J. Indian Chem. Soc.,* 43, 421, 1966.

761. Bundgaard, H. and Johansen, M., Prodrugs as drug delivery systems. IV. N-Mannich bases as potential prodrugs for amide, ureide and other NH-acidic compounds, *J. Pharm. Sci.,* 69, 44, 1980.

762. Risch, N., Langhals, M., and Hohberg, T., Triple (Grob) fragmentation. Retro-Mannich reactions of 1-aza-adamantane derivatives, *Tetrahedron Lett.,* 32, 4465, 1991.

763. Schubert, W. M. and Motoyama, Y., An example of SN₁ cleavage of a sulfide, *J. Am. Chem. Soc.,* 87, 5507, 1965.

764. Kurilo, G. N., Rostova, N. I., and Grinev, A. N., Reaction of Mannich bases with aryldiazonium salts, *Zh. Org. Khim.,* 14, 2627, 1978; *Chem. Abstr.,* 90, 137616, 1979.

765. Möhrle, H. and Schnädelbach, D., Die Darstellung von 2,2-Dimethyl-3-alkylaminopropionaldehyd-hydrochloriden, *Arch. Pharm.,* 308, 783, 1975.

766. Andrisano, R., Della Casa, C., and Tramontini, M., On the reactivity of Mannich bases. IX. Action of diphenylethylene and dimethylbutadiene on some β-aminoketones, *Ann. Chim. Rome,* 57, 1073, 1967; *Chem. Abstr.,* 68, 87098, 1968.

767. Messinger, P. and Gompertz, J., Sulfone als chemische transportformen germicidwirkender Stoffe. 9 Mitt. Sulfonylderivate von N-Mannich-base, *Arch. Pharm.,* 311, 35, 1978.

768. Ogura, K., The Mannich base of troponoid and its application. VII. The synthesis of 5-formyl-derivative of 3-isopropyltropolone, *Bull. Chem. Soc. Jpn.,* 35, 420, 1962.

769. Hennig, H. and Pesch, W., Zur Synthese von Diaza-bicyclo-nonan-dicarbonsäureestern, *Arch. Pharm.,* 307, 569, 1974.

770. Johansen, M. and Bundgaard, H., Pro-drugs as drug delivery systems. XIII. Kinetics of decomposition of N-Mannich bases of salicylamide and assessment of their suitability as possible pro-drugs for amines, *Int. J. Pharm.,* 7, 119, 1980; *Chem. Abstr.,* 94, 162641, 1981.

771. Bundgaard, H. and Johansen, M., Hydrolysis of N-Mannich bases and its consequences for the biological testing of such agents, *Int. J. Pharm.,* 9, 7, 1981; *Chem. Abstr.,* 95, 138592, 1981.

772. Weitzel, G., Schneider, F., Seynsce, K., and Finger, H., Further tumor inhibiting compounds. II. Cytostatic effects of N-C-N compounds, *Z. Physiol. Chem.,* 336, 107, 1964; *Chem. Abstr.,* 61, 4864, 1964.

773. Werner, W., Jungstand, W., Gutsche, W., and Wohlrabe, K., Struktur-wirkungs- beziehungen bei Mannich Basen mit und ohne Stikstofflostgruppen und einigen von β-Aminoketonen abgeleiteten Reduktions Produkten auf grund eines Cancerrostatica-3-sulfinrests mit Transplantationstumoren, *Pharmazie,* 32, 341, 1977.

774. See also survey in Ref. 19.

775. Yamada, M., Ohara, M., Sakuramoto, Y., and Watanabe, T., Vulcanization accelerators for neoprene rubber, *Nippon Gomu Kyokaishi,* 42, 202, 1976; *Chem. Abstr.,* 85, 47895, 1976.

776. Pavlenko, N. I., Marshtupa, V. P., and Baranov, S. N., Introduction of aminomethyl groups into heterocyclic CH-acid molecules, *Dopov. Akad. Nauk Ukr. RSR Ser. B,* 64, 1980; *Chem. Abstr.,* 94, 30627, 1981.

777. Farberov, M. I. and Mironov, G. S., Commercial synthesis of carbonyl monomers by the Mannich reaction, *Dokl. Akad. Nauk SSSR,* 148, 1095, 1963; *Chem. Abstr.,* 59, 5062, 1963.

778. Goldmann, S., Stereoselektive Synthese von Alkoholen. V. Einfache regioselective Umwandlung von Alkyl-ketonen in Allylalkohole, *Synthesis,* 640, 1980.

779. Grass, J. L., A direct synthesis of α-methylene ketones, *Tetrahedron Lett.,* 2111, 1978 and A facile entry to vinylketones, *Tetrahedron Lett.,* 2955, 1978.

780. de Solms, S. J., N,N,N,N-tetramethylmethane diamine. A simple effective Mannich reagent, *J. Org. Chem.,* 41, 2650, 1976.

781. Akhrem, A. A., Kamernitskii, A. V., Reshetova, I. G., and Chenyuk, K. Y., Transformed steroids. LVII. Use of the Mannich reaction for increasing the length of a side chain of steroids, *Izv. Akad. Nauk SSSR Ser. Khim.,* 1633, 1973; *Chem. Abstr.,* 79, 115768, 1973.

782. Koechel, D. A. and Rankin, G. O., Diuretic activity of Mannich base derivatives of ethacrynic acid and certain ethacrynic acids analogues, *J. Med. Chem.,* 21, 764, 1978.

783. Faure, R. and Mattioda, G., Synthèse et stabilité des vinylcétones en position 2 du benzofuranne, *Bull. Soc. Chim. Fr.,* 3059, 1973.

784. Ward, F. E., Garling, D. L., and Buckler, R. T., Antimicrobial 3-methyleneflavanones, *J. Med. Chem.,* 24, 1073, 1981.

785. Moriconi, E. J. and Stemniski, M. A., Effect of α-methyl substitution in the Beckmann and Schmidt rearrangement of 1-hydrindanones, *J. Org. Chem.,* 37, 2035, 1972.

786. Bensel, N., Marschall, H., and Weyerstahl, P., Darstellung von α-Methylen-γ-lactonen mit 9-Oxabicyclo[6.3.0]undecan-skelett, *Chem. Ber.,* 108, 2697, 1975.

787. Otsuka Pharm. Factory, Inc., Japanese Kokai Tokkyo Koho JP 81 128,776, 1981; *Chem. Abstr.,* 96, 85405, 1982.

788. Longo, A. and Lombardi, P., European Patent Appl. EP 307,134, 1989; *Chem. Abstr.,* 111, 134643, 1989.

789. Schulze, K., Pfüller, U., Tichek, H. B., and Mühlstädt, M., Eine Umlagerung beim alkalischen Abbau von 1-Dialkylaminoalkandiin-(2,4)-methojodiden, *Tetrahedron Lett.,* 2999, 1968.

790. Scott, F. L., MacConaill, R. J., and Riordan, J. C., Beckmann rearrangements of Mannich base and related oximes, *J. Chem. Soc. C,* 44, 1967.

791. Kholodov, L. E. and Yashunskii, V. G., Quinidines. III. Amino- and hydroxymethylation of β-quinidans at the third position, *Khim. Geterotsikl. Soedin.,* 1530, 1970; *Chem. Abstr.,* 74, 53480, 1971.

792. Jaszberenyi, J. C., Petrikovics, I., Gunda, E. T., and Hosztafi, S., The Mannich reaction of cephalosporin sulfoxides and sulfones, *Acta Chim. Acad. Sci. Hung.,* 110, 81, 1982; *Chem. Abstr.,* 98, 16465, 1983.

793. Roth, H. J. and Haupt, M., Dienartige Synthesen von Dihydropyronen durch Acetolyse von Mannich-basen, *Arch. Pharm.,* 308, 241, 1975.

794. Polazzi, J. O., Mannich reaction product of dihydrocodeinone, *J. Org. Chem.,* 46, 4262, 1981.

795. Johansen, O. H., Ottersen, T., and Undheim, K., 1H-2-Benzothiapyran-4(3H)-one in aldol and Mannich reactions, *Acta Chem. Scand.,* B33, 669, 1979.

796. Eiden, F. and Felbermeir, G., 2-Aminomethyl-3-hydroxy-1-thiochromone; Darstellung und Reaktionen, *Arch. Pharm.,* 316, 1034, 1983.

797. Roth, H. J., Schvenke, C., and Dvorak, G., Acetolyse von Mannich Basen. VI. Acetolyse des 2-Piperidinomethyl-cyclopentanons, *Arch. Pharm.,* 298, 326, 1965.

798. Böhme, H. and Clement, B., Mannich-kondensationen mit β-Ketosulfoxiden, *Arch. Pharm.,* 312, 527, 1979; see also *Arch. Pharm.,* 312, 531, 1979.

799. Odinokov, V. N., Luneva, S. E., and Gershanov, F. B., 2,6-Xylenol from 2,6-bis(N,N-dimethylaminomethyl)cyclohexanone, *Zh. Org. Khim.,* 8, 2162, 1972; *Chem. Abstr.,* 78, 42951, 1973.

800. Mühlstädt, M. and Zschunke, A., Zur Aminomethylierung der Heterocycloalkanone-(3), *Chem. Ber.,* 101, 1052, 1968.

801. Viola, A., Collins, J. J., and Filipp, N., Intramolecular pericyclic reactions of acetylenic compounds, *Tetrahedron,* 37, 3765, 1981.

802. Cársky, P., Zuman, P., and Horák V., Fission of activated carbon-nitrogen and carbon-sulfur bonds. V. Polarografic study of elimination of β-morpholino propiophenone, *Collect. Czech. Chem. Commun.,* 29, 3044, 1964.

803. Mollica J. A., Smith, J. B., Nunes, I. M., and Govan, H. G., Kinetics of the decomposition of a Mannich base, *J. Pharm. Sci.,* 59, 1770, 1970.

804. Koshy, K. T. and Michner, H., Kinetic study on hydrolysis of a Mannich base compound, *J. Pharm. Sci.,* 53, 1381, 1964.

805. Natova, L. and Khristova, K., Kinetic study of the decomposition of Mannich bases, *God. Vissh. Khimikotekhnol. Inst. Sofia,* 24, 257, 1978; *Chem. Abstr.,* 96, 19376, 1982. See also *God. Vissh. Khimikotekhnol. Inst. Sofia,* 24, 265, 1978.

806. Ivanov, B. E. and Zheltukhin, V. F., Reaction of Mannich bases with triethylphosphite. V. Mechanism of the triethyl phosphite-ketonic Mannich base reaction, *Izv. Akad. Nauk, SSSR Ser. Khim,* 4022, 1969; *Chem. Abstr.,* 71, 48941, 1969.

807. Andrisano, R., Della Casa, C., and Tramontini, M., Reactivity of Mannich bases. XIII. Mechanism of the reaction between aminomethylnaphthols and benzenethiols, *J. Chem. Soc. C,* 1866, 1970.

808. Higashi, K., Kitamura, T., Fukusaki, Y., and Imoto, E., Reaction of phenolic Mannich bases with phenols, *Kôgyô Kagaku Zasshi,* 61, 1035, 1958; *Chem Abstr.,* 55, 22210, 1961.

809. Angeloni, A. S., Angiolini, L., De Maria, P., and Fini, A., Reactivity of Mannich bases. XI. Mechanism of the reaction of thiophenols with nuclear substituted β-morpholinopropiophenones and with some related quaternary iodides, *J. Chem. Soc. C,* 2295, 1968.

810. Horák V., Michl, J., and Zuman, P., A contribution to the studies of elimination reactions of Mannich bases using polarographic methods, *Tetrahedron Lett.,* 744, 1961.

811. Baciocchi, E. and Schiroli, A., Sidechain reactivity of indole derivatives. The reaction of 3-indolylmethyltrimethylammonium methylsulphate with sodium toluene-p-thiolate, *J. Chem. Soc. B,* 401, 1968.

812. Bogardus, J. B. and Higuchi, T., Kinetics and mechanism of hydrolysis of labile quaternary ammonium derivatives of tertiary amines, *J. Pharm. Sci.,* 71, 729, 1982.

813. Craig, J. C. and Ekwuribe, N. N., Synthesis of α,β-unsaturated aldehydes via 1-aminopropan-1,2-dienes. Mechanistic studies, *Tetrahedron Lett.,* 21, 2587, 1980.

814. Kornet, M. G. and Crider, A. M., Potential long-acting anticonvulsants. II. Synthesis and activity of succinimides containing an alkylating group on N or at the 3 position, *J. Med. Chem.,* 20, 1210, 1977.

815. Nerdel, F., Frank, D., and Lengert, H. J., Quartäre Salze von β-Amino-aldehyden und β-Iod-aldehyde, *Chem. Ber.,* 98, 728, 1965.

816. Miller, R. B. and Smith, B. F., Regiospecific synthesis of α-methylene ketones, *Tetrahedron. Lett.,* 5037, 1973.

817. Mikhailov, M. K., Dirlikov, S. K., and Georgieva, T. P., Synthesis of deuterated methyl-methacrylate CD$_2$=C(Me)COOD$_3$, *Vysokomol. Soedin. Ser. B,* 18, 398, 1976; *Chem. Abstr.,* 85, 94704, 1976.

818. Hahne, F. and Zymalkowski, F., Die Synthese partiell hydrierter Benzo[f]isochinoline als potentielle Analgetika, *Arch. Pharm.,* 312, 472, 1979.

819. Boggs, N. T., Chemical modification of peptides containing γ-carboxyglutamic acid, *J. Org. Chem.,* 47, 1812, 1982.

820. Behare, E. S. and Miller, R. B., A new synthesis of α-methylene-γ-butyrolactones, *J. Chem. Soc. Chem. Commun.,* 402, 1970.

821. Austin, E. M., Brown, H. L., and Buchanan, G. L., The thermal Michael reaction. III. The scope of the reaction, *Tetrahedron,* 25, 5509, 1969.

822. Balasubramanian, K., John, J. P., and Swaminatan, S., Robinson-Mannich annellation of cyclohexane-1,3-dione, *Synthesis,* 51, 1974.

823. Hellmann, H. and Pohlmann, J. L. W., C-Alkylierung.mit α-Dimethylaminomethyl-β-naphthol. II. Kondensationen mit cyclischen 1,3-Diketonen, *Liebigs Ann. Chem.,* 642, 35, 1961; see also *Liebigs Ann. Chem.,* 643, 43, 1961.

824. Eiden, F. and Herdeis, C., Chromon-derivate aus Mannich-basen, *Arch. Pharm.,* 310, 573, 1977.

825. Watanabe, T., Katayama, S., Nakashita, Y., and Yamauchi, M., Regiospecific (biogenetic-type) synthesis of 2-methyl-5H-pyrano[3,2-c][1]benzopyran-4-one, the basic skeleton in citromycetin, *J. Chem. Soc. Chem. Commun.,* 761, 1981.

826. Jacobs, R. T., Wright, A. D., and Smith, F. X., Condensation of monosubstituted isopropylidene malonates with Mannich bases, *J. Org. Chem.,* 47, 3769, 1982.

827. Troxler, F., Präparative Verwendung von Mannich Bases von Hydroxy-indolen als Alkylierungsmittel, *Helv. Chim. Acta,* 51, 1214, 1968.

828. Schmidt, A. and Brunetti, H., p-Hydroxybenzylierung von Carbanionen mit chinonmethideliefernden Verbindungen, *Helv. Chim. Acta,* 59, 522, 1976.

829. Ravoux, J. P. and Decombe, J., Contribution à la chimie du ferrocène. I. Réaction de l'iodométhylate de N-diméthylaminométhylferrocène avec des composés à hydrogène mobile en présence de DMF, *Bull. Soc. Chim. Fr.,* 146, 1969.

830. Okuda, T., Alkylation with Mannich bases. VII. Transaminomethylation of N-Mannich bases of theophylline and benzimidazole with indole, *Takugaku Zasshi,* 80, 208, 1960; *Chem. Abstr.,* 54, 13141, 1960.

831. Kamal, V., Qureshi, A. A., and Ahmad, I., Some alkylation reactions of Mannich bases in aqueous medium. Synthesis of some mono- and polyindolyl compounds, *Tetrahedron,* 19, 681, 1963.

832. Ponomarev, G. V., Porphyrine. XIII. Reaction of meso-(dimethylaminomethyl)-etioporphyrine with nucleophiles in the presence of zinc acetate, *Khim. Geterotsikl. Soedin.,* 943, 1980; *Chem. Abstr.,* 94, 65651, 1981.

833. Kline, R. H., U.S. Patent 3,867,467, 1975; *Chem. Abstr.,* 82, 172045, 1975.

834. Messinger, P. and Kusuma, K., Alkylierung der aktiven Methylen-gruppe von Sulfonen mit Mannich Basen, *Synthesis,* 565, 1980.

835. Stephen, J. F., U.S. Patent 3,891,673, 1975; *Chem. Abstr.,* 83, 179139, 1975.

836. Balasubramanian, K. K. and Selvaraj, S., Studies on o-quinonemethides. IV. Alkylation of phenolic Mannich bases with aryl phenacyl sulfones, *Synthesis,* 138, 1980.

837. Mikhailov, V. I., Sholle, V. D., Kagan, E. S., and Rozantsev, E. G., Reaction of Mannich bases with 2,2,6,6-tetramethylpiperidone enamines, *Izv. Akad. Nauk,* 1639, 1976; *Chem. Abstr.,* 85, 159836, 1976.

838. Von Strandtmann, M., Cohen, M. P., and Shavel, J., Carbon-carbon alkylation of enamines with Mannich bases. II. New synthesis of pyran-containing fused ring systems, *J. Org. Chem.,* 30, 3240, 1965.

839. Suzuki, K. and Sekiya, M., A novel method for the α-phenylthiomethylation of carbonyl compounds, *Synthesis,* 297, 1981.

840. Boschetti, E., Molho, D., Chabert, J., Grand, M., and Fontaine, L., Nouvelles coumarines à activité anticoagulante, *Chim. Ther.,* 7, 20, 1972.

841. Bravo, P. and Ticozzi, C., Alkylation of 4-hydroxycoumarins by ketone Mannich bases, *Synthesis,* 894, 1985.

842. Andreani, F., Andrisano, R., Della Casa, C., and Tramontini, M., Reactivity of Mannich bases. XII. Ring alkylation of aromatic and heterocyclic compounds, *J. Chem. Soc. C,* 1157, 1970.

843. Suzuki, K. and Sekiya, M., Novel alkyl- and phenylthiomethylation of aromatic compounds, *Chem. Lett.,* 1241, 1979.

844. Kamal, A. and Asadullah, Alkylation reaction of the Mannich bases. II. 1-(N-Dimethylamino)-3-keto-5-phenylpent-Δ^4-ene and 1-(N,N-dimethylamino)-3-keto-5-phenylpentane, in aqueous medium, *Pak. J. Sci. Ind. Res.,* 9, 316, 1966; *Chem. Abstr.,* 68, 39401, 1968. See also *Pak. J. Sci. Ind. Res.,* 9, 340, 1966; *Chem. Abstr.,* 68, 39405, 1968.

845. Decodts, G., Wakselman, M., and Vilkas, M., Alkoylation de composés indoliques par les hydroxyméthyl- aminométhyl- et halométhyl-phénols, *Tetrahedron,* 26, 3313, 1970.

846. Roth, B., Strelitz, J. Z., and Rauckman, B. S., 2,4-Diamino-5-benzylpyrimidines and analogues antibacterial agents. II. C-Alkylation of pyrimidines with Mannich bases and application to the synthesis of trimetoprim and analogues, *J. Med. Chem.,* 23, 379, 1980; see also *J. Med. Chem.,* 23, 384 and 535, 1980.

847. Eiden, F. and Wendt, R., Pyrano[3,2-c]-chinoline, *Arch. Pharm.,* 309, 70, 1976.

848. Suvorov, N. N., Velezheva, V. S., Vampilova, V. V., and Gordeev, E. N., Indole derivatives. XII. Alkylation of barbituric acids by Mannich bases, *Khim. Geterotsikl. Soedin.,* 515, 1974; *Chem. Abstr.,* 81, 37532, 1974.

849. McEvoy, F. J. and Allen, G. R., A general synthesis of 3-(substituted benzoyl)-3-substituted alkanoic acids, *J. Org. Chem.,* 38, 4044, 1973.

850. Julia, M. and Lallemand, J. Y., Réactions électrophiles en série indolique, *Bull. Soc. Chim. Fr.,* 2046, 1973.

851. Shoemaker, G. L. and Keown, R. W., Mannich bases of nitroparaffins. Steric effect of the aminogroups on carbon alkylation, *J. Am. Chem. Soc.,* 76, 6374, 1954.

852. Gardner, P. D. and Rafsanjani, H. S., Reaction of phenolic Mannich bases methiodides and oxides with various nucleophiles, *J. Am. Chem. Soc.,* 81, 3364, 1959.

853. Pennie, J. T. and Bieber, T. I., Migration of the ferrocenyl-CH$_2$ group from N- to aromatic C, *Tetrahedron Lett.,* 3535, 1972.

854. Pollak, I. E. and Grillot, G. F., On the mechanism of formation of p-aminobenzyl-aryl-sulfides, selenides or sulfones by the acid-catalyzed condensation of aromatic amines with formaldehyde and arenethiols, selenols or sulfinic acids, *J. Org. Chem.,* 32, 3101, 1967.

855. Stetter, H., Schmitz, P. H., and Schreckenberg, M., Über die katalysierte Reaktion von Aldehyden mit Mannich-basen, *Chem. Ber.,* 110, 1971, 1977.

856. Winberg, H., Vries, T., Pouwer, K., Havinga, E. E., and Meijer, E. W., Conducting polymers via a novel polymerization reaction, in *New Aspects Org. Chem. I, Proc. 4th Int. Kyoto Conf.,* 1989, 343; *Chem. Abstr.,* 114, 207839, 1991.

857. Von Strandtman, M., Cohen, M. P., Puchalski, C., and Shavel J., Jr., Reaction of phosphoranes with Mannich bases. Synthesis of α-substituted-β-arylacrylic acids via the Wittig reaction, *J. Org. Chem.,* 33, 4306, 1968.

858. Hellmann, H. and Pohlmann, J. L. W., C-Alkylierung mit α-Dimethylaminomethyl-β-naphthol. I. Kondensationen mit 1,3-Diketonen, *Liebigs Ann. Chem.,* 642, 28, 1961.

859. Asherson, J. L. and Young, D. W., A general and practicable synthesis of polycyclic heteroaromatic compounds. 1. Use of putative quinolone-quinone-methide in the synthesis of polycyclic heteroaromatic compounds, *J. Chem. Soc. Perkin Trans. 1,* 512, 1980; see also *J. Chem. Soc. Perkin Trans. 1,* 522, 1980.

860. Deutsch Gold- und Silber-Scheideaustalt vorm. Roessler, French Patent 1,477,040, 1967; *Chem. Abstr.,* 68, 12708, 1968.

861. Craig, J. C., Johns, S. R., and Moyle, M., Amine exchange reactions. Mannich bases from primary aliphatic amines and from amino acids, *J. Org. Chem.,* 28, 2779, 1963.

862. Gogte, V. N., Mukhedkar, V. A., and Tilak, B. D., Synthesis of heterocyclic compounds. XXI. Dihydro- and tetrahydro-quinolines and their methiodides, *Indian J. Chem.,* 15B, 774, 1977.

863. Boev, V. I., Ferrocenylmethylation and some reactions of the reaction products, *Zh. Obshch. Khim.,* 48, 1594, 1978; *Chem. Abstr.,* 89, 180137, 1978.

864. Glushkova, L. V., Belova, S. Y., Kiro, Z. B., Kutimova, G. V., Efimov, A. A., and Tyutereva, A. F., Synthesis of 1-(3',5',-di-tert-butyl-4'-hydroxybenzyl)-4-alkyl(aryl)thiosemicarbazides and their antioxidant activity, *Zh. Prikl. Khim. Leningrad,* 51, 1852, 1978; *Chem. Abstr.,* 89, 197095, 1978.

865. Cignarella, G., Occelli, E., Cristiani, G., Paduamo, L., and Testa, E., Bicyclic homologs of piperazine, VI. Synthesis and analgesic activity of 3-substituted-8-propionyl-3,8-diazabicyclo[3.2.1]octanes, *J. Med. Chem.*, 6, 764, 1963.

866. Hadlington, M., Rockett, B. W., and Nelhans, A. Unsymmetrically disubstituted ferrocenes. I. Synthesis of 1,2-disubstituted ferrocenes by metallation and nucleophilic substitution reactions, *J. Chem. Soc. C,* 1436, 1967.

867. Hansen, J. F., Szymborski, P. A., and Vidusek, D. A., Formation of N,N-dialkylhydroxylamines in the oximation of some Mannich bases, *J. Org. Chem.*, 44, 661, 1979.

868. Stephen, J. F., U.S. Patent 3,941,746, 1976; *Chem. Abstr.*, 84, 165668, 1976.

869. Böhme, H. and Fuchs, G., Über Darstellung und Umsetzung von Formamidomethylaminen, -sulfiden end -sulfonen, *Chem. Ber.*, 103, 2775, 1970.

870. Rane, D. F., Fishman, A. G., and Pike, R. E., A new application of Mannich bases, *Synthesis,* 694, 1984.

871. Strekowski, L., Alkylation of 2,4-dioxopyridines with Mannich bases, *Bull. Akad. Pol. Sci.*, 21, 257, 1973; *Chem. Abstr.*, 79, 53252, 1973.

872. Aratani, T., Gonda, T., and Nozaki, H., Asymmetric lithiation of ferrocenes, *Tetrahedron*, 26, 5453, 1970.

873. Craig, J. C., Moyle, M., and Johnson, L. F., Amine exchange reactions. Mannich bases from aromatic amines, *J. Org. Chem.*, 29, 410, 1964.

874. Mistryukov, E. A., Aronova, N. I., and Kucherov, V. F., A new method of synthesis of N-alkyl-4-piperidones, *Izv. Akad. Nauk SSSR, Otd. Khim. Nauk,* 932, 1961; *Chem. Abstr.*, 55, 27310, 1961.

875. Hassan, M. M. A. and Casy, A. F., PMR studies of the reaction between methiodides of substituted 4-piperidones and primary bases, *Tetrahedron*, 26, 4517, 1970.

876. Gottarelli, G., Researches on the reactivity of 9-bispidinones, *Tetrahedron Lett.*, 2813, 1965.

877. Scott, F. L., Houlihan, S. A., and Fenton, D. F., Mechanism of pyrazoline formation from the reactions of substituted hydrazines and Mannich bases, *J. Chem. Soc. C,* 80, 1971.

878. Tzschach, A. and Kellner, K., Arsenorgano-Verbindungen. XIX. Umsetzungen von Mannich-basen mit sekundären Phosphinen und Arsinen, *J. Prakt. Chem.*, 314, 315, 1972.

879. Marr, G. and White, T. M., Organometallic derivatives. VI. Synthesis and reactivity of some (ferrocenylmethyl)-phosphines, *J. Chem. Soc. Perkin Trans. 1,* 1955, 1973.

880. Petrov, K. A., Chauzov, V. A., and Pokatun, V. P., γ-Oxophosphine oxides and diphosphine oxides, *Zh. Obshch. Khim.*, 53, 541, 1983; *Chem. Abstr.*, 99, 53858, 1983.

881. Ivanov, B. E., Zheltukhin, V. F., and Valitova, L. A., Mechanism of the reaction of triethylphosphite with Mannich bases, in *Khim. Primen. Fosfororg. Soedin., Tr. 4th Konf.,* 1972, 204; *Chem. Abstr.*, 78, 135260, 1973.

882. Gross, H., Seibt, H., and Keitel, I., Mono- und Bisphosphono-derivate des 2,6-Di-tert-butyl-p-kresols, *J. Prakt. Chem.*, 317, 890, 1975.

883. Ivanov, B. E., Krokhina, S. S., Ryzhkina, I. S., Gaidai, V. I., and Smirnov, V. N., Reaction of diethyl thiophosphite with phenolic Mannich bases, *Izv. Akad. Nauk,* 615, 1979; *Chem. Abstr.*, 91, 20605, 1979.

884. Hellmann, H. and Schumacher, O., Quartäre Phosphoniumsalzen aus tertiären Phosphinen und quartären Ammoniumsalzen, *Liebigs Ann. Chem.*, 640, 79, 1961.

885. Watanabe, S. and Ueda, T., Introduction of substituents to the 7(8)-position of 7-deazaadenosine (tubercidin): conversion to toyocamycin, *Nucleic Acids Symp. Ser.*, 8, 21, 1980; *Chem. Abstr.*, 94, 175411, 1981.

886. Marr, G., Moore, R. E., and Rockett, B. W., Unsymmetrically disubstituted ferrocenes. IV. The synthesis and reactivity of 2(N,N-dimethylaminomethyl)ferroceneboronic acid, *J. Chem. Soc. C,* 24, 1968.

887. Roth, H. J. and Brandes, R., Acetolysis of Mannich bases. VII. Synthesis and properties of certain Mannich bases of theophylline, 8-bromotheophylline and theobromine, *Arch. Pharm.*, 298, 765, 1965.

888. Boschetti, E., Molho, D., Aknin, J., Fontaine, L., and Grand, M., Analeptic activity of some Mannich bases of 7-hydroxycoumarins and determination of their structure, *Chim. Ther.,* 403, 1966.

889. Russkikh, V. V. and Fokin, E. P., α-Aminoalkylation of quinone derivatives. I. Mannich reaction with hydroxyanthraquinones, *Z Org. Khim.*, 7, 371, 1971; *Chem. Abstr.*, 74, 125256, 1971.

890. Böhme, H. and Köhler, E., Carbonsäureester von α-Dialkylamino-alkanolen, *Angew. Chem.*, 72, 523, 1960.

891. Böhme, H. and Hartke, K., Über die Spaltung von Aminalen und α-Dialkylamino-äthern mit Carbonsäurehalogeniden, *Chem. Ber.*, 93, 1305, 1960.

892. Strating, J. and Van Leusen, A. M., Chemisry of α-diazo-sulfones, *Recueil,* 81, 966, 1962.

893. Rosowsky, A., Papathanasopoulos, N., Lazarus, H., Foley, G. E., and Modest, E. J., Cysteine scavengers. II. Synthetic α-methylenebutyrolactones as potential tumor inhibitors, *J. Med. Chem.,* 17, 672, 1974.

894. Andrisano, R., Angeloni, A. S., and Tramontini, M., Ricerche sulla reattività delle basi di Mannich. Nota VI. Azione di alacuni reattivi tiolici su fenil- e stiril-cheto-basi, *Ann. Chim. Rome,* 55, 1093, 1965; *Chem. Abstr.,* 64, 11117, 1966.

895. Abdullaeva, F. A. and Mamedov, F. A., Synthesis of new organosulfur derivatives of shielded phenols, *Tezisy Dokl. Nauchn.,* 140, 1974; *Chem. Abstr.,* 86, 5067, 1977.

896. Gadaginamath, G. S. and Siddappa, S., Synthesis of 3-benzoyl-5-hydroxy-2-phenyl-4-isogramines and 4-arylthiomethyl-5-hydroxy-2-phenylindoles, *J. Indian Chem. Soc.,* 53, 17, 1976.

897. Reese, C. B. and Sanghvi, Y. S., Conversion of 2'-deoxyuridine into thymidine and related studies, *J. Chem. Soc. Chem. Commun.,* 877, 1983.

898. Kuliev, A. M., Aliev, S. R., Mamedov, F. M., and Movsum-Zade, M., Synthesis of aminomethyl derivatives of 2-hydroxy-5-tert-alkylthiophenols and their cleavage by thiols, *Zh. Org. Khim.,* 12, 426, 1976; *Chem. Abstr.,* 84, 150277, 1976.

899. Kreutzkamp. N. and Deicke, G. Modellversuche über die Einwirkung von S-Dialkylaminomethyl-dithiourethanen auf CH-acide Verbindungen, *Arch. Pharm.,* 306, 321, 1973.

900. Matolcsy, G. and Bordás, B., Die Reaktion von Keton-Mannich-basen mit Dithiocarbamaten und mit Schwefelkohlenstoff, *Chem. Ber.,* 106, 1483, 1973.

901. Messinger, P. and Greve, H., Kondensationsreaktionen aktivierter Sulfinsäuren mit Mannich-basen, *Arch. Pharm.,* 311, 280, 1978.

902. Ahuja, R. R., Bhole, S. I., Bhongle, N. N., Gogte, V. N., and Natu, A. A., Optical induction. III. Some mechanistic studies on the reaction between thiophenol and α,β-unsaturated ketones using quinine-quinidine catalysts, *Indian J. Chem.,* 21B, 299, 1982.

903. Avold, H., German Patent 2,614,875, 1977; *Chem. Abstr.,* 88, 74194, 1978.

904. Kreutzkamp, N., Oei, H. Y., and Peschel, H., Über die Umsetzung von Schwefelkohlenstoff mit Mannich-basen, *Arch. Pharm.,* 304, 649, 1971.

905. Fitton, A. O. and Qutob, M., Studies in the dithiocarbamate series. V. The reactions of some N-(4-hydroxybenzyl)-piperidines and -pyrrolidines with carbonyl sulphide, *J. Chem. Soc. Perkin Trans. 1,* 2660, 1972.

906. Pereslegina, N. S., Kuz'mina, G. N., Markova, E. I., and Sanin, P. I., Alkylhydroxybenzyl dialkyldithiocarbamates antioxidizing agents for hydrocarbons, *Neftekhimiya,* 26, 563, 1986; *Chem. Abstr.,* 107, 175623, 1987.

907. Messinger, P. and Greve, H., Synthese symmetrischer Sulfone and Natrium-hydroxymethansulfinat und Mannich-basen, *Synthesis,* 259, 1977.

908. Corbel, B. and Paugam, J. P., An improved synthesis of 2-alkynyl sulfides, *Synthesis,* 882, 1979.

909. Andrisano, R., Angeloni, A. S., and Tramontini, M., Ricerche sulla reattività delle basi di Mannich. Nota IV. Azione di alcuni reattivi nucleofili (anilina, p-metil-tiofenolo) su stiril-cheto-basi, *Ann. Chim. Rome,* 55, 652, 1965; *Chem. Abstr.,* 63, 9854, 1965.

910. Dimmock, J. R., Smith, L. M., and Smith, P. J., The reaction of some nuclear substituted acyclic conjugated styryl ketones and related Mannich bases with ethanethiol, *Can. J. Chem.,* 58, 984, 1980.

911. Messinger, P. and Gompertz, J., Darstellung von Sulfonyl-keton-Mannich-basen, *Arch. Pharm.,* 308, 737, 1975.

912. Chao, H. S. I., The reduction of Mannich bases and their derivatives by tri-n-butyltin hydride, *Synth. Commun.,* 18, 1207, 1988.

913. Carrington, T. R., Long, A. G., and Turner, A. F., Compounds related to the steroid hormones. Part VIII. The Mannich reaction with 3- and 20-oxo-steroids, *J. Chem. Soc.,* 1572, 1962.

914. Maruyama, K., Tobimatsu, T., and Naruta, Y., Synthesis of methyl substituted chromanol. An analogue of vitamin K, *Bull. Chem. Soc. Jpn.,* 52, 1143, 1979.

915. Zacharova, N. V., Liakumovich, A. G., Michurov, Y. L., and Shalimova, A. S., British Patent 1,512,941, 1978; *Chem. Abstr.,* 90, 22571, 1979.

916. Yamada, K., Itoh, N., and Iwakuma, T., 'Onepot' conversion of Mannich bases via quaternary ammonium salts into the corresponding methyl compounds with sodium cyanoborohydride in HMPA, *J. Chem. Soc. Chem. Commun.,* 1089, 1978.

917. Marr, G., Moore, R. E., and Rockett, B. W., 2,7- And 2,10-disubstituted biferrocenyls, *Tetrahedron Lett.,* 2517, 1968.

918. Uchida, H. and Sato, K., Japanese Kokai Tokkyo Koho JP 61,106,530, 1986; *Chem. Abstr.*, 106, 4645, 1987.

919. Previc, E. P., U.S. Patent 3,461,172, 1969; *Chem. Abstr.*, 71, 101520, 1969.

920. Kagan, E. S., Mikhailov, V. I., Pavlikov, V. V., Shapiro, A. B., Sholle, V. D., and Rozantsev, E. G., Unusual product of aminomethylation, *Izv. Akad. Nauk SSSR Ser. Khim.*, 2187, 1978; *Chem. Abstr.*, 90, 6213, 1979.

921. Sinhababu, A. K. and Borchardt, R. T., Selective ring C-methylation of hydroxybenzaldehydes via their Mannich bases, *Synth. Commun.*, 13, 677, 1983.

922. Jones, S. S., Reese, C. B., and Ubasawa, A., A convenient synthesis of 5-methyluridine from uridine, *Synthesis*, 259, 1982.

923. Kadin, S. B., Monomethylation of aromatic amines via sodium borohydride. Mediated carbon-nitrogen bond cleavage, *J. Org. Chem.*, 38, 1348, 1973.

924. Galantay, E., Bacso, I., and Coombs, R. V., The preparation of α-allenic alcohols. Acetylene → allene 'homologization', *Synthesis*, 344, 1974.

925. Azerbaev, I. N., Erzhanov, K. B., Omarova, T. A., Iksanov, Z. A., and Lelyuk, M. I., Aminomethylation of acetylenic alcohols by N,N-bis(ethoxymethyl)aniline, *Izv. Akad. Nauk Kaz. SSR Ser. Khim.*, 25, 58, 1975; *Chem. Abstr.*, 84, 43781, 1976.

926. Poplevskaya, I. A., Kondaurov, G. N., Abdullin, K. A., Shipunova, L. K., Chermanova, G. B., and Kabiev, O. K., Synthesis and antineoplastic properties of acetylene containing bis(2-chloroethyl)amines, *Tr. Inst. Khim. Nauk Akad. Kaz. SSR*, 52, 1980; *Chem. Abstr.*, 94, 120781, 1981.

927. Biere, H. and Redmann, U., Antimikrobiell Wirksame Nitro-perhydropyrido[1,2-a][1,4]-diazepine, *Eur. J. Med. Chem.*, 11, 351, 1976.

928. Daruwala, A. B., Gearien, J. E., Dunn, W. J., Benoit, P. S., and Bauer, L., Amino-ketones. Synthesis and some biological activities in mice of 3,3-dialkyl-1,2,3,4-tetrahydro-4-quinolines and related Mannich bases, *J. Med. Chem.*, 17, 819, 1974.

929. Mann, N., Back, W., and Mutschler, E., Reduktion von Mannich Basen of Monoacylbiphenyl-Reihe. 2 Mitt. Über potentielle Analgetika *Arch. Pharm.*, 306, 67, 1973.

930. Welch, W. M., Plattner, J. J., Stratten, W. P., and Harbert, C. A., Analgesic and tranquilizing activity of 5,8-disubstituted-2-aminomethyl-3,4-dihydronaphthalene, *J. Med. Chem.*, 21, 257, 1978.

931. Fernandez, J. A, Bellate, R. A., Deliwala, C. V., Dadkar, N. H., and Sheth, U. K., Synthesis and central nervous system activity of new piperazine derivatives, *J. Med. Chem.*, 15, 417, 1972.

932. Dimmock, J. R., Hamon, N. W., Noble, L. M., and Wright, D. E., Nuclear-substituted styryl ketone analogs: Effects on neoplasms, microorganisms, and mitochondrial respiration of tumors and normal cells, *J. Pharm. Sci.*, 68, 1033, 1979; see also *J. Pharm. Sci.*, 67, 1536, 1978.

933. Arens, A. and Zicmanis, A., Aminoacetylene dicarbonyl compounds, *Khim. Atsetilena, Tr. 3rd Vses. Konf.*, 1972, 157; *Chem. Abstr.*, 79, 42087, 1973.

934. See also survey in Ref. 591.

935. Granitzer, W. and Stütz, A., Stereoselective trans-Reduktion tert. Propargylamine mit DBAH zu (E)-Allylaminen, *Tetrahedron Lett.*, 3145, 1979.

936. Berger, J. C., Teller, S. R., Adams, C. D., and Guggenberger, L. J., An unusual stereospecific reduction of some basic side-chain substituted indoles, *Tetrahedron Lett.*, 1807, 1975.

937. Viswanathan, N. and Gokhale, U. B., Mannich bases of 9-phenyl-1,2,3,4-tetrahydrocarbazole, *Indian J. Chem. B*, 22B, 121, 1983.

938. Raviña, E., Montañés, J. M., Seco, M. C., Calleja, J. M., and Zarzosa, M., Synthesis and potential antiparkinsonism activity of some tertiary substituted β-aminoalcohols containing a benzofuran nucleus, *Chim. Ther.*, 8, 182, 1973.

939. Jones, G., Maisey, R. F., Somerville, A. R., and Whittle, B. A., Substituted 1,1-diphenyl-3-aminoprop-1-enes and 1,1-diphenyl-3-aminopropanes as potential antidepressant agents, *J. Med. Chem.*, 14, 161, 1971.

940. Iorio, M. A., and Casy, A. F., Synthesis, configuration, and dehydration of some 1-alkyl-and aralkyl-3-methyl-4-o-tolylpiperidin-4-ols, *J. Chem. Soc. C*, 135, 1970.

941. Casy, A. F., Harper, N. J., and Dimmock, J. R., Tertiary alcohols and related compounds derived from 2-dimethylaminomethyl cyclohexanone, *J. Chem. Soc.*, 3635, 1964.

942. Duchon-d'Engenières, M., Maldonado, J., Avril, J. L., Miocque, M., Raynaud, G., Thomas, J., and Pourrias, B., Synthèse et étude pharmacologique d'alcools acétiléniques à fonction amine et ammonium quaternaire, *Eur. J. Med. Chem.*, 9, 59, 1974.

943. Mayrargue, J., Duchon-d'Engenières, M., and Miocque, M., Réactivité des alcools acétyléniques. XI. Synthèse et stéréochimie des dialkylaminométhyl-2 alcynyl-1 cyclohexanol-1, *Bull. Soc. Chim. Fr.*, 133, 1977.

944. Casy, A. F. and Myers, J. L., Some 1,2-diaryl-4-dimethylamino-3-methylbutan-2-ols and derivatives related to propoxyphene, *J. Chem. Soc.*, 4092, 1965.

945. Cavalla, J. F., Marshall, J. P., and Selway, R. A., Compounds derived from the Mannich bases of β-phenylpropiophenone, *J. Med. Chem.*, 7, 716, 1964.

946. Guetté, M. and Lucas, M., Obtention d'aminohydroxyesters par action du réactif de Réformatsky sur les bases de Mannich, *Bull. Soc. Chim. Fr.*, 2759, 1975.

947. Lucas, M. and Guetté, J. P., Addition de réactifs de Réformatsky sur les α- et β-aminocétones. Synthèse de γ- et δ-amino-β-hydroxyesters, *J. Chem. Res.*, 214, 1978.

948. Mladenova, M., Biserkova, M., and Stanchev, S., Addition of lithium reagents of N,N-dialkylsulfonamides to Mannich bases, *Synth. Commun.*, 21, 1555, 1991.

949. Lucas, M. and Guetté, J. P., Induction asymétrique. II. Détermination des configurations rélatives de δ-amino β-hydroxy esters diastéréoisomères, obtenus par la réaction de Réformatsky, *Tetrahedron*, 34, 1675, 1978; see also *Tetrahedron* 34, 1685, 1978.

950. Casy, A. F. and Ison, R. R., Dehydration of alcohols derived from 1-t butyl-3-dimethylaminopropan-1-one: a novel cleavage reaction involving 2,6-dichlorobenzylmagnesium bromide, *Can. J. Chem.*, 48, 1011, 1970.

951. Ison, R. R. and Casy, A. F., Studies of antihystaminics. Dehydration of some substituted 4-aminobutan-2-ols, *J. Chem. Soc. C*, 3048, 1971.

952. Normant, J. F. and Alexakis, A., Carbometallation (C-metallation) of alkynes; stereospecific synthesis of alkenyl derivatives, *Synthesis*, 841, 1981; see, in particular, pp. 845–846.

953. Sahlberg, C. and Claesson, A., Allenes and acetylenes. XXIV. Synthesis of α-allenic amines by organocuprate reactions of acetylenic aminoethers, *Acta Chem. Scand.*, B36, 179, 1982.

954. Sáa, J. M., Llobera, A., Garcia-Raso, A., Costa, A., and Deyá, P. M., Metallation of phenols. Synthesis of benzoquinones by the oxidative degradation approach, *J. Org. Chem.*, 53, 4263, 1988.

955. Omae, I., Organometallic intramolecular-coordination compounds containing a nitrogen donor ligand, *Chem. Rev.*, 79, 287, 1979; see, in particular, p. 301.

956. Stork, G., Ozorio, A. A., and Leong, A. Y. W., N,N-Diethylaminoacetonitrile, a generally useful latent acyl carbanion, *Tetrahedron Lett.*, 5175, 1978.

957. Martin, S. F. and Gompper, R., A facile method for the transformation of ketones into α-substituted aldehydes, *J. Org. Chem.*, 39, 2814, 1974.

958. Broekhof, N. L. J. M. and van der Gen, A., Enamine synthesis using the Horner-Wittig reaction. I. (Aminomethyl)diphenylphosphine oxides, new formyl anion equivalents, *Recueil*, 103, 305, 1984.

959. Russel, G. A., Yao, C. F., Rajaratnam, R., and Kim, B. H., Promotion of electron transfer by protonation of nitrogen-centered free radicals. The addition of radicals to iminium ions, *J. Am. Chem. Soc.*, 113, 373, 1991.

960. Pollak, I. E. and Grillot, G. F., The reaction of Grignard reagents with arylthiomethylarylamines and with tris(phenylthiomethyl)-amines, *J. Org. Chem.*, 32, 2892, 1967.

961. Germon, C., Alexakis, A., and Normant, J. F., Vinyl-copper derivatives. XII. Stereospecific synthesis of allylic amines by aminomethylation of organocopper agents, *Tetrahedron Lett.*, 3763, 1980.

962. Brown, H. C., Kabalka, G. W., Rathke, M. W., and Rogié, M. M., Reaction of organoboranes with Mannich bases. A convenient procedure for the alkylation of cyclic and bicyclic ketones via hydroboration, *J. Am. Chem. Soc.*, 90, 4166, 1968.

963. Roth, H. J., El Raie, M. H., and Schraut, T., Photocyclisierung von 3-Aminoketonen zu 2-Amino-cyclopropanolen-(1) und deren Isomerisierung, *Arch. Pharm.*, 307, 584, 1974.

964. Abdul-Baki, A., Rotter, F., Schrauth, T., and Roth, H. J., Versuche zur photochemischen Synthese ephedrinänlicher Verbindungen. III. Photocyclisierung von 3-Alkylaminopropiophenonen zu 2-Alkylamino-cyclopropanolen und deren Isomerisierung, *Arch. Pharm.*, 311, 341, 1978.

965. Petrosyan, L. M., Gevorgyan, G. A., Engoyan, A. P., and Mndzhoyan, O. L., Aminomethylation of 1,5-diphenyl-1,5-pentanedione, *Zh. Org. Khim.*, 20, 608, 1984; *Chem. Abstr.*, 101, 23071, 1984.

966. Harbert, C. A., Plattner, J. J., Welch, W. M., Weissman, A., and Kenneth-Koe, B., Neuroleptic activity in 5-aryltetrahydro-γ-carbolines, *J. Med. Chem.*, 23, 635, 1980.

967. Sacquet, M. C., Fargeau-Bellasoued, M. C., and Graffe, B., Synthèse de 2-pyrimidinyl-phénols et de 2-pyrazolylphénols, *J. Heterocycl. Chem.*, 28, 667, 1991.

968. Afsah, E. M., Hammouda, M., and Hamama, W. S., Pictet-Spengler reactions of tryptamine and tryptophan with cycloalkanones and ketonic Mannich bases, *Monatsh. Chem.*, 116, 851, 1985.

969. Saint-Ruf, G. and Bourgeade, J. C., Nouvelles isatines N-substituées et leurs dérivés d'intérêt pharmacologique, *Chim. Ther.*, 8, 447, 1973.

970. Epsztein, R. and Le Goff, N., Synthesis of conjugated triendiamines through isomerization of α,ω-diaminononadiynes, *Tetrahedron Lett.*, 1965, 1981.

971. Kruglikova, R. I., Sotnichenko, T. V., Shingareeva, A. G., and Unkovskii, B. V., Synthesis of α-acetylenic alcohol N-phenylcarbamates and their intramolecular cyclization into 4-alkylidene-2oxazolidinones, *Zh. Org. Khim.*, 17, 649, 1981; *Chem. Abstr.*, 95, 61710, 1981.

972. Unkovskii, B. V., Mel'nikova, A. A., Zaitseva, M. G., and Malina, Y. F., Spatial structure and stereochemistry of synthesis of 1-alkyl-4-phenyl-3-benzoyl-4-piperidones, *Zh. Org. Khim.*, 2, 1501, 1966; *Chem. Abstr.*, 66, 46298, 1967.

973. Uchino, M., Studies on the Mannich reaction. II. On the formation of piperidine derivatives, *Bull. Chem. Soc. Jpn.*, 32, 1012, 1959.

974. Von Thiele, K., Posselt, K., and Von Bebenburg, W., Neue Piperidin Derivate aus herzwirksamen β-Aminoketonen. Synthese und Struktur eines Metaboliten von Oxyefedrin, *Arzneim. Forsch.*, 18, 1263, 1968.

975. Chen, G., Xu, X., and Zhao, D., Mannich reaction with arylamine as amine component. V, *Huaxue Xuebao*, 44, 846, 1986; *Chem. Abstr.*, 106, 213715, 1987.

976. Short, J. H. and Ours, C. W., Use of aminoacids in the Mannich reaction, *J. Heterocycl. Chem.*, 12, 869, 1975.

977. Aversa, M. C., Bonaccorsi, P., and Giannetto, P., Chemical behaviour of N-(2-hydroxybenzyl)anthranilic acids in the presence of acyclic anhydrides, *J. Heterocycl. Chem.*, 26, 1383, 1989.

978. Gotz, M. and Grozinger, K., 3-Hydroxy sydnone imines, *Tetrahedron*, 27, 4449, 1971.

979. Yashunskii, V. G., Gorkin, V. Z., Mashkovskii, M. D., Altshuler, R. A., Veryovkina, I. V., and Kholodov, L. E., Synthesis and pharmacological effects of some alkyl-, aryl-, and aralkylsydnonimines, *J. Med. Chem.*, 14, 1013, 1971.

980. Regniert, G., Canevari, R., and Laubie, M., German Offen. 2,241,991, 1973; *Chem. Abstr.*, 78, 159672, 1973.

981. Gonzales Trigo, G., Galvez Ruano, E., and Menendez Aguirre, C., Spiro heterocyclic derivatives. XV. Some N-methyl-N'-alkyl (or aralkyl) -3,7-diazabicyclo σ[3.3.1]nonan-9-ones and some spiro-5'-hydantoin derivatives of 3,7-diazabicyclo[3.3.1]nonanes, *An. Quim.*, 75, 894, 1979; *Chem. Abstr.*, 93, 26411, 1980.

982. Melika, Y. V., Smolanka, I. V., and Staninets, V. I., Bromocyclization of aminomethyl ethers of 2-cyclohexen-1-ol and cinnamic alcohol, *Ukr. Khim Zh.*, 39, 799, 1973; *Chem. Abstr.*, 79, 115481, 1973.

983. Coyle, J. D., Smart, L. E., Challiner, J. F., and Haws, E. J., Photocyclization of N-(dialkylaminomethyl) aromatic 1,2-dicarboximides, *J. Chem. Soc. Perkin Trans. 1*, 121, 1985.

984. Coyle, J. D. and Bryant, L. R. B., Photochemical cyclization of 3-N-(dialkylaminomethyl)imidazole-2,4-diones to 1,3,7-triazabicyclo[3.3.0]octanes, *J. Chem. Soc. Perkin Trans. 1*, 531, 1983.

985. Coyle, J. D., Newport, G. L., and Harriman, A., Nitrogen substituted phthalimides: Fluorescence, phosphorescence and the mechanism of photocyclization, *J. Chem. Soc. Perkin Trans. 2*, 133, 1978.

986. Roth, H. J. and Schwarz, D., Photocyclisierung von Phthalimid-Mannich Basen, *Arch. Pharm.*, 308, 631, 1975; see also *Arch. Pharm.*, 308, 218, 1975.

987. Coyle, J. D., Synthesis of 2,4-benzodiazepines from phthalimide Mannich bases, *Synthesis*, 403, 1980.

988. Brugidou, J. and Christol, H., Condensation diénique des bases de Mannich des cétones, *Bull. Soc. Chim. Fr.*, 1693, 1966.

989. Andrisano, R., Angeloni, A. S., Del Moro, F., and Tramontini, M., Ricerche sulla reattività delle basi di Mannich. V. Decomposizione in mezzo aprotico e sintesi dieniche, *Ann. Chim. Rome*, 55, 968, 1965; *Chem. Abstr.*, 64, 6444, 1966.

990. Andrisano, R., Angeloni, A. S., and Tramontini, M., Richerche sulla reattività delle basi di Mannich. II. Sintesi dieniche su alcune stiril-cheto-basi, *Ann. Chim. Rome*, 55, 143, 1965; *Chem. Abstr.*, 63, 13128, 1965.

991. Rollin, P., Autocondensation des cyclanones catalysée par le sodium métallique. II. Autocondensation de cyclohexanones α-substituées, *Bull. Soc. Chim. Fr.*, 1806, 1973.

992. Mailard, J., Delaunay, P., Langlois, M., Jolly, R., Morin, R., Manuel, C., and Mazmanian, C., Antiinflammatoires dérivés de l'acide phénylacétique. VI. Dérivés oxygénés du (cyclohexyl-4-phényl)-2-éthanol, *Eur. J. Med. Chem.*, 14, 511, 1979; see also *Eur. J. Med. Chem.*, 12, 161, 1977.

993. House, H. O. and Trost, B. M., By-products of the Robinson annellation with cyclohexanone, cyclopentanone, and cyclopentane-1,2-dione, *J. Org. Chem.*, 30, 2513, 1965.

994. Haynes, N. B. and Timmons, C. J., Cyclisation of 2-methyl-2-(4-methyl-3-oxypentyl)-cyclohexanone, *J. Chem. Soc. C*, 224, 1966.

995. Denys, G., Gureviciene, J., Macionyte, V., Bernotas, R., and Cekuoliene, L., Synthesis and reactivity of N-(β-acylethyl)amino-pyridines and aminoquinolines, *Zh. Org. Khim.*, 13, 199, 1977; *Chem. Abstr.*, 86, 155484, 1977.

996. Scott, P. L. and MacConaill, R. J., The establishment of oxime anchimerism using isotopic techniques, *Tetrahedron Lett.*, 3685, 1967.

997. Epsztajn, J., Hahn, W. E., and Tosik, B. K., Cycloalkanes condensed with heterocyclic rings. XIV. Synthesis of 6-methyl-2,3-cycloalkanopyridines, *Rocz. Chem.*, 44, 431, 1970; *Chem. Abstr.*, 72, 132478, 1970.

998. Kröhnke F., The specific synthesis of pyridines and oligopyridines. *Synthesis*, 1, 1976; see, in particular, p. 15.

999. Nietsch, K. H. and Troschütz, R., Heterocyclen via Mannichbasen. 8 Mitt. 7,8-Dihydro-1,6-naphthyridin-5(6H)-on, *Arch. Pharm.*, 318, 175, 1985.

1000. Tramontini, M., Richerche sulla reattività delle basi di Mannich. Nota VII. Sintesi di stirilchinoline e loro proprietà spettroscopiche UV, *Ann. Chim. Rome*, 55, 1154, 1965; *Chem. Abstr.*, 64, 6610, 1966.

1001. Gogte, V. N., Salama, M. A., and Tilak, B. D., Synthesis of nitrogen heterocyclics. VI. Stereochemistry of hydride transfer in acid catalysed disproportionation of 3,4-disubstituted 1,2-dihydroquinolines, *Tetrahedron*, 26, 173, 1970.

1002. Andreani, F., Andrisano, R., Salvadori, G., and Tramontini, M., Reactivity of Mannich bases. XIV. Synthesis of phenylbenzoquinoline and styrylbenzoquinoline derivatives and their photoreactivity, *J. Chem. Soc. C*, 1007, 1971.

1003. Gutsulyak, B. M. and Petrovskii, R. S., Deposited doc., 1980, SPSTL 902; *Chem. Abstr.*, 97, 109849, 1982.

1004. Novitskii, Z. L. Turov, A. V., and Kornilov, M. Y., Structure of cyclization products of secondary amines with formaldehyde and cyclohexanone in the presence of perchloric acid, *Zh. Org. Khim.*, 17, 429, 1981; *Chem. Abstr.*, 94, 208684, 1981.

1005. Henin, H. and Pete, J. P., As easy transformation of ketones into α,α-epoxymethylketones by hydrogen peroxide oxidation of Mannich base derivatives, *Synthesis*, 895, 1980.

1006. Messinger, P. and Greve, H., Cyclische Sulfone aus Mannich-basen und Natrium-hydroxymethan-sulfinat, *Arch. Pharm.*, 310, 674, 1977.

1007. Roper, J. M. and Everly, C. R., Direct synthesis of spiro[5.5]undeca-1,4,7-trienones from phenols via a quinone methide intermediate, *J. Org. Chem.*, 53, 2639, 1988.

1008. Bilgic, O. and Young, D. W., A general and practicable synthesis of polycyclic heteroaromatic compounds. III. Extension of the synthesis of 'quinone-methides' of naphthalene, phenanthrene, and benzene, *J. Chem. Soc. Perkin Trans. 1*, 1233, 1980.

1009. Omae, I., Organometallic intramolecular-coordination compounds containing a nitrogen donor ligand, *Chem. Rev.*, 79, 287, 1979; see, in particular, p. 301.

1010. Bladé-Font, A. and de Mas Rocabayera, T., Synthesis of dihydrobenzofurans from phenolic Mannich bases and their quaternized derivatives, *J. Chem. Soc. Perkin Trans. 1*, 841, 1982.

1011. Cadonà L. and Dalla Croce, P., A convenient synthesis of 2-acyl- or 2-aroyl-substituted 2,3-dihydrobenzofurans and 1,2-dihydronaphtho[2,1-b]furans, *Synthesis*, 800, 1976.

1012. Ivanov, B. E. and Khismatullina, L. A., Reaction of hydrochloric acid and Mannich methiodide phenol base with thiethyl phosphite, *Izv. Akad. Nauk SSSR. Ser. Khim.*, 2150, 1968; *Chem. Abstr.*, 70, 20164, 1969.

1013. Ivanov, B. E., Samurina, S. V., Ageeva, A. B., Valitova, L. A., and Bel'skii, V. E., Reaction of amidophosphites with phenolic Mannich bases, *Zh. Obshch. Khim.*, 49, 1973, 1979; *Chem. Abstr.*, 92, 76024, 1980.

1014. Molho, D., Condensation des bases de Mannich phénoliques avec les dérivés à H mobile. I. Nouvelle syntèse d'(arylalkyl)-3-hydroxy-4-coumarines, *Bull. Soc. Chim. Fr.*, 1417, 1961.

1015. Balasubramanian, K. K. and Selvaraj, S., Novel reaction of o-phenolic Mannich bases with α-chloroacrylonitrile, *J. Org. Chem.*, 45, 3726, 1980.

1016. Mahajan, J. R. and Araùjo, H. C., Synthesis of medium and macrocyclic benzo- and naphthoketolactones. Oxidation of 2,3-polymethylene benzo- and naphthopyrans, *Synthesis,* 111, 1976.

1017. Kagan, E. S., Mikhailov, V. I., Sholle, V. D., Smirnov, V. A., and Rozantsev, E. G., Alkylation of enamines of 2,2,6,6-tetramethyl-4-oxopiperidine and 2,2,6,6-tetramethyl-4-oxo-piperidine-1-oxy by Mannich bases from β-naphthol, *Izv. Adad. Nauk SSSR Ser. Khim.,* 1668, 1978; *Chem. Abstr.,* 89, 197367, 1978.

1018. Leonte, M. V. and Georgescu, M. A., α-Aminoalkylation. III. Synthesis of some new chromon derivatives trrough condensation of Mannich bases with maleic anhydride, *Bull. Univ. Galati,* 1, 89, 1980; *Chem. Abstr.,* 93, 132330, 1980.

1019. Balasubramanian, K. K. and Selvaraj, S., Studies in phenolic Mannich bases. Reaction with acetylenes, *Tetrahedron Lett.,* 851, 1980.

1020. Von Strandtmann, M., Cohen, M. P., and Shavel, J., Reaction of phenolic Mannich bases with imines and oximes. Synthesis of fused ring systems containing 1,3-oxazines, *J. Heterocycl. Chem.,* 6, 429, 1969.

1021. Arct, J., Jakubska, E., and Olszewska, G., Conversion of Mannich phenol bases. II. Synthesis of 2-thioxo-2H-3,4-dihydro-1,3-benzoxazine derivatives, *Synthesis,* 314, 1977.

1022. Slocum, D. W., Rockett, B. W., and Hauser, C. R., Ring metallation of dimethylaminomethyl ferrocene with butyllithium and condensations with electrophilic compounds, *J. Am. Chem. Soc.,* 87, 1241, 1965.

1023. Pauson, P. L., Sandhu, M. C., and Watts, W. E., Ferrocene derivatives. XV. New routes to symmetrically disubstituted ferrocenes, *J. Chem. Soc., C,* 251, 1966.

1024. Greenhill, J. V. and Ramli, M., A base exchange reaction between a 1,2-diketo-bis Mannich base and o-phenylenediamine, *Tetrahedron Lett.,* 4059, 1973.

1025. Werner, W., Jungstand, W., Gutsche, W., Wohlrabe, K., Römer, W., and Tresselt, D., Cancerostatisch Wirksame 6,6a,7,8,13,13a-Hexahydro-1-benzopyrano[4,3-b]-1,5-benzodiazepin-disastereoisomere, *Pharmazie,* 34,394, 1979.

1026. Taylor, E. C. and Garcia, E. E., A new purine synthesis. *J. Am. Chem. Soc.,* 86, 4720, 1964.

1027. Hocker, J., Diehr, H. J., and Merten, R., German Offen. 2,325,927, 1974; *Chem. Abstr.,* 82, 125135, 1975.

1028. Cingolani, G. M., Gualtieri, F., and Pigini, M., Researchers in the field of antiviral compounds. Mannich bases of 3-hydroxycoumarin, *J. Med. Chem.,* 12, 531, 1969.

1029. Raines, S. and Kovacs, C. A., Synthesis and pharmacology of a series of substituted 2-aminomethylpyrroles, *J. Med. Chem.,* 13, 1227, 1970.

1030. Cichra, D. A. and Adolph, H. G., Nitrolysis of dialkyl tert-butylamines, *J. Org. Chem.,* 47, 2474, 1982.

1031. Yamashita, M., Ito, K., Terao, Y, and Sekyia, M., N-[N-Nitrosoalkylamino)methyl]carbamates as new and convenient diazoalkane-generating agents, *Chem. Pharm. Bull. Tokyo,* 27, 682, 1979.

1032. Noll, B., Schreiner, G., Willecke, B., Hille, G., and Hartmann, M., East German Patent DD 229,275, 1985; *Chem. Abstr.,* 106, 34574, 1987.

1033. Saheki, Y., Kimura, M., and Negoro, K., Menshutkin reaction of Mannich bases derived from p-cresol and antimicrobial activities of their quaternary ammonium salts, *Nippon Kagaku Kaishi,* 1123, 1978; *Chem. Abstr.,* 89, 163167, 1978.

1034. Jemison, R. W., Laird, T., Ollis, W. D., and Sutherland, I. O., Base catalysed rearrangements involving ylide intermediates. I. The rearrangements of diallyl- and allylpropynyl ammonium cations, *J. Chem. Soc. Perkin Trans. 1,* 1436, 1980; see also *J. Chem. Soc. Perkin Trans. 1,* 1450 and 2033, 1980.

1035. House, H. O. and Müller, H. C., Decarboxylation and deuterium exchange in some ketone systems, *J. Org. Chem.,* 27, 4436, 1962.

1036. De Stevens, G. and Halamandaris, A., Stereochemical and polar effects in bromination of androstan-3-one Mannich bases, *Experientia,* 17, 297, 1961.

1037. De Stevens, G., U.S. Patent 3,092,621, 1963; *Chem. Abstr.,* 60, 604, 1964.

1038. Kurbanov, A. L., Sirlibaev, T. S., and Kultaev, K. K., Synthesis of unsaturated halogen-containing amines from phenylacetylene, *Uzb. Khim. Zh.,* 31, 1986; *Chem. Abstr.,* 107, 197640, 1987.

1039. Lutz, W. B., Lazarus, S., and Meltzer, R. I., New derivatives of 2,2,6,6-tetramethylpiperidine, *J. Org. Chem.,* 27, 1695, 1962.

1040. Natova, L. and Zhelyazkov, L., Experimental determination of the intermediate product in a Mannich reaction, *God. Vissh. Khimikotekhnol. Inst. Sofia,* 24, 213, 1978; *Chem. Abstr.,* 95, 220039, 1981; see also *Chem. Abstr.,* 95, 220042 and 220043, 1981.

1041. Andrisano, R., Angeloni, A. S., and Gottarelli, G., The oximation of cis-1-alkyl-3,5-diphenyl piperidin-4-ones, *Tetrahedron,* 30, 3827, 1974.

1042. Heindel, D. and Molnar, J., Synthesis and antimalarial activity of amodiaquine analogs, *J. Med. Chem.,* 13, 156, 1970.

1043. Cannon, J. G., Johnson, L. E., Long, J. P., and Heinz, S., 1,6-Diammonium-2,4-hexadiyne analogs of hexamethonium, *J. Med. Chem.,* 17, 355, 1974.

1044. Abidov, U. A., Gapurov, A., Makhsumov, A. G., and Askaraliev, M., Some reactions of propargyl ethers of 2,4-dibromo-phenol, *Zh. Prikl. Khim. Leningrad,* 50, 1371, 1977; *Chem. Abstr.,* 87, 134242, 1977.

1045. Cox, B. G., De Maria, P., Fini, A., and Hassan, A. F., Intramolecular catalysis in the enolisation of β-piperidino propiophenone and its methiodide derivative, *J. Chem. Soc. Perkin Trans. 2,* 1351, 1981.

1046. Unterhalt, B. and Leiblein, F., Nitraminen. XI. Umsetzungen mit Chlormethyl-alkylnitraminen, *Arch. Pharm.,* 311, 879, 1978.

1047. Nacef, B., Daudon, M., and Pinatel, H., Solid bifunctional (nitrogen IV cation) gem chloro nitroso derivatives of Mannich bases, *Synth. Commun.,* 7, 153, 1977.

1048. Scott, F. L., Hegarty, A. F., and McConaill, R. J., Cyclisation versus elimination reactions in a neighbouring group system, *Tetrahedron Lett.,* 1213, 1972; see also *Tetrahedron Lett.,* 1217, 1972.

1049. Ferrand, G., Maffrand, J. P., Eloy, F., and Ferrand, J. C., Synthèse et propriétés pharmacologiques de dérivés aminothiazoliques, *Eur. J. Med. Chem.,* 10, 549, 1975.

1050. Cameron, D. W. and Cracknell, R. H., Some piperidinomethyl quinols and quinones, *Aust. J. Chem.,* 29, 1163, 1976.

1051. Saa, J. M., Llobera, A., and Deya, M., Fremy's salts promoted oxidative degradation of p-hydroxybenzylamines and p-hydroxybenzamides. A novel approach to p-quinones, *Chem. Lett.,* 771, 1987.

1052. Möhrle, H. and Miller, C., Über Mannich Basen. VI. Mannichbasen 2,6-disubstituierter Phenole, *Arch. Pharm.,* 306, 552, 1973.

1053. Lin, A. J., Shansky, C. W., and Sartorelli, A. C., Synthesis and antineoplastic activity of hydroquinone dialdehydes, *J. Med. Chem.,* 17, 558, 1974; see also *J. Med. Chem.,* 23, 627, 1980.

1054. Schlögl, K. and Walser, M., Oxidation von Ferrocen-Mannich Basen mit MnO$_2$ zu Formylferrocenen, *Tetrahedron Lett.,* 5885, 1968.

1055. Shono, T., Matsumura, Y., Hayashi, J., Usui, M., Yamane, S. I., and Inoue, K., Electroorganic chemistry. LXVI. Electrochemical oxidation of animals and enamines using a mediator system, *Acta Chem. Scand.,* B37, 491, 1983.

1056. Voronkov, M. G., Vlasova, N. N., and Pestunovic, A. E., U.S.S.R. Patent 643,507, 1979; *Chem. Abstr.,* 90, 168728, 1979.

1057. Möhrle, H. and Lappenberg, M., Über Mannich Basen. XII. Hydroxylaminomethylierung von 2-Naphthol, *Chem. Ber.,* 109, 1106, 1976.

1058. Craig, J. C., Ekwuribe, N. N., and Gruenke, L. D., Novel rearrangement of prop-2-ynyl N-oxides to hydroxylamine O-allenyl ethers: mechanistic studies, *Tetrahedron Lett.,* 4025, 1979.

1059. Möhrle, H. and Gundlach, P., Nachbargruppeneffect von Phenolen, *Tetrahedron,* 27, 3695, 1971.

1060. Roth, H. J., Schumann, E., George, H., and Assadi, F., Photokemische Reaktionen von Mannich Basen aromatischer Amine, *Tetrahedron Lett.,* 30, 3433, 1968.

1061. Roth, H. J. and Assadi, F., Photoreaktion von Tetralon- und Indanon-Mannich Basen, *Arch. Pharm.,* 303, 732, 1970.

1062. Roth, H. J. and Schumann, E., Darstellung von ω-Aminosäuren aus Cycloalkanon-Mannichbasen. I. Photokemische Reaktionen mit Mannichbasen, *Arch. Pharm.,* 300, 948, 1967.

Chapter III

Mannich Bases in Macromolecular Chemistry

The present chapter deals with the application of Mannich aminomethylation or Mannich bases to the synthesis and modification of macromolecular compounds. As summarized in Fig. 150, a remarkable number of different combinations exist, as the Mannich reaction enables us (a) to perform polymerizations by using bifunctional substrate and amine as well as (d) to functionalize polymeric derivatives behaving, alternatively, as substrate or amine components of Mannich synthesis. On the other hand, the manifold reactivity of Mannich bases makes it possible (b) to produce polymers by amino group replacement with bifunctional nucleophiles or to polymerize suitable moieties (e.g., double bonds) present in the base. Furthermore (e), macromolecular compounds can be subjected to amino group replacement as well as to various other reactions given by Mannich bases. Finally (c), crosslinked derivatives are obtained from oligomeric or polymeric products through any of the above mentioned methods.

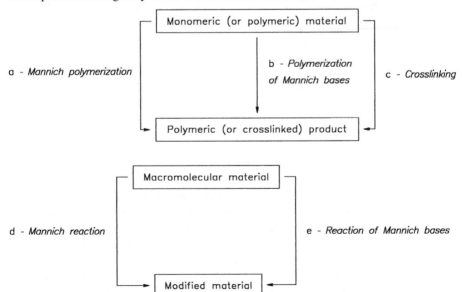

Fig. 150. The chemistry of Mannich bases involved in the synthesis (a–c) and modification (d,e) of macromolecular compounds.

Individual macromolecular derivatives having specific applications in industry are also mentioned in Chap. V, the present chapter being dedicated to the treatment of paths (a)–(e) of Fig. 150. A thorough survey,[1] along with some review papers on particular topics,[2,3] is present in the relevant literature on the subject.

A Mannich Polymerization

When a Mannich reaction is carried out between a substrate containing at least two active hydrogen atoms and a primary or a bis-secondary amine, a polycondensation takes place with production of a polymeric derivative. The polycondensation can also occur when both an NH group and one active hydrogen atom are present in the same molecule. The reaction product is thus characterized by the presence of the methylene moiety, which is derived from the formaldehyde, and forms the polymer backbone, with the consequent possibility of polymer degradation by deamination or deaminomethylation; both of these reactions are typical of Mannich bases (Chap. II, A).

The first kind of polycondensation follows the scheme shown in Fig. 151, involving monomers of type A-A, B-B (Table 31), with formation of the polymeric products **375** and **376**. The monomer mixture generally consists of the usual three components of Mannich synthesis, although in some cases preformed reagents of type X—CH$_2$—N<, such as **382** in Table 31, are used.

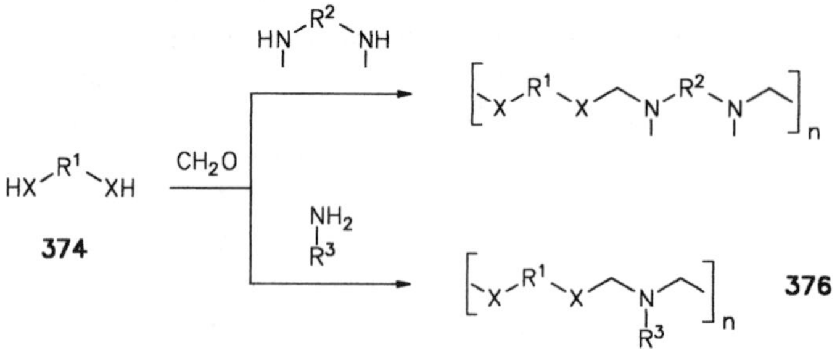

Fig. 151. The Mannich polymerization of monomers of type A-A, B-B.

The two reactive hydrogen atoms of the substrate also may be linked to the same carbon atom, as in the case of alkyl ketones. Other classes of substrates employed in polymerization include phenols, heterocyclic derivatives, etc. Nitrogen-containing substrates, mainly amides and arylamines, are also used. However, as the aromatic ring of arylamines is activated toward electrophilic reactions, products generated by C-aminomethylation of the ring are to be expected along with the polymer formed by N-aminomethylation of the aminic substrate. The amine reagents employed (**377–382** in Table 31) are quite varied; piperazine (**380**) is the most used among bis-amines. Polymeric derivatives having structure **375** are also obtained[18,19] by reaction with primary bis-amines, such as isophorone-diamine, meta-xylilendiamine, etc., in the molar ratio substrate/formaldehyde/amine 1:2:1. Since these last polymeric Mannich bases (e.g., **383**) contain secondary amino groups, they are successfully employed as hardeners for epoxy resins.[18]

Table 31

Three-Component Mannich Polymerization Employing Monomers of Types A-A, B-B, and Formaldehyde (Fig. 151)

Bifunctional Amine

Primary Amines

Alkyl—NH_2 (alkyl groups containing
NR_2, OH, COOH, included)

377

Ar—NH_2

378

Bis-Amines

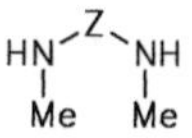

HN—$\overset{Z}{\frown}$—NH
  | |
  Me   Me

($Z = C_{0,2-6}$ chain)

379

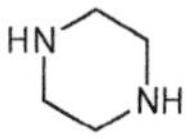

380

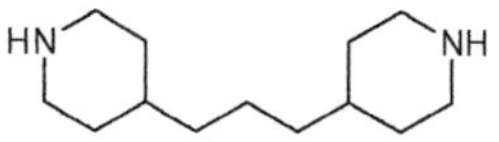

381

Preformed Mannich Reagents

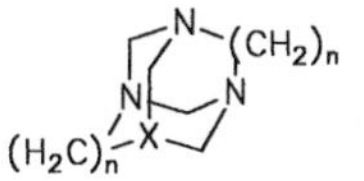

($n = 1, 2$; $X = N, P$)   **382**

Substrate (**374**)[a]	Bifunctional Amine	Polymer	Reference
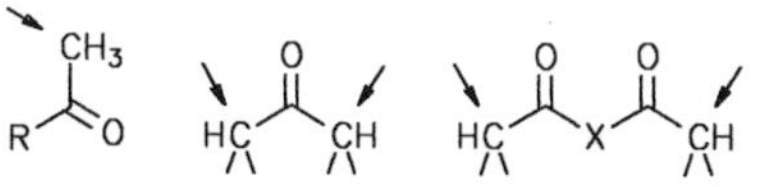	380, 381 377, 378	375 376	4-6 7,8
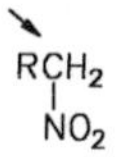 $R\overset{\displaystyle CH_2}{\underset{\displaystyle NO_2}{\mid}}$	379	375	9
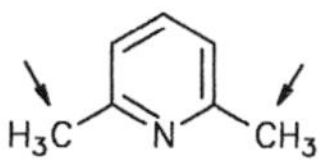	380	375	8
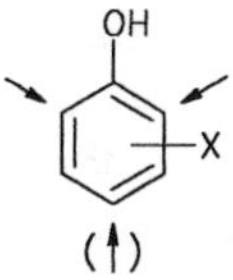 (X = H, Alkyl, OH,...) (↑)	379-381 377, 382	375 376	4,7,10 11-13

It is finally worth mentioning that guanidine, melamine, and similar derivatives are frequently used as amine reagents or, sometimes, as preformed reagents.[20–22] The above reactions are related to the analogous, even more commonly applied, amidomethylations (Chap. I, A.2) involving urea derivatives; all these compounds have similar chemical features and applications.

Table 31

(continued)

structure			
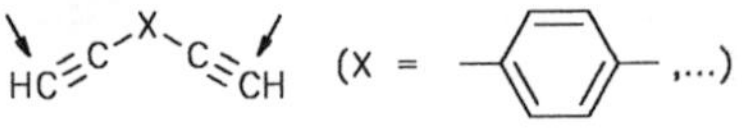 $HC\!\equiv\!C\text{-}X\text{-}C\!\equiv\!CH$ (X = benzene ring, ...)	380 378	375 376	14,15 15
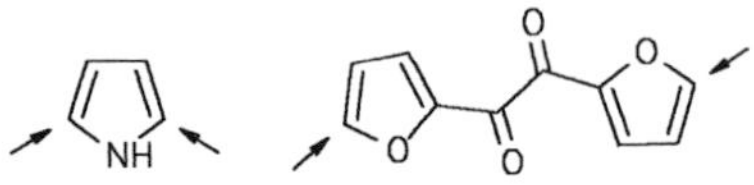 pyrrole (NH), difuranyl diketone	380, 381 377	375 376	6,7 7
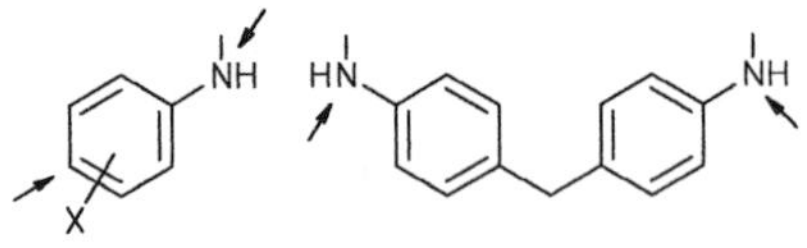 aniline derivatives (X = H, Me,...)	380 377	375 376	4,5,7,16 8
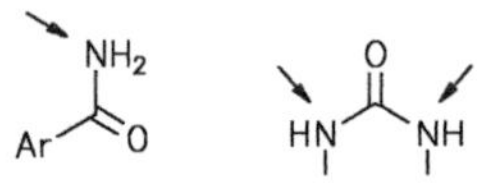 $Ar\text{-}C(=O)\text{-}NH_2$, urea (HN–C(=O)–NH)	381 382	375 376	8 11,17

a Arrow indicates the most likely position of attack by the reagent

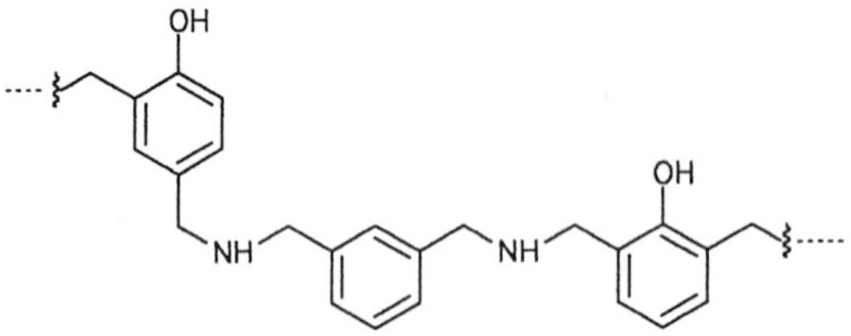

383

The second kind of polycondensation based on the Mannich reaction includes monomers of type A-B (**384**, Fig. 152), requiring equimolar amounts of formaldehyde in order to produce the polymeric derivatives **385**.

As the possibility of concurrent cyclization (Chap. I, C.4) of monomers **384** has to be avoided for a successful polymerization, the most suitable starting materials are arylamino derivatives, which undergo C-aminomethylation on the activated site of the aromatic ring. However, when multiple reactive sites are present in the aromatic ring, branched derivatives and even cross-linked products are likely to be produced.[23] Polymers **386**, obtained from *p*-amino-benzoic acid[24] as well as analogous products given by aminophenols or mixtures of different arylamines,[25] have been described.

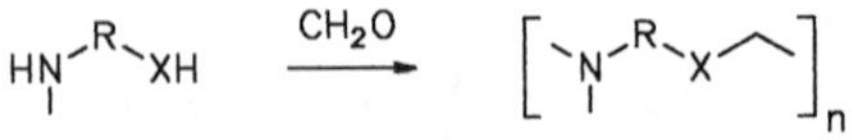

−R−X− = Arylene moieties

Fig. 152. The Mannich polymerization of monomers of type A-B.

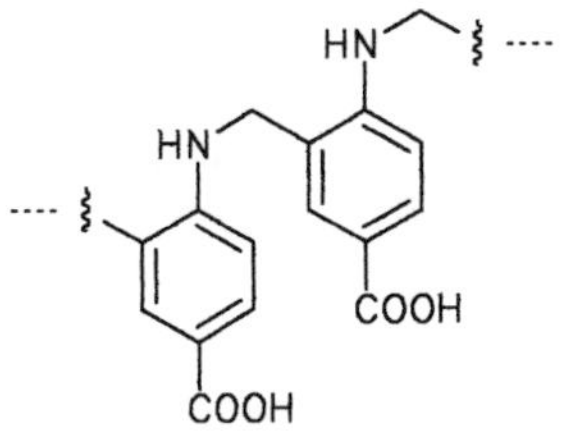

386

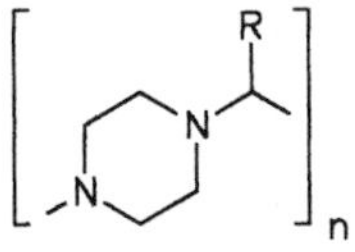

387

R = H, Ph

The synthesis of polymers **387**, starting from piperazine and aldehydes,[26] may also be included in this class of reactions, since piperazine behaves simultaneously as a substrate and as an amine of the Mannich reaction, just like a monomer of type A-B.

B Polymerizations Involving Mannich Bases

Polymeric products derived from reactions on Mannich bases may be prepared according to the following routes:

1. Polymerization of functional groups located in the substrate or the amine moiety of the Mannich base, which, however, is not involved as such, or polymerization by amino group replacement with suitable reagents, thus producing polymers having the methylene moiety inserted into the backbone (Sec. B.1)
2. Chemical modification of the Mannich base producing monomeric derivatives (mainly acrylic and vinylic), which are subsequently subjected to polymerization by the usual methods (Sec. B.2)

Many of the above reactions are also applied to the synthesis of crosslinked derivatives (see below).

B.1 Mannich Bases as Monomers

The occurrence in the substrate (R^1) or amine moiety of Mannich bases **388** (Fig. 153) of suitable groups makes it possible to obtain polymers of types **389** or **391**, respectively.

Co-units constituted by a Mannich base may also be present to a limited extent, as modifiers or residual initiators[27] (**390** and **392**), in the macromolecular chain originated by other different monomeric species (Z). In addition, when the polymerizable moieties are located in both the R^1 and R^2 residues of the Mannich base, polymers of structure **393** are formed, along with branched as well as crosslinked derivatives.

The synthetic path **a** of Fig. 153 makes it possible to obtain, by polyaddition, variously substituted polyenes (**394**, Fig. 154) starting from styrenic[28,29] or acrylic,[30-35] mainly acrylamidic,[31-35] monomers. In particular, the vinyl-β-arylamino propiophenones are prepared from the corresponding Mannich base by amino group replacement and are used in the synthesis of polymers **390** via copolymerization with styrene.[28] Analogous derivatives are obtained by copolymerization with acrylamides.[32,35]

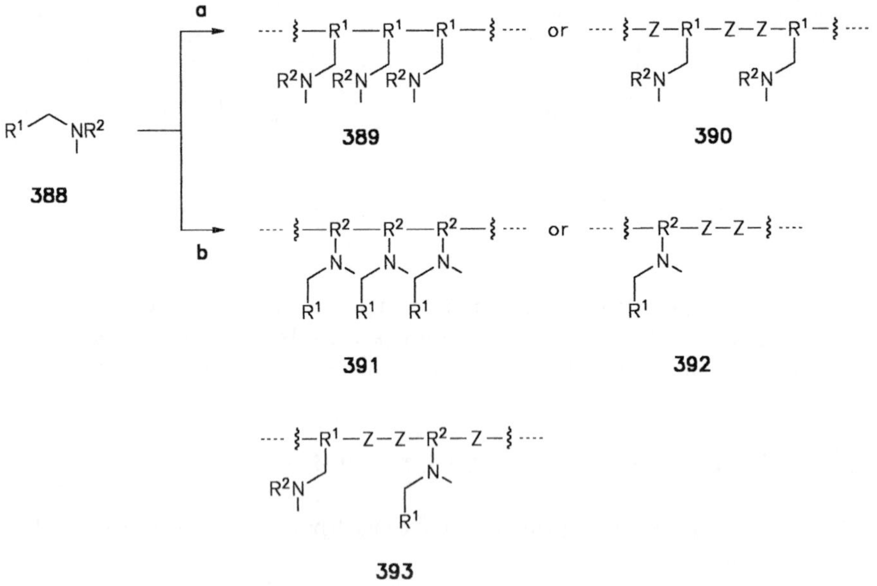

Fig. 153. The polymeric products from Mannich bases containing polymerizable moieties.

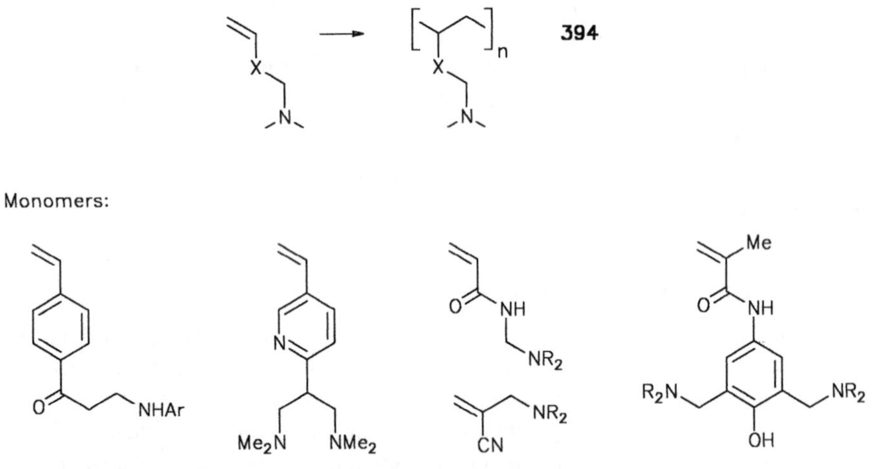

Fig. 154. Polyaddition reactions given by styrenic or acrylic Mannich bases.

Oligomeric products are given by step polymerization, starting from the amino-methyl derivatives of caprolactam (**395**)[36] and bisphenol A (**396**)[37] with epichlorohydrin.

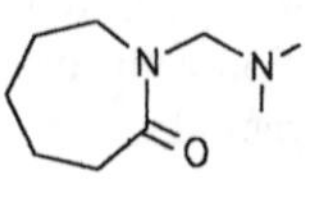

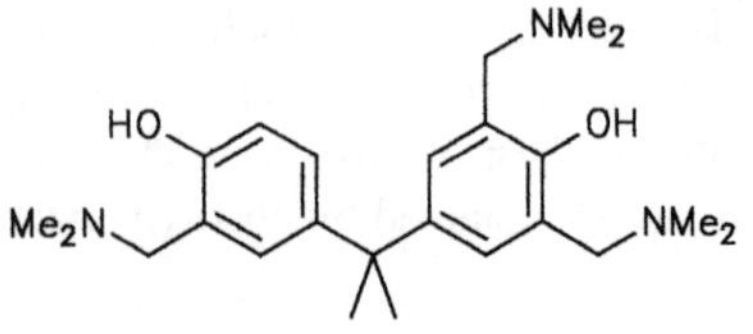

Among the limited examples of polymers originated by path **b** (Fig. 153), the polymerization[38] of diallylamino Mannich bases leading to poly(pyrrolidine)s **397** (Fig. 155) is worth mentioning. The synthesis of ionene polymers of the type **398**, used in water clarification, by reaction between *trans*-1,4-dichloro-2-butene and a phenolic Mannich base,[39] is also interesting.

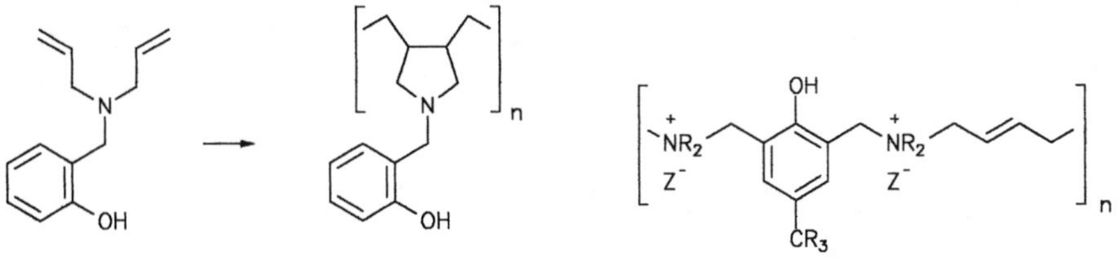

397   **398**

R = Me

Fig. 155.   Polymeric products derived from polymerizations involving the amine moiety of monomeric Mannich bases.

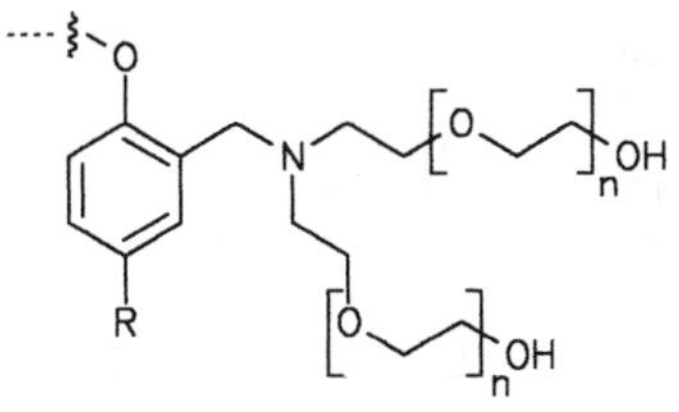

However, Mannich bases are more frequently employed as modifiers of polymeric products, as they are involved in the synthesis of derivatives having structures of type **392** or **393** (Fig. 153), as is evidenced by the epoxy oligomers **399**, obtained from phenolic alkanolamines and ethylene oxide,[40–42] in which the polyether chain involves both the phenolic and hydroxyalkyl moieties.

399

The above products are mostly used in the preparation of crosslinked derivatives (Sec. C.1).

Mannich bases may afford polycondensation with concomitant amine elimination by reaction with nucleophilic reagents. In order to obtain polymeric products through a repetitive and cumulative reaction, both the Mannich base monomer (A-A) and the nucleophilic species (B-B) need to be at least bifunctional or both the aminomethyl- and nucleophilic groups must be present in the same molecule (A-B). Polymers **400** and **401** (Fig. 156) are thus yielded, respectively.

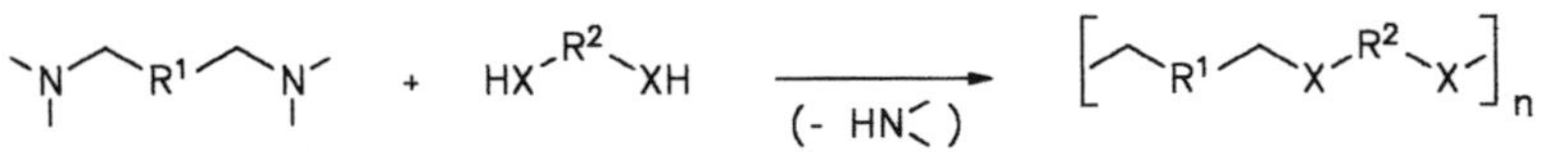

400

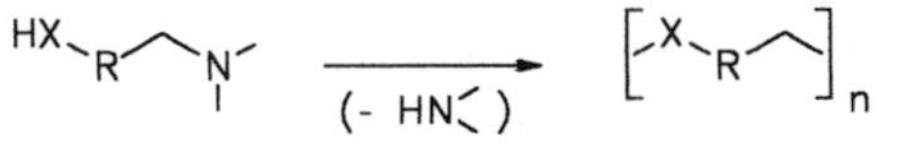

401

Fig. 156.   The polycondensation reactions of Mannich bases.

It should be noted that although the above polymeric derivatives having X = N could in principle be directly obtained by Mannich synthesis involving bifunctional substrate, formaldehyde, and bis-amine, much better results in terms of reaction yield and polymerization degree are given by the replacement method.

Exchange reactions with amines and thiols (X = N, S in Fig. 156) are the most studied (see below); however, interesting C-alkylations,[43] based on the rather unusual replacement by bis-aldehyde (**402**, Fig. 157) to give poly(β-diketone)s **403**, have also been performed.

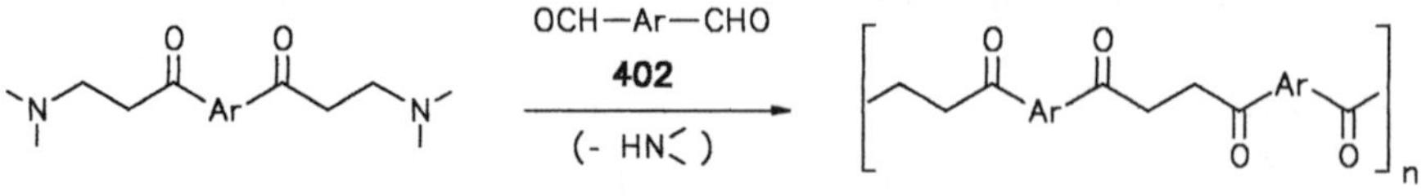

403

Fig. 157. Polycondensation reaction between bifunctional Mannich base and bis-aldehyde.

Various types of bis-Mannich bases (Table 32), ranging from ketonic to phenolic and amidic, are suitable for condensation with bis-amines and bis-thiols. As far as bis-amines are concerned, besides the compounds **379–381** of Table 31, methylpiperazine **404** is used. An even larger number of bis-thiols (**405–409**) are employed in the synthesis of polythioethers.

H_2S — **405**

$HS{-}(CH_2)_n{-}SH$ — **406** (n = 2-6)

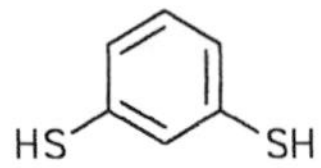

407

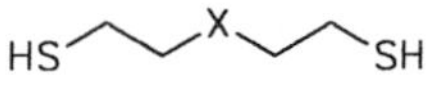

408 (X = O)

409 (X = Piperazine, Alkyl-N,...)

Among the polymers prepared by the above method, polyketoamine **410**[44] and polythioether **411**[57] are mentioned here; it is worth observing that the latter derivatives are often accompanied by variable amounts of macrocyclic compounds formed by competitive cyclization reaction, depending on reaction conditions (concentration of reactants, catalyst, etc.) and type of reacting species.[55,57]

Optically active poly(γ-ketosulfide)s, similarly obtained from chiral bis-thiols, have also been studied.[58,59]

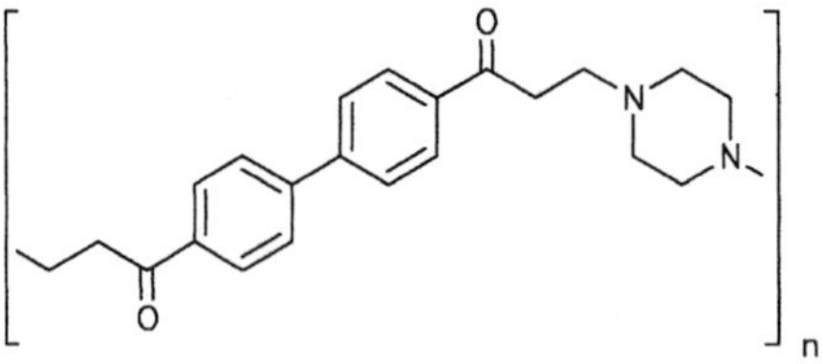

410

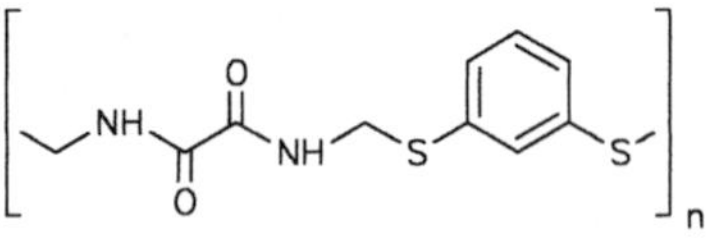

411

Dimethylamino Mannich bases are the preferred starting material, as they release volatile dimethylamine, which is easily removed from the reaction mixture. The corresponding quaternary ammonium salts also have been employed,[46] and good results are obtained by using the hydrochloride of the base in aprotic medium, since the by-product of condensation is in this case an insoluble amine hydrochloride, which is thus subtracted from the equilibrium.[60]

Table 32

Polymerization by Exchange Reaction (Fig. 156) between Mannich Bases and bis-Amines or bis-Thiols to give Polymers 400 (X = N or S, respectively)

Mannich base 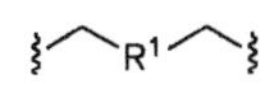	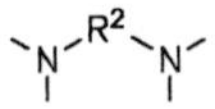 Amine moiety	Ref.	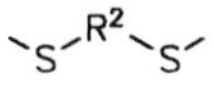 Thiol moiety	Ref.
	$N^{R^2}N$ from:		$S^{R^2}S$ from:	
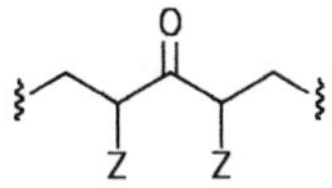 (structure: ketone with Z groups)	379,380,404	44,45	405 406-409	48,49 48,50-52
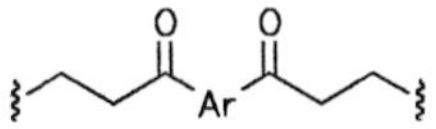 (structure: Ar diketone)	377,379,380,404	6,44,45	405 406-408	53 50,54
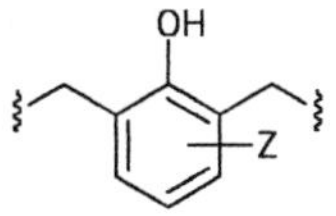 (structure: phenol OH with Z)	379,380	46	406-409	55,56
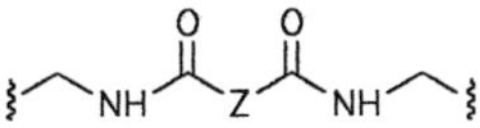 (structure: bis-amide with Z)	379-381,404	47	407	57
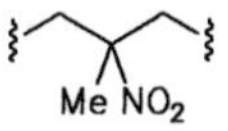 (structure: Me NO$_2$)	379	9		

Monomeric Mannich bases of type A-B, yielding polymers **401** (Fig. 156), are less frequently used.[61] Nevertheless, phenolic Mannich bases affording polymeric derivatives **412** (Fig. 158) are worth mentioning, as they are particularly applied to the production of crosslinked material (Sec. C.1). The reaction is made possible by the release of amine, as occurs, for example, during the baking of epoxy resins.[62,63]

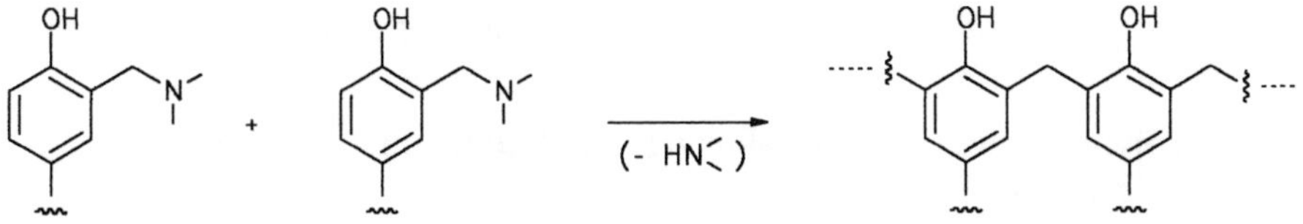

412

Fig. 158. The crosslinking reaction given by aminomethylated phenol derivatives.

B.2   Mannich Bases as Precursors of Monomers

Vinyl,[29,64] acrylic,[65–69] etc.,[71] as well as acrylonitrile[70] monomers **413–415** are readily obtained from Mannich bases. Further examples concerning analogous derivatives are reported in Chap. II, A.2. Deuterated compounds[66,67] are also included among acrylic acid derivatives along with several variously functionalized compounds.

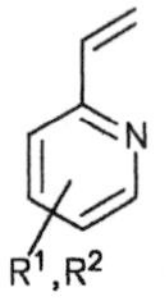

XCH_2

COOR   (R = H, Alkyl)

(X = H, Alkyl, NR'_2, $\underset{\underset{NR'_2}{|}}{CH}$—COOH)

R_2N

CN

(R^1,R^2 = H, Vinyl, Fused ring,...)

413   **414** 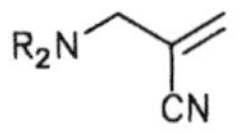   **415**

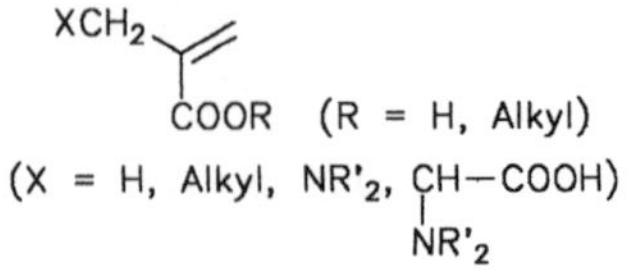

The above monomers are obtained mainly by deamination or deaminodecarboxylation of carboxylic Mannich bases. Acrylic monomers have also been obtained by addition of alkene Mannich bases to acrylonitrile[69] and vinyl monomers by deamination of the Mannich bases of alkylpyridines and alkylquinolines.[29,64]

C   Crosslinked Products from Mannich Bases

The Mannich reaction as well as Mannich bases are frequently involved in the production of crosslinked materials[1] having important applications in industry as resins, coatings, adhesives, etc., as described below in the appropriate chapters on the practical use of Mannich bases. In addition to the inherent complexity of the matter, as far as chemistry and structure of products are concerned, it is necessary to bear in mind that the literature reports, mainly patents, may be rather vague about the chemistry involved in the preparation of the final materials.

As is well known, two main routes can be adopted for obtaining crosslinked derivatives, as schematically depicted in Fig. 159; these include use of polyfunctional oligomers or monomers (route **a**) and reaction of high-molecular-weight linear polymers with crosslinking reagents (route **b**).

Mannich bases can be directly involved as macromolecular components of the above materials, or alternatively, they can participate in the process as catalysts.

C.1   Polymerization and Crosslinking of Polyfunctional Oligomers and Monomers

Almost exclusively, step polymerization involving A-A + B-B, or A-B reagents is employed in the process following route **a** of Fig. 159. As mentioned in connection with Fig. 153, the Mannich base may be the only reacting species yielding the crosslinked product, or it may take part in the reaction as, for example, a comonomer, thus acting as a modifier of the final product. In this context, the main classes of Mannich bases having practical relevance[61] are collected in formulas **416–418**, where the functional groups required for crosslinking are indicated.

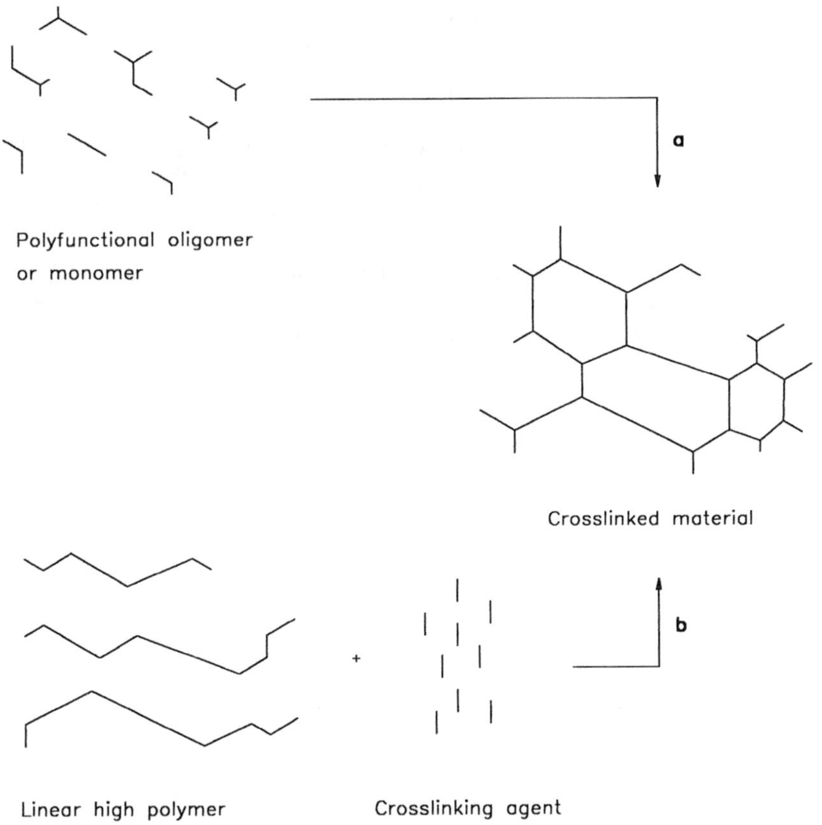

Polyfunctional oligomer
or monomer

Crosslinked material

a

Linear high polymer   Crosslinking agent

b

Fig. 159.   Routes affording crosslinked products.

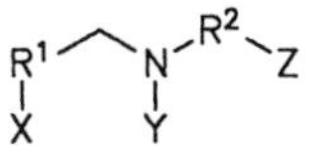

$$R^1 \overset{\displaystyle |}{\underset{\displaystyle X}{}}\!\!\!\frown\!\!\!\overset{\displaystyle |}{\underset{\displaystyle Y}{N}}\!\!\!\frown\!\!\!{}^{R^2}\!\!\!\frown Z \quad (R^1,\ R^2 \text{ also oligomeric moieties})$$

416 (X, Z = OH):   - Oligomers for polyurethanes

417 (Y = H; X and/or Z = NH):   - Reactive hardeners for epoxy resins

418 (R¹-X = substituted phenol with
          free ortho or para positions):   - Hardening of novolacs with urotropine
                                            - Baking of electrophoretic coatings

   Derivatives **416**, rich in hydroxy groups, are very good oligomers in the crosslinking of polyurethanes. Phenolic Mannich bases as well as N- and P-Mannich bases are used to this end. The first group of compounds includes mainly Mannich bases of alkanolamines,[40-42] which are treated with epoxides in order to produce hydroxylated polyfunctional oligomers having structure of type **399** (Sec. B.1). N-Mannich bases are

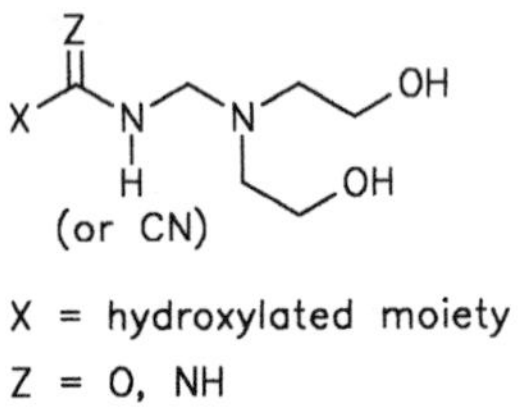

419

X = hydroxylated moiety

Z = O, NH

also consistently represented by a number of products; carbamic acid, cyanoguanidine, and 3-aminopropionamide derivatives aminomethylated with alkanolamines are used.[72–74] These molecules (**419**) may contain, in addition to the hydroxyethylamino moieties, hydroxy groups or other functions bearing active hydrogen capable of reacting with isocyanates.

P-Mannich bases are prepared from phosphonic esters possessing hydroxy groups.[75,76] Representative compounds of this type are **420** (Fig. 160), which give crosslinked polyurethanes of structure **421**, by reaction with diisocyanates.

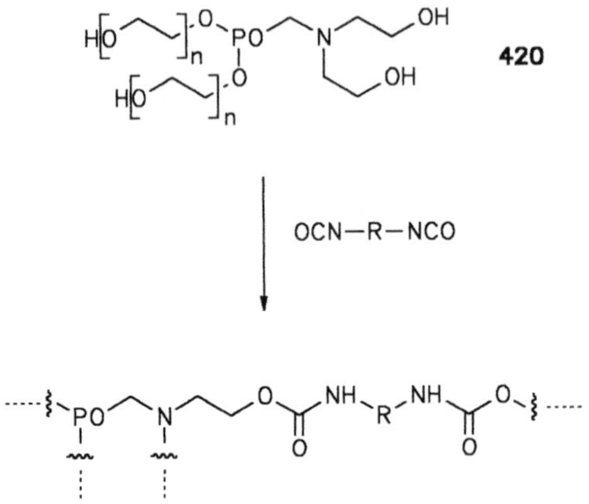

420

OCN—R—NCO

421

Fig. 160.   Crosslinked polyurethanes from polyol P-Mannich bases.

When the Mannich base constitutes only a fraction of the polyol reagent (**390, 392** in Fig. 153), it behaves as a modifier of the final polyurethane resin.

Derivatives of type **417** correspond to polymers **383** (Sec. A), obtained by Mannich polycondensation. These are polyamines frequently employed as curing agents for epoxy resins, due to the high number of reactive NH groups present along the oligomeric chain.[77–79] Thus, crosslinked derivatives having structure **422** can be obtained, even at low temperature, by grafting the epoxide oligomer (chain B), usually deriving from bisphenol and epichlorohydrin, onto the Mannich oligomer (chain A).

The chemical behavior of phenolic compounds **418** in the crosslinking is related to replacement of the amino group by the activated aromatic ring. Other nucleophilic compounds may give this reaction (see below). A significant example is afforded by the reaction of the oligomeric Mannich base **423** (Fig. 161), which, upon being heated to 210° C, releases dimethylamine with formation of the crosslinked resin **424**.[80]

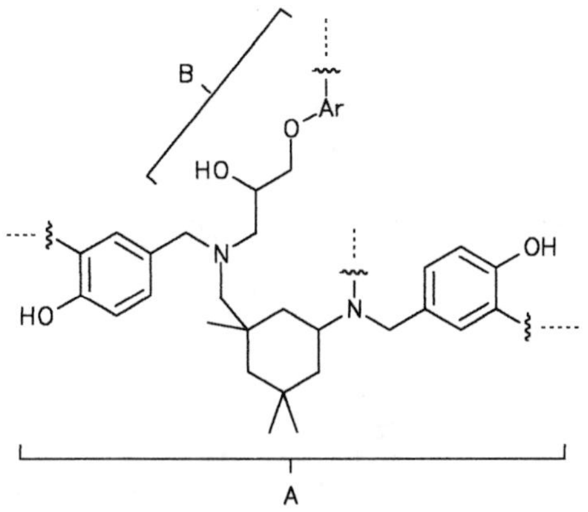

A: Oligomeric phenolic Mannich base from isophorone amine
B: Bisphenol/epichlorohydrin oligomer

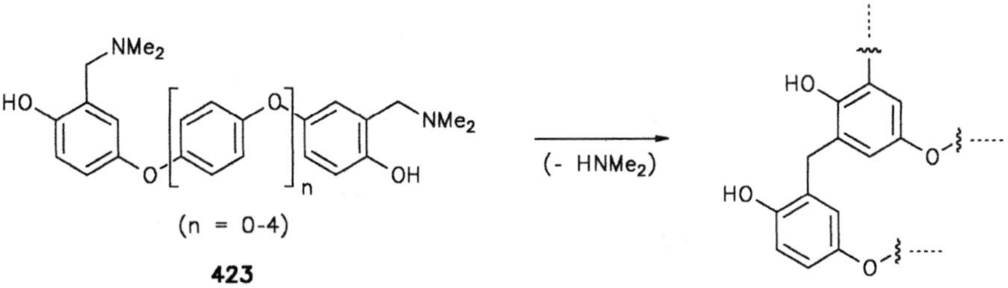

Fig. 161.   Crosslinked products by amino group replacement of oligomeric phenolic Mannich bases.

Two important industrial processes, namely, the curing of phenolic and ureic resins with urotropine[81–84] and similar hardeners[11] or with melamine,[21,85] and the baking of electrophoretic paints,[61] are based on the above reaction. In the former process the crosslinking reaction should take place through the occurrence of both the aminome-thylation reaction and amino group replacement, as indicated in Fig. 158, whereas in the latter process only the replacement reaction should occur, as it is the phenolic Man-nich bases of alkanolamine that are exclusively involved. The undesired release of amine observed during the baking,[62,63,86] due to the amino group replacement by the aromatic ring, can be satisfactorily overcome by introducing into the formulation, for instance, blocked isocyanates capable of reacting with the amine formed.

The use of Mannich bases mixed with other monomers or oligomers as modifiers of phenolic,[87] ureic,[88] epoxy,[89] etc., resins is reported. For instance, crosslinked structures of type **425** may be envisaged for the rapid curing of hydrophilic resins[90] obtained by reaction of acetone/formaldehyde/resorcinol oligomers (moiety A) with aminophenol/formaldehyde oligomers (moiety B).

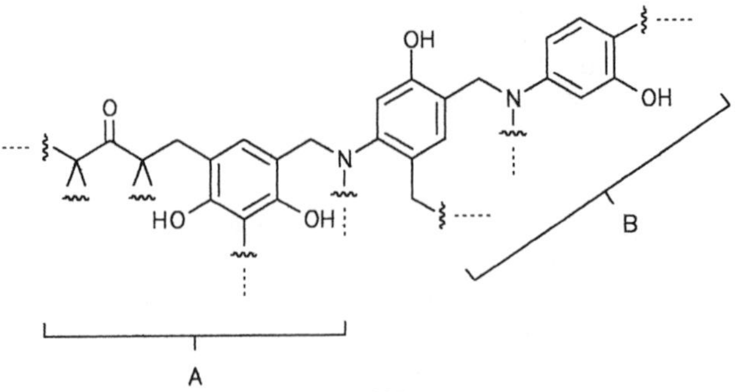

425

Finally, various Mannich bases used as catalysts in the crosslinking of oligomers (see also Chap. V, A.2.), are worth mentioning. Their basic properties are applied in the curing of epoxy oligomers[2] as well as in the production of polyurethanes[91] and, less frequently, in the crosslinking of resols.[92] Compounds of types **426** and **427** are employed mostly for the above purposes.

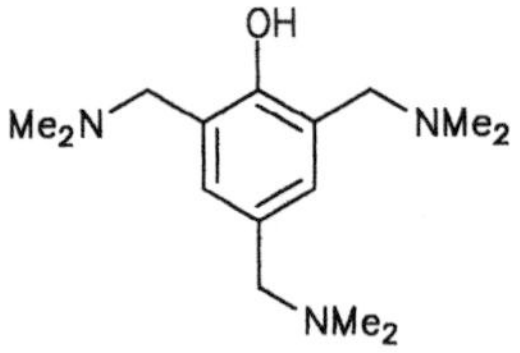

426 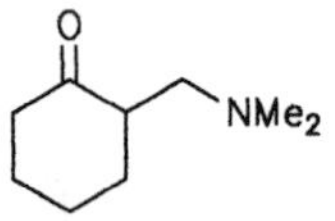   **427**

C.2   Crosslinking of High Polymers

The Mannich reaction has been applied to several crosslinking processes of linear polymers[93–95] through the interaction of di- or polyfunctional amines and formaldehyde with polymers possessing suitable reactive groups (arrows in **428–430**, Fig. 162), such as the α-carboxyketone moiety of the copolymer of allyl acetoacetate with vinyl chloride (**428**),[93] or the amide moiety of polycaprolactam (**430**),[95] for example. Ethylenediamine and aromatic amines, such as benzidine and melamine, are mainly employed as crosslinking reagents.

Melamine, in particular, is used as a hexamethylol derivative, or as the corresponding alkylether,[96,97] in the curing of coatings and adhesives. The Mannich reaction presumably involves hydroxy groups of the alcoholic moieties located on the polymeric substrate (see **429**)[96] thus affording products of type **431**.

The possibility of crosslinking natural macromolecules, such as cellulose and proteic materials, via the Mannich reaction has also been investigated. Cellulose derivatives, however, are mostly subjected to the analogous amidomethylation reaction,[98–100] usually employing bis-methylol amides (urea, oxalyl, adipoyl derivatives, etc.) capable of reacting with the cellulose hydroxy groups.

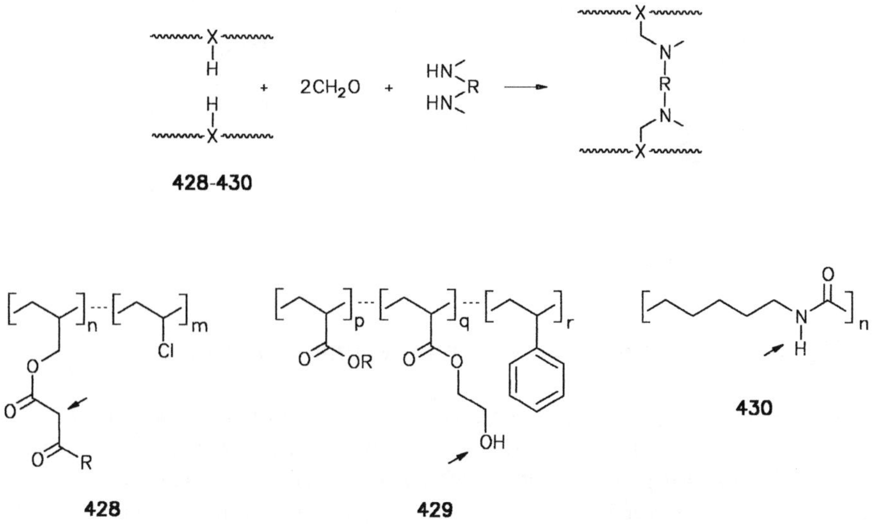

428-430

430

428   **429**

Fig. 162.   The crosslinking of high polymers by Mannich reaction using bifunctional amines.

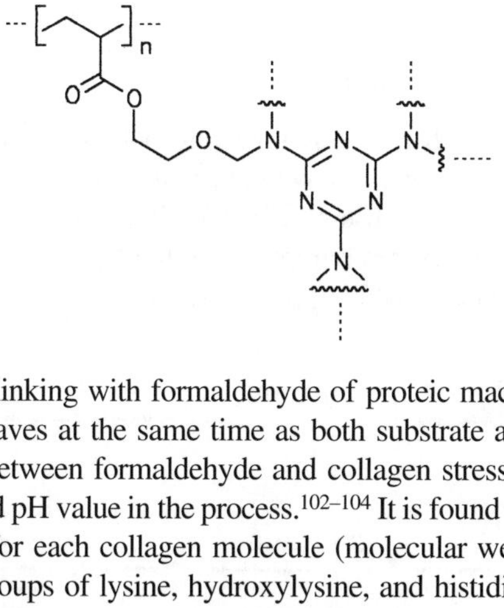

431

In the crosslinking with formaldehyde of proteic macromolecules, such as casein, the polymer behaves at the same time as both substrate and amine reagent.[101] Studies of the reaction between formaldehyde and collagen stress the importance of aldehyde concentration and pH value in the process.[102-104] It is found that the number of methylene bridges formed for each collagen molecule (molecular weight about 10,000) is nearly 15; the amino groups of lysine, hydroxylysine, and histidine present in the macromolecule are suitable for the occurrence of the Mannich reaction.[104,105]

Formaldehyde and polyhydroxy phenols, or their oligomeric derivatives,[106,107] may also lead to crosslinking by reaction with amino groups located in the macromolecular chain, thus producing derivatives with structure **432**.

Finally, certain types of Mannich bases may behave as auxiliaries in the crosslinking process. In particular, the thiolic bases **433** are employed as vulcanization accelerators in the production of tires.[108]

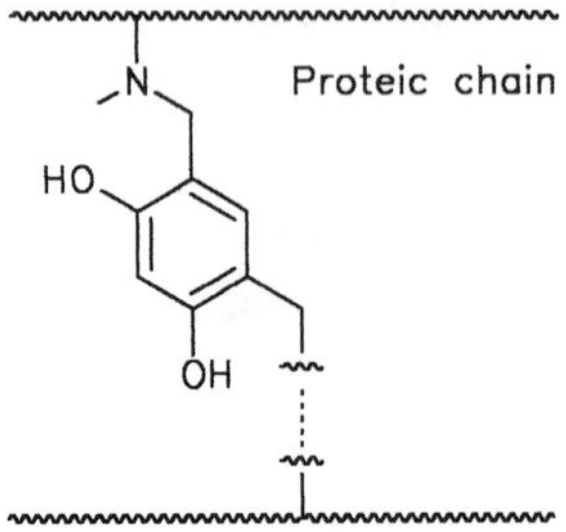

432

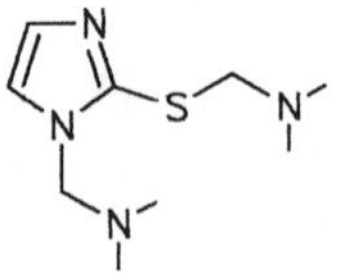

433

D Functionalization of Macromolecular Compounds

In this section we report on reactions, other than the polymerizations described above, that occur on functional groups present on any type of macromolecules and that lead to chemically modified polymers without affecting the chain length and skeleton of the initial material. The reactive groups that allow the macromolecular reactant to behave alternatively as the substrate or the amine reagent of the Mannich reaction, as well as the reactions given by macromolecules possessing Mannich aminomethyl moieties, are therefore discussed.

D.1 Mannich Reaction on Polymers

As depicted in Fig. 163, Mannich reactions involving polymers can be given (a) by polymeric substrates with low-molecular-weight amines or (b) by macromolecular amines with low-molecular-weight substrates. Polymeric amines and substrates interacting to give graft copolymers are rarely reported in the literature. In order to avoid crosslinking, the reacting species should be monofunctional.

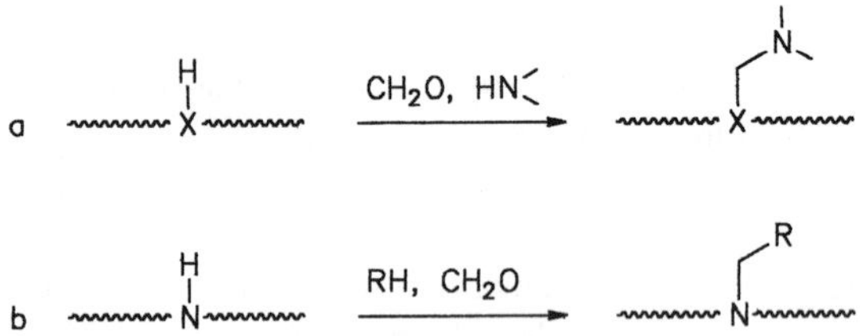

Fig. 163. The Mannich reaction on polymers.

Reactions of type (a) are among the most extensively studied, as they include the synthesis of polyacrylamide Mannich bases, widely employed in water-purification processes.[61] Many other polymeric substrates are, however, successfully subjected to Mannich reaction (Table 33). Moreover, some polymeric substances need to be suitably functionalized in order to undergo the aminomethylation reaction,[61] as reported for polymeric ketones obtained by oxidation of polyenes.[118] Further macromolecular carbonyl substrates could be provided by interesting vinyl monomers purposely designed to give polymers suitable for Mannich reaction.[119]

Macromolecular phenol derivatives have been extensively studied, and the aminomethylation reaction is performed even on crosslinked substrates.[112] Aminomethylation of lignin for the purpose of producing complexing agents should also be included in this group of reactions, as it is likely to occur on the phenolic moieties abundantly present in this material.[120,121]

Table 33

Aminomethylation Reaction Involving Polymers: Polymeric Substrates

Polymeric substrate [a]	Amine	Reference
C-Aminomethylation		
Alkyl Ketones		see text
Phenols (3- and 4-OH derivatives)	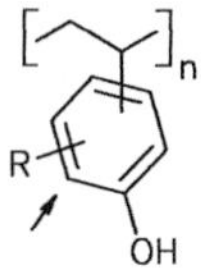 Dimethylamine, Piperidine	109-112
Aza-aromatic alkyl deriv.s	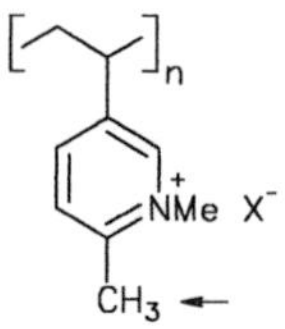 Dimethylamino deriv.s	113
N-Aminomethylation		
Polyacrylamide and copolymers	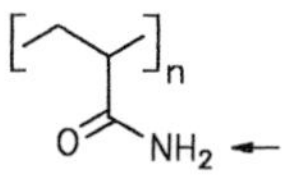 Various amines	61,114
Polyamides	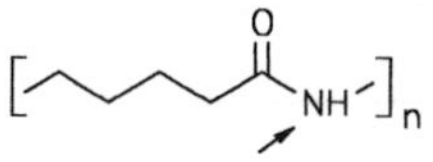 Diethylamines, Arylamines	115,116
Polymaleimide	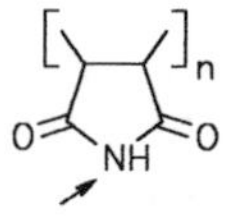 Arylamines	117

[a] Arrow indicates the position of attack

Polymeric N-Mannich bases derived, in particular, from aminomethylation of polyacrylamide[61,122] and its copolymers are the most investigated. The reaction mechanism has also been studied (Chap. I, B.2).[114]

Mannich reaction on nylon is carried out according to a method consisting of the preliminary reaction with formaldehyde/methanol followed by amine condensation on the resultant N-methoxymethyl derivative.[116]

In addition to the abovementioned macromolecular alkyl ketones, it is worth recalling the Mannich reaction on polymers functionalized[61] with phenol, amide, or acetylene moieties;[123] these last give products **434**.

Table 34

Aminomethylation Reaction Involving Polymers: Polymeric Amines

Polymeric amine	Substrate [a]	Reference
Polyalkylene amines		
e.g., 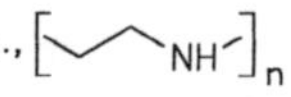 $\left[\text{CH}_2\text{CH}_2\text{NH}\right]_n$	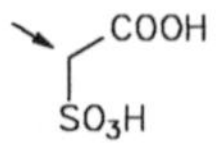 COOH / SO$_3$H	124,125
	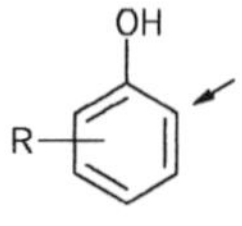 OH / R— benzene	61,126-128
	HPO(OH)$_2$	129,130
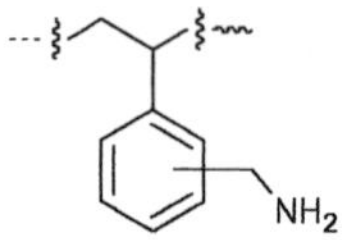 (crosslinked) polystyrene—CH$_2$NH$_2$	HPO(OH)$_2$	131
Collagen	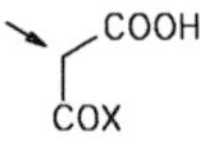 COOH / COX (X = OH, NH$_2$) 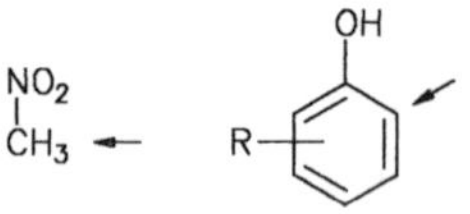	NO$_2$, CH$_3$ ← R— OH benzene	132-134
Chitosan		see text

[a] Arrow indicates the position of attack

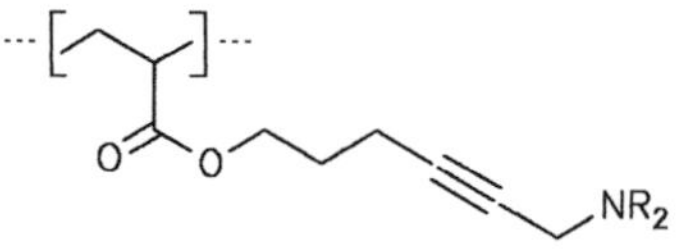

434

Reactions of type (b) (see Fig. 163) enable CH$_2$-R groups to be attached onto macromolecular substances (Table 34).

Oligomeric alkyleneamines, mainly ethyleneamines having fewer than five to ten repeating units, are studied particularly, due to their relevance as detergents for lubricating oils.[61] The substrates employed as reactants are mostly alkylphenols with molecular weights approaching 1000; however, substrates alkylated with polyenes having molecular weights up to 5000 to 10,000 are also reported. In this last case the Mannich reaction would constitute one of the rare examples of grafting between polymers, although the polymeric amine reactant has an appreciably reduced size. Other phenolic substrates such as 8-hydroxyquinoline make it possible to obtain interesting macromolecules (**435**) with complexant properties.[126]

Substrates containing acidic groups are also employed with polyalkyleneamines with the aim of introducing anionic groups as pendant side chains, thus yielding polymeric products with surface activity, sequestering agents, etc. Substrates involved in Mannich reaction with polymeric amine are sulfonyl acetic acid or phosphorous acid, and this last species is also used with cross-linked poly(aminostyrene)s (Table 34). Interesting S- and P-amidomethylations are performed with a similar purpose by using polymeric amides, such as polyacrylamide or nylon-6, with substrates of the bisulfite or phosphorous acid type.[135–137]

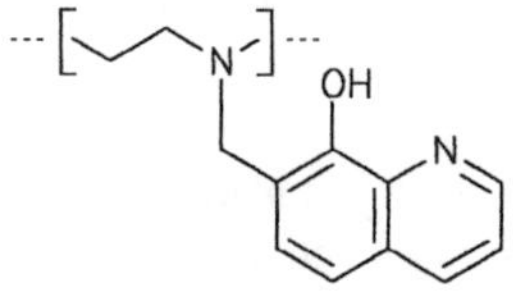

As far as natural macromolecules are concerned, collagen has been allowed to react with various substrates (Table 34) containing activated alkyl groups or phenolic groups, whereas chitosan undergoes reaction with formaldehyde only, due to the simultaneous presence of hydroxy and amino groups in the macromolecule, with production of the oxazolidine ring **436**. Chitosan, however, does not turn out to be a good trapper of formaldehyde.[138]

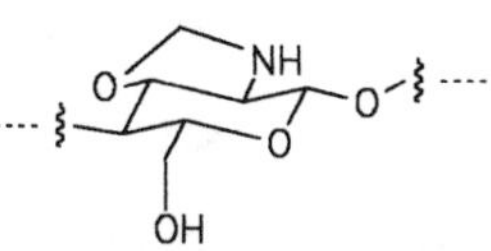

435

436

D.2 Reactions of Polymeric Mannich Bases

Due to the presence of the same moieties as in low-molecular-weight Mannich bases, some polymeric substances prepared by Mannich synthesis or by functionalization exhibit analogous chemical reactivity, thus making it possible to carry out modifications of their properties in order to improve their practical application. In particular, ionic character, stability to degradation, etc., of the polymer may thus be affected.

The production of ionic derivatives may be performed by sulfonation, as reported for the Mannich bases of poly(hydroxystyrene)s;[139] however, the most frequently adopted method is quaternization with the usual alkylating reagents of the amine nitrogen atom, which yields macromolecules having polycation structure (**437**, Fig. 164), generally much more hydrophilic than the starting material.[140–145]

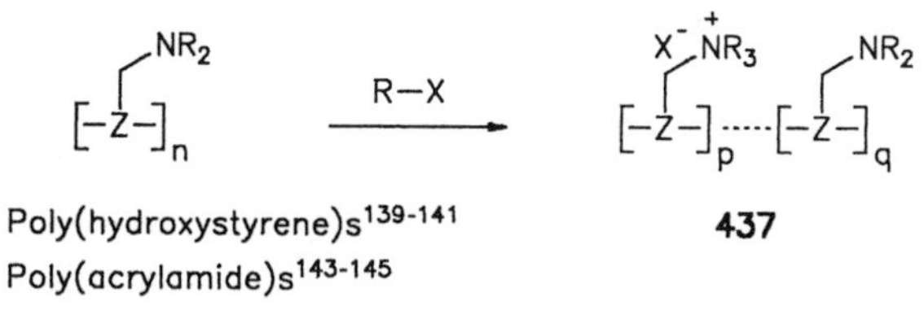

Poly(hydroxystyrene)s[139-141]
Poly(acrylamide)s[143-145]

437

Fig. 164. The quaternization of polymeric Mannich bases.

The possibility of complex formation by polymeric Mannich bases is widely exploited for practical purposes such as water treatment, metal surface protection, etc. (Chap. V).

Polymeric Mannich bases deriving from nitroalkanes[9] or alkyl ketones[146] (Table 31) can be subjected to reduction with the formation of the corresponding amino or hydroxy derivatives, respectively, for the purpose of improving the stability of the polymer to degradation, as the groups promoting the well-known cleavage reactions given by Mannich bases (Chap. II, A) are thus removed. The stereochemistry of the reduction has been investigated[146] with regard to the synthesis of poly(aminoalcohol) **438**.

A variety of specific reactions are performed on polymeric amidic Mannich bases; for instance, poly(maleimide), after aminomethylation with aromatic amine, is subjected to a coupling reaction with a diazo derivative in order to obtain azo dyes, grafted onto the polymeric chain **439**, which are able to give colored chelates.[117]

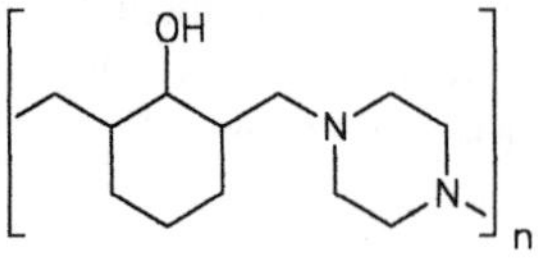

438 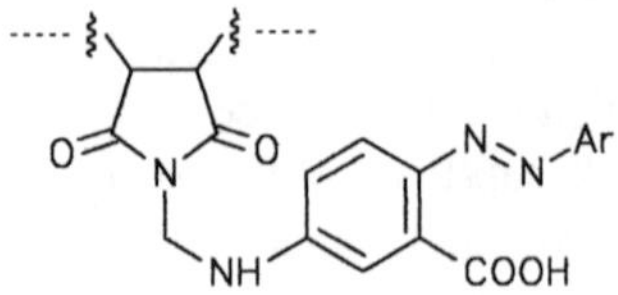   **439**

Finally, worth mentioning is an interesting example of an exchange reaction between the amino group of a Mannich base and a polymeric amine, as it permits the realization of a thus-far scarcely studied, but highly promising, possibility of polymer functionalization by groups having specific properties. The reaction (Fig. 165) consists of the linkage of the antioxidant moiety of the Mannich base **440** to the amino groups located in the main chain of poly(ethyleneimine), with the formation of an antioxidant and emulsifier polymeric product.[147]

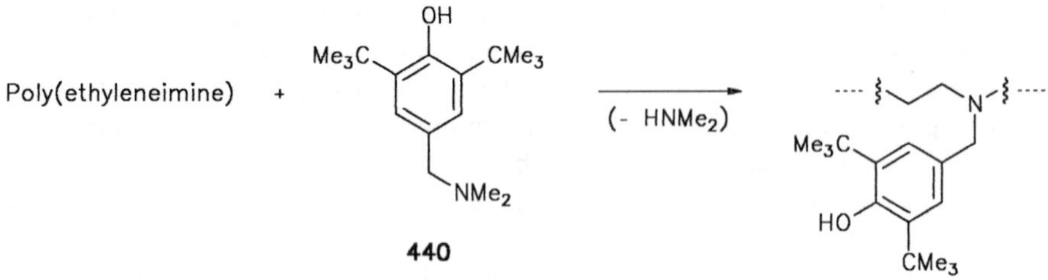

440

Fig. 165.   Functionalization of poly(ethyleneimine) by reaction with phenolic Mannich base.

1. Tramontini, M., Angiolini, L., and Ghedini, N., Mannich bases in polymer chemistry, *Polymer,* 29, 771, 1988.

2. Fedtke, M., Acceleration mechanism in curing reactions involving model systems, *Makrom. Kem. Makrom. Symp.,* 7, 153, 1987.

3. Gotlib, E. M., Voskresenskaya, O. M., Verizhnikov, L. V., Liakumovich, A. G., and Kirpichnikov, P. A., Radical processes in curing of epoxy oligomers with dimethylaminomethyl phenols, *Vysokomol. Soedin. Ser. A,* 33, 1192, 1991; *Chem. Abstr.,* 115, 115577, 1991.

4. Tomono, T., Hasegawa, E., and Tsuchida, E., Polyamine polymers from active hydrogen compounds, formaldehyde and amines, *J. Polym. Sci. Polym. Chem. Ed.,* 12, 953, 1974.

5. Tsuchida, E. and Tomono, T., Japanese Kokai Tokkyo Koho 75 51,593, 1975; *Chem. Abstr.,* 83, 115397, 1975.

6. De Voe, R. J. and Mitra, S., Polymeric Mannich bases as photosensitizers, *Polymer Prepr. Am. Chem. Soc. Div. Polym. Chem.,* 29, 522, 1988.

7. Tsuchida, E. and Tomono, T., Polyamine polymers from pyrrole, formalin and amines by use of the Mannich reaction, *J. Polym. Sci. Polym. Chem. Ed.,* 11, 723, 1973.

8. Tsuchida, E. and Hasegawa, E., Polyamine oligomers obtained from Mannich polymerization, *J. Polym. Sci. Polym. Lett. Ed.,* 14, 103, 1976.

9. Butler, G. B. and Hong, S. H., Preparation of water-soluble polymers via the Mannich reaction, *J. Macromol. Sci. Rev. Macromol. Chem.,* 24, 919, 1987.

10. Tsuchida, H. and Banno, M., Japanese 77 13,551, 1977; *Chem. Abstr.,* 89, 25108, 1978.

11. Deutsche Texaco A.G., Austrian Patent 327,561, 1976; *Chem. Abstr.,* 85, 33945, 1976.

12. Kostyuchenko, V. M., Panov, E. P., Shvarts, A. G., Prushanskaya, N. A., Vasilevich, N. Y., and Fateeva, G. V., German Offen. 2,612,975, 1977; *Chem. Abstr.,* 88, 24105, 1978.

13. Hendrickson, Y. G., Abbot, A. D., and Rothert, K., German Offen. 2,113,916, 1971; *Chem. Abstr.,* 76, 16416, 1972.

14. Akopyan, L. A., Ovakimyan, E. V., Tsaturyan, I. S., and Matsoyan, S. G., Formation of acetylenic polyamines by polycondensation using the Mannich reaction, *Vysokomol. Soedin. Ser. B,* 13, 467, 1971; *Chem. Abstr.,* 75, 152126, 1971.

15. Akopyan, L. A. and Ovakimyan, E. V., Synthesis of acetylenic polyamines by the Mannich reaction, *Arm. Khim. Zh.,* 26, 743, 1973; *Chem. Abstr.,* 80, 121365, 1974.

16. Tsuchida, E. and Tomono, T., Japanese. Kokai Tokkyo Koho 74 11,983 1974; *Chem. Abstr.,* 82, 58494, 1975.

17. Reinhardt, R. M., Daigle, D. J., and Kullman, R. M. H., U.S. Patent Appl. 614,994, 1975; *Chem. Abstr.,* 85, 178935, 1976.

18. Albini, L. and Leoni, R., Induritori a funzionalità amminica, *Pitture Vernici,* 61, 23, 1985; *Chem. Abstr.,* 103, 142828, 1985.

19. Jex Co., Ltd., Japanese Kokai Tokkyo Koho JP 82 25,813, 1982; *Chem. Abstr.,* 97, 78952, 1982.

20. Berge, A., Lien, M., Mellegaard, B., and Ugelstad, J., Aminoformaldehyde resins. Analysis by application of a Mannich-type reaction, *Angew. Makromol. Chem.,* 46, 171, 1975.

21. Samoilenko, T. G., Potapov, E. E., Tutorskii, I. A., and Dogadkin, B. A., Thermal condensation of the hexamethylolmelamine-resorcinol system, *Kauch. Rezina,* 22, 1976; *Chem. Abstr.,* 84, 122423, 1976.

22. Horiguchi, S., Nakamura, M., and Nakajima, K., Japanese 73 02,930, 1973; *Chem. Abstr.,* 80, 121738, 1974.

23. Saunders, K. J., *Organic Polymer Chemistry,* 2nd ed., Chapman & Hall, London, 1988, 356; see also Sec. 1.4.8 in Tramontini, M., Advances in the chemistry of Mannich bases, *Synthesis,* 703, 1973.

24. Chaplin, M. F. and Kennedy, J. F., Magnetic immobilised derivatives of enzymes, *Carbohydr. Res.,* 50, 267, 1976.

25. Petro, J., German Offen. 3,002,318, 1980; *Chem. Abstr.,* 93, 168869, 1980.

26. Sandler, S. R. and Delgado, M. L., Reinvestigation of the reaction of piperazine with aldehydes, *J. Polym. Sci.,* 7, 1373, 1969.

27. Leonte, M., Florea, T., Isbasciu, M., Pope, A., and Ionita, G., Romanian Patent RO 87,062, 1985; *Chem. Abstr.,* 105, 155097, 1986.

28. Kamogawa, H., Kubota, K., and Nanasawa, M., Syntheses of polymerizable aromatic Mannich bases by means of amine exchange reaction, *Bull. Chem. Soc. Jpn.,* 51, 561, 1978.

29. Kagan, E. S. and Ardashev, B. I., Mannich reaction with 2-methyl-5-vinylpyridine, *Khim. Geterotsikl. Soedin.,* 1066, 1968; *Chem. Abstr.,* 70, 77740, 1969.

30. Stevens, J., German Offen. 1,965,872, 1970; *Chem. Abstr.,* 73, 99407, 1970.

31. McDonald, C. J., European Patent Appl. 2,254, 1979; *Chem. Abstr.*, 91, 193884, 1979.

32. Barabas, E. S. and Grosser, F., U.S. Patent 3,929,739, 1975; *Chem. Abstr.*, 84, 91002, 1976.

33. Cascaval, A., Budeano, C., Ofenberg, H., Nicolaescu, T., Antohe, N., and Ionescu, E., Romanian Patent RO 80,000, 1982; *Chem. Abstr.*, 99, 175229, 1983.

34. Nitto Chem. Ind. Co., Ltd., Japanese Kokai Tokkyo Koho JP 60,104,054, 1985; *Chem. Abstr.*, 103, 177957, 1985.

35. Ritter, H. and Rodewald, S., Synthesis of 3′,5′-bis(morpholinomethyl)-4′-hydroxy- and 3′,5′-bis(4-methyl-1-piperazinylmethyl)-4′-hydroxy methacrylamide. Copolymerization and metal ion binding properties of monomers and copolymers, *Makromol. Chem.*, 187, 801, 1986.

36. Meyer, H. R., Reaction products of caprolactam and polycaprolactam with CH_2O and $COCl_2$, *Kunstst. Plast.*, 3, 160, 1956; *Chem. Abstr.*, 52, 11781, 1958.

37. Kyoritsu Org. Ind. Res. Lab., Japanese Kokai Tokkyo Koho JP 58,153,506, 1983; *Chem. Abstr.*, 100, 86653, 1984.

38. Hodgkin, J. H. and Allan, R. J., Cyclopolymerization. XIII. Cyclopolymerization of diallylaminomethylphenols, *J. Macromol. Sci. Chem.*, A11, 937, 1977.

39. Tai, W. T., U.S. Patent 4,038,318, 1977; *Chem. Abstr.*, 87, 136864, 1977.

40. Case, L. C., U.S. Patent 3,671,470, 1972; *Chem. Abstr.*, 77, 89367, 1972.

41. Thorpe, D., European Patent Appl. EP 237,270, 1987; *Chem. Abstr.*, 108, 38588, 1988.

42. Kan, P. T. Y., Cenker, M., and Patton, J. T., Jr., German Offen. 2,546,185, 1976; *Chem. Abstr.*, 85, 124953, 1976.

43. Winberg, H., Vries, T., Pouwer, K., Havinga, E. E., and Meijer, E. W., Conducting polymers via a novel polymerization reaction, *New Aspects Org. Chem. 1*, Proc. 4th Int. Kyoto Conf. 1989, 343; *Chem. Abstr.*, 114, 207839, 1991.

44. Andreani, F., Angeloni, A. S., Angiolini, L., Costa Bizzarri, P., Della Casa, C., Fini, A., Ghedini, N., Tramontini, M., and Ferruti, P., Poly(β-aminoketone)s by polycondensation of bis(β-dialkylaminoketone)s with bis-amines, *J. Polym. Sci. Polym. Lett. Ed.*, 19, 443, 1981.

45. Angeloni, A. S., Ferruti, P., Laus, M., Tramontini, M., Chiellini, E., and Galli, G., The Mannich bases in polymer synthesis. 4. Further studies on synthesis and characterization of poly(β-aminoketone)s containing potentially mesogenic groups, *Polymer Commun.*, 24, 87, 1983.

46. Ghedini, N., Della Casa, C., Costa Bizzarri, P., and Ferruti, P., Mannich bases in polymer synthesis. 8. Synthesis of poly[(alkylimino)alkylene(alkylimino)methylene (2-hydroxy-5-methyl-1,3-phenylene, methylene]s, *Makromol. Chem. Rapid Commun.*, 5, 181, 1984.

47. Angeloni, A. S., Ferruti, P., and Laus, M., Mannich bases in polymer synthesis. 7. Poly(N-aminomethyleneamides) by polycondensation reaction of bis Mannich bases of dicarboxyamides with bis secondary amines, *Polymer Commun.*, 25, 119, 1984.

48. Ferruti, P., Angiolini, L., Ghedini, N., Manaresi, P., and Pilati, F., Italian Patent Appl. 23166 A/83, 1982.

49. Robert, W. M., U.S. Patent 2,820,022, 1958; *Chem. Abstr.*, 52, 5032, 1958.

50. Andreani, F., Angiolini, L., Costa Bizzarri, P., Della Casa, C., Ferruti, P., Ghedini, N., Tramontini, M., and Pilati F., The Mannich bases in polymer synthesis. V. Synthesis and characterization of poly(β-ketothioethers) from aliphatic and aromatic bis(β-dialkylaminoketones) and bis-thiols, *Polymer Commun.*, 24, 156, 1983.

51. Andreani, F., Angeloni, A. S., Angiolini, L., Costa Bizzarri, P., Della Casa, C., Ferruti, P., Fini, A., Ghedini, N., Manaresi, P., Pilati, F., and Tramontini M., Italian Patent Appl. 20305 A/82, 1982.

52. Ranucci, E. and Bignotti, F., New basic multifunctional polymers. VIII. Poly(aminothioether-ketone)s, *Chim. Ind.*, 72, 799, 1990.

53. Robert, W. M., U.S. Patent 2,783,216, 1957; *Chem. Abstr.*, 51, 12545, 1957.

54. Andreani, F., Costa Bizzarri, P., Della Casa, C., Ferruti, P., and Ghedini, N., Italian Patent Appl. 22180 A/82, 1982.

55. Costa Bizzarri, P., Della Casa, C., Ferruti, P., Ghedini, N., Pilati, F., and Scapini, G., Mannich bases in polymer synthesis. VI. Sulphur containing polymers by exchange reaction between 2,6-bis(dimethylaminomethyl)-4-methylphenol and dithiols, *Polymer Commun.*, 25, 115, 1984.

56. Ranucci, E. and Ferruti, P., New basic multifunctional polymers. VII. A poly(aminothioether-phenol) by polycondensation of 2,2'-(1,4-piperazinediyl)diethanethiol with 2,6-bis(dimethylaminomethyl)-4-methylphenol, *Chim. Ind.*, 72, 795, 1990.

57. Laus, M., Angeloni, A. S., Tramontini, M., and Ferruti, P., The Mannich bases in polymer synthesis. 9. Sulphur containing polymers by polycondensation reaction of bis Mannich bases of dicarboxyamides with 1,3-dimercaptobenzene, *Polymer Commun.*, 25, 281, 1984.

58. Angiolini, L., Carlini, C., and Salatelli, E., Sulfur containing optically active polymers. I. Synthesis and chiroptical properties of poly(γ-ketosulfide)s derived from (−) (2R,3R)-1,4-dimercapto-2,3-butanediol, *Makromol. Chem.*, 193, 2883, 1992.

59. Angiolini, L., Carlini, C., and Salatelli, E., Sulfur containing optically active polymers. II. Synthesis and chiroptical properties of poly(γ-ketosulfide)s containing the (2R,3R)-1,4-dimercapto-2,3-butanediolbutyraldehydeacetal moiety, *Polymer*, 34, 3778, 1993.

60. Angiolini, L., Ghedini, N., and Tramontini, M., The Mannich bases in polymer synthesis. 10. Synthesis of poly(β-ketothioethers) and their behaviour towards hydroperoxide reagents, *Polymer Commun.*, 26, 218, 1985.

61. See survey in Ref. 1.

62. Diefenbach, H., Hoppe, K., and Streitberger, H. J., German Offen. 2,751,499, 1979; *Chem. Abstr.*, 91, 41002, 1979.

63. Kempter, F. E., Hartmann, H., and Gulbins, E., German Offen. 2,711,425, 1978; *Chem. Abstr.*, 89, 199207, 1978.

64. Kagan, E. S. and Ardashev, B. I., Synthesis of vinyl derivatives of quinoline by Mannich reaction, *Khim. Geterotsikl. Soedin.*, 701, 1967; *Chem. Abstr.*, 68, 114408, 1968.

65. Carpenter, C. R. and Clovis, J. S., U.S. Patent 4,146,735, 1979; *Chem. Abstr.*, 91, 57789, 1979.

66. Mikhailov, M. K., Dirlikov, S. K., and Georgieva, T. P., Synthesis of deuterated methyl-methacrylate CD_2=$C(Me)COOD_3$, *Vysokomol. Soedin. Ser. B*, 18, 398, 1976; *Chem. Abstr.*, 85, 94704, 1976.

67. Mikhailov, M. K., Dirlikov, S. K., Peeva, N. N., and Georgieva, T. P., Synthesis of poly(methylmethacrylate-α-CD_3), *Vysokomol. Soedin. Ser. B*, 16, 724, 1974; *Chem. Abstr.*, 82, 90247, 1975.

68. Miller, R. B. and Smith, B. F., General synthesis of α-substituted acrylic esters, *Synth. Commun.*, 3, 359, 1973.

69. Boothe, J. E., Sharpe, A. J., and Noren, G. K., Synthesis and polymerization of 3-acrylamido-3-methyl butyl trimethyl ammonium chloride, *Polym. Prepr. Am. Chem. Soc. Div. Polym. Chem.*, 16, 649, 1975; *Chem. Abstr.*, 86, 30095, 1977.

70. Miller, R. B. and Smith, B. F., General synthesis of α-substituted acrylonitriles, *Synth. Commun.*, 3, 413, 1973.

71. Farberov, M. I. and Mironov, G. S., Commercial synthesis of carbonyl monomers by the Mannich reaction, *Dokl. Akad. Nauk SSSR*, 148, 1095, 1963; *Chem. Abstr.*, 59, 5062, 1963.

72. Mueller, E. and Thomas, H., German Offen. 2,246,108, 1974; *Chem. Abstr.*, 81, 121719, 1974.

73. Thomas, H. and Mueller, E., German Offen. 2,223,427, 1974; *Chem. Abstr.*, 80, 134172, 1974.

74. Sellet, L., German Offen. 1,965,562, 1970; *Chem. Abstr.*, 73, 100165, 1970.

75. Lankro Chemicals, Ltd., French Patent 1,470,166, 1967; *Chem. Abstr.*, 67, 117741, 1967.

76. Lester, F., U.S. Patent 3,309,342, 1967; *Chem. Abstr.*, 67, 117738, 1967.

77. Kansai Paint Co., Ltd., Japanese Kokai Tokkyo Koho JP 58,187,462, 1983; *Chem. Abstr.*, 100, 122881, 1984.

78. Marten, M. and Godau, C., German Offen. DE 3,624,314, 1988; *Chem. Abstr.*, 108, 206358, 1988.

79. Mine, S., Japanese Kokai Tokkyo Koho JP 02,269,782, 1990; *Chem. Abstr.*, 114, 187648, 1991.

80. Vorob'ev, Y. P., Sergeev, V. A., Korshak, V. V., and Danilov, V. G., Thermoreactive polymers from bisphenols having poly(p-phenylene oxide) structure and formaldehyde, *Vysokomol. Soedin. Ser. A*, 9, 1763, 1967; *Chem. Abstr.*, 67, 100461, 1967.

81. Kopf, P. W. and Wagner, E. R., Formation and cure of novolacs: NMR study of transient molecules, *J. Polym. Sci. Polym. Chem. Ed.*, 11, 939, 1973.

82. Saunders, K. J., *Organic Polymer Chemistry*, 2nd ed., Chapman & Hall, London, 1988, 326, 335.

83. Ogasawara, K. and Date, H., Japanese Kokai Tokkyo Koho 76 16,399, 1976; *Chem. Abstr.*, 85, 124958, 1976.

84. Kunitomi, K., Japanese Kokai Tokkyo Koho 79 50,097, 1979; *Chem. Abstr.*, 91, 92437, 1979.

85. Takakura, M. and Minoshima, N., Japanese Kokai Tokkyo Koho 79 120,659, 1979; *Chem. Abstr.*, 92, 77364, 1980.

86. Diefenbach, H., Hoppe, K., and Streitberger, H. J., German Offen. 2,759,659, 1980; *Chem. Abstr.*, 94, 5022, 1981.

87. Unitika, Ltd., Japanese Kokai Tokkyo Koho JP 57,197,040, 1982; *Chem. Abstr.*, 98, 127407, 1983.

88. Petersen, H., Syntheses of cyclic ureas by α-ureidoalkylation, *Synthesis*, 243, 1973; see, in particular, p. 251.

89. Streitberger, H. J., Lessmeister, P., and Zdahl, N., German Offen. 2,923,589, 1980; *Chem. Abstr.*, 93, 96970, 1980.

90. Freeman, H. G., Gillern, M., and Smith, H. A., U.S. Patent 3,947,425, 1976; *Chem. Abstr.*, 85, 47584, 1976.

91. Bakirova, I. N., Zenitova, L. A., and Kirpichnikov, P. A., Deposited doc., 1981, SPSTL 425 Khp-D81; *Chem. Abstr.*, 98, 35822, 1983.

92. Laqua, A., Holtschmidt, U., and Schedlitzki, D., German Offen. 2,363,357, 1975; *Chem. Abstr.*, 83, 148305, 1975.

93. Farbwerke Hoechst A.-G., Neth. Appl. 6,609,044; *Chem. Abstr.*, 67, 44405, 1967.

94. Popov, K., Production of new polymer products with flocculating properties by the Mannich amino methylation of prehydrolized ternary acrylonitrile-methyl methacrylate-sodium vinylsulfonate copolymers, *Tr. Vodosnabdyavane, Kanaliz. Sanit. Tekh.*, 12, 119, 1977; *Chem. Abstr.*, 87, 185004, 1977.

95. Theil, I. and Constantinescu, A. C., Romanian Patent RO 62,930, 1977; *Chem. Abstr.*, 98, 72952, 1983.

96. Kolish, B. L., Senseman, W. K., and Konker, C. H., U.S. Patent 3,978,015, 1976; *Chem. Abstr.*, 85, 179135, 1976.

97. Saunders, K. J., *Organic Polymer Chemistry*, 2nd ed., Chapman & Hall, London, 1988, 351–356.

98. Petropavlovskii, G. A. and Kotel'nicova, N. E., Crosslinking of low molecular weight of hydroxyethyl cellulose films with formaldehyde and its derivatives, *Cell. Chem. Technol.*, 11, 551, 1977; *Chem. Abstr.*, 88, 138134, 1978.

99. Vasil'eva, G. G., Petropavlovskii, G. A., and Simanovic, I. E., Study of the solid-phase crosslinking of ethylcellulose, *Zh. Prikl. Khim. Leningrad*, 52, 1833, 1979; *Chem. Abstr.*, 91, 176857, 1979.

100. Vasil'eva, G. G. and Petropavlovskii, G. A., Effect of the molecular weight of methylcellulose on its crosslinking and on the properties of crosslinked films, *Zh. Prikl. Khim. Leningrad*, 53, 2591, 1980; *Chem. Abstr.*, 94, 48999, 1981.

101. Saunders, K. J., *Organic Polymer Chemistry*, 2nd ed., Chapman & Hall, London, 1988, 213.

102. Sugiura, M., Shinbo, T., Kikkawa, M., and Toyoda, H., Membrane potential and ionic permeability of formaldehyde-tanned collagen membranes, *Nippon Nogei Kagaku Kaishi*, 48, 493, 1974; *Chem. Abstr.*, 82, 100090, 1975.

103. Chirita, G., Chirita, A., Bulacovschi, J., and Chisalita, D., Reactions of aldehydes with the functional groups of some amino acids of the collagen structure. II. *J. Am. Leather Chem. Assoc.*, 69, 572, 1974; *Chem. Abstr.*, 82, 100097, 1975.

104. Kubota, M. and Kitada, I., Studies on the bridging reaction of collagen molecules with aldehydes. III. Bridging reaction with formaldehyde and ultraviolet absorption of bridging compounds, *Hikaku Kagaku*, 27, 137, 1981; *Chem. Abstr.*, 96, 144875, 1982.

105. Feairheller, S. H., Taylor, M. M., Gruber, H. A., Mellon, E. F., and Filachione, E. M., Some properties of collagen modified by the Mannich reaction, *J. Polym. Sci. C*, 24, 163, 1967.

106. Hannigan, M. V. and Windus, W., The tanning action of a hemiacetal in combination with resorcinol, *J. Am. Leather Chem. Assoc.*, 60, 581, 1965; *Chem. Abstr.*, 64, 3875, 1966.

107. Gringras, L. and Sjostedt, G., Crosslinking of gelatin with formaldehyde in the presence of some polyhydroxybenzenes, *Angew. Makromol. Chem.*, 42, 123, 1975.

108. Safaev, A., Kadyrov, A., Saidaliev, Z. G., and Afanas'ev, G. V., Deposited doc., 1974, VINITI 751; *Chem. Abstr.*, 86, 189803, 1977.

109. Maruzen Oil Co., Ltd., Japanese Kokai Tokkyo Koho JP 82 102,906, 1982; *Chem. Abstr.*, 97, 198881, 1982.

110. Ferruti, P. and Bettelli, A., Linear, high molecular weight poly(2-alkyl-4-vinyl-6-(dialkylamino-methyl)-phenols) and poly((2,6 bis-dialkylaminomethyl)-4-vinylphenols), *Polymer*, 184, 1972.

111. Maruzen Oil Co., Ltd., Japanese Kokai Tokkyo Koho 80 54,302, 1980; *Chem. Abstr.*, 93, 95935, 1980.

112. Fujiwara, H., Takahashi, A. and Sekiya, M., Japanese Kokai Tokkyo Koho 77 58,087, 1977; *Chem. Abstr.*, 87, 69413, 1977.

113. Evdokimov-Skopinskii, A. N., Zubakova, L. B., Gandurina, L. V., Osokina, M. P., Zhovnirovskaya, A. B., and Buyanovskaya, T. V., U.S.S.R. Patent 460,283, 1975; *Chem. Abstr.*, 83, 60424, 1975.

114. McDonald, C. J. and Beaver, R. H., The Mannich reaction of polyacrylamide, *Macromolecules*, 12, 203, 1979.

115. Horn, J., Albert, W., Batz, H. G., Nelboeck, M., and Hochstetter, M., German Offen. 2,708,018, 1978; *Chem. Abstr.*, 90, 18773, 1979.

116. Nowak, Z. and Szczepaniak, M., Diethylaminomethylation of N-methoxymethyl-polycaprolactam and properties of products, *Polimery Warsaw*, 27, 380, 1982; *Chem. Abstr.*, 99, 38895, 1983.

117. Agency of Ind. Sci. and Tech., Japanese Kokai Tokkyo Koho JP 57,162, 726, 1982; *Chem. Abstr.*, 98, 126868, 1983.

118. West, C. T., U.S. Patent 4,170,562, 1979; *Chem. Abstr.*, 92, 44562, 1980.

119. Ruggeri, G., Bianchi, M., and Aglietto, M., Polymers with oxime groups: reactivity of ketoaromatic side chains towards hydroxylamine, *Chim. Ind.*, 68, 97, 1986.

120. Lin, S. Y. and Hoo, L. H., U.S. Patent 4,728,728, 1988; *Chem. Abstr.*, 108, 188732, 1988.

121. Brezny, R., Paszner, L., Micko, M., and Uhrin, D., The ion-exchanging lignin derivatives prepared by Mannich reaction with amino acids, *Holzforschung*, 42, 369, 1988; *Chem. Abstr.*, 111, 9088, 1989.

122. Krebbs, R. F., Marek, P. J., and Phillips, K. G., U.S. Patent 4,405,728, 1983; *Chem. Abstr.*, 99, 213096, 1983.

123. D'Alelio, G. F., U.S. Patent 3,562,236, 1971; *Chem. Abstr.*, 74, 88460, 1971.

124. Dytyuk, L. T. and Samakaev, R. K., U.S.S.R. Patent 761,494, 1980; *Chem. Abstr.*, 93, 221485, 1980.

125. Oleinikov, A. N., U.S.S.R. Patent 829,637, 1981; *Chem. Abstr.*, 95, 43949, 1981.

126. Balakin, V. M., Glukhikh, V. V., Litvinets, Y. I., Validuda, G. I., and Pogudina, L. K., Synthesis and properties of water-soluble complexing polymers. II. Synthesis and physicochemical properties of high-molecular weight Mannich bases with 7-aminomethylene-8-hydroxyquinoline groups, *Zh. Obshch. Khim.*, 48, 2782, 1978; *Chem. Abstr.*, 90, 169030, 1979.

127. Crawford, J., British Patent 1,519,955, 1978; *Chem. Abstr.*, 90, 139988, 1979.

128. Org. Chem. Dep. du Pont, Mannich condensation products, *Res. Discl.*, 144, 58, 1976; *Chem. Abstr.*, 87, 8336, 1977.

129. Kautsky, G. J., U.S. Patent 3,738,937, 1973; *Chem. Abstr.*, 79, 137279, 1973.

130. Balakin, V. M., Balakin, S. M., and Tesler, A. G., Deposited doc., 1977, VINITI 1359; *Chem. Abstr.*, 90, 187670, 1979.

131. Balakin, V. M., Tesler, A. G., and Vydrina, T. S., Study of the reactions of polymers containing primary or secondary aminogroups. V. Study of the reaction of cros-slinked styrene copolymers containing benzylamine groups with phosphorous acid and formaldehyde, *Izv. Vyssh. Uchebn. Zaved., Khim. Khim. Tekhnol.*, 22, 1279, 1979; *Chem. Abstr.*, 92, 164568, 1980.

132. Feairheller, S. H., Taylor, M. M., and Filachione, E. M., Chemical modification of collagen by the Mannich reaction, *J. Am. Leather Chem. Assoc.*, 62, 408, 1967; *Chem. Abstr.*, 68, 3928, 1968; see also *J. Am. Leather Chem. Assoc.*, 62, 398, 1967.

133. Happich, M. L., Windus, W., and Showell, J. S., Tanning with tris(hydroxymethyl)nitromethane and resorcinol, *J. Am. Leather Chem. Assoc.*, 65, 135, 1970; *Chem. Abstr.*, 72, 112818, 1970.

134. Ranganathan, S., Bose, S. M., and Nayudamma, Y., Studies on the formation of Mannich linkages involving phenolic compounds with reference to the interaction of collagen with certain tanning agents, *Bull. Central Leather Res. Inst. Madras*, 5, 447, 1959; *Chem. Abstr.*, 53, 23027, 1959.

135. Kruglova, V. A., Annenkov, V. V., Zaitseva, I. V., Kalabina, A. V., and Mirskova, A. N., Synthesis, study, and chemical reactions of copolymers of acrylamide with 2-trichloromethyl-4-methylene-1,3-dioxolane, *Vysokomol. Soedin. Ser. B*, 25, 852, 1983; *Chem. Abstr.*, 100, 86172, 1984.

136. Fong, D. W., U.S. Patent 4,795,789, 1989; *Chem. Abstr.*, 110, 174027, 1989.

137. Falewicz, P., Polyfunctional derivatives of polyamides, *Pr. Nauk Inst. Technol. Nieorg. Nowozow. Miner. Politech. Wroclaw.*, 33, 62, 1987; *Chem. Abstr.*, 108, 95072, 1988.

138. Dutkiewicz, J., Some aspects of the reaction between chitosan and formaldehyde, *J. Macromol. Sci. Chem.*, A20, 877, 1983.

139. Fujiwara, H., Sekiya, M., and Suzuki, H., Japanese Kokai Tokkyo Koho 74 53,281, 1974; *Chem. Abstr.*, 81, 170491, 1974.

140. Fujiwara, H., Sekiya, M., and Suzuki, H., Japanese Kokai Tokkyo Koho 74 53,282, 1974; *Chem. Abstr.*, 81, 136995, 1974.

141. Fujiwara, H., Sekiya, M., and Suzuki, H., German Offen. 2,327,661,1973;*Chem. Abstr.*, 80, 121730, 1974.

142. Hunter, W. E. and Sieder, T. P., U.S. Patent 4,049,606, 1977; *Chem. Abstr.*, 87, 202365, 1977.

143. Kyoritsu Yuki Co., Ltd., Japanese Kokai Tokkyo Koho, 80 84,498, 1980; *Chem. Abstr.*, 93, 169975, 1980.

144. Phillips, K. G., Ballweber, E. G., and Hurlock, J. R., U.S. Patent 4,179,424, 1979; *Chem. Abstr.,* 92, 77414, 1980.

145. Mori, H., Sato, S., and Nakanishi, Y., Japanese Kokai Tokkyo Koho JP 61,166,803, 1986; *Chem. Abstr.,* 105, 191861, 1986.

146. Angeloni, A. S., Ferruti, P., Tramontini, M., and Casolaro, M., The Mannich bases in polymer synthesis. III. Reduction of poly(β-aminoketone)s to poly(γ-aminoalcohol)s and their N-alkylation to poly(γ-hydroxy quaternary ammonium salt)s, *Polymer,* 23, 1693, 1982.

147. Volkov, R. N. and Sigov, O. V., U.S.S.R. Patent SU 979,390, 1982; *Chem. Abstr.,* 98, 108327, 1983.

Chapter IV

Mannich Bases in the Chemistry of Natural Products

The chemistry of Mannich bases is implied in important spontaneous reactions involving natural products[1] such as the alkaloids,[2] where a Mannich-type condensation is a key reaction leading to the biogenesis of the final product, as well as in interesting synthetic methods adopted in the laboratory preparation of natural molecules or of structurally related models. The chemical modification of natural compounds and the synthesis of labeled derivatives for biological studies also constitute relevant applications connected with chemistry of Mannich bases.

In the present chapter the matter is classified according to type of natural product and is related to the amount of literature produced on each topic rather than to its biological relevance. The following groups of products are distinguished:

- Alkaloids
- Steroid hormones
- Natural macromolecules and related compounds (proteins and amino acids, lignin, etc.)
- Pharmacologically active substances and vitamins
- Miscellaneous compounds (porphyrins, terpenoids, etc.)

The discussion on these substances is essentially limited to their chemical behavior (synthetic methods, reactivity) as the biochemical aspects are beyond the scope of the present volume.

A Alkaloids

Mannich reactions occur frequently in the chemistry of alkaloids. Many key steps in biogenesis are Mannich-type condensations (see p. 353 in Ref. 2), and several synthetic methods applied to these compounds and related derivatives are often based on Mannich reaction.[3–6]

Most of the applications of Mannich base chemistry are undoubtedly devoted to the synthesis of alkaloids, although some other studies, such as those dealing with the racemization of intermediates in the synthesis of vincamines[7] or the production of derivatives for pharmacological purposes, are also reported in the literature.

A.1   Mannich Bases in the Synthesis of Alkaloids

The aminoalkylation reactions applied to the synthesis of alkaloids are mostly cyclizations of the type described in Chap. I, C.4,5, usually involving CH substrates such as
alkyl ketones. The analogous reactions on alkenes are frequently accompanied by more
or less relevant modifications of the substrate molecule (Chap. I, C.5). Activated aromatic
substrates, particularly phenols and their alkylether derivatives, are also subjected to
Mannich reaction.

Reactions on Activated Alkyl Groups

Aminoalkylation, mainly carried out on alkylketones (**441–445**, Fig. 166), is largely used
and mentioned among the classical synthetic methods in alkaloid chemistry (see pp. 361
and 363 in Ref. 2). In a few cases, for example with dihydrocodeinone, the reaction is
uncontrollable and proceeds up to deamination, followed by dimerization of the vinylketone produced.[8]

Usually, a Mannich reaction involves substrate and amine moieties present in the
same molecule, thus producing intramolecular cyclization. The aldehyde moiety may
be linked to the above reactive system before cyclization takes place, as shown in
Fig. 166, which summarizes the final step of the synthetic routes leading to ring
formation.

The chemical structure of the intermediate alkyleneimmonium ions **442, 444**, and
445 is here depicted only in order to indicate the group affording aminoalkylation and
is not intended to suggest a mechanism of the reaction, although in many cases the
electrophilic attack of this ion on the substrate moiety is assumed. The alkyleneimmonium ion may be generated not only in the usual way, namely, by condensation
between amine and aldehyde, but also by tautomerism or other equilibria favored by
the acidic medium,[11,16] or by CN^- elimination from α-cyanoimines.[9] In some circumstances, the treatment with aldehyde initially gives O-aminomethylation, with formation
of a cyclic intermediate containing the —O—CH_2—N< moiety, by reaction of a hydroxy group suitably located in the molecule[12] or by involvement of the $COCH_3$ group
in a ketal intermediate.[10] In any case, the expected C-Mannich base is finally produced
by proper treatment of the above unstable species.

The cyclization reaction is also successfully performed through amidomethylation,
by using formylated amine.[16]

As represented in Fig. 166, ring closure to γ-piperidone may occur, starting both
from precursors **442** and aminoketones **443** with the suitable aldehyde. The synthesis
of aconitine-type diterpene alkaloids[17] takes place similarly (Fig. 167), the only difference being the presence in the reactive site of an allylic carbon atom (**446**) instead of
the alkyl group in α position to the carbonyl.

Reactions on Alkenes

When the substrate moiety in Mannich aminomethylation is alkene, the reaction may
often produce double bond migration as well as rearrangement involving the whole
molecular skeleton. The former occurrence concerns the aminomethylation of cyclic

441

R^1 = Ar; $R^{2(*)}$ = Chiral group inducing asymmetry: Piperidine alkaloids (see Fig. 39 and Chap. I, C3)

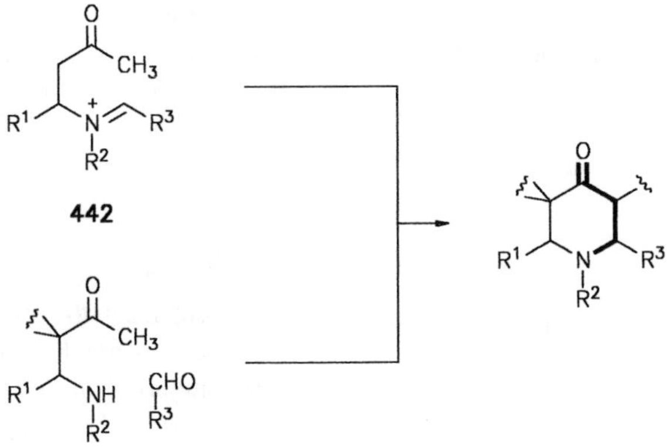

442

443

442: R^1,R^3 = ring: Adaline alkaloids[9]. R^1,R^2 and R^2,R^3 = rings: Coccinelline and Myrrhine alkaloids[10,11]

443: R^1,R^2 = ring; R^3 = H, Ar: Allopumiliotoxin, Lythraceae alkaloids[12,13]. R^1,R^2 = H; R^3 = Ar: Uleine alkaloids[14]

444

R^1,R^3 and R^2,R^4 = rings: Lycopodium alkaloids[15]

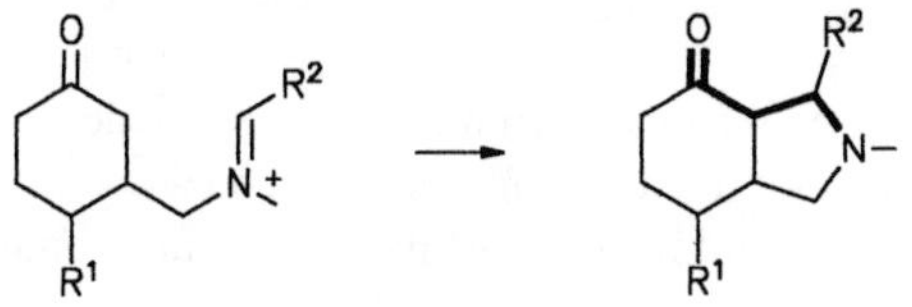

445

R^1,R^2 = ring: Gelsemine alkaloids[16]

Fig. 166. Intermediate steps in the alkaloid synthesis affording β-aminoketone structures (bold lines).

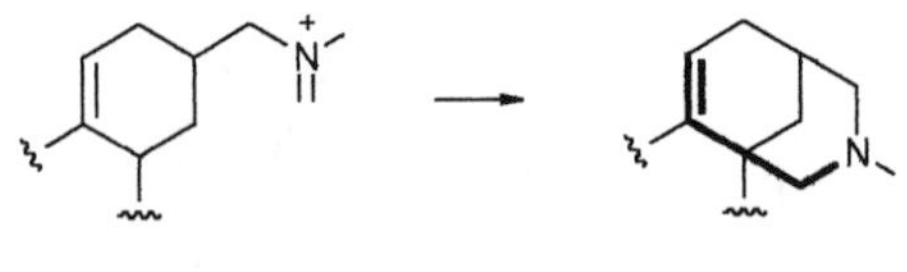

446

Fig. 167. Synthesis of aconitine-type diterpene alkaloids by intramolecular aminomethyla-
tion leading to a β-aminoalkene (bold
lines) derivative.

alkenes devoid of specific activating groups (see also Fig. 65, Chap. I), such as derivatives **447–449** (Fig. 168), giving rise to polycyclic products upon migration of the
double bond toward the most stable position. In some cases (e.g., **447**) racemization of
the final product, attributed to cleavage
of the σ bond connecting the alkyleneimmonium ion to the unsaturated
moiety, is observed.[18]

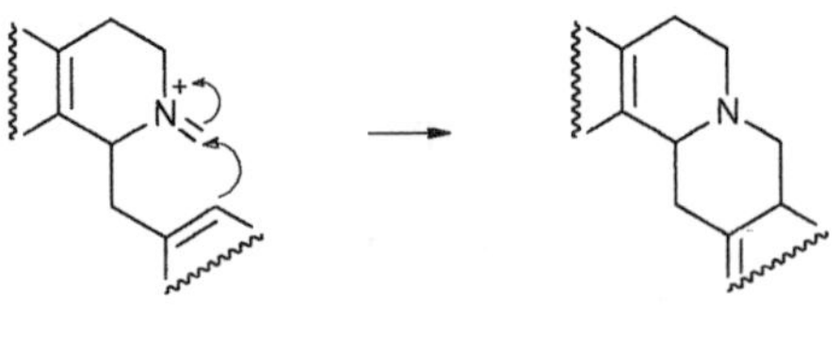

In the reaction on the analogous
systems **450** (Fig. 169), the double
bond remains in the original position;[22]
however, an amidomethylation is actually involved in this case, the cyclization taking place only when R or R′
is equal to carbonyl group.

Among Mannich aminomethylations on alkenes leading to rearranged
products, the tandem cationic aza-Cope
rearrangement–Mannich cyclization
(Figs. 65 and 68, Chap. I) is undoubtedly the most important. The reaction
makes it possible to obtain nitrogen-
containing five-membered rings having
an acyl group in the β position with
respect to the heteroatom, starting from
3-hydroxy-4-amino-1-butene derivatives (**451**, Fig. 170) and aldehyde, or
from species capable of giving alkyleneimmonium ions.

By the above method, the synthesis of interesting alkaloids, such as
meloscine (**452**, Fig. 170)[23] or deriva-

447   Emetine, Yohimbine alkaloids[18,19]

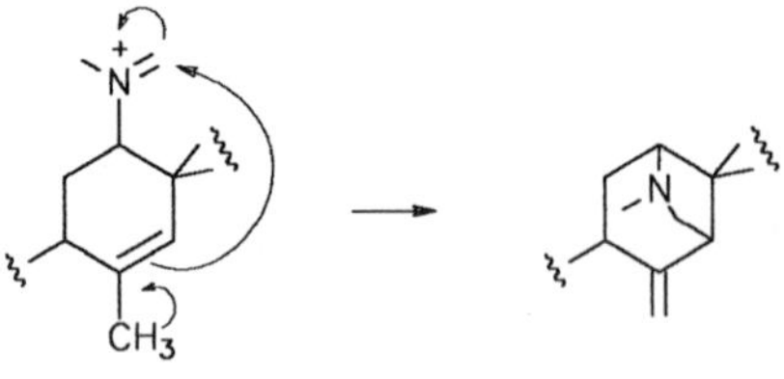

448   Gelsemine-type alkaloids[20]

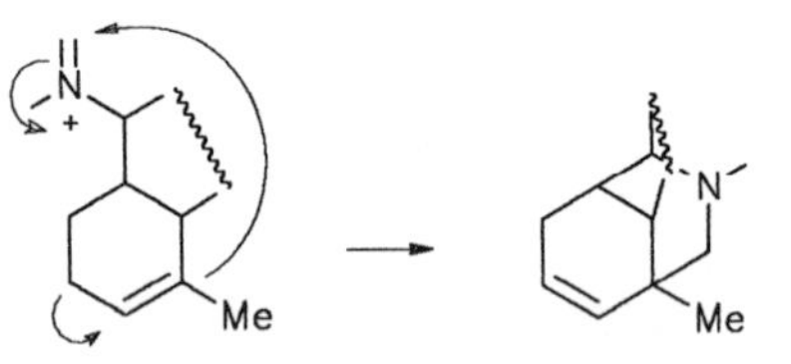

449   Aconitine-type alkaloids[21]

Fig. 168. Intramolecular aminomethylation of alkene
moiety in the alkaloid synthesis.

tives of epipretazetine,[24] aspidospermine,[25] etc., is achieved. In these reactions, a
cyanomethyl derivative of the amine reagent, which can be considered as the Mannich
base of hydrogen cyanide, is employed[24] as aminomethylating agent. The

450 (R, R' = CH₂, CO)

Fig. 169. Intramolecular amidomethylation in alkaloid synthesis.

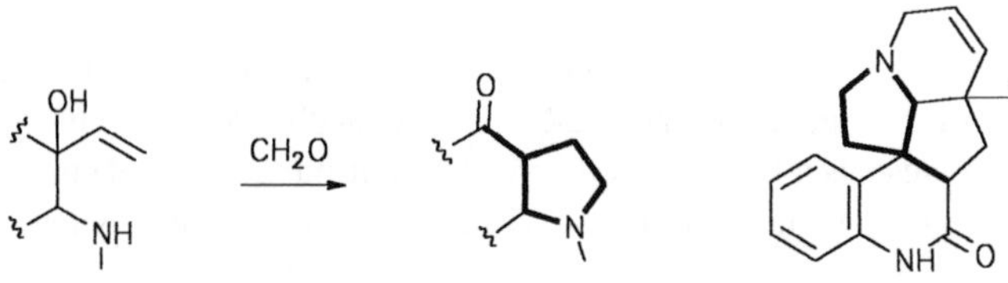

451   **452**

Fig. 170. Tandem aza-Cope rearrangement–Mannich cyclization (bold lines) in alkaloid synthesis.

deamination of this last species (see **38**, Chap. I, A.4), occurring with elimination of the CN⁻ ion, produces the desired immonium cation, intermediate of the cyclization reaction.

Reactions on Aromatic Rings

Mostly phenols and phenolethers are subjected to Mannich reaction, with heteroaromatic substrates also employed. The first group of compounds has been thoroughly investigated (p. 374 in Ref. 2) mainly as far as intramolecular cyclization with the hydroxy- or alkoxy-activated aromatic ring of the tetrahydroisoquinoline-derived methyleneimmonium salt is concerned (**453**, Fig. 171).

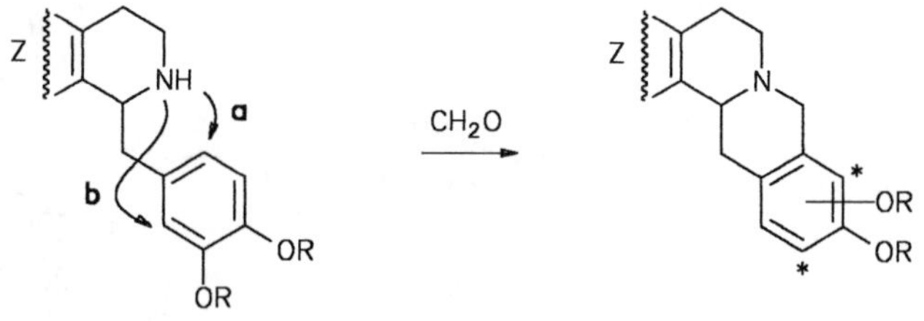

453

R = H, Alkyl; Z =

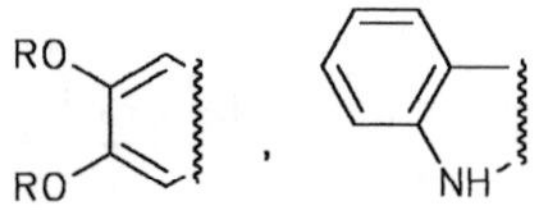

Fig. 171. Intramolecular aminomethylation in the synthesis of protoberberine alkaloids.

The site of cyclization in the above compounds is affected by the ortho-para orienting effect (**a** or **b** attack) by the hydroxy or alkoxy substituents (Fig. 52, Chap. I). A number of alkaloids have been synthesized by this method, including sculerine,[26] coreximine,[27] kikemanine,[28] discretine[29] and other protoberberine derivatives,[30-32] yohimbanes (from tetrahydroazacarbazoles),[33] etc.

Complications may however arise in this type of synthesis, besides the regioselectivity involved in the cyclization step, due mainly to the possibility of cleavage of the bond linking the heterocycle ring to the benzylic carbon atom during the formation of the methyleneimmonium ion. In this way, rupture of the molecule,[31] heterocyclic ring enlargement (see Fig. 69, Chap. I), and the abovementioned racemization[18] may take place.

The heterocyclic ring, also, has been subjected to aminomethylation, as reported for the reaction of the furane substrate **454** (Fig. 172) with dimethylamine and formaldehyde, which represents a key step in the synthesis of muscarines.[5] This is one of the few examples in alkaloid chemistry of the Mannich reaction not involving cyclization.

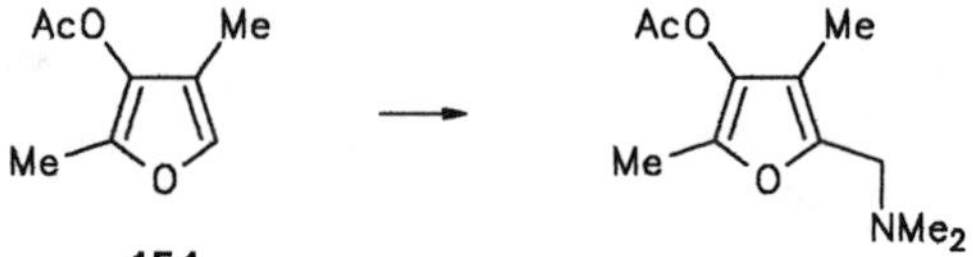

Fig. 172.   Aminomethylation step in the synthesis of muscarines.

Miscellaneous Reactions

O-Aminomethyl derivatives are usually described as unstable intermediates affording more stable C-aminomethyl products by deaminomethylation (see above). However, the synthesis of eburnane alkaloids[34] involves as a key step the formation of a methylol derivative of the α-dicarboxyester **455** (Fig. 173), which then leads to the tetrahydropyrane ring **456** by intramolecular O-aminoalkylation reaction with the immonium ion present in the molecule.

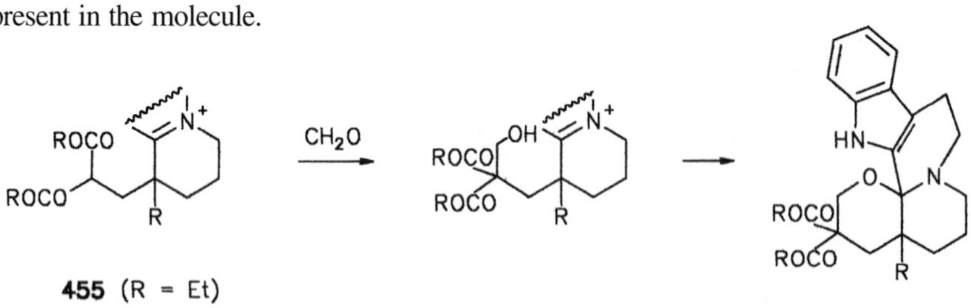

Fig. 173.   O-Aminomethylation step in the synthesis of eburnane alkaloids.

Replacement of the amino group in the Mannich base is also applied to alkaloid synthesis. Thus, ellipticine alkaloids[35] are prepared by deamination of the intermediate indole Mannich base **457** (Fig. 174), which allows ring closure to the hexa-atomic cycle evolving subsequently toward the phenolic moiety (**458**) through the occurrence of tautomeric equilibria.

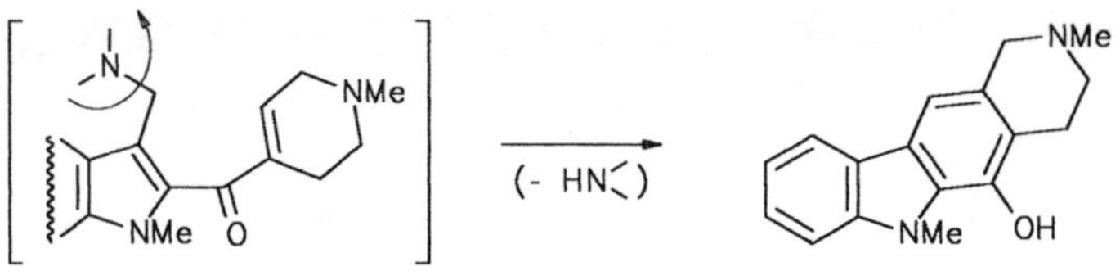

457   **458**

Fig. 174. Intramolecular amino group replacement in the synthesis of ellipticine alkaloids.

A.2 Mannich Reactions on Alkaloids

Mannich reactions on alkaloids employed as substrates or amine reagents are usually done for pharmaceutical purposes. The former case includes aminomethylation of xantine bases, particularly theophilline and its derivatives. The chemoselectivity of the reaction has been investigated by performing the reaction on both the nitrogen and the carbon atom of the imidazole ring[36,37] (see also **68** in Chap. I, C.1).

Good regioselectivity is shown by the Mannich reaction on acetonyl theobromine, which exclusively gives the product **459**, deriving from attack of the amine/formaldehyde reagent on the methyl group of the acetonyl moiety.[38]

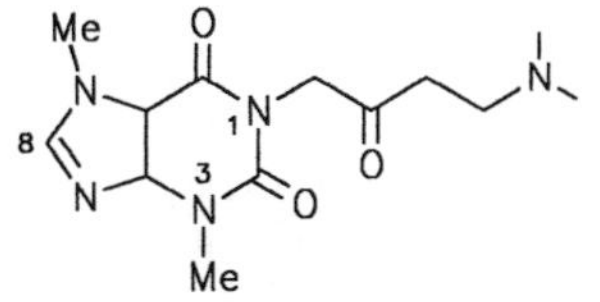

459

In addition to ephedrines (Table 2, Chap. I), other important alkaloids are used as amine reagents for Mannich synthesis. Thus, demethylated nicotine affords aminomethylation of hydrogen cyanide (**460**), employed in studies of the metabolic oxidation of nicotine,[39] and anabasine gives, by reaction with acrylamide, the readily polymerizable monomeric species **461**, the polymeric derivative of which is 50 times less toxic than the original anabasine.[40]

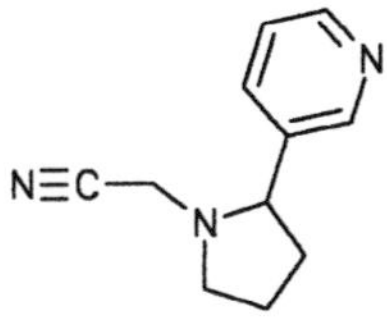

460 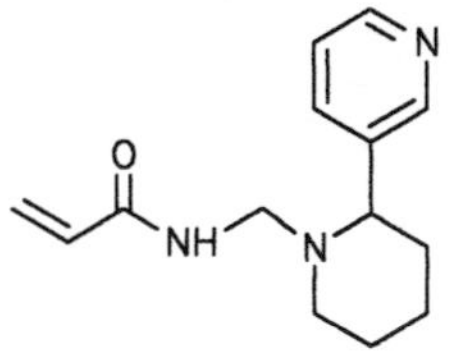   **461**

B. Steroid Hormones

The steroid skeleton is usually subjected to Mannich reaction in order to obtain derivatives possessing basic functionalities. Steroidal aminomethyl compounds are used mostly for the synthesis of compounds having pharmacological activity.

B.1 Mannich Reactions on Steroid Molecules

Steroid molecules may undergo aminomethylation at different positions of the ring-fused system as well as at reactive centers of ring substituents R (**462**). Table 35 lists the possible sites of attack on steroid compounds bearing suitable activating groups.

Table 35

Aminomethylation Reaction on Steroid Compounds

Reaction site [a]	Steroid	Reference
Ring A		
	17-Hydroxy androstanone	41,42
	Cholestanone, synthetic deriv.s	43
	Dihydrocortisone acetate	44
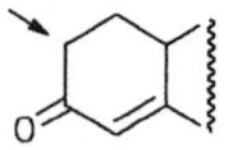	17-Ethynyl hydroxyandrostenone	45
(a) (b)	Prednisone acetate (a)	44
	Cholestadienone (b)	46
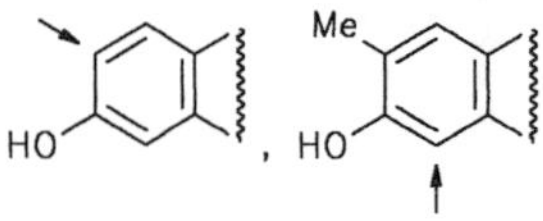 Me	Estrone, Methyl estradiol	47,48
Ring B or D		
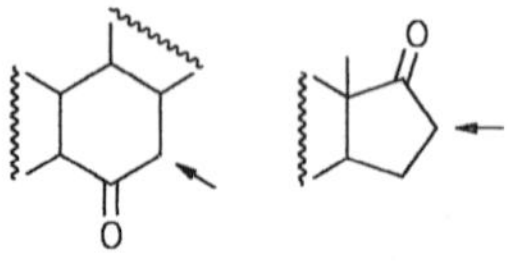	Androstenolone	43,49
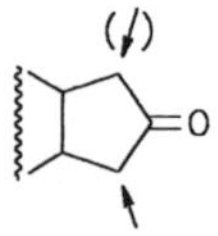 (↓)	16-Ketoandrostane	50

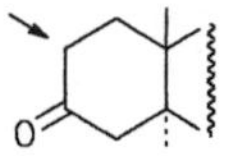

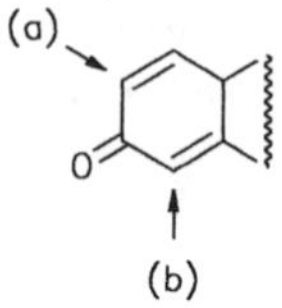

It can be observed that a fair variety of classes of substrates is represented in the table. In addition to the most frequently employed alkyl ketones, particularly cycloalkyl ketones, quinonic, phenolic, and alkyne substrates have been subjected to Mannich

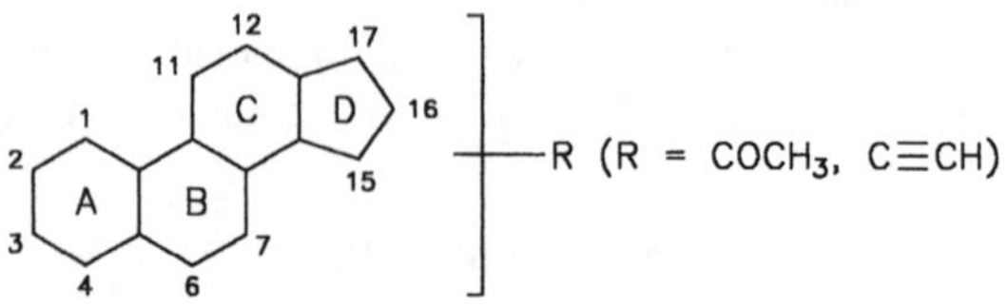

R (R = COCH$_3$, C≡CH)

462

Table 35

(continued)

Exocyclic groups

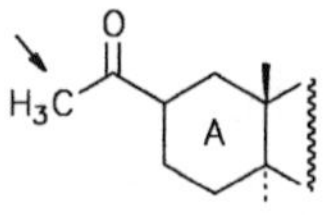	Acetyl-norcholestane and -norpregnane	51
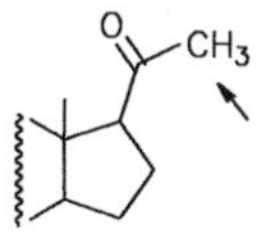	Pregnenones, Pregnenolones, Pregnadienones	52-54
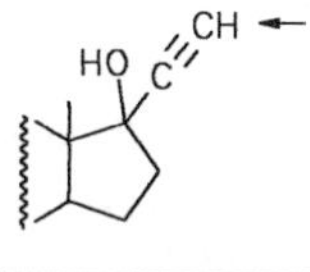	Ethynyl hydroxy-androstenes and -estrenes	55-57

a Arrow indicates the position of attack

reaction. In some cases, different activating groups are present at the same time in the substrate molecule, thus involving chemoselective reactions in addition to regioselective ones.

As far as the amine moiety is concerned, usual amines, such as dimethylamine, are most commonly used. However, in particular situations requiring the preparation of pharmacologically active derivatives, bis(chloroethyl)amine,[47,48,56] aziridine,[48] or norephedrine (this last species for the synthesis of products enhancing coronary blood flow)[53] are employed.

As shown in Table 35, the aminomethyl group can be attached to any fused ring, except ring C. Ring A is the most strongly involved in the reaction, due to the presence of a carbonyl or a phenolic hydroxy group in position 3. Position 2 is usually attacked by the electrophilic reagent, probably due to steric reasons; cholestanedienone gives a predominant reaction only at position 4.[46]

Less abundant examples of aminomethylation on ring B are reported in the literature; most of them are connected with the synthesis of the vinylogous bases described in the following section, which deals with the reactions of steroid Mannich bases. Aminomethyl derivatives at ring C are unknown, and only a few cases of aminomethylation at the penta-atomic ring D are reported. In 16-ketoandrostane a slight preference for attack at the less hindered position 15 is observed.[50]

Two types of substituents linked to the steroid rings are involved in the Mannich reaction, namely, methyl ketone and alkyne residues. The former react on the methyl group, and the latter react chemoselectively on the alkyne hydrogen atom when copper salts are employed as a catalyst.[56]

Although steroid derivatives are in general remarkably reactive toward Mannich synthesis, some cases of unexpected absence of activity of the substrate,[44] as well as the formation of methylene-bis-derivatives R—CH$_2$—R,[41] polyaminomethylated compounds,[44] and mixtures of regioisomers[50] are observed. On the other hand, interesting improvements in the reaction conditions leading to considerably enhanced yields are worthy of mention.[43]

B.2 Reactions of Steroid Mannich Bases

Steroidal aminomethyl derivatives have been subjected to some of the reactions typical of Mannich bases, described in Chap. II, usually with the aim of introducing structural modifications of the molecule by means, in particular, of deamination and hydrogenolysis. The formation of a quaternary ammonium salt (mainly iodomethylate) by N-alkylation of the base[41,57] is also performed to the same end in order to obtain unsaturated deaminated products.[58]

Deamination

Deamination is the most commonly performed reaction of steroid Mannich bases. It is usually carried out by heating the base in the presence of acetic acid/acetic anhydride,[49,54] by column chromatography on silica gel,[44] by heat decomposition of the corresponding hydrochloride or iodomethylate, or under suitably modified reaction conditions leading directly to the unsaturated final product starting from the usual Mannich reactants (Chap. II, A.2).

The production of unsaturated derivatives has been performed by deamination of aminomethyl derivatives at the cycloketonic rings A or D[44,45,49] or of the methyl ketone group in position 17.[54] The same reaction is given with the aminomethyl derivative at position 6 (ring B), starting from progesterone[58] or androstenone derivatives,[59,60] through a vinylogous Mannich base (usually not isolated), which finally yields the ketodiene **463** (Fig. 175).

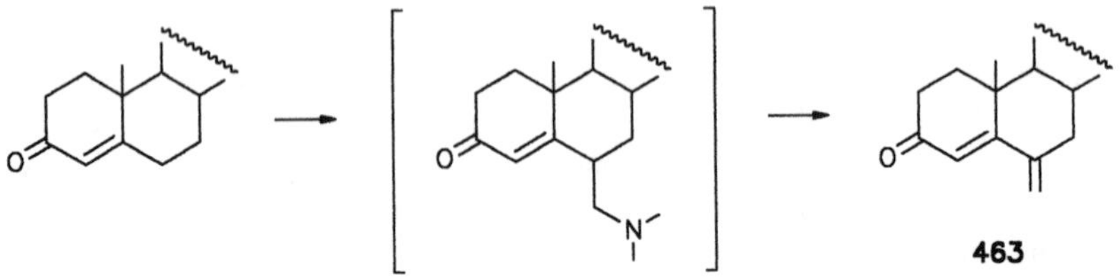

Fig. 175. Deamination reaction of vinylogous Mannich bases of steroid derivatives.

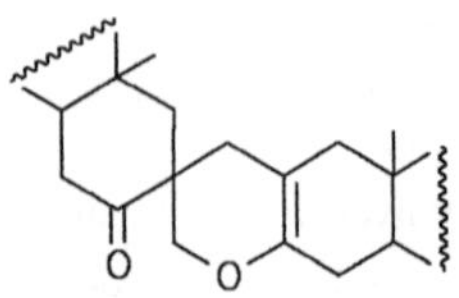

In some cases, deamination is the first step in a synthetic route aimed at forming a methyl group either by hydrogenation (see below) or as a consequence of the establishment of tautomeric equilibria.[44] The production of cyclic derivatives is also pursued.[54] The unsaturated intermediate, however, is seldom isolated and characterized, as it readily dimerizes,[42] giving products of type **464**.

Hydrogenolysis and Other Reactions

Hydrogenolysis of the CH_2—N bond of the Mannich base produces a methyl group, which may be thus introduced at different positions of the steroid rings[44,48,49] or used in order to lengthen the carbon chain linked at position 17 (ring D).[44] The reaction is successfully performed on phenolic[48] as well as ketonic Mannich bases; however, in the latter case it is possible to obtain diastereomeric products, if the aminomethyl group is bonded to an unsaturated, prochiral carbon atom simultaneously undergoing hydrogenation.[46]

In the hydrogenolysis of acetylenic Mannich bases,[57] rearrangement with double bond migration, accompanying the elimination of the amine, may take place with formation of an allene derivative instead of the expected methyl alkyne.

A typical example of hydrogenolysis is given by the synthesis of 2,4-dimethyl estradiol **465**, obtained in good yield from the dimethylamino Mannich base of the corresponding 2-methyl derivative.[48]

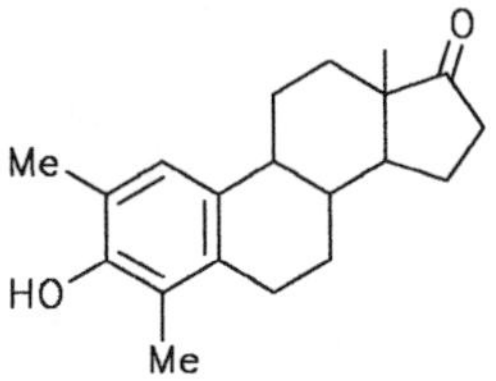

465

Steroidal Mannich bases have also been subjected to acetylation and bromination. Thus, the possibility of esterifying the hydroxy group in position 17 of aminomethylated hydroxypregnenones[52] as well as the regioselectivity of bromination of 2-aminomethyl androstan-3-ones activated by the carbonyl group in position 3[61] have been investigated.

▊C▊ Natural Macromolecules and Related Compounds

Important compounds related to nucleic acids and natural macromolecules such as peptides and polysaccharides, are frequently involved in the chemistry of Mannich bases. Low-molecular-weight compounds connected with the chemistry of the above macromolecules are also considered.

C.1   Compounds Related to Nucleic Acids

Studies of nucleosides and the corresponding heterocyclic bases predominate, although nucleotides are also investigated. The synthesis of these compounds, based on the chemistry of Mannich bases, is reported by few papers, as in the case of the ring closure of imidazole (**364** in Chap. II, E.2); however, the functionalization of nucleosides and nucleotides, using these molecules as substrates or amine reagents in aminomethylation reactions, is more diffusely reported.

Reactions on Nucleosides

Chemo- and regioselectivity is involved in these reactions (see **71** and **82** in Chap. I, C.1,2); C- and N-Mannich bases, starting, respectively, from uracil[62] and 5-fluorocytosine,[63] adenine or other purines,[64,65] can result. Multiple aminomethylation may also occur with purines, the synthesis of monoaminomethylated derivatives, such as **466**, being possible only with weakly basic amines.

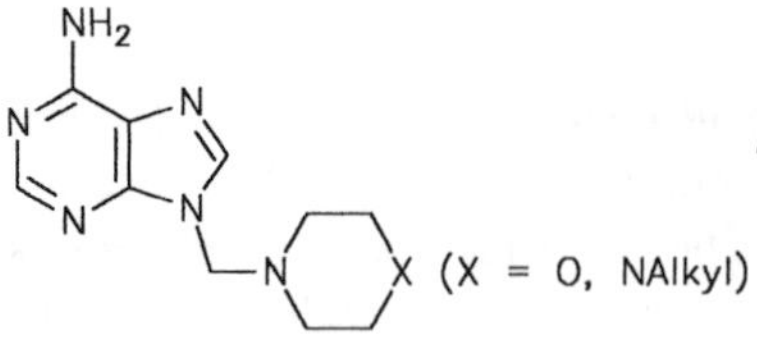

466 (X = O, NAlkyl)

C-Mannich derivatives of uridine **467** and thymidine[66] or N-Mannich bases of cytidine iododerivatives[67] are yielded from nucleosides subjected to aminomethylation. The presence of the glycoside moiety may favor the reaction at C-5 (arrow in **468**) through the formation of a polycyclic structure bearing an electron-attracting group inserted at the adjacent position, C-6.[66]

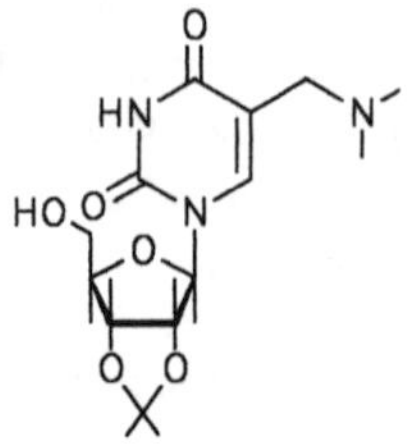

467 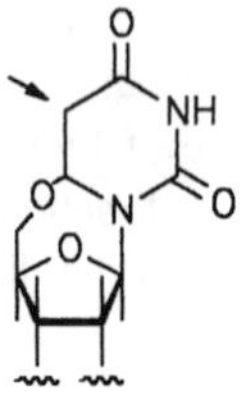   **468**

The synthesis of O- and S-Mannich bases by reaction of nucleosides as amine reagents with formaldehyde and ethanol (**469**)[68] or sodium bisulfite (**470**),[69] respectively, is reported.

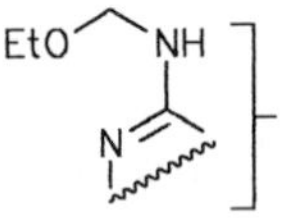

469 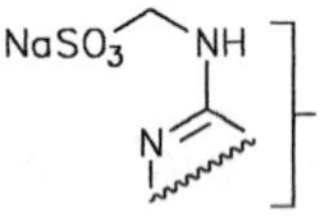   **470**

= Adenosine, Guanosine, Cytidine moiety

Products having the structure **470** containing adenylic, guanylic, etc., acid moieties are analogously obtained from nucleotides.[69]

Reactivity of Mannich Bases of Nucleosides

As far as the reactivity of the above-described Mannich bases is concerned, studies[68] of products **469** indicate a much better resistance to alkaline than to acid hydrolysis; the nature of the nucleoside also plays a role, as the cytidine derivatives are much more stable than the purine derivatives. Similar investigations have been made[65] into the deaminomethylation of Mannich bases of type **466**.

The amino group replacement reaction has been studied with different reagents. In addition to a series of C-nucleophiles (HCN, nitroalkanes, etc.),[70] amines and thiols[71] are employed. The former reagents are esters of amino acids that are hydrolyzed after the exchange reaction, while thiols give thioethers, which may be subjected to hydrogenolysis with cleavage of the CH_2—S bond, thus producing the corresponding methyl derivative. This makes it possible to obtain thymidine (**471**, Fig. 176) from 2′-deoxyuridine.[66]

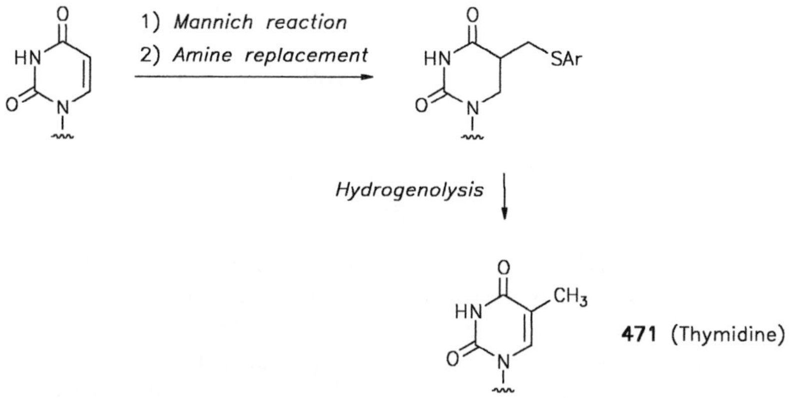

Fig. 176.   Synthesis of thymidine involving the chemistry of Mannich bases.

The CH_3 group can be similarly introduced into compounds of the same type by direct hydrogenolysis of the Mannich base (Fig. 108, Chap. II).

C.2   Peptides

α-Amino acids, in addition to peptides (oligopeptides, proteins, enzymes), are involved in the chemistry of Mannich bases. In particular, the synthesis of a glutamic acid derivative, the γ-carboxyglutamic acid (**473**, Fig. 177), is effected[72] through amino group replacement of **472** by C-alkylation with fully esterified amidomalonic acid followed by hydrolysis and decarboxylation.

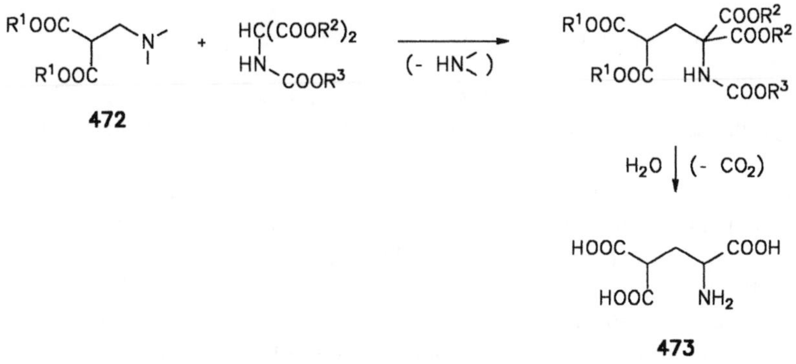

Fig. 177.   Synthesis of γ-carboxyglutamic acid.

A satisfactory method for introducing the amino group in the α position with respect to amidic or carboxylic CO is given by the aminomethylation of hydrogen cyanide followed by partial or total hydrolysis of the nitrile group. This reaction has relevant applications in the field of complexants (Chap. V). A similar way to synthesize peptide derivatives, reported[73] in studies on opioids, consists in the amidomethylation of hydrogen cyanide. The nitrile derivative thus obtained is subsequently hydrolyzed in an acidic medium to give the corresponding amide **474** (Fig. 178).

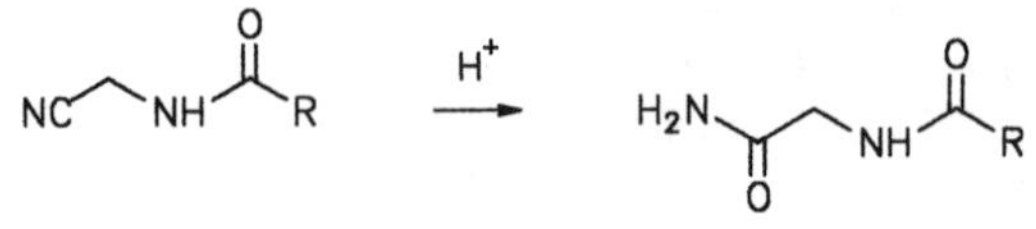

R = Metazocine residue   **474**

Fig. 178.   Synthesis of peptide derivatives.

Some investigations have been made into the use of amino acids as amine reagents in Mannich synthesis. The reaction between formaldehyde and amino acids such as arginine, lysine, and alanine,[74,75] to give the corresponding hydroxymethyl derivatives, occurs spontaneously and depends on the type of amino group. However, it may evolve toward the formation of dimethylene ether-bonded products by the further condensation of two molecules of the hydroxy intermediate. The most varied substrates have been subjected to Mannich reaction with amino acids, as shown by **475**.[76–78] A specific review paper on this matter, containing a large number of examples, is available.[79]

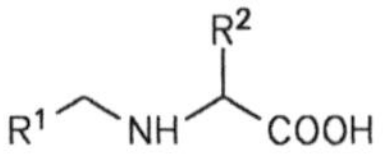

R^1 = (NO$_2$)$_2$CF–, Me

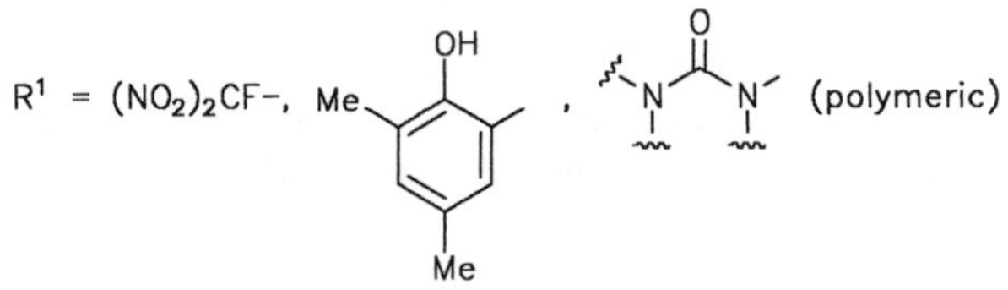

R^2 = H, Me, CH$_2$Ph,...

The application of the Mannich reaction to polypeptides has two main objectives, namely, functionalization and crosslinking of the macromolecules (see also Chap. III, C.2).

The functionalization reaction, which is particularly applied to collagen, is based on the aminomethylation of various substrates[80] (see also Table 34, Chap. III) by employing the NH$_2$ groups of the protein as Mannich amine reagents. Formaldehyde is preferred[81] for crosslinking; however, other agents, such as resols (see **432** in Chap. III, C.2) or those employed with cellulosic materials, are successfully used for proteic macromolecules such as keratin.[82] The resulting crosslinked product should have structure **476** (Fig. 179), which also accounts, to some extent, for the possible polymerization of the bis-methylolamide reagent.

A small number of quite interesting papers have been published on the application of Mannich base chemistry to enzymes, with particular reference to functionalization and immobilization. Thus, phenolic substrates, such as the vanillomandelic acid and analogous derivatives, have been linked to alkyl phosphatase (**477**) by aminomethylation involving the amino groups of the protein as the amine reagent;[83] the possibility of anchoring β-D-glucoxidase on magnetic material by linkage to the polymeric coating created by Mannich polymerization over a metallic surface has also been investigated.[84]

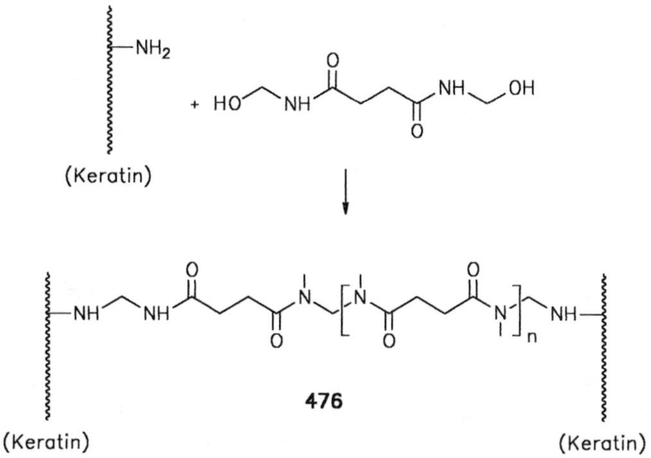

Fig. 179. The crosslinking of keratin.

Indeed, the polymeric coating, in addition to structure **386** (Chap. III, A), possesses primary arylamino groups which can be connected by diazocoupling to the side-chain tyrosine groups of the enzyme to give products **478**.

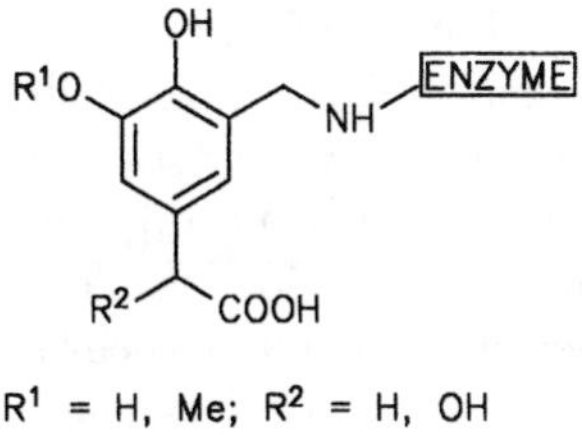

R^1 = H, Me; R^2 = H, OH

477

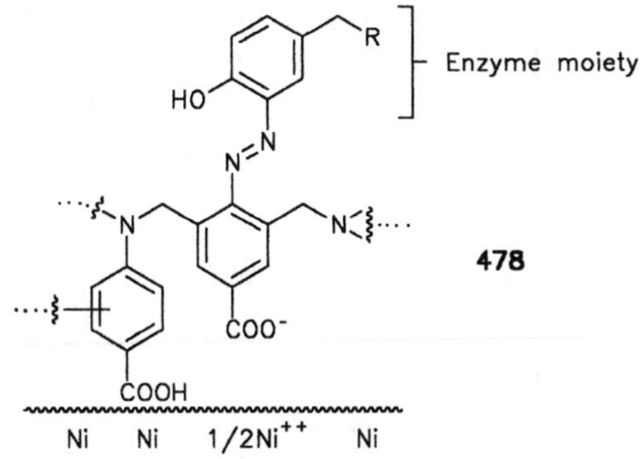

478

As suggested in a paper on the anchorage of hydroquinone over indium-tin oxide electrodes by reaction with the methyleneimmonium derivative of trimethoxysilyl al-kylamine,[85] the method of linking side-chain reactive groups to oxide surface by Mannich reaction could also be usefully applied to enzyme immobilization. Moreover, in a manner similar to the crosslinking of proteic macromolecules shown in Fig. 179, bis-thiols and other analogous polyfunctional substrates could be successfully employed in the place of the methylol bis-amide.

C.3 Polysaccharides, Lignin, and Related Compounds

As far as the chemistry of Mannich bases is concerned, almost exclusively aminomethylation reactions aimed at introducing appropriate modifications into this group of substrates, rather than synthesizing new compounds, are reported in the literature.

The functional groups present in glycosides allow these derivatives to be used as substrates, aldehyde, or as amine reagents of the Mannich reaction. Thus, amide and hydrazide derivatives of D-galacturonic acid are employed[86] as substrates of type **479**, and the use of glucose as an aldehyde reagent (**480**) in combination with polyfunctional amines and various substrates is proposed[87] for the preparation of ion exchange resins. Finally, aminosugars **481** are used as chiral auxiliaries in the asymmetric synthesis of alkaloids or β-amino acids (see Fig. 39, Chap. I).

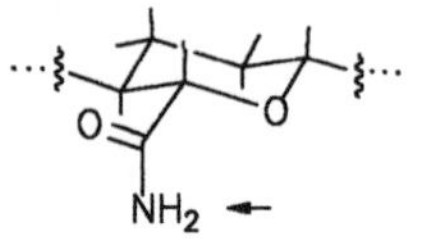

| 479 | 480 | 481 |

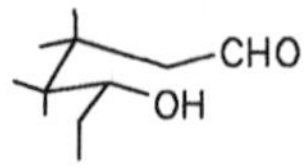

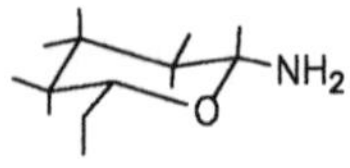

Polysaccharides are also involved, to some extent, in the chemistry of Mannich bases. For instance, acetonic Mannich bases are conveniently combined with starch-based adhesives,[88] probably due to the possibility of interactions such as exchange reactions, between base and polyglycoside. Information is not abundant on the nature of the chemical structures involved in these applications, but the process of crosslinking the cellulose derivatives carried out by means of amidomethylation with bis-methylolamides (Chap. III, C.2) is reasonably clear. Moreover, studies of the reaction of poly(aminosaccharide)s with formaldehyde give grounds for suggesting the presence of oxazolidine structures in equilibrium with open chain methylolamine species (see **436**, Chap. III, D.1), and intermolecularly linked methylenic bridges are presumed to be present in products employed in water desalination.[89]

Lignin is consistently used as a substrate of aminomethylation, as many derived products are prepared for various applications[90,91] (Chap. V). The site of attack by the aminomethylating agent on lignin is not completely understood. The reaction with formaldehyde only has been studied in different mediums.[92] Derivatives of type **482** should be formed in the reaction employing formaldehyde and glycine or similar amino acids as amine reagents when the ortho position with respect to the phenolic hydroxy group is available.[91] In addition, partial O-aminomethylation of the hydroxy groups present in lignin in large quantities cannot be excluded.

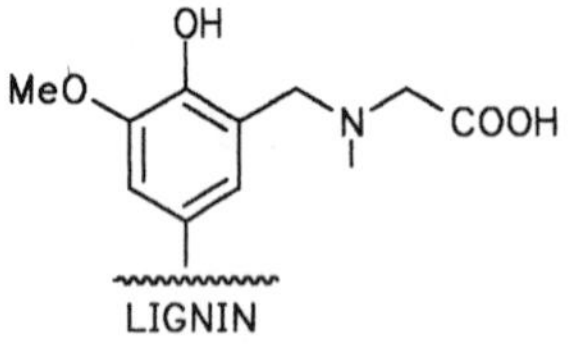

D Pharmacologically Active Substances and Vitamins

The chemistry of Mannich bases has been associated with pharmacological studies since the early investigations by Carl Mannich, and it is still extensively involved in pharmaceutical chemistry, as is demonstrated by the profusion of papers appearing in specialized publications. The applications of this topic are reported in Chap. V, B, and the present section deals only with natural compounds that function as drugs or vitamins.

D.1 Antibiotics

Only a few cases of the synthesis of antibiotics, such as citromicetin, carried out by amino group replacement,[93] and cephalosporin,[94] are mentioned in the literature; most of the reports are dedicated to the use of the antibiotic molecule as a substrate or amine reagent of the Mannich reaction. Chemoselectivity may play an important role with this type of substrate. Thus, rifamycine[95,96] undergoes C-aminomethylation on the unsaturated carbon atom in the α position with respect to the carbonyl rather than on the amide NH (**483**), and tetracycline gives N-Mannich bases **484** instead of the C-aminomethylated product at the phenolic ring.[97–100] Finally, cojic acid is aminomethylated in position 6, vicinal to the hydroxy group (**485**) (see Ref. 568 for Chaps. I and II).

The pharmacological activity of tetracyclines bearing different amino groups has been thoroughly investigated; various amines have been employed, ranging from the bifunctional ones, such as piperazine, which allows the linkage of two tetracycline moieties, to hydrophilic amino acids, which improve the water solubility of the product.

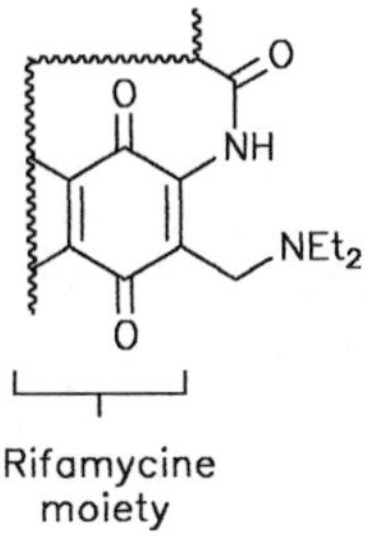

Rifamycine
moiety

483

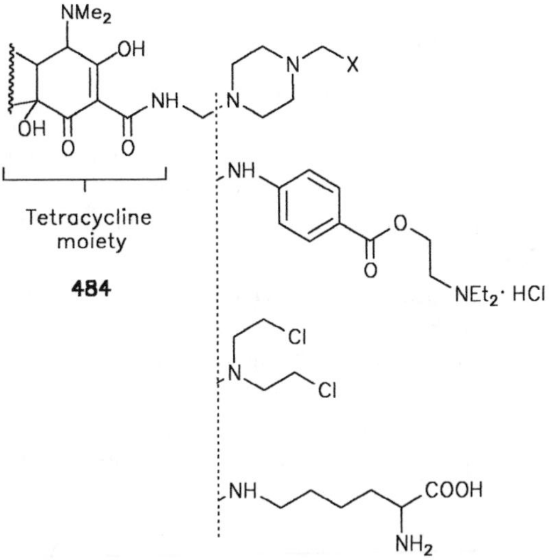

Tetracycline
moiety

484

NEt$_2 \cdot$ HCl

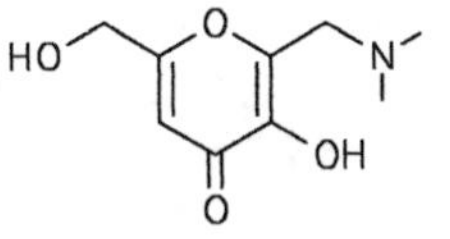

485

Finally, gentamicin is used as an amine in the reaction with various substrates, including arylmethyl ketones **486**, by conversion of the side-chain amino group.[101]

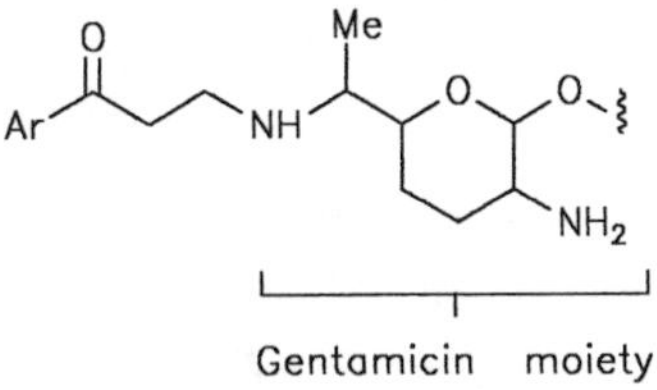

486

Gentamicin   moiety

D.2   Flavones and Related Phenolic Derivatives

As in the case of antibiotics, this group of substrates may be readily functionalized by aminomethylation due to the presence of phenolic rings, which permit preparation of products with basic properties. Flavonoids such as rutin[102,103] and silymarin[104] possess two hydroxylated aromatic nuclei, both of which may undergo aminomethylation, thus preventing the clear detection of the site of attack; however, the products obtained usually maintain the pharmacological activity of the initial material. In the case of arctigenin,[105] aminomethylation can take place only in the ortho position with respect to the hydroxy group, and chrysin (**487**, Fig. 180) reacts at position 8.

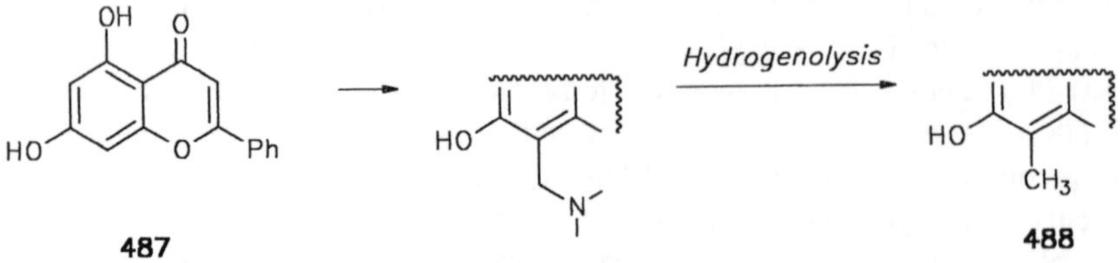

Fig. 180.   Aminomethylation of chrysin in the synthesis of tectochrysin.

The corresponding 8-methyl derivative, tectochrysin **488**, having diuretic properties, is synthesized by hydrogenolysis of the aminomethylated intermediate.[106]

D.3   Vitamins

In addition to the studies of molecules with similarities to natural vitamins, such as the models of tetrahydrofolic acid or the analogs of vitamin K (Refs. 512 and 914 for Chaps. I and II), aminomethylation has been performed on the nicotinamide group of vitamin P, with the formation in good yields of the corresponding N-Mannich base.[107] Improved conversion is obtained, starting from the palladium complex of the vitamin (Fig. 6, Chap. I).

The process, which includes the simultaneous separation and synthesis of vitamin E (α-tocopherol) (Fig. 181), appears particularly interesting; aminomethylation at the phenolic ring of type **489** compounds allows the separation from the aminomethylated material of any vitamin E that may be present and that is unable to undergo Mannich reaction. In a second step, further amounts of vitamin E are provided by hydrogenolysis of the aminomethylated product to the corresponding methyl derivative.[108]

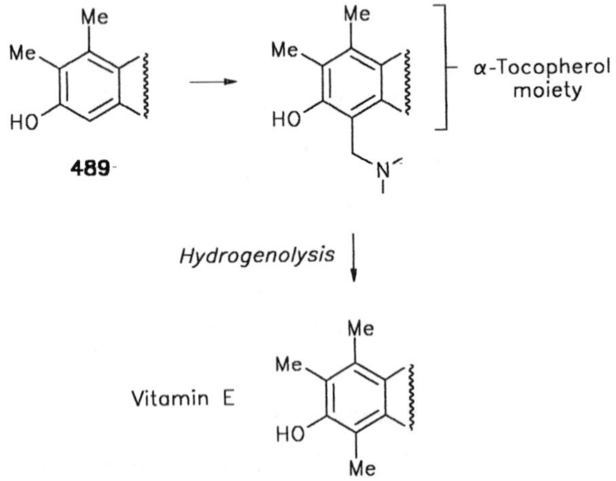

Fig. 181.   Aminomethylation reaction as applied to the separation and synthesis of vitamin E.

E   Miscellaneous Compounds

A few studies of small groups of compounds, such as porphyrins and terpenoids are described here. A brief survey of labeled Mannich bases, particularly useful as tracers in biochemistry or as tools for structure determination, is also given.

E.1   Porphyrins

Porphyrin synthesis and functionalization based on the chemistry of Mannich bases, briefly mentioned in previous chapters, are recalled here. As far as porphyrin synthesis is concerned, studies of biomimetic models of photochemically active reaction centers are worth noting. The synthetic procedure involves amino group replacement of the pyrrole bis-Mannich base with formation of the tetrapyrrole ring of porphyrin (see **360**, Chap. II).

The functionalization of porphyrins is performed by introducing a methyl group into dicyanocobaltheptamethylcorrin through hydrogenolysis of the corresponding aminomethyl derivative (Ref. 176 for Chaps. I and II). An appreciable number of replacement reactions on the amino group of etioporphyrin Mannich bases has also been tried (Ref. 832 for Chaps. I and II).

E.2   Terpenoids

Terpenoids are subjected to aminomethylation in order to introduce a basic moiety into the molecule. The reactive site of the substrate is usually an alkene or alkyl ketone group; however, unsaturated derivatives of type **490** are obtained from terpenes suitably functionalized by the alkyne moiety.[109]

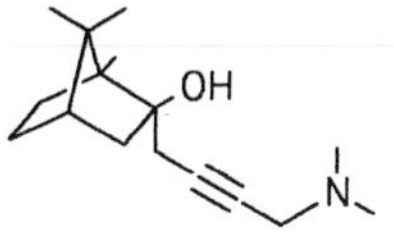

Due to the rather complex structure of terpenoid substrates, chemo- and regioselectivity of the reaction as well as the possibility of rearrangements in the substrate moiety are to be taken into account. Indeed, the widely investigated aminomethylation of alkene derivatives indicates that

490

Mannich reaction is frequently accompanied by changes in the substrate structure, with a particular involvement of the double bond location in the final product (Table 11, Chap. I, C.5).

Among terpenoid substrates bearing the alkyl ketone moiety, 4-caranone[110] and 2-pinene-4-one,[111] in addition to camphor, are worth mentioning. The stereoselectivity of aminomethylation[112] is remarkably affected by the reaction conditions; moreover, the tendency of 2-pinene-4-one, like steroid substrates, to give the vinylogous Mannich base (Table 8, Chap. I, C.1) is important.

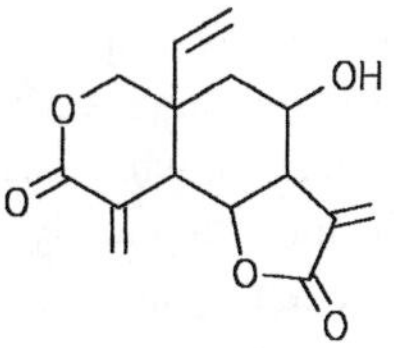

Deaminomethylation of terpenoid Mannich bases is frequently applied (Table 23, Chap. II, A.2) in order to create an unsaturated moiety in the molecule. The reaction has been carried out on the Mannich bases of *cis*-4-caranone, 2-pinene-4-one and sesquiterpene lactones, these last compounds being intermediates in the synthesis[113] of

491

tumor inhibitors vernolepin and vernomenin, which have a structure of type **491**.

Table 36

Isotopically Labeled Mannich Bases

Isotope	Mannich base (* = labeling isotope)	Reference
^{3}H	3-Indole Mannich base precursor of a 6-^{3}H-derivative	114
^{11}B	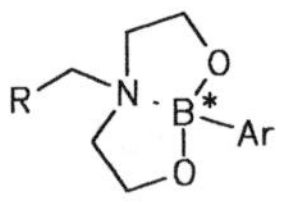 (R = Salicylamide group)	115
^{14}C	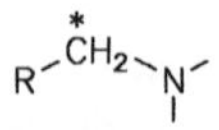 (R = 3-Indole or Alkyne)	114,116
	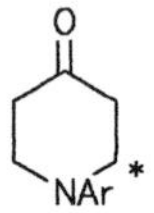	117
	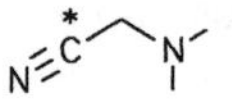	118
^{57}Co	Co-complex of uracil/iminodiacetic acid Mannich base	119
^{131}I	High molecular weight 8-hydroxyquinoline Mannich base bearing absorbed iodine	120

Hydrogenation and organometal addition to the carbonyl group of Mannich bases of camphor and related compounds are also widely reported (see references in Chap. II, C.2 and D.1).

E.3　Labeled Mannich Bases

Labeled Mannich bases of natural compounds or of compounds capable of interacting with biomolecules in the course, for example, of metabolism, such as drugs, prodrugs, etc., are usefully employed as tracers in biological and pharmacological investigations. Table 36 summarizes Mannich bases labeled with isotopes other than the commonly used deuterium, which are reported in the literature.

The most frequently used derivatives are ^{14}C-labeled Mannich bases obtained from ^{14}C-hydrogen cyanide or, to a larger extent, ^{14}CH$_2$O, analogously to the deuterated derivatives prepared with CD$_2$O.

In addition to the isotopically labeled Mannich bases, it is worth mentioning the paramagnetic Mannich base **492**, which represents a very good spin-label for the protein SH group,[121] as well as the cyclic Mannich base deriving from L-DOPA (**493**), which exhibits much enhanced optical rotatory power with respect to the open chain precursor allowing it to be more precisely estimated.[122]

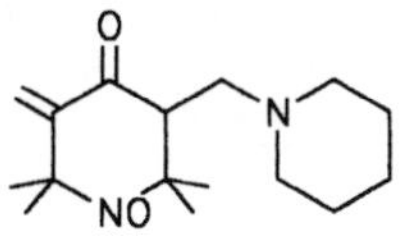

492 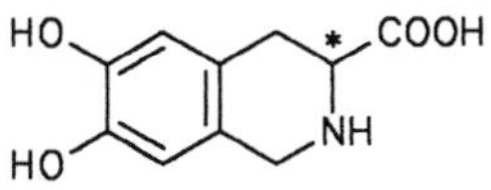   **493**

References for Chapter IV

1. Bolte, M. L., Crow, W. D., and Paton, D. M., Frost hardiness in Eucalyptus grandis: a possible molecular mechanism, *Plant Biol.*, 5, 129, 1987; *Chem. Abstr.*, 107, 195135, 1987.
2. Kametani, T., Honda, T., Fukumoto, K., and Ihara, M., Alkaloids, in *Ullmann's Encyclopedia of Industrial Chemistry*, Vol. A1, 5th ed., Gerhartz, W., Ed., VCH, Weinheim, 1985, 356, 363, 373, and 374.
3. Corey, E. J. and Balanson, R. D., A total synthesis of (±) Poranterine, *J. Am. Chem. Soc.*, 96, 6516, 1974.
4. Kametani, T., Noguchi, I., and Saito, K. Mannich and Eschweiler-Clarke reaction of 1-benzyl-1,2,3,4-tetrahydroisoquinolines. Studies on the synthesis of heterocyclic compounds, *J. Heterocycl. Chem.*, 6, 869, 1969.
5. Hayakawa, Y., Takaya, H., Makino, S., Hayakawa, N., and Noyori, R., Synthesis of 3(2H)-furanones by the iron carbonyl-promoted cyclocoupling reaction of α,α'-dibromo ketones and carboxyamides. A convenient route to muscarines, *Bull. Chem. Soc. Jpn.*, 50, 1990, 1977.
6. Openshaw, H. T. and Wittaker, N., The synthesis of emetine and related compounds. IV. A new synthesis of 3-substituted 1,2,3,4,5,7-hexahydro-9,10-dimethoxy-2-oxo-11bH-benzo[a]quinolizines, *J. Chem. Soc. C.*, 1449, 1963.
7. Feldman, P. L. and Rapoport, H., Synthesis of optically pure Δ^4-tetrahydroquinolinic acids and hexahydroindolo[2,3-a]quinolines from L-aspartic acid. Racemization on the route to vindoline. *J. Org. Chem.*, 51, 3882, 1986.
8. Polazzi, J. O., Mannich reaction product of dihydrocodeinone, *J. Org. Chem.*, 46, 4262, 1981.
9. Gnecco Medina, D. H., Grierson, D. S., and Husson, H. P., 2-Cyano-Δ^3-piperideines X^1: biomimetic synthesis of the Ladyburg alkaloids of the adaline series, *Tetrahedron Lett.*, 24, 2099, 1983.
10. Ayer, W. A. and Furuichi, K., The total synthesis of coccinelline and precoccinelline, *Can. J. Chem.*, 54, 1494, 1976.
11. Ayer, W. A., Dawe, R., Eisner, R. A., and Furuichi, K., A total synthesis of myrrhine, (±)-hippodamine, and (±)convergine, *Can. J. Chem.*, 54, 473, 1976.
12. Overman, L. E. and Goldstein, S. W., Enantioselective total synthesis of allopumiliotoxin A alkaloids 267A and 339B, *J. Am. Chem. Soc.*, 106, 5360, 1984.
13. Hanaoka, M., Ogawa, N., Shimizu, K., and Arata, Y., Synthetic studies on lythracea alkaloids. II. The Mannich reaction of isopelletierine with 3-methoxybenzaldehyde and 3-hydroxybenzaldehyde, *Chem. Pharm. Bull. Tokyo*, 23, 1573, 1975.
14. Büchi, G., Gould, S. J., and Näf, F., Stereospecific synthesis of uleine and epiuleine, *J. Am. Chem. Soc.*, 93, 2492, 1971.
15. Heathcock, C. H., Kleinman, E. F., and Binkley, E. S., Total synthesis of lycopodium alkaloids: (±)-lycopodine, -lycodine, and -lycodoline, *J. Am. Chem. Soc.*, 104, 1054, 1982; see also *J. Am. Chem. Soc.*, 100, 8036, 1978.
16. Earley, W. G., Oh, T., and Overman, L. E., Synthetic studies directed towards gelsemine. Preparation of an advanced pentacyclic intermediate, *Tetrahedron Lett.*, 29, 3785, 1988.
17. Shishido, K., Hiroya, K., Fukumoto, K., and Kametani, T., Highly efficient construction of the 7-axabicyclo[3.3.1]non-2-ene system. An application to the synthesis of the AEF ring system of the aconitine-type diterpene alkaloids, *J. Chem. Res. Synop.*, 100, 1989; *Chem. Abstr.*, 112, 21174, 1990.
18. Guiles, J. W. and Meyers, A. I., Asymmetric synthesis of benoquinolizidines: a formal synthesis of (−)-emetine, *J. Org. Chem.*, 56, 6873, 1991.
19. Meyers, A. I., Miller, D. B., and White, F. H., Chiral and achiral formamidine in synthesis. The first asymmetric route to (−)-yohimbone and an efficient total synthesis of (±)-yohimbone, *J. Am. Chem. Soc.*, 110, 477, 1988.
20. Earley, W. G., Jacobsen, E. J., Meier, G. P., Oh, T., and Overman, L. E., Synthesis studies directed towards gelsemine. A new synthesis of highly functionalized cis-hydroisoquinolines, *Tetrahedron Lett.*, 29, 3781, 1988.
21. Shishido, K., Hiroia, K., Fukumoto, K., and Kametani, T., An efficient and highly regioselective intramolecular Mannich-type reaction: a construction of AEF ring system of aconitine-type diterpene alkaloids, *Tetrahedron Lett.*, 27, 1167, 1986.
22. Speckamp, W. N., Recent developments in alkaloid synthesis, *Recueil*, 100, 345, 1981.
23. Overman, L. E., Robertson, G. M., and Robichaud, A. J., Total synthesis of (±) meloscine and (±) epimeloscine, *J. Org. Chem.*, 54, 1236, 1989.

24. Overman, L. E. and Wild, H., Preparation of functionalized hydroindol-3-ols via tandem aza-Cope rearrangement-Mannich cyclizations. Total synthesis of (±) 6α-epipretazetine and related alkaloids, *Tetrahedron Lett.,* 30, 647, 1989.

25. Overman, L. E., Sworin, M., and Burk, R. M., A new approach for the total synthesis of pentacyclic aspidosperma alkaloids. Total synthesis of dl-16-methoxytabersonine, *J. Org. Chem.,* 48, 2685, 1983.

26. Kametani, T. and Ihara, M., An alternative total synthesis of (±)-scoulerine and (±) tetrahydropalmatine, *J. Chem. Soc. C,* 530, 1967.

27. Kametani, T., Noguchi, I., Nakamura, S., and Konno, Y., Synthesis of heterocyclic compounds. CLXII. Coreximine and related compounds. I, *Yakugaku Zasshi,* 87, 168, 1967; *Chem. Abstr.,* 67, 54309, 1967.

28. Kametani, T., Honda, T., and Ihara, M., A total synthesis of kikemanine, *J. Chem. Soc. C,* 3318, 1971.

29. Kametani, T., Takeshita, M., and Takano, S., Total synthesis of discretine, *J. Chem. Soc. Perkin Trans. 1,* 2834, 1972.

30. Kametani, T., Taguchi, E., Yamaki, K., and Kozuka, A., Syntheses of heterocyclic compounds. DXVII. Syntheses of berberine and related compounds. II Mannich reaction of 1-benzylisoquinoline under various conditions, *Yakugaku Zasshi,* 93, 529, 1973; *Chem. Abstr.,* 79, 79005, 1973.

31. Kametani, T., Matsumoto, H., Satoh, Y., Nemoto, H., and Fukumoto, K., Studies on the synthesis of heterocyclic compounds. DCLXXIX. A stereoselective total synthesis of (±)-ophiocarpine, *J. Chem. Soc. Perkin Trans. 1,* 376, 1977.

32. Chiang, H. C. and Brochmann-Hanssen, E., Total synthesis of (±)-discretamine and (±)-stepholidine, *J. Org. Chem.,* 42, 3190, 1977.

33. Kametani, T., Japanese Kokai Tokkyo Koho 75 24,299, 1975; *Chem. Abstr.,* 83, 114719, 1975.

34. Kreidl, J., Szantay, C., Szabo, L., Farkas Kirjak, M., et al., French Patent Appl. FR 2,648,817, 1990; *Chem. Abstr.,* 115, 71988, 1991.

35. Martinez, S. J. and Joule, J. A., Synthesis of 5-hydroxy-2,6-dimethyl-6H-pirido [4,3-b]carbazole, *J. Chem. Soc. Chem. Commun.,* 818, 1976.

36. Bariana, D. S. and Groundwater, G., Mannich reaction studies in the theophylline series, *J. Heterocycl. Chem.,* 6, 583, 1969.

37. Rida, S. M., Farghaly, A. M., and Ashour, F. A., Mannich bases of theophylline, *Pharmazie,* 34, 214, 1979.

38. Graefe, G., Mannich bases of acetonyltheobromines. II. Synthesis of some 1-(4-dialkylaminobutan-2-onyl)theobromines, *Arzneim. Forsch.,* 17, 1465, 1967.

39. Nguyen, T. L., Gruenke, L. D., and Castagnoli, N., Metabolic oxidation of nicotine to chemically reactive intermediates, *J. Med. Chem.,* 22, 259, 1979.

40. Sikdar, M. B., Babaev, T. M., and Musaev, U. N., Synthesis and polymerization of N,N′-anabasinomethyleneacrylamide, *Uzb. Khim. Zh.,* 42, 1985; *Chem. Abstr.,* 103, 123959, 1985.

41. Stevens, G. and Halamandaris, A., Bases prepared from steroids, *J. Org. Chem.,* 26, 1614, 1961.

42. Mauli, R., Ringold, H. J., and Djerassi, C., Steroids. CXLV. 2-Methylandrostane derivatives. Demonstration of boat form in the bromination of 2-α-methyl-androstan-17β-ol-3-one, *J. Am. Chem. Soc.,* 82, 5494, 1960.

43. Ahond, A., Cavé, A., Kan-Fan, C., and Potier, P., Modification de la réaction de Polonovsky: préparation d'un nouveau réactif de Mannich, *Bull. Soc. Chim. Fr.,* 2704, 1970.

44. Carrington, T. R., Long, A. G., and Turner, A. F., Compounds relate to the steroid hormones. VIII. The Mannich reaction with 3- and 20-oxo-steroids, *J. Chem. Soc.,* 1572, 1962.

45. Manson, A. J., Sjogren, R. E., and Riano, M., A new microbiological steroid reaction, *J. Org. Chem.,* 30, 307, 1965.

46. Gregory, G. I. and Long, A. G., The Mannich reaction with cholesta-1,4-dien-3-one, *J. Chem. Soc.,* 3059, 1961.

47. Pettit, G. R. and Das Gupta, A. K., Actidione acetate nitrogen mustard, *Chem. Ind.,* 1016, 1962.

48. Gandhi, V. S. and Schwenk, E., Mannich reaction of estrone, *J. Indian Chem. Soc.,* 39, 306, 1962.

49. Julian, P. L., Meyer, E. W., and Printy, H. C., Sterols. VI. 16-Methyltestosterone, *J. Am. Chem. Soc.,* 70, 3872, 1948.

50. Heimes, A., Brienne, M. J., Jacques, J., Johnston, D. B. R., and Windholz, T. B., Some reactions of 5-α-A-nor-androstan-3-ones, in *Proc. 2nd Int. Congr. Hormonal Steroids,* Milan, 1966.

51. Papadaki-Valiraki, A., Study of a-norsteroids, *Prakt. Akad. Athenon,* 52, 476, 1977; *Chem. Abstr.,* 92, 181474, 1980.

52. Hirai, S., Harvey, R. G., and Jensen, E. V., The Mannich reaction: improved conditions and application to 20-ketosteroids, *Tetrahedron Lett.*, 1123, 1963.
53. Thiele, K. and Posselt, K., U.S. Patent 3,733, 340, 1973; *Chem. Abstr.*, 79, 32182, 1973.
54. Akhrem, A. A., Kamernitskii, A. V., Reshetova, I. G. and Chenyuk, K. Y., Transformed steroids. LVII. Use of the Mannich reaction for increasing the length of a side chain of steroids, *Izv. Akad. Nauk SSSR Ser. Khim.*, 1633, 1973; *Chem. Abstr.*, 79, 115768, 1973.
55. Burn, D. and Petrov, V., Modified steroid hormones. XXXV. Steroidal ethynylamines, *Tetrahedron*, 20, 2295, 1964.
56. Pettit, G. R. and Saldana, E. I., Antineoplastic agents. XXXVI. Acetilenic carrier groups, *J. Med. Chem.*, 17, 896, 1974.
57. Galantay, E., Bacso, I., and Coombs, R. V., The preparation of α-allenic alcohols. Acetylene → allene 'homologization', *Synthesis*, 344, 1974.
58. Danishefsky, S., Prisbylla, M., and Lipisko, B., Regioselective Mannich reactions via trimethysilyl enol ethers, *Tetrahedron Lett.*, 21, 805, 1980.
59. Longo, A. and Lombardi, P., European Patent Appl. EP 326,340, 1989; *Chem. Abstr.*, 112, 36257, 1990.
60. Longo, A. and Lombardi, P., European Patent Appl. EP 307,134, 1989; *Chem. Abstr.*, 111, 134643, 1989.
61. De Stevens, G., U.S. Patent 3,092,621, 1963; *Chem. Abstr.*, 60, 604, 1964.
62. Wilson, R. S. and Mertes, M. P., A chemical model for thymidylate synthetase catalysis, *J. Am. Chem. Soc.*, 94, 7182, 1972.
63. Koch, S. A. M. and Sloan, K. B., N-Mannich base derivates of 5-fluoro-cytosine: a prodrug approach to improve topical delivery, *Int. J. Pharm.*, 35, 243, 1987; *Chem. Abstr.*, 106, 219506, 1987.
64. Beandes, R. and Roth, H. J., Zur Aminomethylierung von Xantin, Hypoxantin, Adenin, Guanin und Harnsäure, *Arch. Pharm.*, 300, 1000, 1967.
65. Sloan, K. B. and Siver, K. G., The aminomethylation of adenine, cytosine and guanine, *Tetrahedron*, 40, 3997, 1984.
66. Reese, C. B. and Sanghvi, Y. S., Conversion of 2'-deoxyuridine into thymidine and related studies, *J. Chem. Soc. Chem. Commun.*, 877, 1983.
67. Koch, S. A. M. and Sloan, K. B., N-Mannich base pro-drugs of 5-iodo-2'-deoxycytidine as topical delivery enhancers, *Pharm. Res.*, 4, 317, 1987; *Chem. Abstr.*, 107, 223132, 1987.
68. Bridson, P. K., Jiricny, J., Kemal, O., and Reese, C. B., Reactions between ribonucleoside derivatives and formaldehyde in ethanol solution, *J. Chem. Soc. Chem. Commun.*, 208, 1980.
69. Hikoya, H., Yasuhiro, Y., Seiko, Y., Yuriko, Y., Kenichi, T., and Kazuo, N., N-Sulfomethylation of guanine, adenine and cytosine with formaldehyde bisulfite, *Nucleic Acids Res.*, 10, 6281, 1982; *Chem. Abstr.*, 98, 89805, 1983.
70. Badman, G. T. and Reese, C. B., Reactions between methiodides of nucleoside Mannich bases and carbon nucleophiles, *J. Chem. Soc. Chem. Commun.*, 1732, 1987.
71. Reese, C. B. and Sanghvi, Y. S., The synthesis of 5-carboxymethylaminomethyluridine and 5-carboxymethylaminomethyl-2-thiouridine, *J. Chem. Soc. Chem. Commun.*, 62, 1984.
72. Juhasz, A. and Bajusz, S., Synthesis of γ-carboxyglutamic acid and its derivatives via Mannich base condensation, *Int. J. Pept. Protein Res.*, 15, 154, 1980; *Chem. Abstr.*, 92, 181615, 1980.
73. Ramakrishnan, K. and Portoghese, P. S., Synthesis and biological evaluation of a metazocine-containing enkephalinamide. Evidence for nonidentical roles of the tyramine moiety in opiates and opioid peptides, *J. Med. Chem.*, 25, 1423, 1982.
74. Tome, D., Naulet, N., and Martin, G. J., NMR Studies of the reaction of formaldehyde with the aminogroup of alanine and lysine versus pH, *J. Chim. Phys. Chim. Biol.*, 79, 361, 1982.
75. Trézl, L., Rusznák, I., Náray-Szabo, G., Szarvas, T., Csiba, A., and Ludanyi, A., Essential differences in spontaneous reactions of L-lysine and L-arginine with formaldehyde, *Period. Polytech. Chem. Eng. Budapest*, 32, 251, 1988; *Chem. Abstr.*, 112, 99170, 1990.
76. Grakauskas, V. and Baum, K., Mannich reaction of 2-fluoro-2,2-dinitroethanol, *J. Org. Chem.*, 36, 2599, 1971.
77. Blass, J., Préparation et étude d'une base de Mannich cristallisée obtenue par action due formol sur un mélange de thréonine et de diméthyl-2,4-phénol, *Bull. Soc. Chim. Fr.*, 3120, 1966.
78. Amos, H. E., Evans, J. J., Himmelsbach, D. S., and Barton, F. E., *In vitro* stability, *in vivo* hydrolysis, and absorption of lysine and methionine from polymerized amino acid preparations, *J. Agric. Food Chem.*, 28, 1250, 1980; *Chem. Abstr.*, 93, 202990, 1980.

79. Agababyan, A. G., Gevorgyan, G. A., and Mndzhoyan, O. L., Aminoacids as the amine component in a Mannich reaction, *Usp. Khim,* 51, 678, 1982; *Chem. Abstr.,* 97, 24174, 1982.

80. Ramaswamy, T. and Ramaswamy, D., Further structural stabilization of the chrome-collagen compounds, *J. Soc. Leather Technol. Chem.,* 59, 149, 1975; *Chem. Abstr.,* 83, 195285, 1975.

81. Gringas, L. and Sjosted, G., Crosslinking of gelatin with formaldehyde in the presence of some polyhydroxy benzenes, *Angew. Makromol. Kem.,* 42, 123, 1975.

82. Oreal, S. A., Netherlands Patent Appl. 80 03,421, 1980; *Chem. Abstr.,* 95, 67812, 1981.

83. Yoshioka, M., Japanese Kokai Tokkyo Koho JP 62,233,761, 1987; *Chem. Abstr.,* 108, 164415, 1988.

84. Chaplin, M. F. and Kennedy, J. F., Magnetic immobilised derivatives of enzymes, *Carbohydr. Res.,* 50, 267, 1976.

85. Narasimhan, K. and Wingard, L. B., Jr., Immobilization of hydroquinone on electrode surfaces, *J. Mol. Catal.,* 33, 151, 1985.

86. Lapenko, V. L., Potapova, L. B., Slivkin, A. I., and Vasil'eva, E. V., Synthesis and some analogs of D-galacturonic acid amide and hydrazide, *Izv. Vyssh. Uchebn. Zaved. Khim. Khim. Tekhnol.,* 30, 38, 1987; *Chem. Abstr.,* 108, 38260, 1988.

87. Sidney, S., U.S. Patent 2,442,989, 1948; *Chem. Abstr.,* 31, 6162, 1948.

88. Schoenberg, J. E., U.S. Patent 4,009,311, 1977; *Chem. Abstr.,* 86, 123253, 1977.

89. Matsuda, H., Kanki, K., Nakagawa, T., and Koike, M., Japanese Kokai Tokkyo Koho 75 67,287, 1975; *Chem. Abstr.,* 85, 83100, 1976.

90. Lin, S. Y. and Hoo, L. H., U.S. Patent 4,728,728, 1988; *Chem. Abstr.,* 108, 188732, 1988.

91. Brezny, R., Paszner, L., Micko, M., and Uhrin, D., The ion-exchanging lignin derivatives prepared by Mannich reaction with amino acids, *Holzforschung,* 42, 369, 1988; *Chem. Abstr.,* 111, 9088, 1989.

92. Van der Klashorst, G. H., South African Patent ZA 88 05,771, 1989; *Chem. Abstr.,* 112, 38477, 1990.

93. Watanabe, T., Katayama, S., Nakashita, Y., and Yamauchi, M., Regiospecific (biogenetic-type) synthesis of 2-methyl-5H-pyrano[3,2-c][1]benzopyran-4-one, the basic skeleton in citromycetin, *J. Chem. Soc. Chem. Commun.,* 761, 1981.

94. Ratcliffe, R. W. and Christensen, B. G., Total synthesis of β-lactam antibiotics. I. α-Thio-formamido-diethylphosphonoacetates, *Tetrahedron Lett.,* 4645, 1973.

95. Lepetit S.P.A., Belgian Patent 661,982, 1965; *Chem. Abstr.,* 65, 5475, 1966.

96. Korea Inst. of Science and Technology, Japanese Kokai Tokkyo Koho, JP 58 67,693; *Chem. Abstr.,* 99, 70474, 1983.

97. Sobiczewski, V. and Biniecki, S., Preparation of aminomethyl derivatives of tetracyclines by Mannich reaction, *Acta Pol. Pharm.,* 21, 421, 1964; *Chem. Abstr.,* 62, 6446, 1965.

98. Popova, I. V. and Sokolov, L. B., New aminomethyl derivative of tetracycline, *Tr. Leningr. Nauchno Issled Inst. Antibiot.,* 84, 1972; *Chem. Abstr.,* 78, 71740, 1973.

99. Kaplan, M. A., Bradner, W. T., Buckwalter, F. H., and Pindell, M. H., Antitumor activity of N-(β,β'-dichloro-diethylaminomethyl)-tetracycline (tetracycline mustard), *Nature,* 205, 399, 1965; *Chem. Abstr.,* 62, 11030, 1965.

100. de Carneri, I., Coppi, G., Lauria, F., and Logemann, W., Una nuova tetraciclina solubile: la tetraciclina-L-metilenlisina, *Farmaco Ed. Prat.,* 16, 65, 1961; *Chem. Abstr.,* 55, 18990, 1961.

101. Martinez, F., Ibanez, J., and Lazaro, A., Spanish Patent 482,506, 1980; *Chem. Abstr.,* 94, 103778, 1981.

102. Klosa, J. and Voigt, H., German Patent 1,249,878, 1967; *Chem. Abstr.,* 68, 59849, 1968.

103. Mueller, J. and Menard, E., Swiss Patent 484, 097, 1970; *Chem. Abstr.,* 73, 66435, 1970.

104. Bonati, A., Bombardelli, E., and Gabetta, G., British Patent 1,383, 053, 1975; *Chem. Abstr.,* 83, 43342, 1975.

105. Foldeak S., Hegyes, P., and Dombi, G., Extraction of arctigenin and synthesis of its amino derivatives, *Acta Phys. Chem.,* 20, 459, 1974; *Chem. Abstr.,* 83, 58560, 1975.

106. Molho, D. and Gerphagnon, M. C., Méthylation nucléaire au moyen des bases de Mannich. I. Synthèse de la méthyl-8-chrysine et de la strobochrysine, *Bull. Soc. Chim. Fr.,* 604, 1963.

107. Singh, G. B., Chawla, A. S., and Vijjan, V. K., N-Mannich bases of nicotinamide, *Indian J. Pharm.,* 30, 231, 1968; *Chem. Abstr.,* 70, 28790, 1969.

108. Baldwin, W. S., Willging, S. M., and Siegel, B. M., European Patent Appl. EP 159,018, 1985; *Chem. Abstr.,* 104, 95446, 1986.

109. Mariani, E., Longobardi, M., Schenone, P., Bondavalli, F., and Bianchi, C., Acetylenic Mannich bases with terpenoid structure, *Chim. Ther.,* 8, 281, 1973.

110. Lajunen, M. and Krieger, H., Versuche zur Darstellung von α-Methylen-3(10)carenen, *Rapp. Univ. Oulu. Ser. Chem.,* 19, 1985.

111. Krieger, H., Kojo, A., and Oikarinen, A., Aminomethylation of 2-pinen-4-one, *Finn. Chem. Lett.,* 185, 1978; *Chem. Abstr.,* 90, 23283, 1979.

112. McClure, N. L., Dai, G. Y., and Mosher, H. S., exo,endo-3-(Dimethylaminomethyl)-d-camphor: d-camphor Mannich products, *J. Org. Chem.,* 53, 2617, 1988.

113. Danishefsky, S., Schuda, P. F., Kitahara, T., and Etheredge, S. J., Total synthesis of dl-vernolepin and dl-vernomenin, *J. Am. Chem. Soc.,* 99, 6066, 1977.

114. Schreier, E., Radiolabeled peptide ergot alkaloids, *Helv. Chim. Acta,* 59, 585, 1976.

115. Csuk, R., Hoenig, H., and Romanin, C., Stability studies of B-N-heterocycles by ^{11}B-NMR spectroscopy, *Monatsh. Chem.,* 113, 1025, 1982.

116. Fowler, J. S., 2-Methyl-3-butyn-2-ol as an acetylene precursor in the Mannich reaction. A new synthesis of suicide inactivators of monoamine oxidase, *J. Org. Chem.,* 42, 2637, 1977.

117. Nakatsuka, I., Kawahara, K., and Yoshitake, A., Labeling of neuroleptic butyrophenones. III. Synthesis of 2'-amino-4'-fluoro-4[4-hydroxy-4(3-trifluoromethylphenyl)piperidino-2-^{14}C]butyrophenone, *J. Labelled Compd. Radiopharm.,* 18, 495, 1981; *Chem. Abstr.,* 95, 97533, 1981.

118. Masuda, K., Toga, T., and Hayashi, N., Synthesis of 3-morpholino-N-ethoxycarbonyl sydnone imine-5-^{14}C (SIN-10-^{14}C), *J. Labelled Compds.,* 11, 301, 1975; *Chem. Abstr.,* 84, 121730, 1976.

119. Rzeszotarski, W. J., Eckelman, W. C., and Reba, R. C., Application of the Mannich reaction for introducing chelating groups into a biologically active carrier, *J. Labelled Compds.,* 13, 171, 1977; *Chem. Abstr.,* 87, 53192, 1977.

120. Ivannikov, A. T., Popov, B. A., Parfenova, I. M., Balakin, V. M., and Glukhikh, V. V., Study of the effect of high-molecular-weight Mannich bases with 8-OH-quinoline groups on the absorption of radioiodine (^{131}I) during its entry into the digestive tract, *Khim. Pharm. Zh.,* 12, 34, 1978; *Chem. Abstr.,* 88, 148257, 1978.

121. Sholle, V. D., Kagan, E. S., Michailov, V. I., Rozantsev, E. G., Frangopol, P. T., Frangopol., M., Pop, V. I., and Benga, G., A new spin label for sulfhydryl groups in proteins: The synthesis and some applications in labelling of albumin and erythrocyte membranes, *Cent. Inst. Phys. Bucharest RB,* 2, 1980; *Chem. Abstr.,* 93, 182216, 1980.

122. Chaftez, L. and Chen, T. M., Enhancement of optical rotation of levodopa by cyclization, *J. Pharm. Sci.,* 63, 807, 1974.

Chapter V

Practical Applications of Mannich Bases

As we have outlined in the previous chapters, Mannich bases not only are important in research, usually aimed at improving our knowledge of the chemistry of this type of compounds, but they also have considerable practical use over a wide range of other fields. In this connection, a description of the relationships linking chemical structure with the behavior of Mannich bases and their derivatives may conveniently indicate the main features that make these compounds useful for practical purposes.

As we have seen before (Fig. 3, Chap. I, and Fig. 80, Chap. II), the versatile chemistry of Mannich bases offers many opportunities for creating tailor-made molecular structures suited to different needs by virtue of simple and relatively inexpensive procedures that can be readily carried out in industry. In the present chapter, therefore, we shall examine both the chemical structures that provide the properties required and the synthetic strategies that lead to the products best suited to the desired functions. A survey of the main uses of individual Mannich bases according to the type of application is also included.

A Structure/Property Relationships in Mannich Bases and Derivatives

A.1 Chemical Structure

The design of products having, for instance, antioxidant, crosslinking, or anticorrosion abilities, to say nothing of pharmacological applications, calls for a well-defined chemical structure suitably configured in terms of shape and size as well as type, number, and location of the relevant chemical moieties. The predominant molecular conformation also plays an important role.

As far as the shape and size of the molecules are concerned, direct Mannich synthesis or typical reactions of Mannich bases, such as amino group replacement, open up the possibility of producing derivatives that have the chemical structures **I–VII** (Fig. 182). Examples of such derivatives are compounds **494–503**, all selected from products that have properties with useful practical applications. Thus, nonsymmetrical molecules with different A and B functionalities (type **I** in Fig. 182), along with symmetrically disubstituted molecules (type **II**), cyclic or branched compounds (types **III** and **IV**), as well as polymeric or crosslinked derivatives (types **V–VII**), can be produced as the result of appropriate combinations of mono- or polyfunctional A and B reactants.

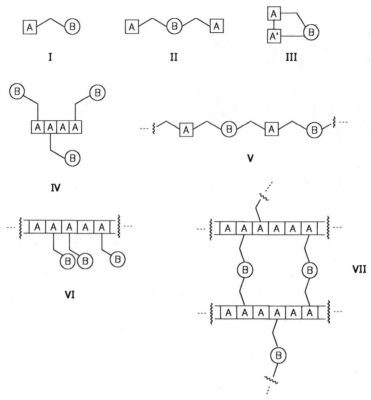

Fig. 182.   The chemical structures of derivatives obtainable by means of Mannich base chemistry. A and B represent different moieties (usually substrate and amine) connected together by a methylene bridge.

Molecules of type **I** are the classical product of reactions between monofunctional reactants. In many cases, they can be considered to be derived from the functionalization of a substrate or, reciprocally, of an amine, in order to attach chemical moieties purposely designed for specific uses, without substantially affecting the essential molecular structure. This is the case, for instance, of the antibiotic substance **494**, which assumes hydrophilic properties upon aminomethylation of the pharmacologically active substrate having an amino acidic residue,[1,2] or of the complexant agent **495**,[3] as well as of the polyacrylamide Mannich base **502**.[4,5]

Symmetrically disubstituted molecules of type **II** are mostly obtained from bifunctional substrates such as bisphenols, cyclic ureids, etc., or bifunctional amines such as ethylenediamine[6] or piperazine. Accordingly, the amino group replacement of monofunctional Mannich bases by bifunctional nucleophiles such as nitroalkanes or bisamides gives derivatives of type **II**, as represented by the heat stabilizer for polymers **496**,[7,8] bearing two antioxidant phenolic moieties per molecule.

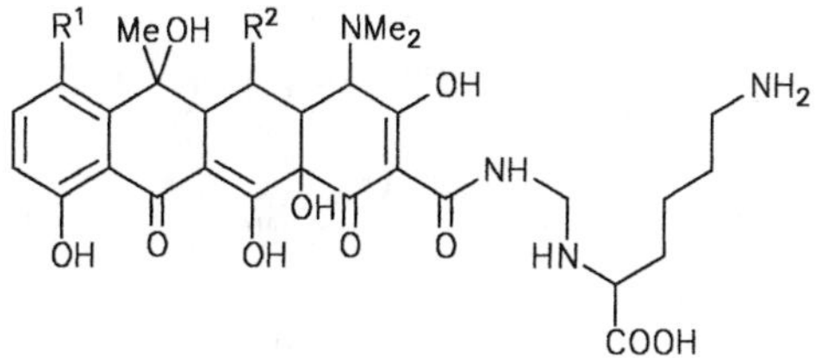

494

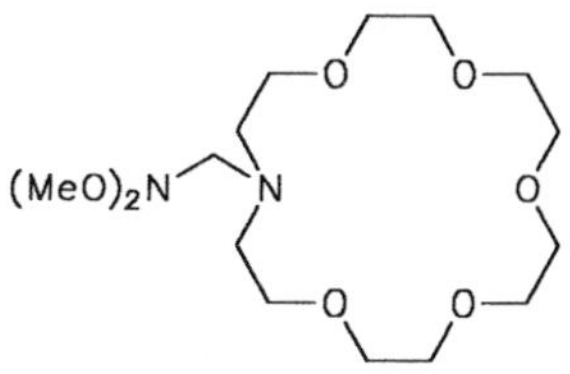

495

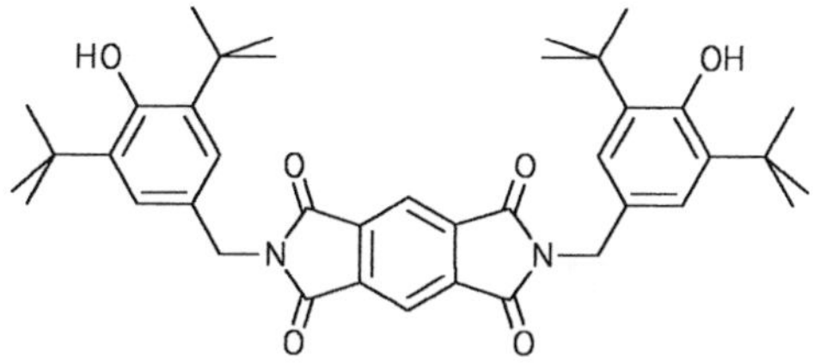

496

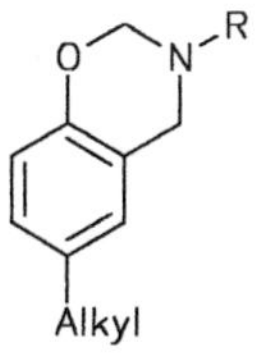

497

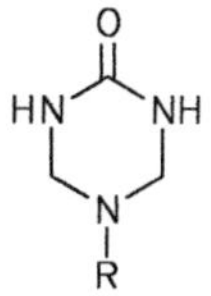

498

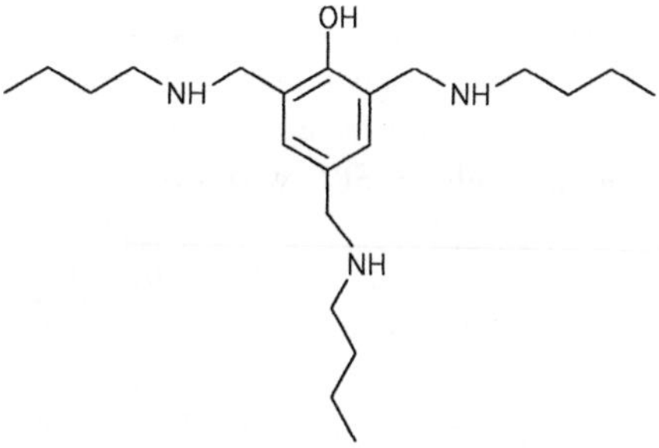

499

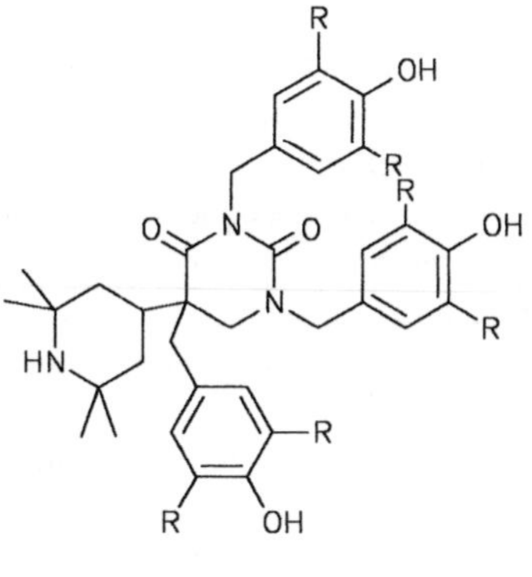

500 (R = CMe₃)

Cyclic derivatives of type **III** include cyclic Mannich bases, such as dihydroben-zoxazines **497**, employed as detergents for lubricating oils,[9,10] and cyclic ureides **498**, precursors of crosslinking agents for fabrics,[11] as well as other cyclic derivatives prepared by conversion of Mannich bases. Macromolecular derivatives of type **IV** are relatively small in size and have branched (star-shaped) structures; they are of considerable importance as, for example, corrosion inhibitors **499**,[12] plastics stabilizers **500**,[13] pre-polymers for epoxy-based electrophoretic paints,[4,14] and polyols in polyurethane synthesis.[15]

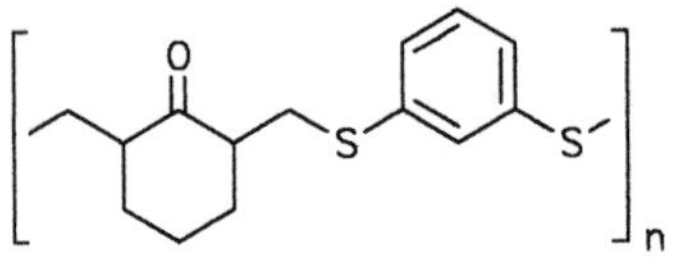

501

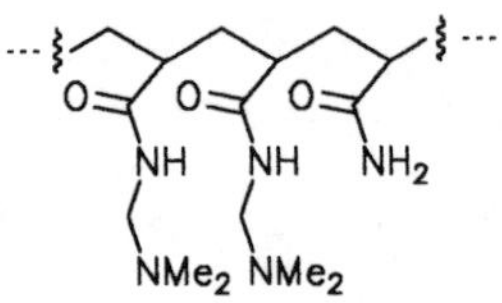

502

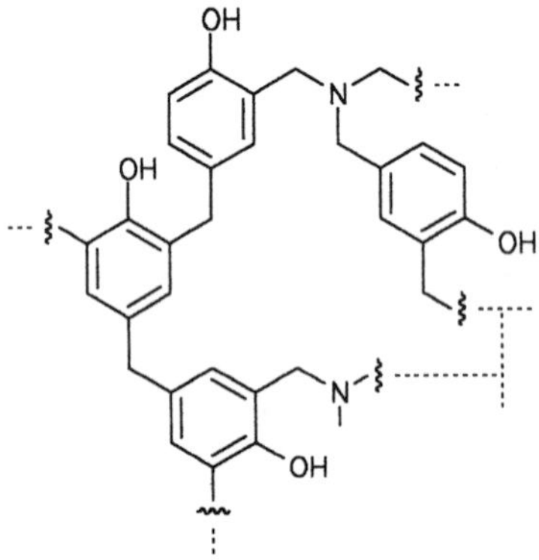

503

Polymeric macromolecules of types **V–VII** are needed for applications requiring enhanced molecular size (see also Chap. III). Thus, linear polymers of type **V** are prepared by Mannich polymerization or from bis-Mannich bases by exchange reaction, for example, with bis-thiols, as in the case of poly(ketosulfide) **501**, which is useful as a high-molecular-weight antioxidant in the processing of plastics.[16,17] Macromolecules of type **VI** are generated by the functionalization of polymers, as shown by **502**, which is obtained by aminomethylation of polyacrylamide and used as flocculant in water treatment.[4,5]

Finally, crosslinked structures of type **VII** may result from the hardening, for instance, of phenolic resins (**503**) through reaction of novolacs with hexamethylenetetramine,[18] which is a preformed aminomethylating agent (see Table 31, Chap. III).

The nature of the chemical moieties present is of course fundamental in determining the possible function of a molecule. Indeed, chemical behavior, such as reactivity, ability to produce ionic species, complexant power, reducing properties, etc., as well as physicochemical features, such as solubility and surface activity, are established by the presence of well-defined groups, as schematically depicted in Table 37; a rigid division between chemical and physicochemical properties exclusively attributable to different chemical moieties can hardly be made, however. Ionogenic groups, for instance, also provide hydrophilicity to the molecule in conjunction with the possibility of ion formation.

Examples of applications determined mainly by chemical properties of the compounds employed are found among resin hardeners **504**,[19] flocculants **505**,[4,20,21] complexants **506**,[22,23] and ionamines **507**,[24,25] while the use of additives for lubricant oils[26] and antistatic agents[27,28] **508–510** is more markedly based on physicochemical behavior.

Table 37

Influence of Relevant Chemical Moieties on the Properties of Mannich Bases

Type	Chemical moiety	Affected properties
a.	Basic groups (as in **504**)	
b.	Ionogenic groups (as in **505, 507, 509**)	mainly chemical
c.	Reactive groups (complexant, hydrolizable, having reducing power, etc. as in **506, 507**)	
d.	Polar-Apolar groups (as in **508 - 510**)	mainly physico-chemical
e.	Hydrophilic-Hydrophobic groups (as in **509, 510**)	

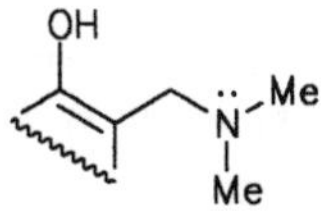

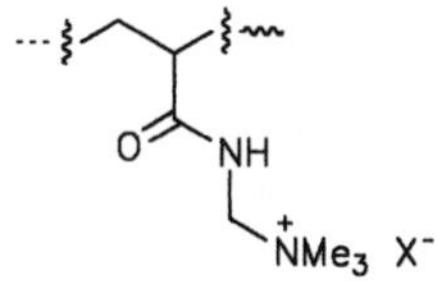

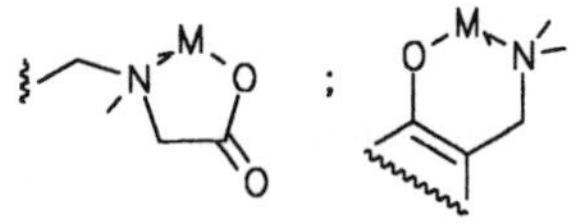

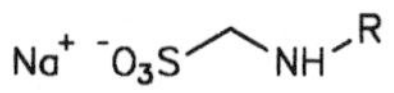

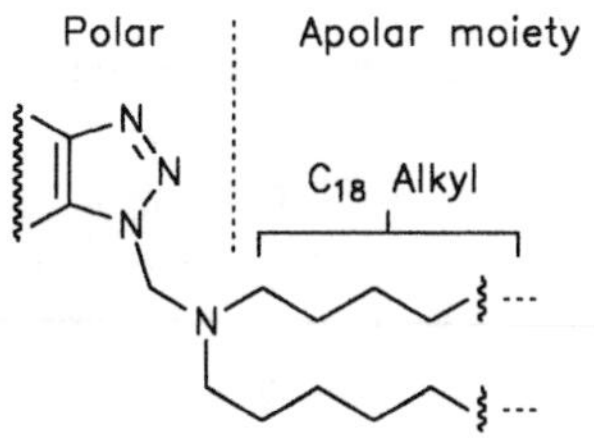

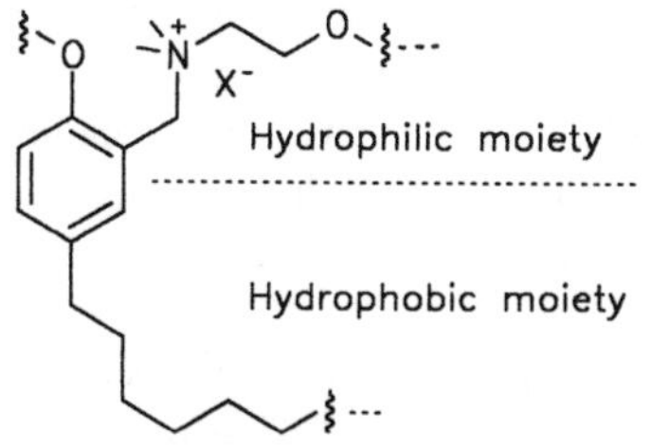

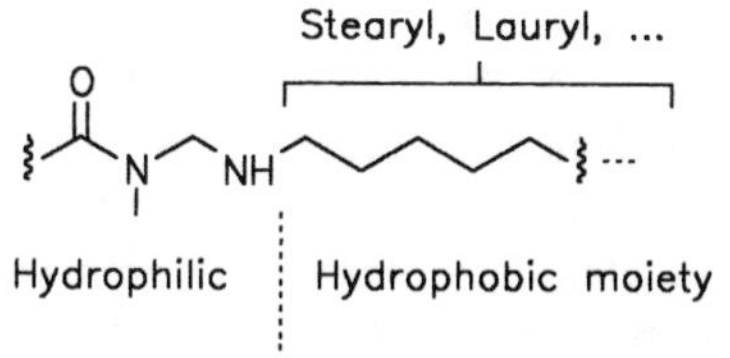

The major feature of the Mannich reaction, namely, the possibility of introducing one or more amine moieties into a substrate, represents a peculiar method for producing substances possessing basic groups (a, Table 37). This allows a potentially large selection of the best-suited amine (Chap. I, A.2), which can be further widened by the possibility of replacing an amino group by another amino compound (Chap. II, B.2). When non-basic analogous structures are required, the replacement reaction can be carried out with amides or other reagents. Direct amidomethylation (Refs. 137–140 for Chaps. I and II) is also frequently performed for many practical applications.[11,29–31] Thiomethylation,[32–34] sulfomethylation, etc., can be similarly performed (Fig. 12, Chap. I).

Due to the possibility of protonation or formation of the corresponding quaternary ammonium salts **505, 509**, the amino group of Mannich bases may constitute an iono-genic moiety (b, Table 37) capable of producing a cationic center in the molecule,[4,20,21,27] although the presence of the quaternary ammonium group may be detrimental to stability (Chap. II, A.2). Cationic species are also afforded by substrates containing N-hetero-cyclic rings, including α-alkyl-pyridines and α-alkyl-quinolines (Chap. I, D). Anionic derivatives may be produced either from the carboxyacid group by employing amino acids as amine reagents or by subjecting alkaline bisulfite (**507**)[24,25] or phosphorous acid[35] to aminomethylation. Readily obtainable precursors of acidic groups are carbox-yesters, alkyl phosphites (Fig. 11, Chap. I), and Mannich derivatives of hydrogen cya-nide.[36] The aldehyde reagent glyoxylic acid also allows introduction of the carboxylate anion (see **28** and Table 3, Chap. I).

Certain reactive properties (c, Table 37), determined by the presence of the ami-nomethyl moiety in Mannich bases, offer the possibility of complex formation (**506**),[22,23] which is frequently exploited in various applications, and of bond cleavage at the meth-ylene moiety adjacent to nitrogen, which allows (e.g., through deamination) the gradual release of amine catalytically active toward macromolecular materials.[37,38] Bond cleavage at the methylene moiety is also profitably applied to dyes[24] and drugs,[25,39–41] as it allows water-soluble compounds of type **507** to lose the hydrophilic sodium al-kylsulfonate group (Fig. 90, Chap. II).

The simultaneous presence in the same molecule of polar and apolar groups (d, Table 37), responsible for attractive or repulsive interactions between molecules, yields surface-active compounds capable of acting at the interface between different substances. Indeed, compounds such as **508**, possessing a highly polar residue, due to the presence of the unsaturated heterocyclic ring, and a large apolar hydrocarbon moiety, are able to interpose themselves at the contact surface between polar and apolar materials,[26] ar-ranging accordingly the relevant groups. Apolar moieties may also behave as "internal plasticizers" when linked to macromolecular materials with a closely packed structure; in this way they can improve, for example, elastic behavior and impact strength.[42]

Similarly, polar hydrophilic groups, capable of readily producing hydrogen bonding or of being easily solvated by water, connected to bulky lipophilic hydrocarbon chains having affinity with organic substances (e, Table 37), are present in compounds such as **509** and **510**, which are used as antistatic agents.[27,28]

In addition to type of functional group, the relative position in the molecule is frequently relevant in order for the expected function to take place. The molecular conformation may also contribute to the occurrence of chemical or physicochemical interactions specifically required for obtaining a particular behavior. Thus, as amino-methylation allows linkage of the amine to the phenolic ring or to the CN group with formation of derivatives having an appropriate distance between the amine nitrogen atom and the hydroxy or the carboxy group (**506**), respectively, it is possible to create chelating agents effective toward various metal ions. Also, the activity of bis-methylolamides employed as crosslinking agents of proteic macromolecules (Fig. 179, Chap. IV) can be affected by the extent of separation between methylol groups that must be adequately related to the distance between the macromolecular chains.

Table 38

Chemical Structure of Mannich Bases and Derivatives Prevalently Used in Practical Applications

Type of Mannich base	Essential chemical structure	Application (relevant Ref.)
C-Mannich bases from:		
Alkyl Ketones ↓	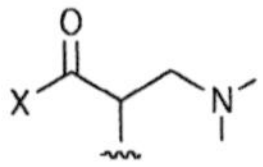 **511**	Pharmaceuticals (43)
γ-Aminoalcohols and further derivatives	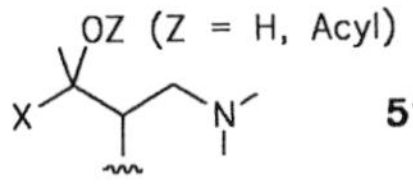 OZ (Z = H, Acyl) **512**	Pharmaceuticals (43)
Phenols	**513** (R = H, CH$_2$NMe$_2$) (R = C$_{8-20}$-Alkyl)	Basic catalysts (44,45) Detergents, Anticorrosion agents (46-48)
	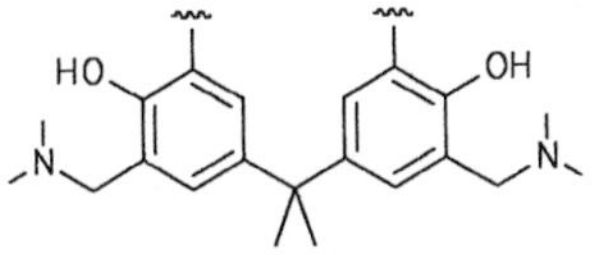 **514**	Prepolymers for epoxy resins (49-51)
	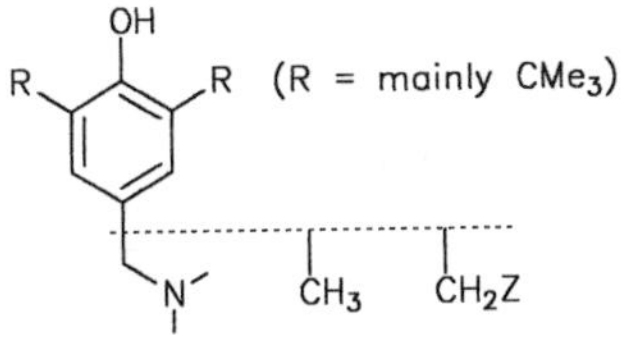 **515** (R = mainly CMe$_3$)	Radical inhibitors (52-56)
Hydrogen cyanide ↓ α-Aminoacids	HOOC—N **516**	Complexants, Sequestering agents (36,57)

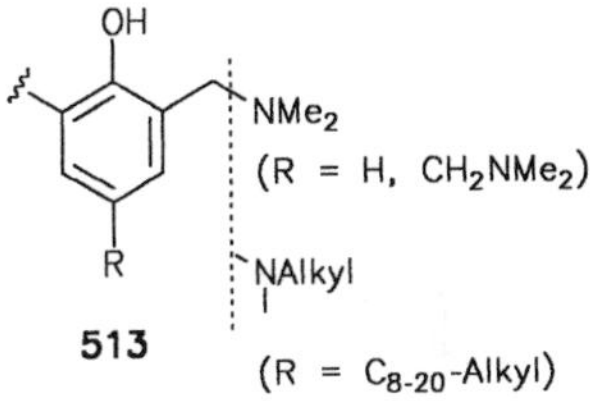

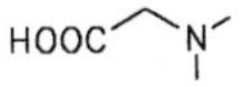

A selected list of the chemical structures of Mannich bases most frequently investigated and applied in industry is summarized in Table 38 along with their applications and relevant references. A more exhaustive treatment of the possibilities offered in this field by the chemistry of Mannich bases is developed in the following sections.

As indicated in the table, almost all the known types of Mannich base, along with an appreciable number of various derivatives (**512, 515, 517**), find practical use.

Table 38
(continued)

N-Mannich bases from:

Amides

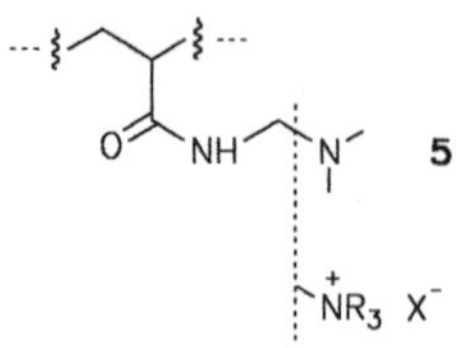

517 Flocculants (58-60)

↓

Quaternary ammonium salts

$^+\!NR_3\ X^-$ Flocculants (20,21)

Benzimidazoles and Benzotriazoles

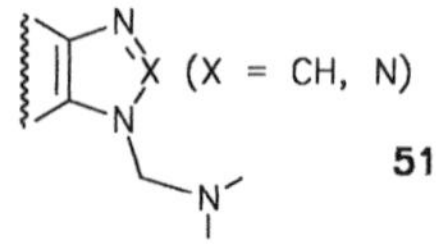

(X = CH, N) Corrosion inhibitors (61-63)

518

O-Mannich bases from:

Melamine (amine reagent)

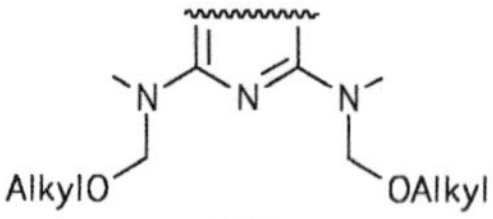

Crosslinking agents (64-66)

519

S-Mannich bases from:

NaHSO₃

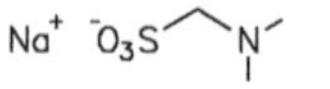

$Na^+\ {}^-O_3S$ **520** Solubilizing, Hydrolyzable groups (24,25)

P-Mannich bases from:

Phosphorous acid and derivatives

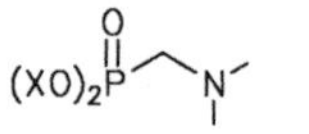

$(XO)_2P$ **521** Complexants (67-69) Pesticides (70-72)

β-Aminoketones **511** are employed mostly as pharmaceuticals, many of them currently in use. Investigations of their properties are of continuing interest in several branches of medicinal chemistry (Fig. 76, Chap. I).[73–77] Moreover, these compounds are precursors of a substantial group of derivatives equally important in pharmacology, namely, the γ-aminoalcohols **512**, which are readily prepared by hydrogenation or addition of organometal to the carbonyl group.[73]

Phenolic Mannich bases **513–515** are quantitatively by far the most important group of compounds having practical application. Variously para-substituted phenols undergoing aminomethylation in the ortho position **513**, range from simple mono- or polydimethylaminomethylated derivatives, used as catalysts for the crosslinking of epoxy resins, to para-alkyl phenolic bases, which are useful as additives in lubricant oils. Moreover, diethanolamino phenolic Mannich bases are precursors of polyols for polyurethanes,[78,79] and piperazino, or amino acidic, bases display selective complexing power.[80–82] Bisphenolic Mannich bases **514** are mainly exploited as prepolymers for electrophoretic epoxy-based coatings, and 2,6-dialkylphenolic bases, as such, or when properly modified via amino group replacement (**515**), give diffusely used radical inhibitors.

C-Mannich bases deriving from hydrogen cyanide make it possible to obtain, upon hydrolysis of the cyano group, α-amino acids **516**, which are employed largely as complexants; the well-known ethylenediaminetetraacetic acid (EDTA) is the main representative of this group of compounds.

Among N-Mannich bases, derivatives of polyacrylamide (**517**) are investigated[58] particularly for their flocculant properties in water treatment. The corresponding quaternary ammonium salts are also used. Benzimidazolic and benzotriazolic Mannich bases **518**, usually with the bulky alkyl group linked to the amine moiety, are used effectively against corrosion in a manner similar to alkylphenols **513**, which have the apolar group linked to the substrate moiety.

Melamine and guanamine afford O-Mannich bases **519** with basicity that is weak, yet higher than that of the arylamines. They behave similarly to urea derivatives as crosslinking agents for macromolecular materials and easily give transaminomethylation with replacement of the O-Alkyl group by residues deriving from hydrogen-active reagents or by reaction with active groups already present in the molecule.

The S-Mannich bases deriving from sodium bisulfite (**520**), as mentioned before, are ionamines possessing hydrophilic properties that can be easily hydrolyzed when the presence of the polar ionic group is unnecessary.

Finally, the P-Mannich bases of phosphorous acid (**521**) are mainly used as complexants, as are α-amino acids **516**, and pesticides, particularly herbicides.

A.2   Functions Performed

As has been pointed out in the preceding section, certain classes of Mannich bases are particularly suited to perform practical functions. The properties required for this purpose are the subject of the present section, which describes the above functions in detail and highlights, whenever possible, structure/property relationships. To clarify further the definition of practical function, it is worth recalling that the same class of products can be employed in different applications; for example, a crosslinking agent may find use in the textile industry as well as in hairdressing preparations or in the cure of resins. Similarly, a radical inhibitor may be used as an antioxidant for food preservation, in the preservation of lubricants, or as a polymerization inhibitor in the handling of unsaturated monomeric substances. Despite the large number of diverse applications, however, some of the more obvious relationships that connect the behavior of a product to its chemical structure, that is, to the shape and size of the molecules, and to type and location of functional groups, can be outlined with reasonable accuracy.

Some prominent properties, such as basicity, ionogenic and complexing power, hydrophilicity, lipophilicity as well as chemical reactivity, can be suitably combined to provide the desired function. The identification of such properties may be readily accomplished in many cases. Thus, for ion-exchange resins, crosslinked or anyhow insoluble products containing ionogenic groups are readily envisaged, while equally relevant properties required for this function, such as polarity and/or complexing capability of the chemical moieties present in the molecule, are more difficult to conceive. The contribution of these last features to the final performance of a product is hardly measurable in the case of anticorrosion agents and, in general, for all the substances that operate at the interface between different phases. Even more complex considerations would apply in the case of molecules, such as drugs and pesticides, acting into a biological environment; this subject would require a treatment completely beyond the scope of the present book.

Primary Functions

If we consider the main properties exhibited by Mannich bases or their derivatives, we can collect into the following few classes the large number of specific applications (see Table 41, below) provided by individual molecules:

- **Complexing agents**—Sequestering agents, some types of ion-exchange resins, etc.
- **Crosslinking agents**—Hardeners, finishing agents for textiles, etc.
- **Ion exchangers**
- **Radical inhibitors**—Antioxidants, polymerization inhibitors, etc.
- **Surface-active agents**—Surfactants, wetting agents, antistatics, etc.

Complexing Agents

The complexing power of a molecule can be exploited either to exhaustively capture or gradually release a particular metal ion, or to produce ion exchange. To this end, the most suitable molecules are chelating agents having two or more properly located moieties capable of behaving as ligands.[23] As amine nitrogen atoms are frequently involved in these functions, the chemistry of Mannich bases can offer valid synthetic solutions that make it possible to connect a substrate to an amine, either of which has chelating power,[22,83–85] or to directly create a complexant species by Mannich reaction. Structures **522–525** are examples of the former type, with systems **523** and **524** grafted onto polymeric chains,[83–85] in which the chelation sites (indicated by the double arrow) are the same as those originally present in the substrate or amine molecule.

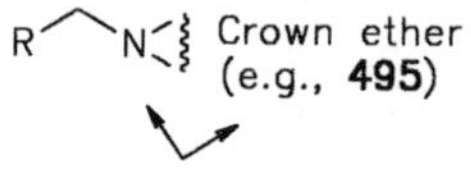

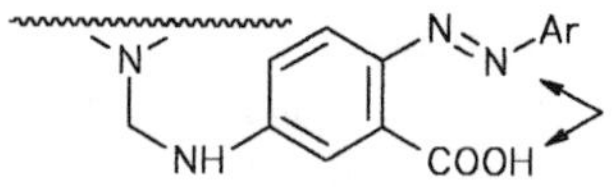

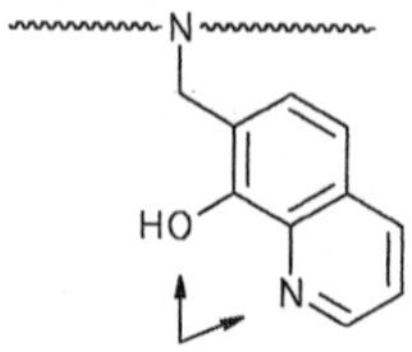

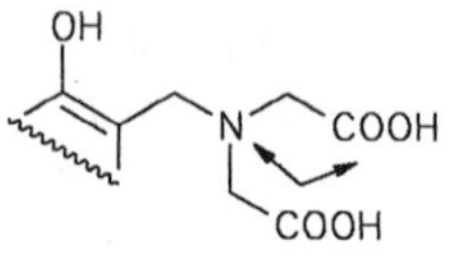

Mannich bases of the latter type usually contain one ligand in the substrate moiety, with the second one provided by the amine nitrogen atom. Products **524** and **525** may be included in this group; under particular conditions, the phenolic hydroxyl can also be involved in the chelating system.[22] However, the most important representatives of this class are products containing the moieties **526–528**.

Table 39

Metal Ions Affording Chelates with Mannich Bases

Metal	Mannich base from:	Reference
UO_2^{II}	Alkyl Ketones (Resin)	95
Co, Mn^{II}	γ-Piperidones (**529**)	91
Cr, Al, Zr	Malonic acid/Collagen	96
Co, Al, Zn, Pb^{II}, Cd	Phenols	22
Cu, Co, Cd, Ca	Phenols	86
Cu, Fe^{III}	Phenols	23,94
Cu, Zn, Cd	Lignin	93,97
Cu, Ni, Co, Mn^{II}	8-Hydroxyquinoline	98
Cu, Co, Ni, Pd	Polymeric Amides (**523**)	84
Ni, Mo	Cyclic Phosphines (**530**)	92

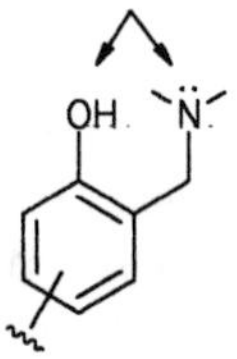

526 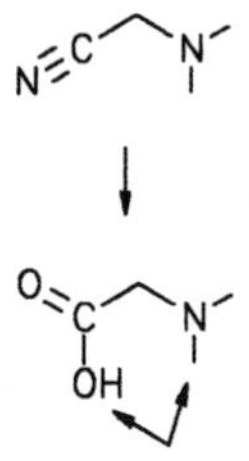   **527** 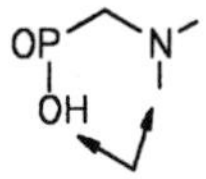   **528**

Phenolic Mannich bases **526**[86] and Mannich bases derived from hydrogen cyanide, precursors of α-amino acids **527** such as EDTA and similar derivatives,[36,87,88] are largely used. The P-Mannich bases **528** may be also inserted into polymeric structures.[89,90]

Structures represented by γ-piperidone **529**[91] and Mannich bases derived from phosphine (**530**)[92] enable us to point out the wide possibility of synthesizing products having the chelating groups most suitably located for the best performance.

Good stability of metal chelate and, connected to this, the selectivity toward different ions, which are both strongly affected by the nature of the Mannich base, are reported.[93–95] Table 39 lists the main metal ions giving complexes with different Mannich bases.

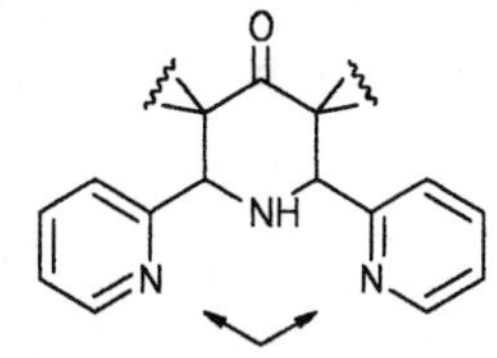

529    **530**

An important feature connected with the structure of the chelating agent is its interaction with the surrounding environment, that is, the aqueous medium, the organic solvent, or both. When the agent has to remain in aqueous solution after chelation, the presence of a strongly hydrophilic group, such as the sulfonic one, or of actively ionogenic groups allowing good solubility even of macromolecular derivatives, is required. In the case of the organic medium, particularly hydrocarbon solvents,[99,100] bulky alkyl groups (**531**) or extended hydrocarbon chains combined with the oligomeric nature of the product **532** make it possible to obtain satisfactory solubility of the chelate. When the complexant operates at the interface between organic and aqueous layers, an analogous function is performed by large apolar moieties anchoring the molecule to the lipophilic phase. Crosslinked macromolecular derivatives are employed when the complexing action has to be exerted by insoluble substances.

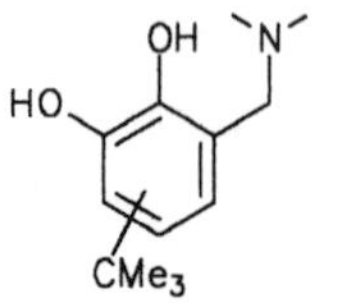

531 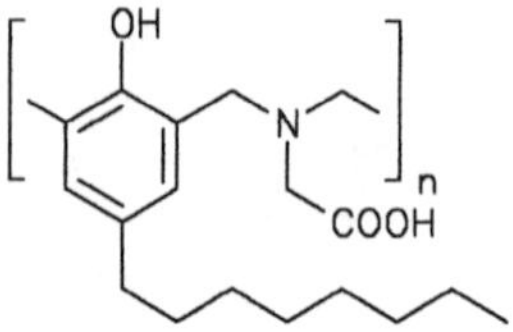   **532**

Similar chelating properties are exhibited by the following agents (see also Ref. 36): scale inhibitors,[89,90] sequestering agents, water softeners, surfactants, peroxide stabilizers,[89] soil antiredeposition agents, well-drilling needs,[35] etc.

Crosslinking Agents

The important process of crosslinking makes it possible to connect sections of macromolecular chains, so as to severely limit or prevent their mutual movement, in order to reduce the possibility of mechanical deformation of the material, obtain insolubility, and establish the desired degree of swelling in selected solvents. Crosslinked products (see Fig. 159, Chap. III) can derive from the polymerization of polyfunctional monomers, or oligomers, as well as from reaction of high-molecular-weight macromolecules with crosslinking agents. The former process is based on the molding of compounds with branched (star-shaped) structures bearing, at each end, chemically reactive groups such as the polyols employed in the production of crosslinked polyurethanes. These prepolymers can be prepared by Mannich reaction[15,101,102] between a phenolic substrate and alkanolamine to give products **533**, which are then usually allowed to react with epoxides to afford polyols having variable chain length (see **399** and Fig. 160, Chap. III). The molding process is carried out in the presence of polyfunctional isocyanates, and the crosslinking degree, directly affecting the mechanical stiffness of the resulting polyurethane, is related to the length of the polyol chains.

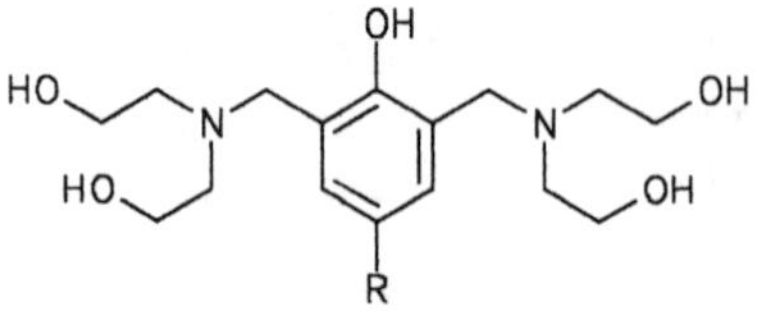

533

Crosslinking of macromolecular derivatives may be performed either in conjunction with the material being manufactured (e.g., in the hardening of epoxy resins with polyfunctional amines) or by treatment of a previously formed material with the appropriate agent (finishing or post-treatment). This last species must have at least two equally, or slightly different, reactive functions and adequate molecular size in order to provide sufficient capability for diffusion into the solid material. The distance between the reactive moieties should also play an important role in crosslinking. This aspect, usually rather neglected, is examined only in analogous thiomethylation reactions.[34,103]

Crosslinking agents affording aminomethylation reactions (Chap. III) range from formaldehyde, employed in the well-known curing process of casein,[104] to preformed aminomethylating agents, including methylolamines[105–107] such as bis-methylolpiperazine **534**, used in photography.[106] The adamantanone derivatives **535** are employed for cotton fabrics.[107] They have a structure similar to urotropine, which is actually the most frequently mentioned among crosslinking agents along with hydroxy- and alkoxymethyl melamine derivatives.[105]

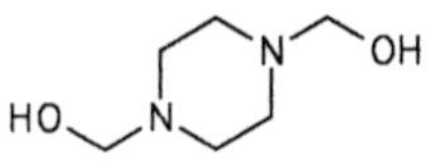

534 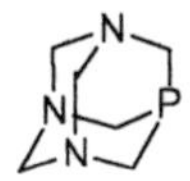   **535** 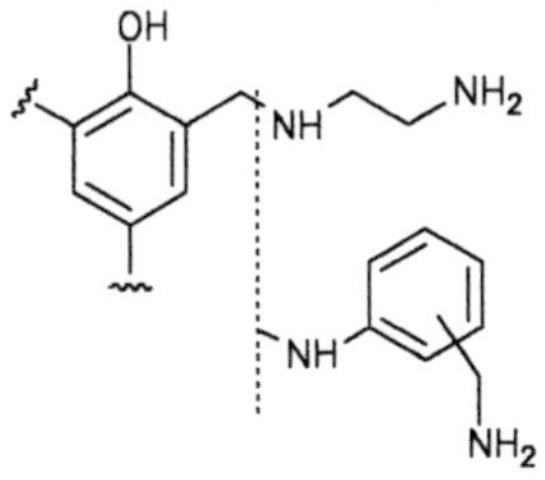   **536**

Bis-methylolamides achieve crosslinking by reaction with the amino groups of proteic macromolecules, for example, in hairdressing preparations (Fig. 179, Chap. IV). The same compounds can afford amidomethylation of materials that lack amino groups, as in the well-known anticrease treatment of cellulosic fibers.[108] Methylol derivatives of the cyclic Mannich bases **498** (Sec. A.1) are similarly used with cellulosic fibers.

Finally, aminomethyl derivatives of type **536** (see also **422**, Chap. III, C.1), having NH groups, are used in the reactive curing of epoxy resins.[109–111]

Ion Exchangers

Ion-exchange resins are used in many practical applications, due to their ability to replace ions present in solution by different ions having an electrical charge of the same sign. Thus, selective or exhaustive water demineralization, purification from undesired ions, recovery of valuable ionic species, etc., can be performed. In addition, ion-exchange resins are employed in the controlled release of microfertilizers or drugs, in dehydration processes, in the absorption of acidic or alkaline volatile substances (e.g., in deodorants),[112] etc. The fundamental properties of these products are therefore insolubility in the liquids submitted to treatment, ionogenicity, and the ability to keep the mobile counterions close to the ionic centers through polar interactions, complex formation, and steric restraints. In this context, the chemistry of Mannich bases may appreciably contribute, in particular, by making it possible to functionalize crosslinked structures or to crosslink polymeric materials for the production of insoluble derivatives (see Chap. III).

Functionalization of crosslinked polymers by Mannich reaction[113] includes mainly polystyrenes and polyacrylics such as styrene/divinyl benzene copolymers **537–539**[114–117] and acrylic ester/divinyl benzene copolymers **540**, respectively.[118] These materials are involved in the reaction as substrate (**539**)[117] or, more frequently, as amine reagent (see also Fig. 163, Chap. III) when the crosslinked product, containing amino groups, is allowed to react with phosphorous acids (**537, 538,** and **540**). Thus, chelating properties are assumed by the resins.

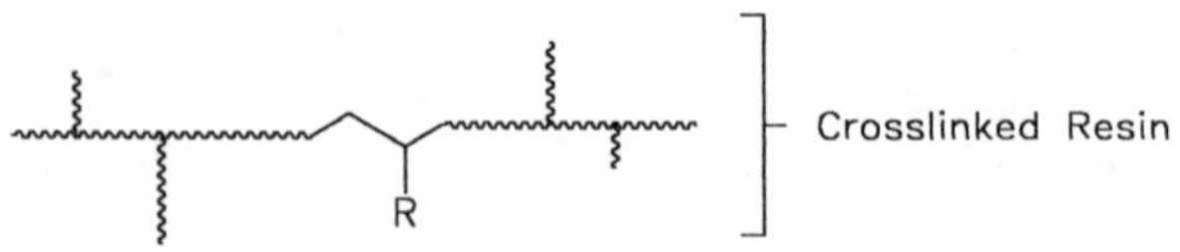

Crosslinked Resin

537   R =

NH⌒PO(OH)$_2$

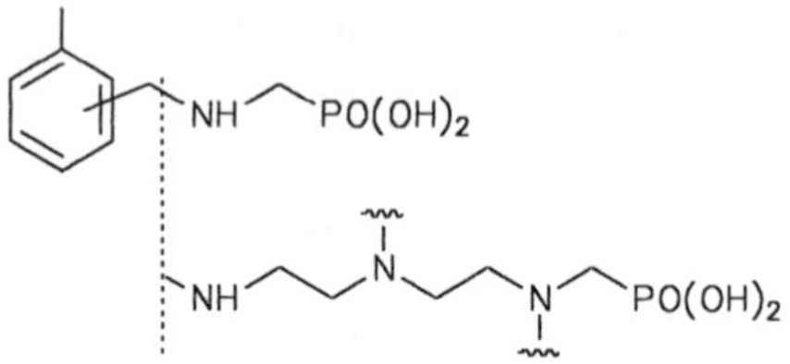

538

539

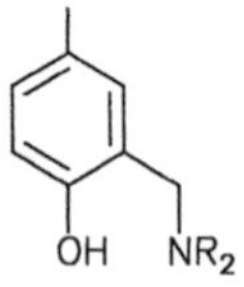

540

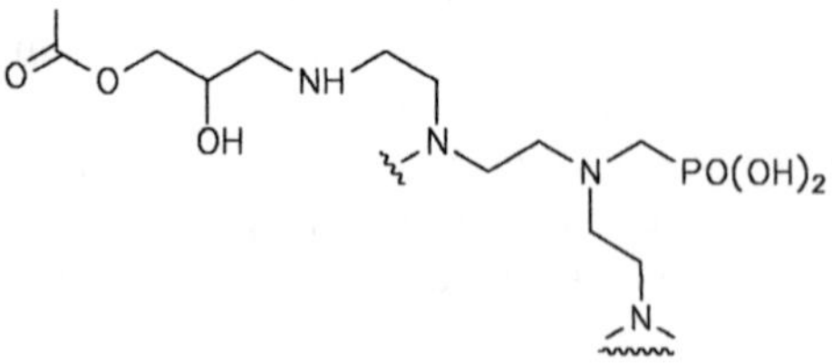

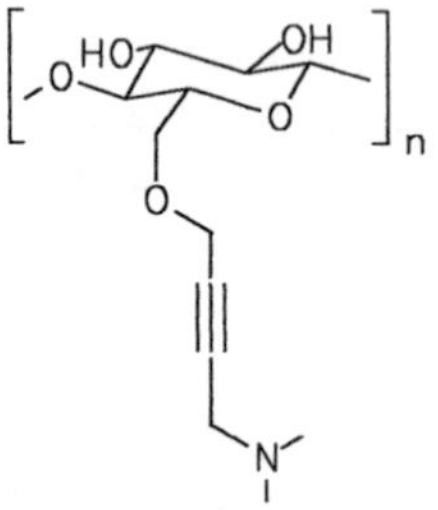

541

In addition to crosslinked resins, insoluble ion exchangers may also be obtained with good results by suitable functionalization of fibers based on cellulose[119,120] or other materials.[121] In this connection, product **541** is a notable example of a Mannich derivative of the propargylic ether of cellulose.[119]

The crosslinking of polymeric materials by Mannich reaction includes the polycondensation of acetone with oligomeric polyalkyleneamines and aldehydes[95,122] (see also **480**, Chap. IV) and, more relevantly, polyamides (**430**, Chap. III) crosslinked with benzidine and formaldehyde. Macromolecular materials such as cellophane and polyglucosamine (Fig. 183), deriving from natural substances, are also subjected to the reaction.

Product **542** is an anion-exchange resin produced from cellophane through treatment with polyethyleneimine (PEI) and formaldehyde, followed by quaternization of the amine nitrogen atom.[123] Polyglucosamine is crosslinked by formaldehyde to give **543**, with methylene bridges interconnecting the amine moieties of different polymeric chains.[124]

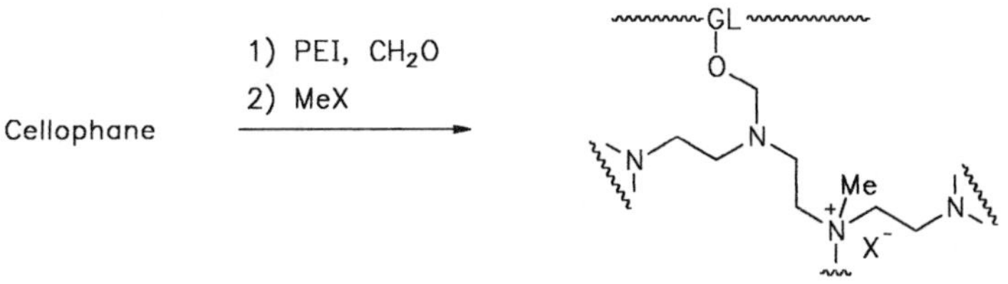

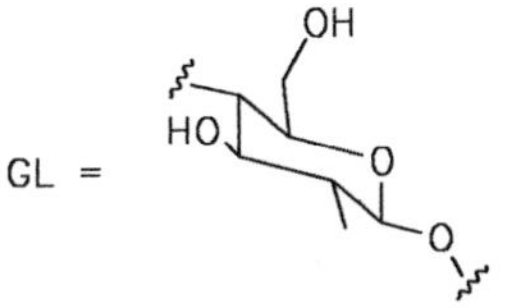

Fig. 183. Mannich bases as ion exchange resins derived from cellulose or poly(glucosamine).

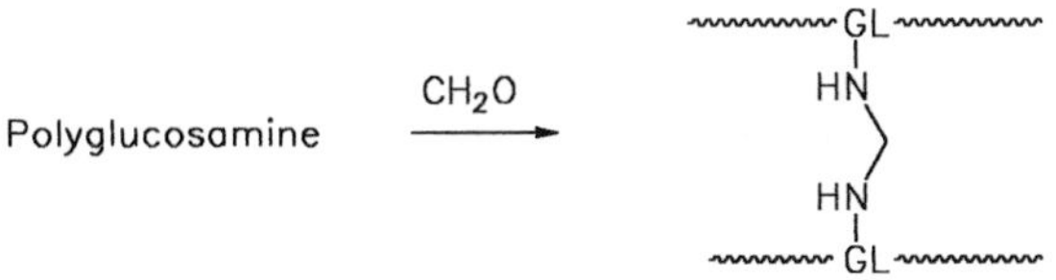

Radical Inhibitors

Radical formation is frequently an occurrence that modifies the chemical structure of many substances with negative effects on their overall performance. Thus, for example, polymeric materials may reduce or lose their mechanical properties, turn yellow, etc.; hydrocarbon fluids (lubricants, fuels) may give rise to precipitation of solids; unsaturated compounds (monomers, etc.) may produce undesired reactions during storage. As a consequence, the need to inhibit radical generation is in many cases of vital importance.

The processes involving the formation of radical species are generally rather complex chain reactions induced by light, heat, etc.; however, the degradative oxidation of hydrocarbon in particular, which is of major relevance, is fundamentally determined by two concomitant mechanisms. The first involves oxygen fixation to the initial radical species, with formation of the peroxide radical R—OO·, followed by hydroperoxide generation due to H· extraction from the hydrocarbon molecule. The second route consists of the initiation by the hydroperoxide R—OOH moieties of radical chain reactions; this can be determined by homolytic cleavage of the reactive O—O bond induced by the light or by the presence of particular metal species having oxido-reductive properties. The possibility of preventing these processes is therefore connected with (a) the supply of hydrogen radicals from sources other than the hydrocarbon molecules, avoiding at the same time the occurrence of chain transfer reactions, (b) the reductive annihilation of hydroperoxide, (c) the protection of the material against light, and (d) the inactivation of metal ions capable of promoting hydroperoxide decomposition.

In this context, the chemistry of Mannich bases offers convenient tools for the synthesis of suitable compounds, mainly based on phenolic derivatives such as those depicted in formulas **544–547**, which can inhibit the formation of radical species.

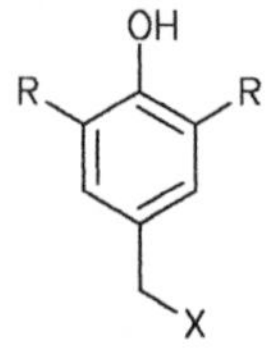

544

(a)

(R = Bulky alkyl groups)

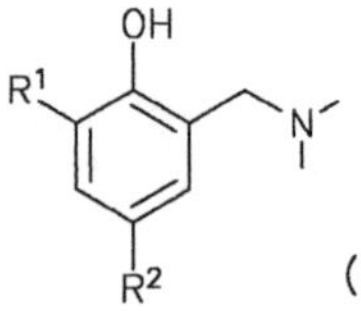

545

(a,d)

(R^1, R^2 = Alkyl, $CH_2N\langle$)

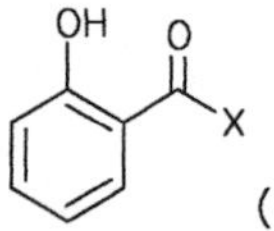

(c)

546

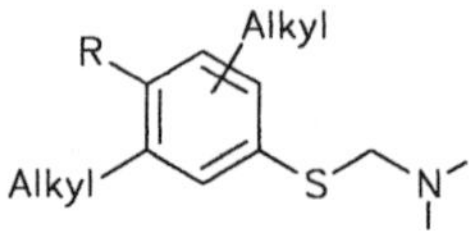

547

(b)

(R = H, OH)

Para-substituted phenols with strongly hindering groups in the ortho positions (**544**; X, see Fig. 184, below) are used as effective inhibitors of radical chain reactions, according to requirement (a), above. They can be prepared by Mannich synthesis or by reaction of aminomethyl derivatives.[125–129] The relationships between structure and antioxidant properties of this class of compounds have been thoroughly investigated.[125,130–132]

Compounds **545** have a similar structure;[132–134] however, the presence of the aminomethyl group in the ortho position should presumably give also the possibility of protection, by chelation, against the action of metal ions[135,136] usually present in the material as catalyst residues, contaminants deriving from corrosion of metal containers, etc. (d). This is indeed confirmed by comparison with analogous compounds having weak or no chelating power.[135,137] Thioethers **547**, obtained by aminomethylation of thiols[138,139] or aminomethylated phosphines,[140] behave as hydroperoxide reducing agents, thus corresponding to requirement (b). Finally, the action of light on hydroperoxide species (c) can be prevented by the addition of compounds having strong ultraviolet (UV) absorption in the relevant spectral regions, such as derivatives of type **546**, which can be appropriately functionalized by Mannich reaction on the ortho position of the phenolic hydroxy group[141] or on the α-position to the carbonyl group (**546**, X = $CH_2CH_2NR_2$).[142]

The synthesis of phenolic antioxidants of type **544**, starting from 2,6-diterbutyl phenol, involves formation of the corresponding para-aminomethyl derivative **548** (Fig. 184), a very good radical inhibitor, which may be used as a precursor of other compounds having similar properties. In addition to the dimethylamino derivative, Mannich synthesis also permits the production of particular amines, such as **551**, which can be grafted onto copolymers containing the glycidyl ether group,[143] thus affording polymeric antioxidants unextractable from the material to be protected, in a manner similar to example **501** of Sec. A.1.

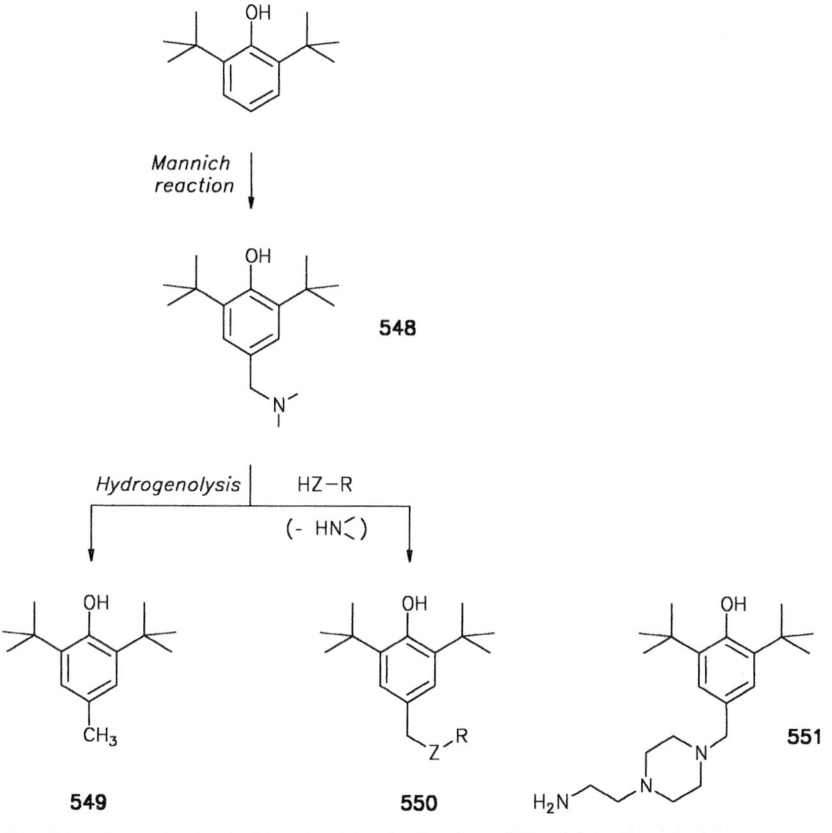

Fig. 184.   Synthesis of antioxidants involving the chemistry of Mannich bases.

The hydrogenolytic reaction leading to ionol **549** has been accurately studied[144,145] (see also Chap. II, C.1), as has the synthesis of compounds **550**. Indeed, the amino group replacement makes it possible to prepare a very large series of derivatives, significant examples of which are represented by **552–555** (i.e., Z in **550** = C, N, P, S).

The presence of large alkyl groups in **552** and **554** is aimed at improving the affinity toward materials or fluids having a hydrocarbon nature[146–148] as well as acting as corrosion inhibitors. Different antioxidant moieties, constituted by hindered phenolic and amino groups, may be connected in the same molecule,[149] as in **553**. Antioxidant groups working at different steps of the radical degradation process may be jointed together,[150] as exhibited by **555**, having in addition a sulfur atom with reductive power toward hydroperoxides. Dithiocarbamates obtainable by insertion of carbon disulfide into Mannich bases (Fig. 106, Chap. II) and other thioethers[151] are also largely employed as antioxidants.

In addition to plastics, elastomers, lubricants, and fuels, the above mentioned inhibitors are applied as protective agents in halogenation[152] and reforming processes[153] as well as in prevention of the polymerization of monomers and unsaturated compounds during storage.[154,155]

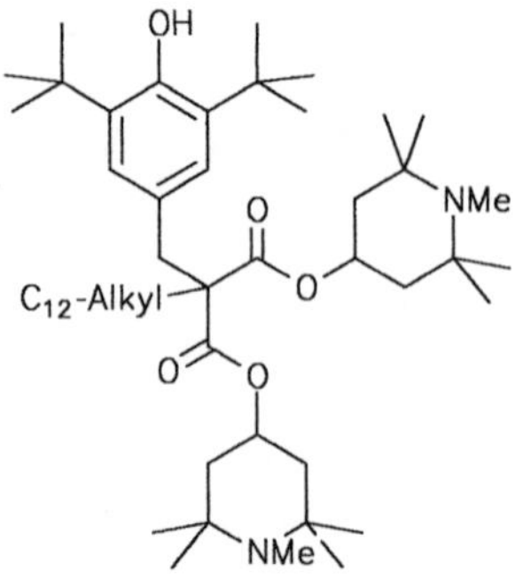

552

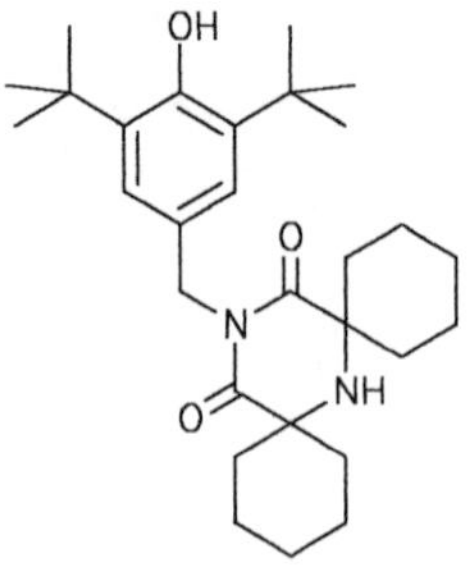

553

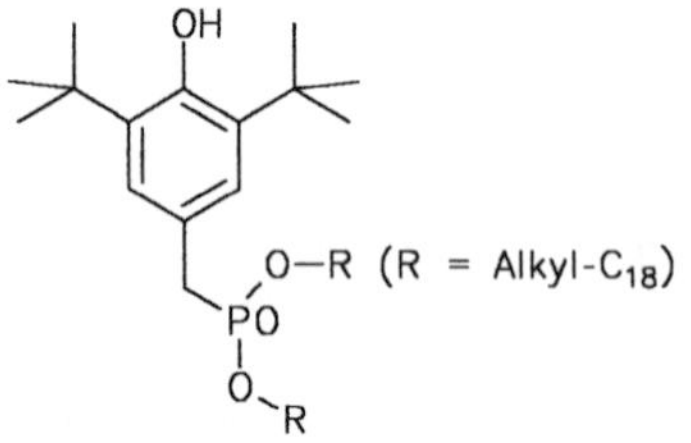

554

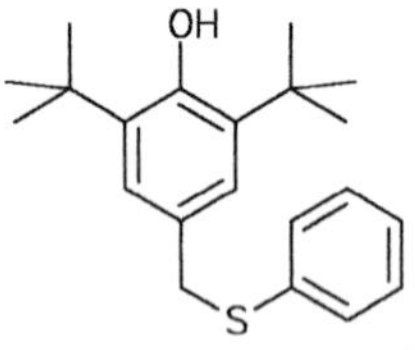

555

Surface-Active Agents

Compounds interacting with the chemical moieties located at the surfaces of different physical phases are employed in the most diverse applications, as they may behave as dispersants in liquid/solid or liquid/liquid systems, antistatics at the solid/air interface, foaming agents (or antifoams) in liquid/air systems, and so on. Their action is essentially based on the presence of groups capable of adequate interaction with two phases of different chemical composition as well as on properly designed molecular shapes and sizes. Thus, the facility of Mannich base chemistry to connect quite different moieties (Sec. A.1) may be helpful in many cases. We can consider the types of compound suitable for this type of application to be divided roughly into two main groups: medium- and small-sized molecules having apolar/hydrophobic moieties combined with polar moieties, and water-soluble ionic macromolecules having the polar groups uniformly ordered along the polymeric chain. The polar moiety of the above compounds is always located in proximity to the methylene group of the Mannich base, as both amine and substrate reagents of Mannich reaction exhibit polar features. The apolar moiety may be present on the amine or the substrate side (Fig. 185).

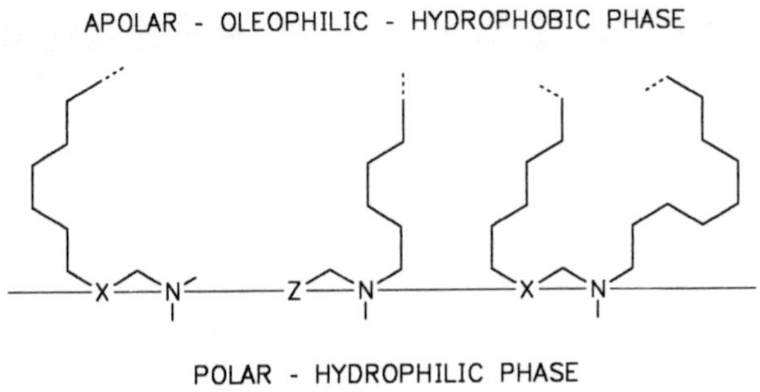

APOLAR - OLEOPHILIC - HYDROPHOBIC PHASE

POLAR - HYDROPHILIC PHASE

Fig. 185. Mannich bases as surface active agents (X, Z = substrate moiety).

Table 40 lists the most frequently reported chemical features of Mannich bases employed as surfactants, along with brief indications of their individual function, although this aspect is hardly specifiable, as several different practical definitions can be given for the same molecule. For example, an antioxidant for lubricating oil may be also named an antisludge compound, due to its capability to reduce the formation of solid precipitate caused by oxidation processes.

Phenolic compounds, particularly para-substituted alkyl phenols, are subjected mainly to aminomethylation to give the ortho-aminomethyl derivatives (**556, 560, 561** in Table 40). As a consequence, the molecule may also acquire chelating properties, as mentioned before, at the same time a polar moiety is introduced. The alkyl moiety may vary appreciably in size and shape, depending on the lipophilicity requirements, ranging from short, linear hydrocarbon chains to branched polymeric structures of considerable size (**556, 560**). From this point of view, the boundary between polymeric and low-molecular-weight compounds appears rather vague; the only difference characterizing the above-defined groups of surface-active agents resides in the presence or absence of ordered distribution of polar groups in the macromolecule. Thus, compounds **561** belong to the second group, despite their not particularly high molecular weight, due to the regular sequences of phenolic groups along the polymeric chain providing enhanced antistripping properties to asphalts in contact with polar surfaces of inorganic materials.

Polyacrylamides **562** are second in importance only to alkyl phenols.[4] Their water solubility as well as ionogenic power are increased by the introduction, through Mannich reaction, of the dimethylamino group, which allows better adhesion to the surface of suspended micelles and hence an improvement of the precipitating action.

Further examples reported in Table 40 are given by the alkyl ketone moieties deriving from oxidation of polyenes, which are subjected to Mannich aminomethylation in order to produce compounds **559** having dispersant properties for lubricating oils, by benzotriazoles **557**, capable of forming, due to physical adsorption, thin layers over the surfaces subjected to friction, and by S-Mannich bases **558**, combining antifriction and antioxidant properties.

Although much less frequently, the amines employed in the production of surface-active Mannich bases may bear lipophilic alkyl groups (**563**). In other cases, they contain groups that may improve the complexant power and/or hydrophilicity, the highest activity being required by antistatic agents **567**, which must be able to retain a thin film of humidity over the surface of the treated material in order to avoid the accumulation of electric charges in limited surface regions.

Table 40

Substrate and Amine Moieties in Surface-Active Mannich Bases

Substrate moiety	Function performed	Reference
Nonpolymeric		
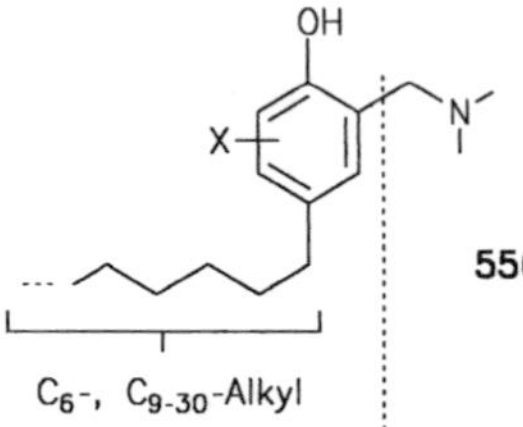 **556** C_6-, C_{9-30}-Alkyl	Oleophilization of minerals	156
	Deposit inhibitor (with **564**)	157
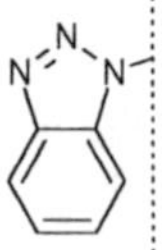 **557**	Antifriction (with **563**)	26
	Corrosion inhibitor	158
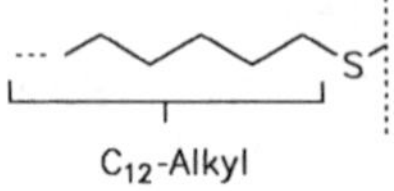 **558** C_{12}-Alkyl	Antifriction, Anticorrosion and Antioxidant (with **563**)	159-161
Polymeric		
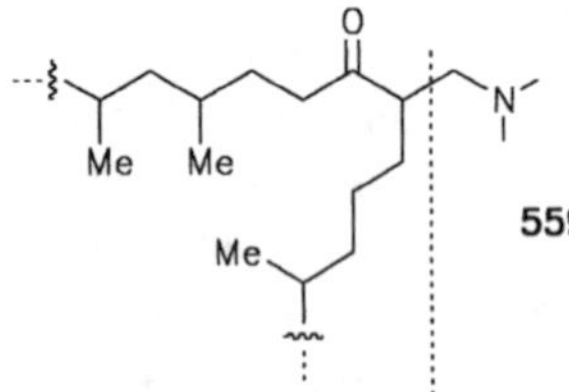 **559**	Dispersant (with **569**)	162,163
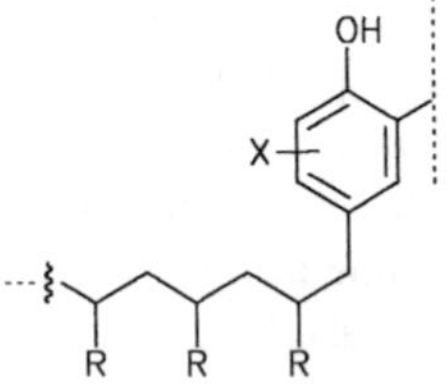 **560** (R = Me, Et, Octyl) $\overline{X}_n$ = 10-150	Detergent, Dispersant (with **565**, **566**)	164-166
	Detergent (with dimethylamine)	167
	Antifriction (with **563**)	168,169

In general, amines of type **565** are more frequently preferred as they expose several sites to Mannich reaction with, for example, phenols. The molar ratio of phenol to amine adopted for this type of synthesis is in fact often less than 2:1. Studies are being conducted into the influence of this molar ratio on the dispersing, antifoulant, etc., properties of the products obtained.[4,6,177]

Finally, aminoalkylsilanes **568**, which generate groups with high affinity to glass surfaces, make it possible to anchor macromolecular materials to fillers made of fiberglass or similar materials.

Table 40

(continued)

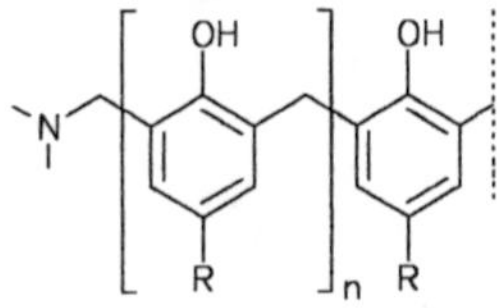

561   Asphalt antistripping   170

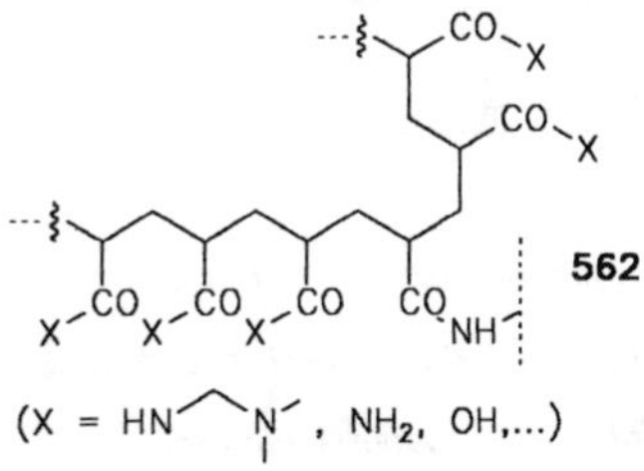

(R = C$_{9-15}$-Alkyl)
$\overline{X}_n$ = 0-30

562   Flocculant (with, mostly, dimethylamine)   4,171,172

(X = HN$\diagdown$N$\diagup$, NH$_2$, OH,...)

Amine moiety

---

563   Antifriction   169

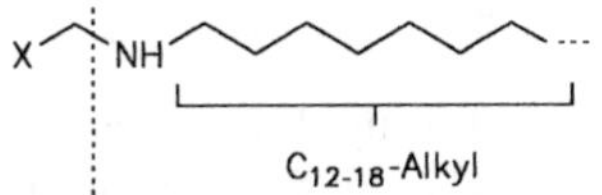

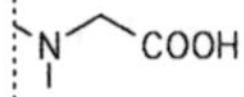

C$_{12-18}$-Alkyl

564   Deposit inhibitor   157

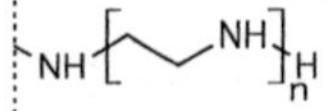

565   4

(n = 0-5)

R (R = Alkyl, Ar)

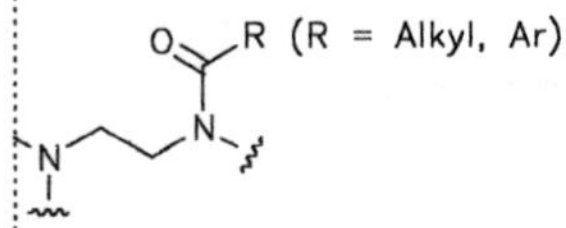

566   Corrosion inhibitor (with **556, 560**)   166,173,174

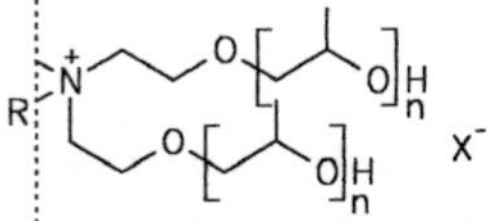

567   Antistatic (with **556**)   27

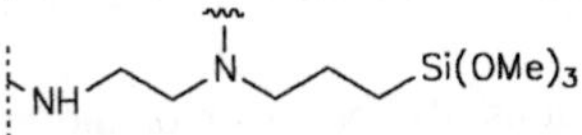

Si(OMe)$_3$ **568**   Adhesion promoter for glass fibers   175,176

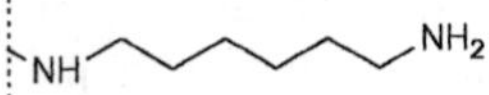

NH$_2$ **569**   162,163

Table 41
Specific Functions Performed by Mannich Bases and Derivatives

• Compounds with Curative (a) or Biocide (b) Activity

  (a): *Pharmaceuticals*

  (b): *Pesticides*

• Constitutive Components (a) or Auxiliary Components (b) of Materials

  (a): *Monomers, Pre-polymers, Polymers, Macromolecules*

  (b): *Catalysts, Crosslinking agents, Hardeners, Functionalizing agents*

• Additives Providing Improved Properties (a) or Reducing Detrimental Properties (b)

  of Materials

  (a): *Adhesion promoters, Detergents, Dyes, Flocculants, Surfactants*

  (b): *Antioxidants, Antistatics, Corrosion inhibitors, Detergents/Dispersants,*

    Flame retardants, Homolytic reaction inhibitors, Scale inhibitors,

    Sequestering agents

Specific Functions

The abovementioned primary features of Mannich bases find many different applications, aimed at satisfying specific needs, and these are grouped according to their final use in Table 41. Further examples are mentioned in Sec. B, which deals with individual branches of industry that are involved to a varying extent with the chemistry of Mannich bases.

Three main classes of compounds are recognized and accordingly listed in Table 41:

- Molecules having biological activity
- Constituents and auxiliaries in the production of synthetic materials
- Additives

Compounds Having Biological Activity

This group includes pharmaceuticals and pesticides, which have therapeutical and biocide action, respectively, on living organisms, along with a few compounds that improve plant germination.[178]

Apart from any medicinal chemistry considerations, the behavior of the aminomethyl group of the Mannich base in connection with the function here described deserves a short comment. In particular, the hydrophilic and/or lipophilic properties of well-known pharmacologically active substances may be conveniently modified in order to improve their diffusion properties, and hence their performance, in the microenvironment in which the molecules act.[2,179,180] Thus, aminomethylation of a drug may lead to a product (**570**, Fig. 186) having better hydrophilic properties due to both the enhancement of polar groups and the possibility of ammonium salt formation. Similarly, formaldehyde/bisulfite condensation on the amino group of a drug[25] (**571**, see also **507**, Sec. A.1) may give rise to water solubility of the product. Alternatively, the lipophilic properties may be improved by aminomethylation with suitably alkyl-substituted amine.[179]

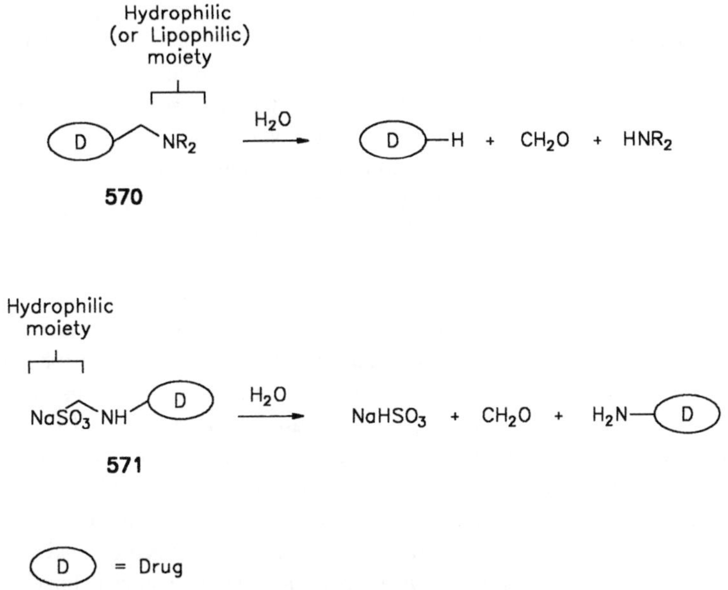

Fig. 186. Physicochemical properties and the deaminomethylation reaction of pharmacologically active Mannich bases.

Reverse Mannich reaction also plays an important role with biologically active molecules, in particular with prodrugs, capable of releasing the pharmacologically active constituent under controlled hydrolytic conditions (Fig. 186), thus improving the overall performance. Several types of Mannich bases are employed, including the polymeric ones,[181] all characterized by the presence of X-aminomethyl groups (X = amide-N, heterocycle-N, sulfonic-S) readily undergoing deaminomethylation (Chap. II, A.1).

In this context, amidic[39,40,182–184] as well as hydroxamic acid[185] Mannich bases of well-known drugs, such as tetracyclines, indometacine, etc., have been investigated along with model compounds. The N-heterocyclic derivatives of theophylline, mercaptopurine, etc.,[41,179] and the bisulfite Mannich bases, particularly interesting as diuretics (see **520**, Table 38),[186,187] have been similarly studied.

As far as the biocide properties are concerned, O-Mannich bases containing extended hydrocarbon chains and used for their combined detergent/germicide action[188,189] are worth mentioning. A good preserving function is also performed by plywood adhesives, based on urea, melamine, and formaldehyde, which are fixed by suitable iodo-derivatives presumably by amino- and amidomethylation reactions.[190,191]

Constitutive Components and Auxiliaries in the Production of Synthetic Macromolecular Materials

Mannich bases are involved in the synthesis of macromolecular derivatives either by directly participating in the chemical structure of the final product as monomers, oligomers, etc., or by contributing as essential auxiliaries, such as crosslinking agents, modifiers, etc. (see Chap. III).

The macromolecular materials having Mannich bases, or their derivatives, as structural components are usually formaldehyde resins, epoxy resins, and polyurethanes.

Indeed, Mannich aminomethylation represents the repetitive-cumulative reaction giving rise to several types of formaldehyde resins; the most relevant of these are melamine resins, which have been abundantly investigated and reported in a quite large series of patents.[192–195] Melamines, as well as guanamines, have also been tested in combination with phenols or urea derivatives, behaving as Mannich substrates in this type of reaction. Further examples of Mannich polymerization involving other amines and acetonic, phenolic, etc., substrates[196–200] are found in Chap. III, A (see also **425**, Chap. III, C.1).

Epoxy resins, mostly employed in the coating industry, are characterized by the presence of a phenolic Mannich base, usually deriving from bisphenol A or alkyl phenols,[4,201–206] linked to an epoxy oligomer, mixed with various crosslinking agents,[201–203,207] and dispersed into aqueous solution in the presence of organic acid. This formulation is laid down by electrophoresis over the material to be coated, acting as the cathode, and finally heat cured.[4] The process requires molecules with chemical structure suitably designed for different applications, and the chemistry of Mannich bases may represent a convenient tool for preparing variously substituted derivatives. In fact, as depicted in Fig. 187, the reactive functions leading to the final crosslinked material may be bound to several different positions of the oligomeric epoxy resin. The Mannich derivatives of bisphenol A (**572**)[201,203–206] and nonylphenol (**573**)[202,205,206] can be considered as the most frequently employed.

Addition of oligomeric epoxide to the Mannich base involves both the phenolic hydroxy group and the alcohol moieties present in the amine residue. On the other hand, the amino group of the base, in conjunction with other amine functions, as in **573**, acts as a catalyst for the reaction and provides, in the presence of acid, the cationic center enhancing the hydrophilic power of the binder and allowing the electrophoretic transfer of the resin. In many cases, semiblocked diisocyanate is added to secondary amino groups for better crosslinking.[206] In order to improve the hydrophilicity of centrifugally coated resins, a sulfonic group may be inserted into the amine residue.[204] The presence of long chained alkyl groups linked to the amine may endanger the possibility of dispersion of the binder into the aqueous medium; however, they act as "internal plasticizers" as they reduce the cohesive interactions, thus allowing an easier mutual sliding of chain sections of the macromolecules, with an improvement in mechanical properties such as elasticity and impact strength in the final material.[42] Finally, the hydroxy groups belonging to the oligomeric epoxy chains can be exploited for crosslinking by reaction with diisocyanates or alkoxymethyl melamines.

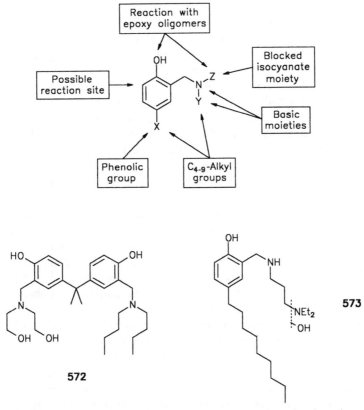

Fig. 187. Features of phenolic Mannich bases used in epoxy electrophoretic coatings.

The Mannich reaction also makes an important contribution to the synthesis of polyurethanes, in particular to the preparation of polyols, which represent the pre-polymerization step of the process. These compounds are predominantly constituted by phenolic Mannich bases; however, derivatives of phosphorous acid and guanamines are also used (**533** and Ref. 15; see also **399**, Fig. 160, and **419** in Chap. III). Usually, the aminomethylation is carried out with diethanolamine on hydroxylated substrates in order to produce molecules having a sufficient number of hydroxy groups available for the successive steps. The resulting Mannich base is then treated with ethylene- and/or propylene-oxide to give pre-polymers which, as such[208–210] or mixed with other polyols,[211] are subjected to reaction with diisocyanate. As the abovementioned polyols obtained by Mannich synthesis are already provided with the basic moieties performing the catalytic function required for polymerization,[208,209,212] the need to add low-molecular-weight catalysts undergoing migration toward the surface of the final material, with production of undesired emissions, is avoided. In addition, it is possible to obtain "internally plasticized" materials (see above) or fire resistance of the product by selecting suitable substrates, for example, ring-alkylated[208,209] or halogenated phenols,[212] respectively.

In contrast to the compounds described above, Mannich derivatives employed as auxiliaries are used in relatively low amounts as curing agents in the crosslinking of polymeric products or as modifiers to improve the final performances of macromolecular materials. In particular, they may work in crosslinking as accelerating or reactive agents, with both functions occurring simultaneously in some cases.

As far as the accelerating behavior is concerned, the main contribution to this catalytic activity is obviously given by the presence of basic amino groups; however, the mobility of the molecules in the reaction mixture may be critical, particularly in a highly viscous medium. Thus, compatibility toward the interacting material,[213] complexant power, and capability to evolve free amine acting as the real catalyst[37,38,214] may be the determining features of such a chemical auxiliary. The most commonly used catalytic agents (see also **426, 427** in Chap. III and Ref. 4) are undoubtedly mono- and polyfunctional phenolic Mannich bases bearing the aminomethyl moieties in ring positions 2, 4, and 6 (**574**).[19,215] The R group may have various chemical structures, including the possibility of covalently bonding the material to be polymerized[216,217] in order to reduce catalyst migration.

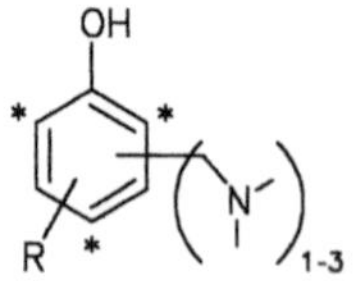

574

Polymerization of epoxy monomers, requiring probably the simultaneous participation of the phenolic hydroxy group, and the production of polyurethanes are the main applications of the abovementioned catalysts. In particular, their role in the trimerization of the isocyanate group[218,219] appears quite relevant and several investigations of the synthesis of epoxy resins[220–222] and polyurethanes[223] are reported. A different type of catalyst is constituted by acetophenone-derived β-aminoketones employed with photocurable resins.[224,225]

The crosslinking action performed by Mannich bases or Mannich reagents relates to three main types of reactions of notable practical interest.

The first type concerns aminomethylation giving rise to the macromolecular network starting from polyfunctional polymeric substrates and/or amines. Hardening of phenolic novolacs[226] and other resins with urotropine belongs to this type of reaction. The process may be carried out in the presence of proteic molecules, which are thus inserted into the cured material.[227,228] Similarly, the crosslinking of collagen or other proteic macromolecules with formaldehyde, or with mixtures of aldehydes, is frequently encountered (see also Chap. III, C.2).[229,230]

The second type of reaction concerns the use of hydroxymethyl or, more frequently, alkoxymethyl derivatives of melamine, that is, a multifunctional amine, in aminomethylation (or transaminomethylation) reactions. Crosslinked epoxy resins are also obtained by reaction with epoxides (Fig. 188).[231,232]

Finally, the Mannich reaction may be applied to the synthesis of reactive amines used as crosslinking agents of epoxy resins. The process requires the availability of molecules possessing more than two NH groups, which are obtained by reaction of polyfunctional substrates with polyfunctional primary amines (oligomeric polyalkyleneamines, diamino cyclohexane, etc.)[233–237] (see also **422**, Chap. III, C).

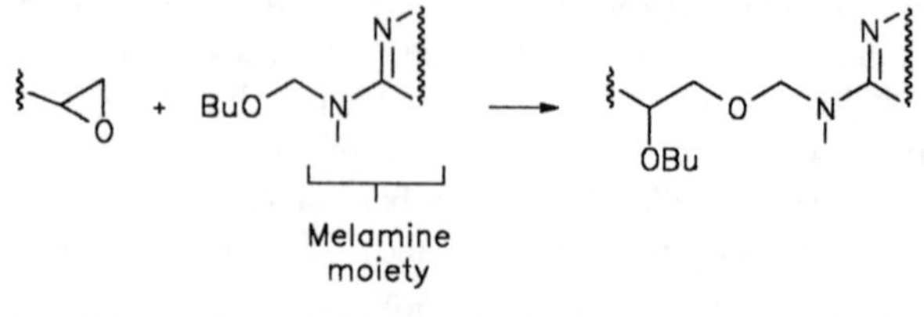

Fig. 188. Aminomethylation reaction involving epoxy groups in the synthesis of crosslinked resins.

As mentioned above, the use of low-molecular-weight additives as modifiers of polymeric materials may release colored, toxic, and/or unpleasant smelling substances from the material surface. As a consequence, it is convenient for the required moieties to be linked to the polymeric backbone, as in the case, for instance, of the internal plasticizers seen previously in the case of epoxy resins (Fig. 187). A very similar behavior is exhibited by polyamine-modified urea-formaldehyde resins,[238] having improved resistance to cyclic stress due to the incorporation of flexible alkylamine moieties. Suitably modified polyols for polyurethanes having flame-retardant properties have been largely investigated, in particular, for polyurethane foams, which offer a quite extended surface to burning. Thus, halogens and phosphorus derivatives are inserted into polyols[239,240] starting from Mannich bases of types **575** or **576**. Many other applications of this type are reported,[212,241,242] including polymeric materials other than polyurethanes.[197]

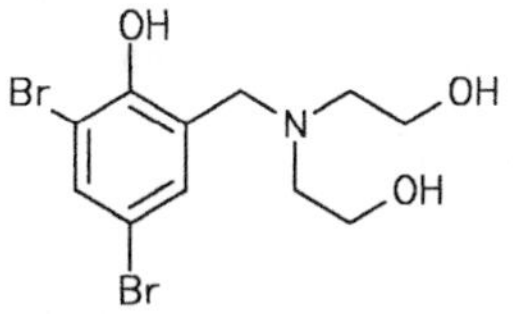

575 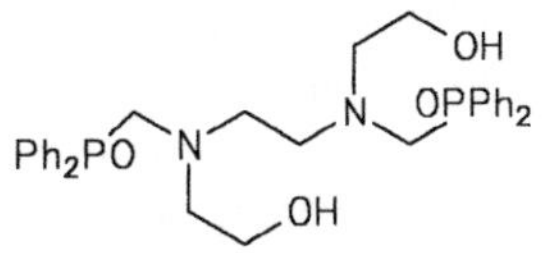   **576**

Further examples of Mannich reactions aimed at the modification of materials range from the use of suitable aldehyde mixtures in the crosslinking of proteins for contact lenses[229] to the modification of resols with arylamines[243,244] for obtaining rapidly curing resins employed as wood adhesives. Some further examples of the functionalization of macromolecular derivatives[190,191] for modifying the final properties of the material are found in Chap. III, D.

Additives

The addition of compounds in relatively small amounts to materials or to widely employed fluids such as fuels, lubricant oils, and even water is fundamentally intended

- to improve the existing positive properties and/or to provide new features
- to reduce the consequences of defects in the material or impurities in fluids and to protect against wear and tear caused by the environment and/or the working conditions

Additives Providing Improved Positive Properties

Adhesion promoters—Their scope is to favor the best adhesion between a macromolecular material and the filler (powder, fiber, etc.) employed in order to provide the material with required mechanical properties. As this function is performed at the contact surface between chemically different solid phases, the adhesion promoters must possess affinity toward both the components, one of them (the filler) having usually high polarity, and capability of giving hydrogen bonding, due to its nature as inorganic salt, fiberglass, cellulosic derivative, etc. Thus, phenolic Mannich bases containing mainly alkanolamino groups (see also Table 40) are well suited to this purpose.[200,245,246] In some circumstances, the amine moiety is supplied by aminosilyl ethers[247] undergoing hydrolysis in the course of manufacturing, thus providing very good adhesion to glass materials. In addition to adhesive properties, these additives must have strong internal cohesion in order to satisfy the requirements. To this end, arylamine/phenol/formaldehyde resins[198,200,243] and polyacrylamides[248] (see section ''Paper,'' below) are preferred.

Detergents—The contribution of Mannich bases to this class of products, generally used as cleansing agents for hand or machine washing, is quite negligible (see section ''Surfactants,'' below). However, the term *detergent/dispersant* also includes compounds employed as additives to organic fluids, mainly hydrocarbons (fuels, lubricating oils), having the function of preventing the deposition of suspended solid particles that lead to undesired scale formation over the metal surfaces of machinery. As the function of these additives is essentially protective, it is described below in the appropriate section.

Dyes—In addition to chromophoric moiety, dyes must possess a suitable chemical structure that favors the dyeing process and avoids, to the greatest possible extent, running of the color from dyed materials. In this context, the chemistry of Mannich bases is variously involved in several aspects, ranging from the synthesis of important indigo dyes[249] to the introduction into the dye of chemical moieties capable of promoting dyeing as well as the provision of auxiliaries and post-treatments[250–252] required for stabilizing the color over the fiber.

The most interesting features that concern the structure/property relationships of Mannich bases are, however, connected with their contribution to the dyeing process (Fig. 189). In particular, the dye may be subjected to aminomethylation (Fig. 189a), thus introducing a cationic moiety (e.g., **577**) that is exploited for providing both solubility into the dyeing bath and affinity toward negatively charged fibers.[253–258]

Aminomethylation of sodium bisulfite (Fig. 189b), by means of primary or secondary amino groups belonging to the dye molecule, is also carried out to obtain fine dispersion of the coloring agent into the bath, which is readily accomplished by deaminomethylation on heating (Fig. 90, Chap. II, and **507**).

Amino group replacement by nucleophilic moieties (OH, NH) present in the fiber molecules (Fig 189c) is also applied to reactive dyeing when Mannich ketobases (e.g., **578**) are employed.[256,259,260]

Finally, the introduction into the dye of the aminomethyl group CH_2NH_2 may be conveniently performed by amidomethylation followed by hydrolysis[257,261] (Fig. 15, Chap. I). The above group allows the subsequent linkage of the well-known reactive moieties halodiazines or triazines to the dye molecule.[261]

Flocculants—The need to separate undesired solid particles finely suspended in aqueous effluents, which is particularly relevant in wastewater treatment, the paper industry, etc.,[4,262–264] requires the use of water-soluble macromolecules, which are mostly ionic in character and capable of interacting with the suspended material. With high-molecular-weight derivatives, the presence of a suitable content of ionogenic moieties as well as their uniform distribution along the polymeric chains are important in order to obtain the best performance of the flocculant, which should not completely cover the surface of the dispersed particles.

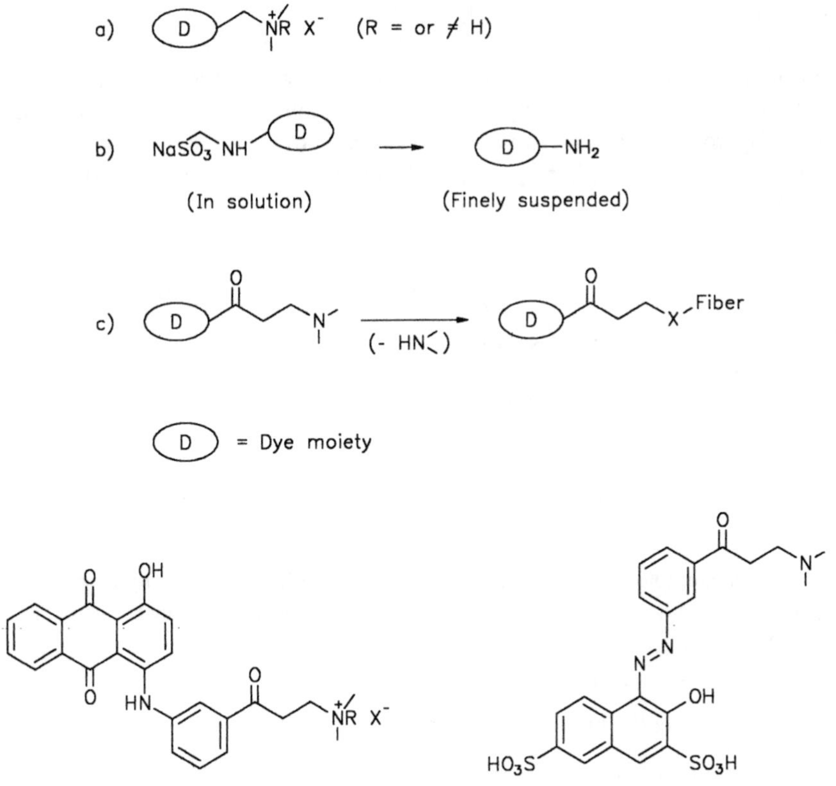

Fig. 189. Mannich bases used as (a) cationic, (b) disperse, and (c) reactive dyes.

Different types of flocculants can be prepared by aminomethylation; the amino-methyl derivatives of polyacrylamide represent the most widely adopted class of compounds, which includes high-molecular-weight flocculants (**502, 517** in Table 38, **562** in Table 40), and has been widely patented. The large amount of research on these products is concerned, in particular, with isoelectric points, the content of functionalized co-units, the extent of amido groups hydrolysis, and the use of copolymers.[4,262–267] Applications of acrylamide copolymers grafted onto starch or other polysaccharides are also considered.[4,268]

Flocculants other than polyacrylamides are oligomeric phenolic Mannich bases or their derivatives, with the only exception being oligomeric urea resins,[269] working with a charge-neutralization mechanism. They include lignin Mannich bases,[270] quaternary ammonium salts (**398**, Chap. III, B), and nonionic products having the phenolic hydroxy group etherified by epoxide.[271,272]

Surfactants—This group includes various bases used for different applications, such as protein dispersants,[273] stabilizers of aqueous pigments,[274] blood preservatives,[275] flotation collectors,[276] etc. Apart from rare cases,[277] they are phenolic derivatives[278,279] (see also Table 40 and **440**, Chap. III, D) having lipophilic power supplied by the presence of variously extended alkyl chains[274,280] and hydrophilicity provided by carboxy groups located in the substrate or amine moiety (Mannich bases of salicylic acid[275] or of amino acids,[273] respectively), or by sulfonic groups, as in the lignin derivative **579**.[281] Hydrophilic properties can also be obtained by attaching polyethyleneglycol chains to the phenolic hydroxy group,[274,280] with formation of nonionic surfactants of type **580**.

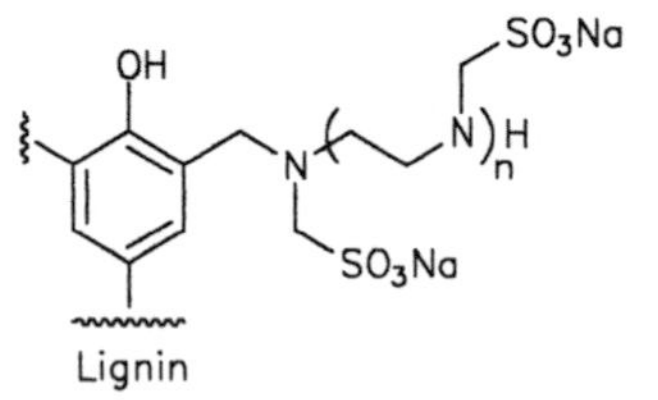

579 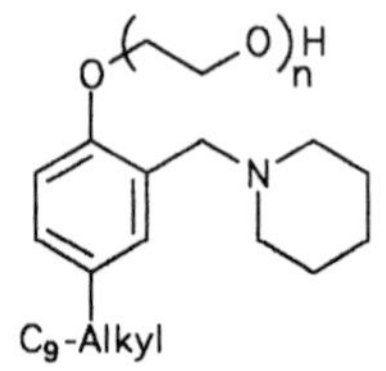   **580**

Further examples of additives providing specific performances (wetting agents, etc.) are mentioned in the next section, which is devoted to individual branches of industrial application.

Additives Reducing the Effects of Unfavorable Properties

Additives performing a protective or corrective action against aging and/or deterioration of performance (see Table 41, above) are grouped as follows:

Antioxidants—The tendency of hydrocarbon chains to undergo oxidative processes through radical mechanism (see section "Radical Inhibitors" above) leads to undesired crosslinking and degradation with loss of important mechanical properties, such as flexibility, elasticity, etc., in solid materials or of relevant features connected with viscosity in fluids. Antioxidants are also employed as heat stabilizers, in particular,[130,146,282–285] in order to inhibit oxidation in the manufacture of plastics, which is usually carried out under strong heating. Some Mannich bases used as antioxidants for rubber exhibit also an antiozone action,[126,286] probably due to the presence of amino groups.

The hydrocarbon chain in fuels, oils, and lubricants is considerably shorter than in polymeric materials and produces, on oxidation, insoluble tars requiring the presence of antioxidants performing antisludge action.[287–289] It is also worth noting that Mannich bases used as antioxidants and detergents are ashless,[290] a property that is of great importance in additives used in this type of application.

Antistatics—In addition to preventing the accumulation of electrical charges over the surface of solid materials, particularly fibers, these auxiliaries are also used as fuel additives.[291] They are actually surface-active agents capable of anchoring their apolar moieties to the material so as to allow the polar groups to retain the humidity required for static-charge dispersion (Fig. 190).

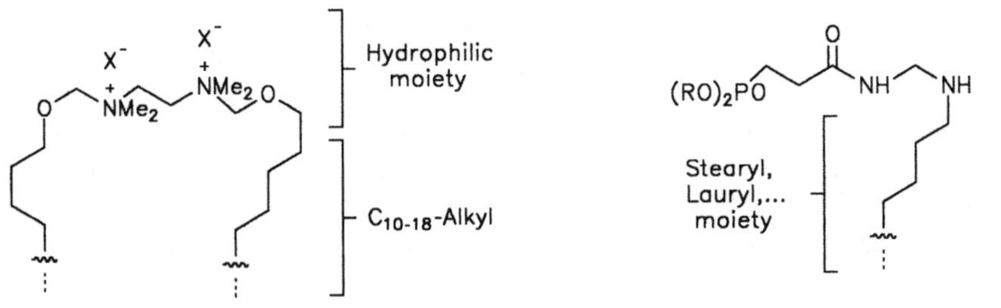

Fig. 190. Mannich bases used as antistatic agents.

To this end, highly hygroscopic quaternary ammonium salts of type **581** are employed.[292,293] Epoxide etherification of hydroxy groups (**567**, Table 40) helps in improving hydrophilicity. Moieties such as EDTA[294] and others,[28,295] having a strong tendency to give hydrogen bonding with water, are also used. Among these last compounds, **582**, in particular, combines the antistatic action with the flame-retardant properties[296] exhibited by phosphorus derivatives. A combined antistatic and softening action is provided by melamine derivatives covering the fiber surface with molecules that have a high content of nitrogen atoms.[295]

Corrosion inhibitors—Their action is based on the ability to cover metal surfaces with a monomolecular layer strongly adhering to the metal so as to form a film impermeable to water and oxygen. These molecules are therefore made up of polar moieties and apolar hydrocarbon groups of variable size according to the type of application. Alkyl phenols[4,297,298] are mostly used, but the Mannich bases of benzotriazoles,[299] thiols[300] (see also **556–558** and **566** in Table 40), and other substrates (alkynes, thiourea, etc.)[301–303] are also employed. Fatty acids, precursors of the corresponding acid salts or of amides by reaction with the amine moiety of Mannich bases (**566**, Table 40),[298,304] also provide hydrophobic groups.

Larger alkyl groups are required when the Mannich base is used as an additive to fuels, lubricants, etc. In some cases the anticorrosion power is afforded by exhaustive aminomethylation of polyfunctional substrates with aliphatic amines, thus obtaining branched basic derivatives (see, e.g., **499**, Sec. A.1) that are able to protect the material against acid corrosion.[137,305,306] Detergent[307] (see also **513**, Table 38) and antioxidant[138,308] actions can be combined with the anticorrosion ability of certain additives to oils and fuels, although the undesired formation of emulsion or foam may be favored.

The protection of metal surfaces by anticorrosion agents[300,301,303] is also applied to heat exchangers,[298] in the hydrogen fluoride cleaning of surfaces,[302] with printing inks,[299] with primers for coatings,[309] etc. A typical derivative having corrosion inhibitor and surfactant properties is represented by **583**, employed on steel surfaces.[310]

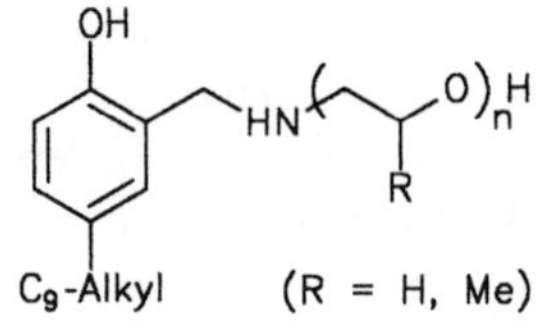

Detergents/dispersants—Scale prevention or removal related to organic fluids (fuels, lubricating oils, etc.) is usually performed by molecules possessing lipophilic as well as polar and amine moieties that have been progressively preferred, as ashless products, to analogous derivatives containing inorganic cations. Two types of Mannich bases deriving from high-molecular-weight alkyl-substituted phenols[311–314] or from partially oxidized polyenes[315,316] are well suited to the purpose (see **556, 559, 560** in Table 40). Cyclic Mannich bases possessing the tetrahydrobenzoxazine ring are also used.[9] Moreover, Mannich bases obtained from aminophenols, subsequently acylated with cyclic anhydrides containing extended hydrocarbon chains, are reported.[317] Antioxidant properties, in addition to detergent ability, are exhibited by acrylic copolymers bearing Mannich bases of type **551** (Fig. 184) grafted onto the polymeric chain.[318]

Flame retardants—Whereas halogen-containing Mannich bases are mainly used as modifiers of macromolecular materials, as mentioned before, phosphorus derivatives are employed as additives. These products are obtained from phosphorous acid or phosphites **584**[319–321] and from other analogous phosphorus-containing compounds.[296] N-Heterocycles, after reaction with melamine and formaldehyde (**585**), are also used as flame retardants.[322]

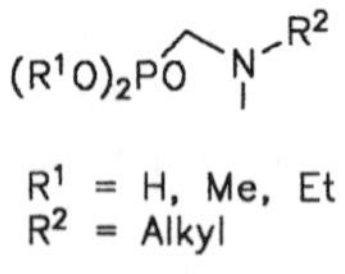

$(R^1O)_2\overset{\text{O}}{\overset{\|}{P}}\frown N\overset{R^2}{\underset{|}{}}$

R^1 = H, Me, Et
R^2 = Alkyl

584

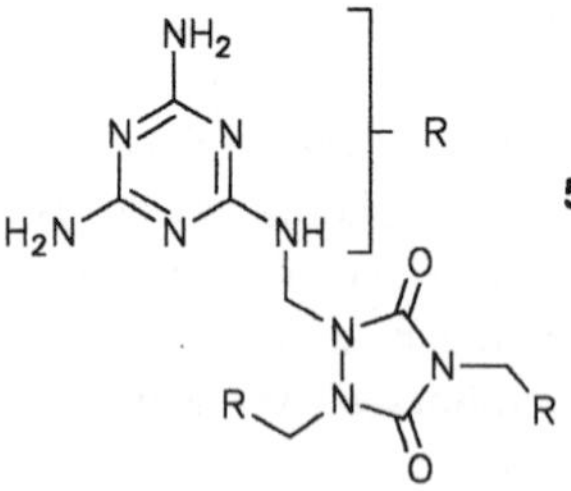

585

Homolytic reaction inhibitors—In addition to degradative oxidation (see section on "Radical Inhibitors", homolytic reactions may take place in industrial processes carried out at temperatures favoring pyrolysis,[153,323] or in the presence of halogens,[152] and during the storage of unsaturated products.[154,324,325]

Usually, para-substituted phenols bearing bulky groups at both ortho positions are employed as additives, except in a few cases,[154] the most frequently used being the Mannich base of 2,6-di-*tert*-butyl phenol **548** and its derivatives,[326] including the materials recovered from its hydrogenolysis[153] to give ionol (Fig. 184).

Sequestering agents/scale inhibitors—Chelating ability toward metal ions is the main feature of these products, which must possess at least two groups located at a relative distance suitable for successfully coordinating ionic species (see **527, 528**). Moreover, they must have good solubility in the reaction medium, both in the free and in the coordinate form, when they are employed as soluble additives. Derivatives of α-amino carboxyacids[327] and aminoalkyl phosphonates[328–331] (**516, 521** in Table 38 and **527, 528**) are mainly employed. Significant examples are given by **586**[327] and by **587** and **588**,[67,331] respectively.

Low-molecular-weight derivatives, as well as oligomeric[90,332] and insoluble cross-linked resins (ion-exchange resins **537–540**), are also reported as sequestering agents in particular applications.[333]

The required water solubility is usually provided by the presence of a high amount of complexant groups having hydrophilic properties. The sulfonic group (**587**) or polyethyleneglycol moieties[330] may be introduced in order to enhance solubility. A combined deflocculating action, particularly useful in water clarification, can be performed by some of these additives.[68,69]

Table 42
Industrial Branches of Application of Mannich Bases and Derivatives

1 - Pharmaceutical Industry and Pesticides Production

2 - Production of Natural (a) and Synthetic (b) Macromolecular Materials

     (a): *Leather, Paper, Textiles*

     (b): *Adhesives/Sealants, Coatings, Elastomers, Foams, Plastics/Resins*

3 - Petroleum Products

     Fuels, Lubricants,...

4 - Miscellaneous Productions

     Analytical reagents, Cosmetics, Dyes/Pigments, Explosives, Food industry, Inks, Photography

5 - Purification (a) and Recovery (b)

     (a): *Water, Gases*

     (b): *Recovery of Industrial Products*

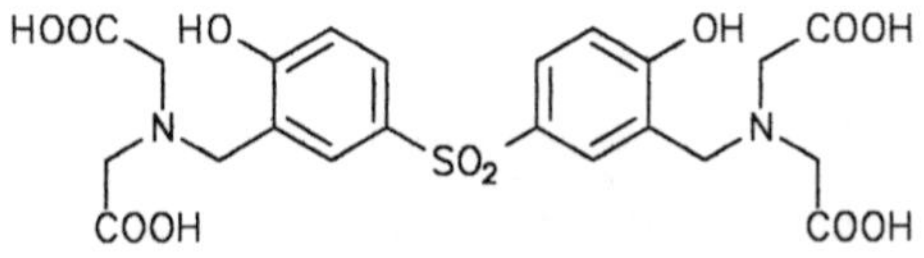

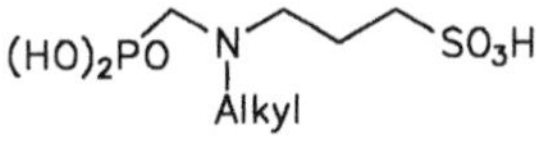

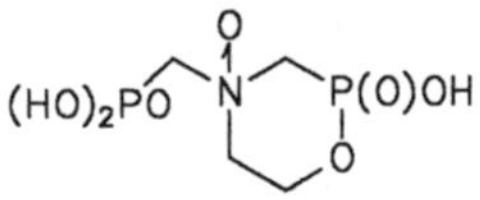

B — Industrial Branches Employing Mannich Bases and Derivatives

As reported in Table 42, Mannich bases are present to a variable extent in many products of the organic chemical industry. In particular, they are frequently the active principle of pharmaceuticals and pesticides or an essential constituent of macromolecular synthetic materials. In this latter field, as well as in the production of petroleum products, in the treatment of wastes, etc., they mainly provide important additives and auxiliaries.

B.1. The Pharmaceutical Industry and Pesticides Production

The chief contribution to the development of the chemistry of Mannich bases has un-doubtedly come from the pharmaceutical research that began with the early studies by

Table 43

Mannich Bases Used as Pharmaceuticals

Essential structure	Tradename (activity[a])	Reference[b]
C-Mannich bases		
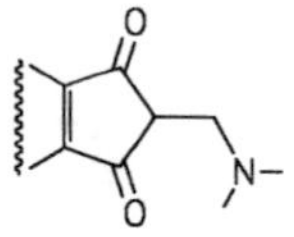	*Oxazidione (aCO)*	334 (6882)
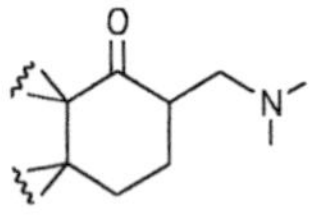	*Molindone (aPS)* *Pimeclone (RS)*	334 (6142) 334 (7399)
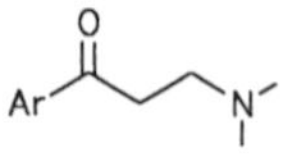	*Dyclonine (AN)* *Propipocaine (AN)* *Falicain (AN)* *Eprazinone (aT)* *Oxyfedrine (aAN)*	334 (3453) 334 (7843) 335 (3872) 334 (3577) 334 (6915)
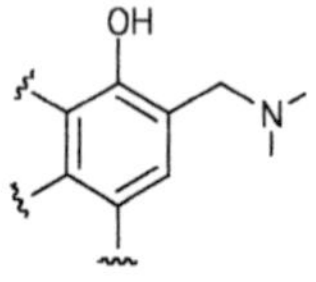	*Amodiaquine (aM)* *Amopyroquine (aM)* *Bialamicol (aM)* *Clamoxyquin (aM)* *Rytmol (aAR)*	334 (602),336 337 (II, 359) 334 (1217) 335 (2316) 335 (8066)
	*Pargyline (aHT)* *Tremorine* *Oxotremorine*	337 (III, 302) 334 (9498) 334 (6904)
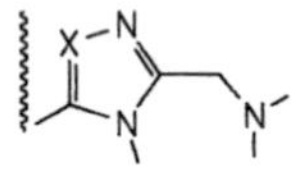 (X = C, N)	*Clemizole (aHY)* *Adinazolam (aD)*	334 (2346) 338
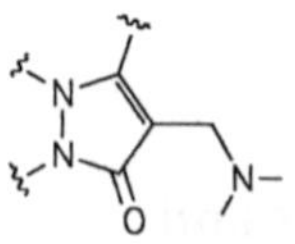	*Morazone (AA,aPY)*	334 (6176),342

Carl Mannich at the beginning of this century. Indeed, the huge amount of papers published on the matter along with the number of commercially available drugs (Tables 43 and 44) provide significant proof of the relevance of such compounds in this branch of industry.

Pharmaceuticals made from Mannich bases are reported in Table 43, where almost the whole range of the different classes is represented. C-Mannich bases, in particular cyclic and acyclic β-aminoketones, predominate, although N-Mannich bases, especially amide derivatives, are also well represented.

Mannich bases used as precursors of drugs, such as Amoxicillin,[340] Benetazon,[341] Morazone[342] (see also Chap. IV, D, and **511, 512** in Table 38), as well as those listed in Table 44, are also present in large numbers.

	Table 43
	(continued)

N-Mannich bases

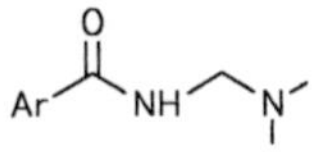	Morphazinamide (aB)	334 (6182)
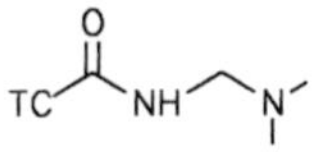 (TC = Tetracycline moiety)	Guamecycline (aB) Lysinemethyl-TC (aB) Roli-TC (aB)	334 (4468) 334 (5500) 334 (8236)
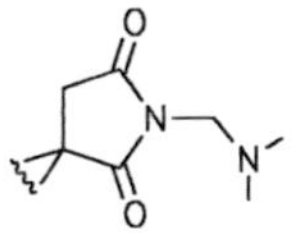	Morussimide (aCV)	186 (IV, 361)
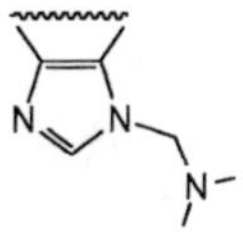	N-Morpholinomethyl- Theophylline (D)	334 (6196)
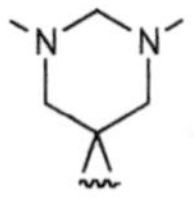	Hexetidine (aF)	334 (4624)
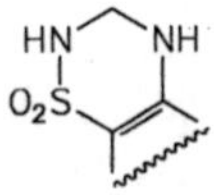	Hydrochlorothiazide (D) Hydroflumethiazide (D)	334 (4704) 334 (4716)

S-Mannich bases

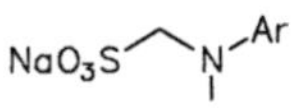	Dipyrone (AA) Sulfamipyrine (aPY) Sulfoxone (aL)	334 (3358) 334 (8896) 339 (A6, 195)
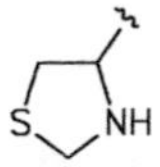	Timonacic (C)	334 (9375)

[a] AA = analgesic; AN = anesthetic; aAN = antianginal; aAR = antiarrhythmic; aB = antibacterial; aCH = anticholinergic; aCO = anticoagulant; aCV = anticonvulsant; aD = antidepressant; aF = antifungal; aHT = antihypertensive; aHY = antihystaminic; aL = antileprosyl; aM = antimalaric; aPS = antipsycotic; aPY = antipyretic; aT = antitussive; B = bronchodilator; C = choleretic; D = diuretic; RS = respiratory stimulant; V = vasodilator

[b] Merck's Index Number or Volume and page in parentheses

Various synthetic methods (Table 44) are applied to the modification of Mannich bases in order to obtain pharmacologically active derivatives. The carbonyl group of β-aminoketones is frequently involved in the reaction, as is demonstrated by the elegant synthesis of Phenindamine **589**[346] reported in Fig. 191 (see also Fig. 124, Chap. II).

Table 44

Derivatives of Mannich Bases Used as Pharmaceuticals

Essential structure (method of synthesis)	Tradename (activity[a])	Reference [b]
(Deamination of β-aminoketones)	Ethacrinic acid (D)	334 (3669)
(Amine replacement in aminomethyl indole by nitroalkane - Reduction)	Bucindolol (β-blocking drug)	343
(Carbonyl reduction of β-aminoketones)	Eprozinol (B) Halofantrine (aM) Mepiperphenidol (aCH)	334 (3578) 339 (A6, 209) 335 (5686)
(Organometal addition to carbonyl group of β-aminoketones)	Tramadol (AA)	334 (9485),344
R OH (R = Alkyl, Ar') (Organometal addition to carbonyl group of β-aminoketones)	Biperiden (aCH) Clobutinol (aT) Clophedianol (aT) Cycrimine (aCH) Procyclidine (aCH) Tiemonium iodide (aCH) Tridihexethyl (aCH) Trihexyphenidyl (aCH)	334 (1246) 335 (2334) 334 (2053) 334 (2763) 334 (7770) 334 (9363) 334 (9577) 334 (9607)
(Acylation of the above aminoalcohols)	Propoxyphene (AA)	334 (7851)

Promising trends in the research in this field include some interesting developments suggested both by review papers[347] (see also the Introduction) and by recent reports on β-aminoketones,[73–77] alkyne Mannich bases with radioprotective properties,[348] and benzotriazole Mannich bases.[349]

In addition to the studies dealing with the synthesis of new molecules, the modification of known drugs is receiving continued attention. In this connection, natural[2,350,351] (see also Chap. IV) as well as synthetic molecules (Pargyline,[352] Niridazole,[353] etc.) have been subjected to investigation. The prodrugs[354] described above (Fig. 186) are also worth mentioning.

Table 44		
(continued)		

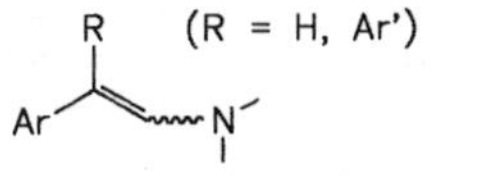

(Dehydration of aminoalcohols) (R = H, Ar')

	Cinnarizine (aHY)	334 (2308),345
	Pyrrobutamine (aHY)	334 (8023)
	Tinofedrine (V)	334 (9378)
	Triprolidine (aHY)	334 (9661),346

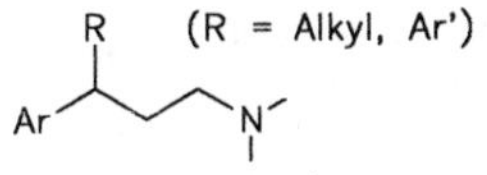

(Reduction of aminoalcohols) (R = Alkyl, Ar')

	Gamfexine (aB)	337 (III, 1041)
	Pheniramine (aHY)	334 (7198)
	Prenylamine (V)	334 (7744)

589 (see text) | *Phenindamine* (aHY) | 334 (7195),346 |

[a,b] As in Table 43

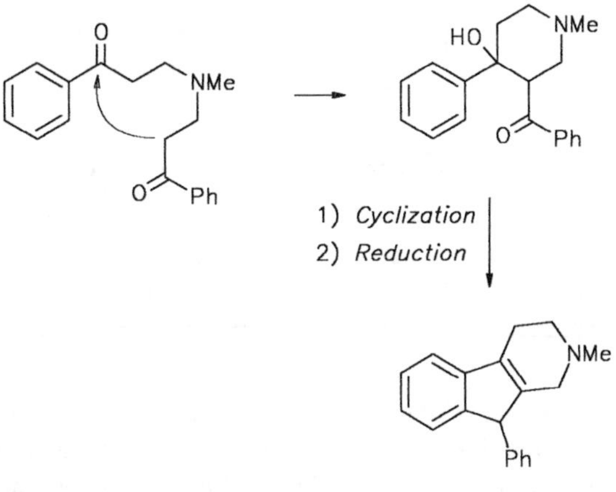

1) Cyclization
2) Reduction

589

Fig. 191.   Synthesis of Phenindamine from ketonic Mannich base.

Mannich bases and derivatives used as pesticides cover three essentially different fields, namely, insecticides and insect repellants, which act against noxious insects; fungicides and virucides, which are frequently employed as agrochemicals; and herbicides. As in the case of pharmaceuticals, almost all classes of bases are represented, in some instances as precursors of the active pesticide molecule[355–358] (Fig. 5, Chap. I). Mannich bases of type X—CH$_2$—N< (X ≠ C) are mainly used (Table 45), probably due to their high reactivity, which makes these derivatives suitable to interact with the metabolic mechanisms of the species to be destroyed.

Phenolic Mannich bases (**590**, Table 45) are the most common among insecticides, but cyclic and acyclic O-Mannich bases[376,377] and phosphorous acid Mannich bases[71] are also significantly employed.

Table 45

Mannich Bases Having Biocide Activity

Activity	Essential structure	Reference
Insecticides and Insect Repellants	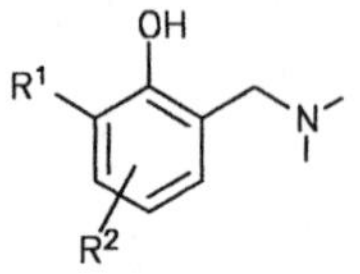 $(R^1 = Me, OMe, COOH,...$ $R^2 = H, Cl,...)$ **590**	359-362
Fungicides and Virucides	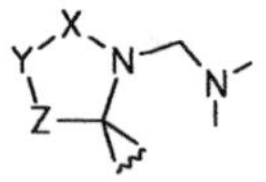 $(X - Z = C, N, S)$ **591**	363-368
	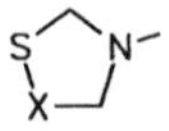 $[X = CH_2, N(Alkyl)-CS]$ **592**	369-371
Herbicides and Plant Growth Regulators	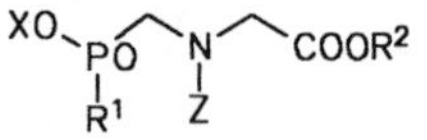 $(R^1 = OX, Alkyl;$ $R^2 = H, Alkyl)$ **593** $(X = H, SiPh_2tert.Bu, PBu_4;$ $Z = H, OH, Alkyl)$	70,72,358, 372-375

Among fungicides and virucides, N-Mannich bases deriving from five-membered heterocyclics such as pyrroles, pyrazoles, benzothiazoles, oxa- and thiadiazoles, tetrazoles (**591**), as well as from six-membered heterocyclics (e.g., pyrimidines),[378] are frequently used. S-Aminomethyl derivatives (**592**) are usually cyclic Mannich bases, while C-derivatives are prepared from cyclohexanone,[379,380] phenols,[381,382] alkynes[383] (see also Table 1, Chap. I), and amides, these last being used as wood antibacterials.[191]

Herbicides are mainly aminomethylphosphonic acids (**593**); amidic N-Mannich bases are also represented.[384,385]

B.2   Production of Natural and Synthetic Macromolecular Materials

It is mainly the leather, paper, and textile industries, especially those involved in the production of cellulosic and proteic fibers, that are interested in the use of Mannich bases as auxiliaries or additives for the treatment of natural macromolecules (see also Chap. IV, C).

Leather—Several investigations, relating to leather-working processes, are dedicated to the crosslinking of collagen. In particular, experiments have been carried out using formaldehyde or other polyfunctional precursors of Mannich bases (see **432**, Chap.

III, C.2, and "Collagen" in Table 34, Chap. III). Studies of phenolic Mannich bases used as agents in the treatment of leather are worthy of note.[386,387]

Paper—In addition to the wood industry, where different types of aminomethyl derivatives are produced by treatment of lignin[270,281] (see also **482**, Chap. IV, C.3, and **579**), the cellulose industry[21] and, in particular, the paper manufacturing process frequently requires products derived from Mannich base chemistry as processing aids or additives. The former help in the process as retention,[263] dispersion,[388] filtration,[263,268] and drainage[262,264,389] aids; the latter improve the performance of the final product as sizing agents[20,264,388] or strength additives.[20,262,268,390,391] Usually, Mannich bases of homopolymeric polyacrylamide,[4] and of its derivatives that are copolymeric[262–264] or grafted onto natural macromolecules,[268,388,391] are employed. Their action depends on the size of the molecule as well as on the ionogenic properties assumed after Mannich reaction. Nitrogen quaternization combined with concomitant hydrolytic processes may also contribute to the performances[4] (see also **502, 517** in Table 38, **562** in Table 40).

More specific additives aimed at improving the wet strength,[392] for electroconductive coatings,[393] biostatic agents,[380] dyes,[257] etc., are based on different compounds.

Textiles—The chemical treatment of natural and synthetic fibers is concerned with functionalization reactions having the purpose of obtaining improvements in both the manufacturing process and in the final product. Improvements in dyeing are obtained through pre-[394,395] or post-treatments[250–252,396] aimed at stabilizing the color, as described in Sec. A.2.

Anticrease treatments consist in the controlled functionalization and crosslinking of the proteic or cellulosic macromolecules of fibers. This topic has been already discussed (Sec. A.2, Chap. III, C.2, and Fig. 179, Chap. IV), but a very accurate study[11] of N,N'-bis-methylol ureides (**594** and **595**), which stresses the importance of a well-balanced hydrophilic power among these additives, is worth mentioning. The appropriate choice of either of the above products, or the use of a mixture of both, in fact permits a satisfactory optimization of the results.

594   **595**

Fiber wetting,[397] in relation to antistatic properties[27,28,398] (**567**, Table 40), is obtained by additives constituted by the Mannich bases of phosphorous acid, which also serve as fireproofing agents.[296,319,399]

Production of macromolecular materials—The industrial production of synthetic macromolecular materials requiring the application of Mannich bases is reported in Table 46. Some of them (adhesives, coatings, foam plastics, and resins) use Mannich bases or their derivatives as structural components of the material, whereas all the listed branches are concerned with the use of important additives (mainly antioxidants) or auxiliaries such as basic catalysts and accelerators. Specific functions are performed, for instance, by agents improving the adhesion of photopolymerizable paints[409] and by accelerator-modified adhesion promoters for rubber-to-wire adhesion.[418]

Table 46

Mannich Bases and Derivatives in the Production of Macromolecular Materials

Branch	Function performed	Material	Reference
Adhesives	Component/Modifier	Phenol-based Starch-based Various (Melamine, Polyacrylamide, Polyurethanes)	198,200,243 400,401 195,248,386
	Catalyst	Epoxy	215
Coatings	Component	Epoxy (electro- phoretic) Various	4,402-405, a 232,406
	Accelerator, Curing agent	Epoxy	407,408, a
	Adhesion improver, Pigment	-	409,410
Elastomers	Vulcanization accelerator	Natural and synthetic rubber	411-415
	Antioxidant	-	416,417, b
	Adhesion promoter	-	175,418, c
Foams	Component/Modifier	Polyurethanes Collagen-based	a 230
	Processing aid (stabilizer)	Polyurethanes	419,420
Plastics	Antioxidant	Polyenes, mainly	421
	Accelerator	Polyenes, Polyacrylics,...	422,423
	Plasticizer	-	424
Resins	Component/Modifier	Melamine- and Urea-based	a
	Catalyst	Melamine-based, Polyurethanes	425-427
	Adhesion promoter	-	176, c
Sealants	Catalyst	Polyurethanes, Epoxy	428-431

a see also *Constitutive components* in Sec. A.2 and Chap. III, B, C
b see also "Radical Inhibitors" and "Antioxidants" in Sec. A.2
c see also **568**, Table 40

B.3   Petroleum Products

Mannich bases are present as auxiliaries and additives throughout the whole working process of the petroleum industry, from crude oil extraction up to the final products. Indeed, they are used as oleophilizers in drilling fluids[35,156] or in the treatment of crude oil, such as desalting[432] and prevention of paraffin deposits.[433] Mannich bases and derivatives are, moreover, employed as antifoulants[6,434,435] in heat exchangers[434] as well as demulsifiers[271] and antifoaming agents;[436] all these functions are based on their surface activity properties (Sec. A.2).

A much more substantial use of Mannich bases is observed, however, with the primary products of petroleum, namely, fuels and lubricating oils.

Fuels—The antioxidant and detergent functions of Mannich bases are the ones mainly exploited.[307,437–440] They are also claimed to assist in improving storage stability,[439] as metal deactivators,[438] and in reducing soot ignition temperatures.[440]

Lubricating oils and similar products—Modern mechanical engineering requires lubricants capable of more and more sophisticated performances and hence needing a large number of additives suited to the most varied applications. Many of these requirements may be satisfied by properly designed Mannich bases, in particular when the drawback represented by their polar/basic features, which make them water extractable, can be partially overcome by using the corresponding carbonated salts.[441]

In addition to the compounds described in previous sections ("Antioxidants," "Corrosion Inhibitors," "Detergents"), more specific additives include friction-reducing agents[442] (see also Table 40), working as surface-active species affording chemisorption at the metal surface with formation of a boundary layer. Mannich bases, moreover, are inserted into the polymeric structure (see **559**, Table 40) of viscosity index improvers,[315,316,443,444] which serve to reduce the dependence of oil viscosity on temperature. Other polymeric additives, such as the pour-point depressants, may contain grafted aminomethyl groups[318] or be mixed with Mannich bases acting as dispersing agents.[445,446]

Functions similar to those performed with lubricating oils are reported for the additives of greases[160,297,447] and dielectric oils,[448] whereas the Mannich bases used as additives to asphalts for pavements serve as antistripping agents (**516**, Table 40).[449]

B.4 Miscellaneous Products

In addition to the major branches of application described above (Table 42), the versatile chemistry of Mannich bases makes it possible to synthesize various products for specific uses.

Analytical reagents—Besides the diagnostic reagents represented by the labeled compounds described in Chap. IV, E.3, titration indicators and colorimetric reagents of the complexons series, based on phenolic Mannich bases ortho-aminomethylated with α-amino acids, are reported to be able to produce stable complexes (see also "Complexing Agents"). Various phenols (**596**),[450–452] including alizarin derivatives **597**,[453,454] have been investigated to this end.

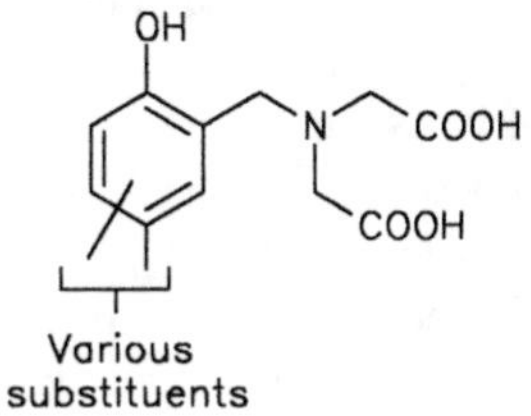

Various
substituents

596

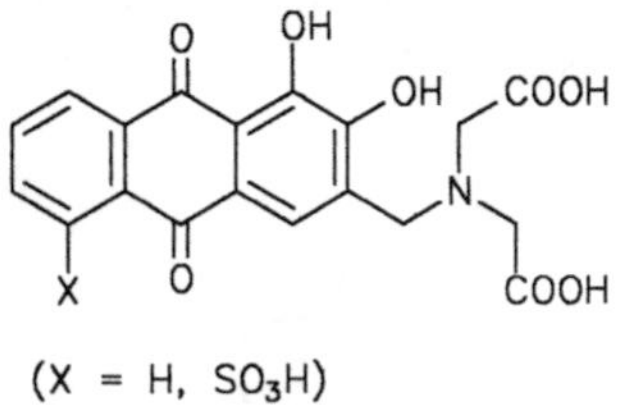

(X = H, SO₃H)

597

Moreover, macromolecular indicators having the ortho-carboxy azoic[455] or phenolphthalein[456] moiety have been prepared by amino- or amidomethylation of polymeric products.

Cosmetics—Hair preparations based on the crosslinking of keratine (see **432**, Chap. III, C.2, and Fig. 179, Chap. IV)[457] as well as deodorants for hygienic care, produced by polycondensation of phenols with formaldehyde and arylamines, are reported.[458]

Dyes—Mannich bases are involved in the synthesis of indigo, of reactive and disperse dyes (Sec. A.2), and in several pre- and post-treatments of textiles (Sec. B.2). A few reports deal with pigments[274] and paper dyes (see "Paper").

Explosives and propellants—Although some diazo derivatives capable of producing instantaneous or gradual gas evolution are prepared from Mannich bases,[459] the synthesis of nitro-derivatives is more relevant. Essentially two means of obtaining Mannich bases bearing NO_2 groups are followed (Fig. 192): by Mannich reaction on nitroalkanes (**a**) or by nitration of aminomethyl derivatives (**b**). Polynitro-derivatives of type **598** are usually prepared by combining these methods, that is, through Mannich reaction to give a cyclic nitro-derivative followed by nitration.[460,461]

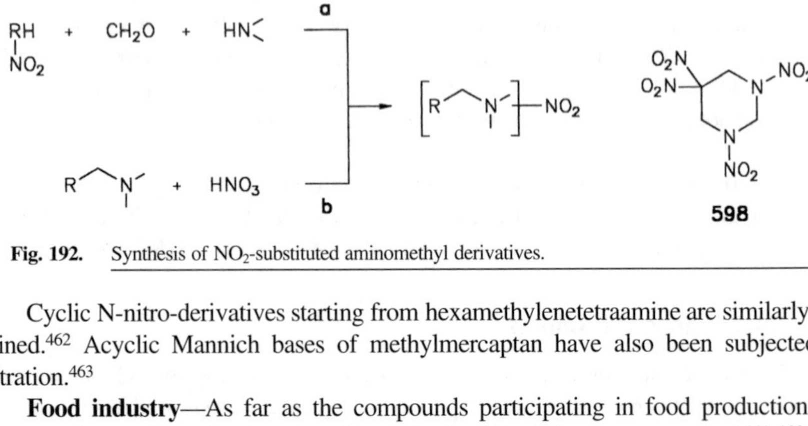

Fig. 192.   Synthesis of NO_2-substituted aminomethyl derivatives.

Cyclic N-nitro-derivatives starting from hexamethylenetetraamine are similarly obtained.[462] Acyclic Mannich bases of methylmercaptan have also been subjected to nitration.[463]

Food industry—As far as the compounds participating in food production are concerned, some investigations into the synthesis of food flavors are reported.[464-466] Key steps of the process are represented by the preparation of cycloketones bearing methyls as ring substituents, carried out by hydrogenolysis of precursor aminomethyl derivatives. The Mannich synthesis is also applied to amino acid preparations for animal nutrition.[467]

Surface-active agents[265,273] and antioxidants derived from gallic acid,[468] used as protein dispersants and food preservatives, respectively, are employed as additives.

Inks—Mannich bases derived from naphthol[278] and benzotriazole[299] are additives providing dispersant and anticorrosion properties for intaglio printing and ballpoint pen inks, respectively.

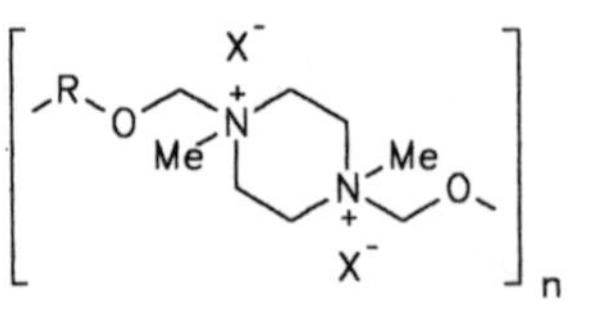

Photography—Several types of Mannich bases and derivatives are employed in this field. Crosslinking agents derived from piperazine (**534**, Sec. A.2) are hardeners for gelatin photographic emulsions, and O-aminomethyl derivatives of piperazine, used under the form of polymeric quaternary ammonium salts **599**, are useful as sensitizers for photographic emulsions.[469]

Melamine resins containing aromatic sulfonic acid groups can be used as a mordanting layer in color photography,[470] and finally, complexing agents of the carboxymethyl-poly(ethyleneimine) type are components of the processing bath.[471]

B.5   Purification and Recovery

The treatment of industrial effluents for purposes of purification and recovery is of great economic and environmental importance and requires specific auxiliary materials well suited to carry out the process efficiently. A contribution to this field is also given by Mannich bases, particularly in wastewater treatment and in the production of ion-exchange resins.

Water treatment—A wide range of patents covers the synthesis and application of polyacrylamide Mannich bases[4,472,473] used as flocculants for wastewater (see also "Surface-active agents" and "Flocculants") and applied occasionally to the treatment of solid wastes.[60] Other cationic polymers, such as polyacrylamide Mannich derivatives grafted onto natural polymers,[4,474] as well as various phenolic derivatives (e.g., **398** in Chap. III, B) are also used for the same purpose. The sequestering action of compounds having chelating properties[67,68] is exploited in order to avoid the formation of undesired inorganic precipitates (see also "Complexing agents" and "Sequestering agents").

Ion-exchange resins are employed in water-desalting membranes,[475] in the depuration from fluoride ions,[116] in uranium recovery (Table 39), etc. (see also "Ion exchangers," Sec. A.2).

Treatment of gaseous effluents—Polymeric[476,477] as well as low-molecular-weight[478] Mannich bases are employed in the treatment of gaseous effluents, particularly in desulfurization processes. Even the aminomethylation reaction on sulfur-containing effluents may be applied to the same end.[479]

Recovery of industrial by-products—Mannich reactions may help in the conversion of process by-products or residues into more valuable materials, as is evidenced in the cellulose industry, where the recovery of Ca-lignin sulfonate[480] involves employing the aminomethylation reaction to give the Mannich bases of lignin (**482** in Chap. IV and **579** in Sec. A.2), which are then used for various applications.[270,281]

Still bottoms deriving from ionol production (Fig. 184, Sec. A.2) are exploited as polymerization inhibitors in catalytic reforming,[153] and the residues of alkylphenol production, after aminomethylation, are useful as adhesion additives for bitumens.[449] Finally, corrosion inhibitors are obtained from the by-products of polycaprolactam production,[481] and shapable materials are similarly produced from the keratin of pig bristles.[227]

■ References for Chapter V

1. de Carneri, I., Coppi, G., Lauria, F., and Logemann, W., Una nuova tetraciclina solubile: la tetraciclina-L-metilenlisina, *Farmaco Ed. Prat.*, 16, 65, 1961; *Chem. Abstr.*, 55, 18990, 1961.

2. Lepetit SpA, Netherlands Patent Appl. 6,144,012, 1965; *Chem. Abstr.*, 63, 13180, 1965.

3. Rudchenko, V. F., Ignatov, S. M., Chervin, I. I., Nosova, V. S., and Kostyanovskii, R. G., Asymmetric nitrogen. XLVII. Geminal systems. XXXI. Dialkoxyamines: synthesis, hydroxy- and aminomethylation, NMR spectra, configurational stability, *Izv. Akad. Nauk SSSR Ser. Khim.*, 1153, 1986; *Chem. Abstr.*, 106, 175847, 1987.

4. Tramontini, M., Angiolini, L., and Ghedini, N., Mannich bases in polymer chemistry, *Polymer*, 29, 771, 1988.

5. Imada, T., Masafuda, H., and Minoda, S., Japanese Kokai Tokkyo Koho JP 61 91,240, 1986; *Chem. Abstr.*, 105, 191852, 1986.

6. Roling, P. V., Niu, J. H. Y., and Reid, D. K., European Patent Appl. EP 267,715, 1988; *Chem. Abstr.*, 109, 95881, 1988.

7. Dexter, M., Knell, M., Klemchuk, P., and Stephen, J. F., German Offen. 2,248,306, 1973; *Chem. Abstr.*, 79, 19700, 1973.

8. Stephen, J. F., German Offen. 2,248,339, 1973; *Chem. Abstr.*, 79, 19738, 1973.

9. Stover, W. H., U.S. Patent 4,025,316, 1977; *Chem. Abstr.*, 89, 113760, 1978.

10. Adamek, M., Bobulova, P., Glask, S., Gregor, F., and Kovacikova, A., Czechoslovakian Patent 133,091, 1969; *Chem. Abstr.*, 73, 87933, 1970.

11. Petersen, H. A., Cross-linking with formaldehyde-containing reactants, in *Handbook of Fiber Science and Technology, Vol. 2: Chemical Processing of Fibers and Fabrics: Functional Finishes, Part A*, Lewin, M. and Sello, S. B., Eds., Dekker, New York, 1983; see, in particular, paragraphs 3, 5, and 6.

12. Abdullaev, G. K., Agamolieva, E. A., Abasova, N. A., and Mamedov, I. A., Condensation of phenols with propyl- and butylamines in the presence of paraformaldehyde, *Azerb. Neft. Khoz.*, 53, 35, 1973; *Chem. Abstr.*, 79, 42077, 1973.

13. Rasberger, M. and Carrer, F., German Offen. 2,730,503, 1978; *Chem. Abstr.*, 88, 137447, 1978.

14. Lawrenz, D., Schupp, H., Schwerzel, T., Oslowski, H. J., and Heimann, U., European Patent Appl. EP 319,834, 1989; *Chem. Abstr.*, 112, 22445, 1990.

15. Wu, D. and Wu, Y., CN 1,032,344, 1989; *Chem. Abstr.*, 113, 80172, 1990.

16. Andreani, F., Angiolini, L., Costa Bizzarri, P., Della Casa, C., Ferruti, P., Ghedini, N., Tramontini, M., and Pilati F., The Mannich bases in polymer synthesis. V. Synthesis and characterization of poly(β-ketothioethers) from aliphatic and aromatic bis(β-dialkylaminoketones) and bis-thiols, *Polym. Commun.*, 24, 156, 1983.

17. Andreani, F., Angeloni, A. S., Angiolini, L., Costa Bizzarri, P., Della Casa, C., Ferruti, P., Fini, A., Ghedini, N., Manaresi, P., Pilati, F., and Tramontini M., Ital. Patent Appl. 20305 A/82, 1982.

18. Saunders, K. J., *Organic Polymer Chemistry*, 2nd ed., Chapman & Hall., London, 1988, 335–336.

19. Gotlib, E. M., Voskresenskaya, O. M., Verizhnikov, L. V., Liakumovich, A. G., Kirpichnikov, P. A., and Ivanov, B. E., Use of phenolic Mannich bases for curing of epoxy compounds, *Plast. Massy*, 28, 1987; *Chem. Abstr.*, 108, 151466, 1988.

20. Kyoritsu Org. Ind. Res. Laboratory, Japanese Kokai Tokkyo Koho JP 82,101,097, 1982; *Chem. Abstr.*, 97, 218353, 1982.

21. Mahanta, D. and Rahman, A., Control of wet-end chemistry by synthetic organic polymers: Adsorption of cationic PAA onto cellulose pulp, *Indian J. Chem. A*, 25A, 825, 1986.

22. Murakami, S., Ogura, K., and Yoshino, T., Equilibria of complex formation between bivalent metal ions and 3,3'-bis[N,N'-bis(carboxymethyl) aminomethyl]-o-cresolsulfonphthalein, *Bull. Chem. Soc. Jpn.*, 53, 2228, 1980.

23. Hodgkin, J. H., New Mannich base ligands, *Aust. J. Chem.*, 37, 2371, 1984.

24. Anon., *Colour Index*, Vol. 4, Society of Dyers and Colourists, Bradford, England, 1971, 4044, 4554, 4556.

25. Kamiya, T., Tanaka, K., Teraji, T., and Henmi, K., Japanese Kokai Tokkyo Koho 75,112,395, 1975; *Chem. Abstr.*, 84, 59500, 1976.

26. Nippon Oil Co., Japanese Kokai Tokkyo Koho JP 81,163,195, 1981; *Chem. Abstr.*, 96, 184008, 1982.

27. Sellet, L., U.S. Patent 3,790,606, 1974; *Chem. Abstr.*, 80, 108175, 1974.

28. Jaeger, H., German Offen. 2,556,481, 1976; *Chem. Abstr.*, 85, 143299, 1976.

29. Crook, J. W., Nicholson, E. S., and Sharma, V. R., British Patent 1,384,055, 1975; *Chem. Abstr.*, 83, 180306, 1975.

30. Petropavlovskii, G. A., Vasil'eva, G. G., and Simanovich, I. E., Chemical crosslinking of methyl cellulose, *Zh. Prikl. Khim. Leningrad,* 51, 2336, 1978; *Chem. Abstr.,* 90, 55646, 1979.

31. Schupp, E., Loch, W., Osterloh, R., and Ahlers, K., German Offen. DE 3,311,512, 1984; *Chem. Abstr.,* 102, 80432, 1985.

32. Dronov, V. I. and Nikitin, Y. E., Thioalkylation reactions, *Usp. Khim.,* 54, 941, 1985; *Chem. Abstr.,* 103, 214237, 1985.

33. Meier, H. R., European Patent Appl. EP 165209, 1985; *Chem. Abstr.,* 105, 6307, 1986.

34. Angiolini, L., Carlini, C., Ghedini, N., and Tramontini, M., Crosslinked sulphur-containing polymers. II. Reaction of acetone with formaldehyde and bis-thiols, *Polymer,* 31, 353, 1990.

35. Leighton, J. C., Iovine, C. P., and Carmine, P., U.S. Patent 4,707,306, 1987; *Chem. Abstr.,* 108, 132459, 1988.

36. Hart, J. R. and Grace, W. R., Ethylenediaminetetraacetic acid and related chelating agents, in *Ullmann's Encyclopedia of Industrial Chemistry,* Vol. A10, 5th ed., Gerhartz, W., Ed., VCH, Weinheim, 1987, 95.

37. David, C. B. and Earl, E. P., U.S. Patent 2,950,262, 1960; *Chem. Abstr.,* 55, 2182, 1961.

38. Yamada, M., Ohara, M., Sakuramoto, Y., and Watanabe, T., Study of vulcanization of chloroprene rubber with amine salts of 3,5-dinitrobenzoic acid and 2,6-substituted 4-dimethylaminomethyl phenols, *Nippon Gomu Kyokaishi,* 49, 202, 1976; *Chem. Abstr.,* 85, 47895, 1976.

39. Johansen, M. and Bundgaard, H., Decomposition of rolitetracycline and other N-Mannich bases and of N-hydroxymethyl derivatives in the presence of plasma, *Arch. Pharm. Chemi Sci. Ed.,* 9, 40, 1981.

40. Johansen, M. and Bundgaard, H., Prodrugs as drug delivery systems. XXIV. N-Mannich bases as bioreversible lipophilic transport forms for ephedrine, phenethylamine and other amines, *Arch. Pharm. Chemi Sci. Ed.,* 10, 111, 1982.

41. Sloan, K. B., Sherertz, E. F., and McTiernan, R. G., The effect of structure of Mannich bases prodrugs on their ability to deliver theophylline and 5-fluorouracil through hairless mouse skin, *Int. J. Pharm.,* 44, 87, 1988.

42. Schupp, E. and Kempter, F. E., German Offen. DE 3,146,640, 1983; *Chem. Abstr.,* 99, 89691, 1983.

43. Burger, A., *Burger's Medicinal Chemistry,* 4th ed., Wolff, M. E., Ed., John Wiley & Sons, New York, 1981, part III; see, in particular, pp. 417 (Table 45.1) and 942.

44. Gupta, P., Nast, R., and Windemuth, E., German Offen. 2,452,532, 1976; *Chem. Abstr.,* 85, 78923, 1976.

45. Gotlib, E. M., Voskresenskaya, O. M., Khamitov, I. K., Verizhnikov, L. V., and Liakumovich, A. G., *Plast. Massy,* 58, 1989; *Chem. Abstr.,* 112, 21690, 1990.

46. Dorer, C. J., German Offen. 2,452,662, 1975; *Chem. Abstr.,* 83, 150194, 1975.

47. Gordash, Y. T. and Grechko, A. N., Study of products of the reaction of technical-grade alkylsalicylic acids with formaldehyde and ammonia, *Neftekhimiya,* 23, 399, 1983; *Chem. Abstr.,* 99, 73438, 1983.

48. Kulieva, K. N., Namazova, I. I., Ismailova, N. D., and Dorokhina, I. V., Detergent dispersing additives based on high-molecular-weight alkylphenols, *Khim. Tekhnol. Topl. Masel,* 3, 1988; *Chem. Abstr.,* 108, 97491, 1988.

49. Schupp, E., Osterloh, R., Loch, W., and Ahlers, K., Ger. Offen. DE 3,422,457 1985; *Chem. Abstr.,* 104, 150916, 1986.

50. Paar, W., German Offen. DE 3,634,483, 1987; *Chem. Abstr.,* 107, 156469, 1987.

51. Lawrenz, D., Schupp, H., Schwerzel, T., Oslowski, H. J., and Heimann, U., European Patent Appl. EP 351,618, 1990; *Chem. Abstr.,* 112, 218859, 1990.

52. Patel, D. K., Shah, K. H., and Krishnan, V., A new class of protective agents for general purpose rubber vulcanisates, in *10th Prog. Paper and Rubber Conf.,* 127, 1978; *Chem. Abstr.,* 93, 169383, 1980.

53. Mardanov, M. A., Veliev, K. G., Akhmed-Zade, D. A., Alekperova, N. G., Mirzabekova, K. A., and Adzhamova, S. G., Synthesis and study of aminoalkylphenol compounds as jet fuel additives, *Sb. Tr. Inst. Neftekhim. Protsessov Akad. Nauk Az. SSR,* 13, 105, 1982; *Chem. Abstr.,* 99, 90708, 1983.

54. Gershanov, F. B., Liakumovich, A. G., Michurov, Y. I., Pantukh, B. I., Rutman, G. I., Sobolev, V. M., Grinberg, A. A., Gurvich, Y. A., Zacharova, N. V., and Nafikova, A. M., U.S. Patent 3,946,086, 1976; *Chem. Abstr.,* 85, 46190, 1976.

55. Amery, A., Crook, J. W., and Sharma, V. R., British Patent 1,363,233, 1974; *Chem. Abstr.,* 82, 73905, 1975.

56. Rasberger, M., U.S. Patent 4,198,334, 1980; *Chem. Abstr.,* 93, 73028, 1980.

57. Refined Products Corp., British Patent 767,162, 1957; *Chem. Abstr.,* 51, 13911, 1957.

58. See also survey in Ref. 4.

59. Seiko Chemical Industry Co., Ltd., Japanese Kokai Tokkyo Koho JP 58,164,633, 1983; *Chem. Abstr.,* 100, 140188, 1984.

60. Matsumoto, S., Teraishi, S., Watanabe, J., Sato, M., and Shiratori, I., Japanese Kokai Tokkyo Koho JP 62,250,008, 1987; *Chem. Abstr.,* 108, 76130, 1988.

61. Castrol Ltd., British Patent 1,061,904, 1967; *Chem. Abstr.,* 67, 73608, 1967.

62. Popplewell, A. F. and Clark, D. R., German Offen. 2,601,719, 1976; *Chem. Abstr.,* 86, 93000, 1977.

63. Sung, R. L., U.S. Patent 4,376,635, 1983; *Chem. Abstr.,* 98, 218654, 1983.

64. Kelly, D. P., Melrose, G. J. H., and Solomon, D. H., Thermosetting vinyl and acrylic co-polymers, *J. Appl. Polym. Sci.,* 7, 1991, 1963.

65. Nakai, M., Japanese Kokai Tokkyo Koho 79,112,993, 1979; *Chem. Abstr.,* 92, 77350, 1980.

66. Horvath, S. K., U.S. Patent 4,415,681, 1983; *Chem. Abstr.,* 100, 70018, 1984.

67. Redmore, D. and Welge, F. T., U.S Patent 4,085,134, 1978; *Chem. Abstr.,* 89, 109960, 1978.

68. Redmore, D., U.S. Patent 3,720,498, 1973; *Chem. Abstr.,* 78, 159670, 1973.

69. Dersa, S. A., Spanish Patent 411,113, 1976; *Chem. Abstr.,* 85, 108770, 1976.

70. Franz, J. E., U.S. Patent 4,084,953, 1978; *Chem. Abstr.,* 89, 109961, 1978.

71. Kuwahara, M. and Mutsukado, M., Japanese Kokai 77,100,452, 1977; *Chem. Abstr.,* 88, 121376, 1978.

72. Miller, W. H., Reitz, D. B., and Pulver, M. J., U.S. Patent 4,657,705, 1987; *Chem. Abstr.,* 107, 59256, 1987.

73. Von Thiele, K., Posselt, K., Offermanns, H., and Thiemer, K., Neue zerebral Wirksame basische Dithienyl-verbindungen, *Arzneim. Forsch.,* 30, 747, 1980.

74. Cagniant, P., Kirsch, G., Wierzbicki, M., Lepage, F., Cagniant, D., Loebenberg, D., Parmegiani, R., and Scherlock, M., Synthesis, antifungal and antibacterial activity of aminoalkyl ketones in sulfur heterocyclic series, *Eur. J. Med. Chem.,* 15, 439, 1980.

75. Dimmock, J. R., Erciyas, E., Raghavan, S. K., and Kirckpatrick, D. L., Evaluation of the cytotoxicity of some Mannich bases of aceto-phenone against the EMT6 tumor, *Pharmazie,* 45, 755, 1990.

76. Dimmock, J. R., Ercyias, E., Bigam, G. E., Kirkpatrick, D. L., and Duke, M. M., Intra-molecular cyclization and cytotoxicities of some Mannich bases of styryl ketones, *Eur. J. Med. Chem.,* 24, 379, 1989.

77. Meindl, W., Laske, R., and Böhm, M. Tu-morhemmende β-Aminoketone, *Arch. Pharm.,* 320, 730, 1987.

78. Upjohn Co., British Patent 1,188,897, 1970; *Chem. Abstr.,* 73, 35032, 1970.

79. McDaniel, K. G. and Speranza, G. P., U.S. Patent 4,383,102, 1983; *Chem. Abstr.,* 99, 39286, 1983.

80. Huitink, G. M., Isocein: a new fluorescent reagent for calcium, *Anal. Chim. Acta,* 70, 311, 1974.

81. Badrinas Vancells, A., Swiss Patent 614,696, 1979; *Chem. Abstr.,* 92, 215062, 1980.

82. Hodgkin, J. H., Linear aminophenol polymers using the Mannich reaction, *J. Polym. Sci. Polym. Chem. Ed.,* 24, 3117, 1986.

83. Roska, A., Klavins, M., and Zicmanis, A., High-molecular-weight catalysts in organic synthesis. XIX. New method of synthesis of polymer supported crown ethers, *Latv. PSR Zinat. Akad. Vestis Kim. Ser.,* 458, 1988; *Chem. Abstr.,* 110, 192345, 1989.

84. Agency of Ind. Sci. and Tech., Japanese Kokai Tokkyo Koho JP 57, 162726, 1982; *Chem. Abstr.,* 98, 126868, 1983.

85. Balakin, V. M., Glukhikh, V. V., Litvinets, Y. I., Validuda, G. I., and Pogudina, L. K., Synthesis and properties of water-soluble complexing polymers. II. Synthesis and phys-icochemical properties of high-molecular weight Mannich bases with 7-aminomethy-lene-8-hydroxyquinoline groups, *Zh. Obshch. Khim.,* 48, 2782, 1978; *Chem. Abstr.,* 90, 169030, 1979.

86. Ritter, H. and Rodewald, S., Synthesis of 3',5'-bis(morpholinomethyl)-4'-hydroxy- and 3',5'-bis(4-methyl-1-piperazinomethyl)-4'-hydroxymethyl acrylanilide. Copolymeriza-tion and metal ion binding properties of monomer and copolymer, *Makromol. Chem.,* 187, 801, 1986.

87. Singer, J. J., European Patent Appl. EP 85,277, 1983; *Chem. Abstr.,* 99, 212156, 1983.

88. Lin, S. Y. and Hoo, L. H., U.S. Patent 4,728,728, 1988; *Chem. Abstr.,* 108, 188732, 1988.

89. Sommer, K. and Raab, G., U.S. Patent 4,098,814, 1978; *Chem. Abstr.,* 90, 23247, 1979.

90. Quinlan, P. M., U.S. Patent 4,035,412, 1977; *Chem. Abstr.,* 87, 102436, 1977.

91. Haller, R., Zur Kenntnis substituierter 3,7-Diaza- und 3-Oxa-7-aza-bicyclo-[3.3.1]-nonanone, *Arzneim. Forsch.,* 15, 1327, 1965.

92. Märkl, G., Jin, G. Yu, and Schoerner, C., Chiral aminomethylphosphines and aminomethyldiphosphines, *Tetrahedron Lett.,* 1409, 1980.

93. Brezny, R., Schraml, J., Cermak, J., Micko, M. M., and Paszner, L., Specific ion-exchangers based on lignins and their characterization by silicon-29 NMR spectroscopy, *Tappi,* 73, 199, 1990; *Chem. Abstr.,* 113, 61535, 1990.

94. Hodgkin, J. H., PCT Int. Appl. 81 00,848, 1981; *Chem. Abstr.,* 95, 81026, 1981.

95. Maura, G. and Rinaldi, G., Resine poliamminocarboniliche per il recupero di uranio da soluzioni acquose, *Chim. Ind.,* 69, 77, 1987.

96. Feairheller, S. H., Taylor, M. M., and Filachione E. M., Chemical modification of collagen by the Mannich reaction, *J. Am. Leather Chem. Assoc.,* 62, 408, 1967; *Chem. Abstr.,* 68, 3928, 1968; see also *J. Am. Leather Assoc.,* 62, 398, 1967.

97. Brezny, R., Paszner, L., Micko, M., and Uhrin, D., The ion-exchanging lignin derivatives prepared by Mannich reaction with amino acids, *Holzforschung,* 42, 369, 1988; *Chem. Abstr.,* 111, 9088, 1989.

98. Goyal, M. and Chaturvedi, K. K., Potentiometric study of the complexes of Mannich base with a few bivalent metals, *Res. J. Sci. Devi Ahilya Vishwavidyalaya Indore,* 9, 11, 1987; *Chem. Abstr.,* 107, 243945, 1987.

99. Scanlon, P. M. and Young, E. R., U.S. Patent 4,387,244, 1983; *Chem. Abstr.,* 99, 141280, 1983.

100. Roling, P. V., Reid, D. K., and Niu, J. H. Y., U.S. Patent 4,894,139, 1990; *Chem. Abstr.,* 113, 43772, 1990.

101. Austin, T. H., U.S. Patent 4,371,629, 1983; *Chem. Abstr.,* 99, 6526, 1983.

102. Marugg, J. E., Gansow, M. A. P., and Thoen, J. A., U.S. Patent 4,883,826, 1989; *Chem. Abstr.,* 112, 159695, 1990.

103. Angiolini, L., Carlini, C., Ghedini, N., and Tramontini, M., Cross-linked sulfur containing polymers: reaction of poly(β-ketosulfide)s with formaldehyde and bis-thiols, *Polymer,* 30, 564, 1989.

104. Saunders, K. J., *Organic Polymer Chemistry,* 2nd ed., Chapman & Hall, London, 1988, 213.

105. Endo, T., Japanese 73 17,757, 1973; *Chem. Abstr.,* 80, 134174, 1974.

106. Zavlin, P. M., Rodnyawskaya, E. R., Shvchik, N. D., Levit, N. V., and Naidis, F. B., U.S.S.R. Patent SU 1,318,983, 1987; *Chem. Abstr.,* 108, 46759, 1988.

107. Reinhardt, R. M., Daigle, D. J., and Kullman, R. M. H., U.S. Patent Appl. 614,994, 1975; *Chem. Abstr.,* 85, 178935, 1976.

108. Petersen, H. A., Cross-linking with formaldehyde-containing reactants, in *Handbook of Fiber Science and Technology, Vol. 2: Chemical processing of fibers and fabrics: Functional Finishes, Part A,* Lewin, M. and Sello, S. B., Eds., Dekker, New York, 1983; see, in particular, paragraphs 3, 5, and 6.

109. Goeke, U. and Richter, M., German Offen. 2,823,682, 1979; *Chem. Abstr.,* 92, 95135, 1980.

110. Becker, W., Hubner, H., and Marten, M., European Patent Appl. EP 42,617, 1981; *Chem. Abstr.,* 96, 144629, 1982.

111. Becker, W. and Marten, M., German Offen. DE 3,124,370, 1982; *Chem. Abstr.,* 98, 108991, 1983.

112. Jex Co., Ltd., Japanese Kokai Tokkyo Koho JP 82 25,813, 1982; *Chem. Abstr.,* 97, 78952, 1982.

113. Balakin, V. M., Balakin, S. M., and Tesler, A. G., Deposited doc., 1977, VINITI 1359; *Chem. Abstr.,* 90, 187670, 1979.

114. Klipper, R. M., Heller, H., Lange, P. M., Friedrich, W., and Mitschker, A., European Patent Appl. EP 355,007, 1990; *Chem. Abstr.,* 113, 41932, 1990.

115. Sabrowski, E., Schwachula, G., Bachmann, R., Kochmann, W., and Broman, H. G., East German Patent 276,486, 1990; *Chem. Abstr.,* 115, 160552, 1991.

116. Kanesato, M., Yokoyama, T., and Suzuki, T., Japanese Kokai Tokkyo Koho JP 01 94,948, 1989; *Chem. Abstr.,* 111, 120323, 1989.

117. Fujiwara, H., Takahashi, A., and Sekiya, M., Japanese Kokai Tokkyo Koho 77 58,087, 1977; *Chem. Abstr.,* 87, 69413, 1977.

118. Maeda, H. and Yoshida, H., Preparation of macroreticular chelating resins containing aminomethylphosphonic acid groups from poly(glycidyl methacrylate) beads and their adsorption capacity, *Kenkyu Hokoku Kumamoto Kogyo Daigaku,* 14, 153, 1989; *Chem. Abstr.,* 111, 174778, 1989.

119. Rashidova, S. S. and Valiev, A. K., Aminomethylation of cellulose propargyl ether, *Uzb. Khim. Zh.,* 44, 1977; *Chem. Abstr.,* 88, 24406, 1978.

120. Dimov, K., Dimitrov, D., Terlemezyan, E., Semkova, M., and Bandova, M., Some possibilities of obtaining ion-exchanging fibers based on cellulose, *Cellul. Chem. Technol.*, 14, 665, 1980; *Chem. Abstr.*, 94, 122958, 1981.

121. Nishiyama M., On the preparation of ion exchange fibers containing amino and carboxyl groups and paper therefrom, *Kami Pa Gikyoshi*, 34, 421, 1980; *Chem. Abstr.*, 93, 48365, 1980.

122. Samborskii, I. V., I'in, V. A., Grachev, L. L., Chetverikov, A. F., and Gorbarenko, A. N., U.S.S.R. Patent 260,885, 1979; *Chem. Abstr.*, 92, 7392, 1980.

123. Suszer, A., U.S. Patent 3,714,010, 1973; *Chem. Abstr.*, 78, 137416, 1973.

124. Matsuda, H., Kanki, K., Nakagawa, T., and Koike, M., Japanese Kokai Tokkyo Koho 75, 67,287, 1975; *Chem. Abstr.*, 85, 83100, 1976.

125. Terekin, S. V., Vishnyakova, T. P., Golubeva, I. A., and Glebova, E. V., Study of phenolic antioxidants during the stabilization of fuel T-7, *Khim. Tekhnol. Topl. Masel*, 11, 1978; *Chem. Abstr.*, 89, 132066, 1978.

126. Takehisa, M., Watanabe, T., Imamura, K., and Iwata, M., Japanese 72, 09,258 1972; *Chem. Abstr.*, 77, 141203, 1972.

127. Brindell, G. D. and Macander, R. F., U.S. Patent 4,038,327, 1977; *Chem. Abstr.*, 87, 118802, 1977.

128. Schmidt, A., Peterson, J. B., and Dexter, M., U.S. Patent 4,044,019, 1977; *Chem. Abstr.*, 89, 111291, 1978.

129. Rasberger, M. and Gegner, E., German Offen. 2,933,206, 1980; *Chem. Abstr.*, 93, 28964, 1980.

130. Bruk, Y. A., Zolotova, L. V., and Borodulina, M. Z., Shielded phenols. VIII. Stabilization of polyolefins by 3,5-di-tert-butyl-4-hydroxy-benzylideneamines and N-substituted 3,5-di-tert-butyl-4-hydroxybenzylamines, *Zh. Prikl. Khim. Leningrad*, 48, 2734, 1975; *Chem. Abstr.*, 84, 106459, 1976.

131. Jiráčková, L. and Pospisil, J., Antioxidative activity of phenols containing substituents with heteroatoms O, S and N in the oxidation of tetralin, *Collect. Czech. Chem. Commun.*, 40, 2800, 1975.

132. Abdullaeva, F. A. and Mamedov, F. N., Phenol derivatives as oil stabilizers, *Neftekhimiya*, 18, 124, 1978; *Chem. Abstr.*, 89, 27167, 1978.

133. O'Shea, F. X., U.S. Patent 3,686,312, 1972; *Chem. Abstr.*, 77, 141024, 1972.

134. Layer, R. W., European Patent Appl. EP 42,589, 1981; *Chem. Abstr.*, 96, 143989, 1982.

135. Watanabe, T., Ikeda, S., Imamura, K., and Iwata, M., Substituted alkyl derivatives as antiozonants. IV. Antidegradant efficiency of benzylamino and aminomethylphenol derivatives, *Nippon Gomu Kyokaishi*, 44, 279, 1971; *Chem. Abstr.*, 75, 37544, 1971.

136. Kline, R. H., U.S. Patent 3,935,160, 1976; *Chem. Abstr.*, 84, 136997, 1976.

137. Gordash, Y. T., Zhurba, A. S., Sopkina, A. S., Chermenin, A. P., Perekrest, A. N., Marusyak, O. V., and Bereza, S. N., Use of Mannich bases and their Ba salts as additives for lubricating oils, *Khim. Tekhnol. Topl. Masel*, 22, 1974; *Chem. Abstr.*, 83, 134611, 1975.

138. Kurbanov, K. B., Mamedova, Z. A., and Akhmedov, S. T., Study of aminomethylation of cyclohexyl-p-dimethyl-thiophenols with different amines and formaldehyde, *Tezisy Resp. Nauchn. Konf. Molodych Uch. Khim. Azerb.*, 86, 1974; *Chem. Abstr.*, 87, 5553, 1977.

139. Spivack, J. D. and Pastor, S. D., European Patent Appl. EP 119,160, 1984; *Chem. Abstr.*, 102, 46816, 1985.

140. Khardin, A. P., Tuzhikov, O. I., Prikhid'ko, T. I., and Khokhlova, T. V., U.S.S.R. Patent SU 582,262, 1977; *Chem. Abstr.*, 88, 75180, 1978.

141. Allen, C. F. and Van Allan, J. A., U.S. Patent 2,763,657, 1956; *Chem. Abstr.*, 51, 3666, 1957.

142. Kuliev, A. M., Sardarova, S. A., Guseinov, M. S., and Kadyrov, M. S., Stabilization of polyethylene with some β-aminoketones, *Plast. Massy*, 74, 1978; *Chem. Abstr.*, 89, 111071, 1978.

143. Liu, C. S., Clarke, D. J., and Grina, L. D., U.S. Patent 4,780,230, 1988; *Chem. Abstr.*, 110, 79112, 1989.

144. Gershanov, F. B., Liakumovich, A. G., Mitsarov, Y. I., Pautukh, B. I., Rutman, G. I., Sobolev, V. M., Grinberg, A. A., Gurvich, Y. A., Zakharova, N. V., and Nafikova, A. M., German Offen. 2,053,181, 1972; *Chem. Abstr.*, 77, 101169, 1972.

145. Zakharova, N. V., Liakumovich, A. G., Michurov, Y. I., and Shalimova, Z. S., U.S. Patent 4,122,287, 1978; *Chem. Abstr.*, 90, 87014, 1979.

146. Rasberger, M., German Offen. 2,647,452, 1977; *Chem. Abstr.*, 87, 118757, 1977.

147. Parekh, M. G., U.S. Patent 4,263,232, 1981; *Chem. Abstr.*, 95, 98021, 1981.

148. Tsibul'skaya, L. V., Vasil'kevich, I. M., Ostroverkhov, V. G., and Seleznenko, L. V., *Nefteperera. Neftekhim. Kiev*, 37, 52, 1989; *Chem. Abstr.*, 112, 182558, 1990.

149. Ramey, C. E. and Luzzi, J. J., German Offen. 2,314,115, 1973; *Chem. Abstr.,* 80, 96764, 1974.

150. Bruk, Y. A. and Zolotova, L. V., U.S.S.R. Patent SU 437,758, 1974; *Chem. Abstr.,* 82, 31125, 1975.

151. Abdullaeva, F. A., Synthesis of screened phenols, *Azerb. Khim. Zh.,* 66, 1973; *Chem. Abstr.,* 83, 79881, 1975.

152. Distillers Co., Ltd., French Patent 1,355,292, 1964; *Chem. Abstr.,* 61, 4214, 1964.

153. Danil'yan, T. D., Rogacheva, O. V., Akhmetov, S. A., Varfolomeev, D. F., Makhov, A. F., Usmanov, R. M., Bezmel'nitsin, A. G., Surkov, V. D., and Malysheva, N. Y., U.S.S.R. Patent, SU 1,490,142, 1989; *Chem. Abstr.,* 111, 177738, 1989.

154. Roling, P. V., U.S. Patent 4,912,247, 1990; *Chem. Abstr.,* 113, 7007, 1990.

155. Mukhina, M. V., Zolotova, L. V., Komarov, P. S., and Bruk, Y. A., U.S.S.R. Patent SU 618,399, 1978; *Chem. Abstr.,* 89, 155961, 1978.

156. Ostroverkhov, V. G., Primak, R. G., Lukashevich, K. N., Tsebenko, V. A., and Mysak, A. E., *Neftepererab. Neftekhim. Kiev,* 24, 25, 1983; *Chem. Abstr.,* 99, 90698, 1983.

157. Harrison, J. J., European Patent Appl. EP 240,291, 1987; *Chem. Abstr.,* 108, 78530, 1988.

158. Sung, R. L. and Zoleski, B. H., U.S. Patent 4,278,553, 1981; *Chem. Abstr.,* 95, 118196, 1981.

159. Horodysky, A. G. and Kaminski, J. M., U.S. Patent 4,394,278, 1983; *Chem. Abstr.,* 99, 107952, 1983.

160. Horodysky, A. G. and Kaminski, J. M., U.S. Patent 4,402,842, 1983; *Chem. Abstr.,* 99, 161265, 1983.

161. Horodysky, A. G. and Kaminski, J. M., U.S. Patent 4,486,321, 1984; *Chem. Abstr.,* 102, 98182, 1985.

162. Serres, C. and Schaffhausen, J. G., U.S. Patent 4,424,317, 1984; *Chem. Abstr.,* 101, 24777, 1984.

163. Song, Y. S. and Basalay, R. J., European Patent Appl. EP 285,088, 1988; *Chem. Abstr.,* 110, 41833, 1989.

164. Karll, R. E. and Lee, R. J., U.S. Patent 4,384,138, 1983; *Chem. Abstr.,* 99, 178848, 1983.

165. Zakupra, V. A., Struzhko, V. L., Petrenko, L. M., and Zhurba, A. S., Study of selectivity of separation of Mannich base-type additives by liquid chromatography, *Neftepererab. Neftekhim. Kiev,* 30, 26, 1986; *Chem. Abstr.,* 105, 136632, 1986.

166. Lundberg, R. D. and Gutierrez, A., European Patent Appl. EP 304,175, 1989; *Chem. Abstr.,* 111, 10101, 1989.

167. Otto, F. P., U.S. Patent 3,649,229, 1972; *Chem. Abstr.,* 76, 143212, 1972.

168. Horodysky, A. G. and Gemmill, R. M., European Patent Appl. EP 182,940, 1986; *Chem. Abstr.,* 105, 46129, 1986.

169. Horodysky, A. G. and Gemmill, R. M., U.S. Patent 4,787,996, 1988; *Chem. Abstr.,* 110, 138508, 1989.

170. Kostusyk, J. L., PCT Int. Appl. WO 88 08,017, 1988; *Chem. Abstr.,* 110, 178566, 1989.

171. Kozakiewicz, J. J. and Huang, S. Y., U.S. Patent 4,956,399, 1990; *Chem. Abstr.,* 114, 83189, 1991.

172. Nakra, G., Kane, J. P., and Sortwell, E. T., European Patent Appl. EP 405,712, 1991; *Chem. Abstr.,* 114, 170585, 1991.

173. Korotushenko, T. P., Sukhoverkhov, V. D., Gordash, Y. T., and Grechko, A. N., Production of ashless additives from technical alkylsalicylic acids, *Neftepererab. Neftekhim. Kiev,* 24, 22, 1983; *Chem. Abstr.,* 99, 90697, 1983.

174. Anderson, R. L. and Chamot, E., European Patent Appl. EP 132,383, 1985; *Chem. Abstr.,* 102, 134823, 1985.

175. Atwell, E. C., British Patent 1,175,651, 1969; *Chem. Abstr.,* 72, 56511, 1970.

176. Plueddemann, E. P., U.S. Patent 3,554,952, 1971; *Chem. Abstr.,* 74, 64967, 1971.

177. Evstaf'ev, V. P., Ivanova, E. A., Fufaev, A. A., Trofimov, G. A., Shor, G. I., and Bondarenko, A. P., Synthesis and functional properties of high-molecular-weight Mannich bases, *Khim. Tekhnol. Topl. Masel,* 27, 1977; *Chem. Abstr.,* 87, 186888, 1977.

178. Kozlovskaya, T. F. and Gudriniece, E., *Latv. PSR Zinat. Akad. Vestis Khim. Ser.,* 619, 1988; *Chem. Abstr.,* 110, 212168, 1989.

179. Saab, A. N., Sloan, K. B., Beall, H. D., and Villanueva, R., Effect of aminomethyl (N-Mannich base) derivatization on the ability of S^6-acetyloxymethyl-6-mercaptopurine prodrug to deliver 6-mercaptopurine through hairless mouse skin, *J. Pharm. Sci.,* 79, 1099, 1990.

180. Bonati, A., Bombardelli, E., and Gabetto, B., British Patent 1,383,053, 1975; *Chem. Abstr.,* 83, 43342, 1975.

181. Ferruti, P., Angeloni, A. S., Scapini, G., and Tanzi, M. C., New oligomers and polymers as drug carriers, in *Recent Advances in Drug Delivery Systems,* Anderson, J. M., and Kim, S. W., Eds., Plenum, London, 1984, 71–74.

182. Bundgaard, H. and Johansen, M., Prodrugs as drug delivery systems. XIX. Bioreversible derivatization of aromatic amines by formation of N-Mannich bases with succinimide, *Int. J. Pharm.*, 8, 183, 1981.

183. Bundgaard, H. and Johansen, M., Prodrugs as drug delivery systems. X. N-Mannich bases as novel pro-drug candidates for amides, imides, urea derivatives, amines and other NH-acidic compounds. Kinetics and mechanisms of decomposition and structure-reactivity relationships, *Arch. Pharm. Chem. Sci. Ed.*, 8, 29, 1980.

184. Bundgaard, H., Johansen, M., Stella, V., and Cortese, M., Prodrugs as drug delivery systems. XXI. Preparation, physicochemical properties and bioavailability of a novel water-soluble prodrug type for carbamazepine, *Int. J. Pharm.*, 10, 181, 1982; *Chem. Abstr.*, 97, 11708, 1982.

185. Sloan, K. B., European Patent Appl. EP 39,051, 1981; *Chem. Abstr.*, 96, 104087, 1982.

186. Runti, C., *Fondamenti di Chimica Farmaceutica*, Vols. 2 and 4, Lint, Trieste, 1973, 213 and 361, respectively.

187. Neumann, M. G. and De Groote, R. A. M. C., Reaction of sodium hydroxymethanesulfonate with substituted anilines, *J. Pharm. Sci.*, 67, 1283, 1978.

188. Th. Goldschmidt A.-G., British Patent 771,635, 1957; *Chem. Abstr.*, 51, 12515, 1957.

189. Komkov, I. P. and Pankratov, V. A., Surface active salts of quaternary ammonium bases with antimicrobial action, *Zh. Prikl. Khim. Leningrad*, 43, 1371, 1970; *Chem. Abstr.*, 73, 100287, 1970.

190. Sato, T., Yamamoto, H., and Hashizume, T., Japanese Kokai Tokkyo Koho JP 02,120,001, 1990; *Chem. Abstr.*, 113, 174381, 1990.

191. Yoshitomi Pharmaceutical Industries, Ltd., Japanese Kokai Tokkyo Koho JP 82 92,076, 1982; *Chem. Abstr.*, 97, 184270, 1982.

192. Saunders, K. J., *Organic Polymer Chemistry*, 2nd ed., Chapman & Hall, London, 1988, 351–356.

193. Kamata, K., Suzuki, T., Yamaguchi, M., Saito, J., Mitsuishi, T., and Waki, H., Japanese Kokai Tokkyo Koho JP 61,268,713, 1986; *Chem. Abstr.*, 107, 135952, 1987.

194. Roh, J. K., Higuchi, M., and Sakata, I., Curing behaviour and bonding properties of thermosetting resin adhesives. V. Cocondensation reaction of phenol with melamine, *Mokuzai Gakkaishi*, 36, 42, 1990; *Chem. Abstr.*, 113, 41891, 1990.

195. Lin, J., Chen, W., and Hu, B., Preparation of water-soluble melamine-formaldehyde resin and its characteristics, *Fujian Shifan Daxue Xuebao*, 6, 51, 1990; *Chem. Abstr.*, 115, 50886, 1991.

196. Takhirov, M. K., Abbaskhanov, N. A., Korotin, M. M., Vorosova, T. G., and Solomatov, V. I., Polymeric composition, *Otkrytiya Izobret. Prom. Obratztsy Tavarnye Znaky*, 63, 1984; *Chem. Abstr.*, 101, 39526, 11984.

197. Farbwerke Hoechst A.-G., Netherlands Patent Appl. 6,609,044; *Chem. Abstr.*, 67, 44405, 1967.

198. Schreiber, H., German Offen. 2,323,936, 1973; *Chem. Abstr.*, 81, 38305, 1974.

199. He, B., Huang, W., Guo, X., and Yu, Y., Catalysis by condensation polymers containing salicylic acid for hydrolysis of p-nitrophenyl acetate in aqueous solution, *Gaofenzi Tongxun*, 416, 1982; *Chem. Abstr.*, 99, 21670, 1983.

200. Spoor, H., German Offen. 2,320,536, 1974; *Chem. Abstr.*, 82, 99155, 1975.

201. Sekmakas, K. and Shah, R., U.S. Patent 4,396,732, 1983; *Chem. Abstr.*, 99, 124179, 1983.

202. Paar, W. and Hoenel, M., European Patent Appl. EP 213,626, 1987; *Chem. Abstr.*, 107, 135973, 1987.

203. Lawrenz, D., Schupp, E., and Schwerzel, T., European Patent Appl. EP 304,854, 1987; *Chem. Abstr.*, 111, 80020, 1989.

204. Demmer, C. G., French Patent Appl. FR 2,523,980, 1983; *Chem. Abstr.*, 100, 70012, 1984.

205. Paar, W. and Gmoser, J., Austrian Patent AT 382,633, 1987; *Chem. Abstr.*, 107, 156465, 1987.

206. Paar, W. and Pampouchidis, G., Austrian Patent AT 392,075, 1991; *Chem. Abstr.*, 115, 31233, 1991.

207. Vianova Kunsthanz A.-G., Japanese Kokai Tokkyo Koho JP 62 68861, 1987; *Chem. Abstr.*, 107, 135972, 1987.

208. Klein, H. P. and Speranza, G. P., U.S. Patent 4,404,121, 1983; *Chem. Abstr.*, 99, 196318, 1983.

209. Speranza, G. P. and Klein, H. P., European Patent Appl. EP 134,338, 1985; *Chem. Abstr.*, 103, 72093, 1985.

210. Texaco Development Corp., Belgian Patent BE 905,005, 1986; *Chem. Abstr.*, 107, 60027, 1987.

211. Asano, A., Odaka, H., and Tanabe, K., Japanese Kokai Tokkyo Koho JP 60,166,315, 1985; *Chem. Abstr.*, 104, 34863, 1986.

212. Jefferson Chemical Co., Inc., British Patent Amended 1,002,272, 1969; *Chem. Abstr.,* 76, 25975, 1972.

213. Bratychak, V. A., Bychkov, V. A., and Puchin, V. A., Peroxide aniline-formaldehyde oligomers, *Plast. Massy,* 15, 1984; *Chem. Abstr.,* 101, 8113, 1984.

214. Yamada, M., Ohara, M., Sakuramoto, Y., and Watanabe, T., Vulcanization accelerators for neoprene rubber, *Nippon Gomu Kyokaishi,* 42, 202, 1976; *Chem. Abstr.,* 85, 47895, 1976.

215. Kazaryan, G. A., Shirinyan, A. A., Airapetov, Y. S., and Sarkysian, E. S., U.S.S.R. Patent SU 992,543, 1983; *Chem. Abstr.,* 98, 144744, 1983.

216. Sorokin, M. F., Shteinpress, A. B., Shode, L. G., and Zuev, V. V., Coating compositions made from epoxy and dimethylaminomethylated p-alkylphenol-formaldehyde resins, *Lakokras. Mater. Ikh Primen.,* 12, 1973; *Chem. Abstr.,* 79, 54955, 1973.

217. Andrews, C. M., Bull, C. H., Demmer, C. G., and Rolfe, W. M., European Patent Appl. EP 351,365, 1990; *Chem. Abstr.,* 113, 7535, 1990.

218. Lockwood, R. J., Reymore, H. E., and Thompson, E. J., U.S. Patent 3,896,052, 1975; *Chem. Abstr.,* 83, 180430, 1975.

219. Selivanov, A. V., Zenitova, L. A., Bakirova I. N., and Kirpichnikov, P. A., Catalysis of trimerization of phenylisocyanate in the presence of phenolic Mannich bases, *Kinet. Katal.,* 29, 586, 1988; *Chem. Abstr.,* 110, 94429, 1989.

220. Fedtke, M., Acceleration mechanism in curing reactions involving model systems, *Makrom. Kem. Makrom. Symp.,* 7, 153, 1987.

221. Kirpichnikov, P. A., Il'yasov, A. V., Kadirov, M. K., Nefed'ev, E. S., Liakumovich, A. G., Verizhnikov, L. V., and Gotlib, E. M., Role of radical processes in epoxy oligomer hardening by 2,4-bis[(dimethylamino)methyl] phenol in the presence of quinone imines, *Izv. Akad. Nauk SSSR Ser. Khim.,* 2824, 1986; *Chem. Abstr.,* 107, 176596, 1987.

222. Voskresenskaya, O. M., Gotlib, E. M., Verezhnikov, L. V., and Liakumovich, A. G., Effect of benzoyl peroxide on curing of epoxy oligomers with phenolic Mannich bases, *Izv. Vyssh. Uchebn. Zaved. Khim. Khim. Tekhnol.,* 32, 87, 1989; *Chem. Abstr.,* 112, 180514, 1990.

223. Bakirova, I. N., Zenitova, L. A., and Kirpichnikov, P. A., Synthesis and properties of SKU-OM-type polyurethane elastomers, *Kauch. Rezina,* 7, 22, 1985; *Chem. Abstr.,* 104, 51854, 1986.

224. Yamamiya, S., Abe, Y., Kanno, T., and Takezawa, N., Japanese Kokai Tokkyo Koho 79 37,182, 1979; *Chem. Abstr.,* 91, 41039, 1979.

225. Mitra, S. and DeVoe, R. J., U.S. Patent 4,791,045, 1988; *Chem. Abstr.,* 111, 15343, 1989.

226. Saunders, K. J., *Organic Polymer Chemistry,* 2nd ed., Chapman & Hall., London, 1988, 326 and 335.

227. Sontag, D., Mellentin, J., Bretshneider, W., Fickel, K., and Schubert, W., East German Patent 85,184, 1971; *Chem. Abstr.,* 78, 59271, 1973.

228. Hinterwaldner, R., Casein and its importance in the paper and coating industry. VIII. Modification and additives. VI, *Coating,* 15, 32, 1982; *Chem. Abstr.,* 96, 201283, 1982.

229. Battista, O. A., U.S. Patent 4,416,814, 1983; *Chem. Abstr.,* 100, 56894, 1984.

230. Gagniard, P. and Vinard, D., French Patent Appl. FR 2,545,088, 1984; *Chem. Abstr.,* 102, 205743, 1985.

231. Saunders, K. J., *Organic Polymer Chemistry,* 2nd ed., Chapman & Hall., London, 1988, 429.

232. Seknakas, K. and Shah, R., U.S. Patent 4,525,510, 1985; *Chem. Abstr.,* 104, 35614, 1986.

233. Smith, H. A., U.S. Patent 4,115,365, 1978; *Chem. Abstr.,* 90, 55678, 1979.

234. Takai, Y., Yamamoto, J., Yamazaki, K., and Shibahara, Y., Japanese Kokai Tokkyo Koho 80 27,364, 1980; *Chem. Abstr.,* 93, 9643, 1980.

235. Hitachi, Ltd., Japanese Kokai Tokkyo Koho 81 82,815 1981; *Chem. Abstr.,* 95, 204938, 1981.

236. Waddill, H. and Speranza, G. P., European Patent Appl. EP 222,512, 1987; *Chem. Abstr.,* 107, 177202, 1987.

237. Nishimura, T., Mine, S., and Kasuya, T., m-Xylenediamine Mannich type epoxy resin curing agent, *Purasuchikkusu,* 41, 113 and 76, 1990; *Chem. Abstr.,* 114, 25223, 1991.

238. Ebewele, R. O., River, B. H., Myers, G. E., and Koutsky, J. A., Polyamine-modified urea-formaldehyde resins. II. Resistance to stress induced by moisture cycling of solid wood joints and particleboard, *J. Appl. Polym. Sci.,* 43, 1483, 1991.

239. Thorpe, D., European Patent Appl. EP 237,270 1987; *Chem. Abstr.,* 108, 38588, 1988.

240. Ishizuka, Y. and Komakibara, I., Japanese Kokai Tokkyo Koho 78 47,095, 1978; *Chem. Abstr.,* 90, 168730, 1979.

241. Clubley, B. G., Dellar, R. J., Buszard, D. L., and Richardson, N., European Patent Appl. EP 149,480, 1985; *Chem. Abstr.*, 104, 34886, 1986.

242. Speranza, G. P., Brennan, M. E., and Grigsby, R. A., Jr., U.S. Patent 4,681,965, 1987; *Chem. Abstr.*, 108, 38975, 1988.

243. Freeman, H. G., Baxter, G. F., and Allan, G. G., U.S. Patent 3,920,613, 1975; *Chem. Abstr.*, 84, 75047, 1976.

244. Tiedeman, G. T., U.S. Patent 3,784,514, 1974; *Chem. Abstr.*, 81, 26564, 1974.

245. Eilerman, G. E., U.S. Patent 3,876,405, 1975; *Chem. Abstr.*, 83, 98917, 1975.

246. Anderson, D. W. and Neubert, T. C., U.S. Patent 4,210,475, 1980; *Chem. Abstr.*, 93, 151478, 1980.

247. Plueddemann, E. P., German Offen. 2,002,420, 1970; *Chem. Abstr.*, 73, 78030, 1970.

248. Dombrovskii, A. V., Shkol'nik, Y. S., and Shkol'nik, R. S., U.S.S.R. Patent SU 170,601, 1965; *Chem. Abstr.*, 63, 10140, 1965.

249. Keim, W. and Röper, M., Acylation and alkylation, in *Ullmann's Encyclopedia of Industrial Chemistry*, Vol. A1, 5th ed., Gerhartz, W., Ed., VCH, Weinheim, 1985, 185; see, in particular, p. 212.

250. Ohme, R. and Schmitz, E., East German Patent 39,618, 1964; *Chem. Abstr.*, 64, 19852, 1966.

251. Chavan, R. B. and Langer, M. H., Sublimation transfer printing of polyester/cotton blends, *Text. Res. J.*, 58, 51, 1988.

252. Noll, B., Schreiner, G., Willecke, B., Hille, G., and Hartmann, M., East German Patent DD 229,275, 1985; *Chem. Abstr.*, 106, 34574, 1987.

253. Braun, W., Weissauer, H., and Waechter, R., British Patent 857,391, 1960; *Chem. Abstr.*, 55, 11869, 1961.

254. Braun, W., Waechter, R., and Weissauer, H., German Patent 1,132,269, 1962; *Chem. Abstr.*, 59, 12954, 1963.

255. Farbwerke Hoechst A.-G., French Patent 1,363,216, 1964; *Chem. Abstr.*, 62, 10572, 1965.

256. Matsui, K., Otaguro, K., Asaumi, E., and Fujimura, T., Dyes containing a dimethylaminomethyl group, *Yuki Gosei Kagaku Kyokai Shi*, 26, 75, 1968; *Chem. Abstr.*, 70, 38867, 1969.

257. Patsch, M. and Ruske, M., German Offen. DE 3,117,956, 1982; *Chem. Abstr.*, 98, 199840, 1983.

258. Havlíckova, L., Antrachinonfarbstoffe. XVI. Methylierung, Hydroxymethylierung und Aminomethylierung in der 1-Aminoantrachinon-reihe, *Collect. Czech. Chem. Commun.*, 38, 2003, 1973.

259. Andrisano, R., Baroncini, L., and Tramontini, M., Ricerche sulla reattività delle basi di Mannich. III. Su alcuni coloranti reattivi contenenti il gruppo β-ammino-etil-chetonico, *Chim. Ind.*, 47, 173, 1965.

260. Yamase, I., Kuroki, N., and Konishi, K., Reactive dyes. XIX. Reactive dyes containing ω-(dimethylamino)propiophenone groups, *Kogyo Kagaku Zasshi*, 98, 1713, 1965; *Chem. Abstr.*, 64, 11350, 1966.

261. Jaeger, H., German Offen. 2,264,698, 1974; *Chem. Abstr.*, 82, 45059, 1975.

262. Showa Denko K. K., Japanese Kokai Tokkyo Koho JP 59 30,996, 1984; *Chem. Abstr.*, 101, 40122, 1984.

263. Uchida, T. and Ogawa, M., Japanese Kokai Tokkyo Koho JP 60 151,400, 1985; *Chem. Abstr.*, 104, 70565, 1986.

264. Kurokawa, A., Toki, H., Suzuki, Y., and Yodoya, T., Japanese Kokai Tokkyo Koho JP 63 54,402, 1988; *Chem. Abstr.*, 109, 95024, 1988.

265. Kawaguchi, H., Hoshino, H., and Ohtsuka, Y., Modification of polymer latex particles, *Proc. 28th IUPAC Macromol. Symp.*, 609, 1982.

266. Kawaguchi, H., Hoshino, H., Amagasa, H., and Ohtsuka, Y., Modifications of a polymer latex, *J. Colloid Interface Sci.*, 97, 465, 1984.

267. Kyoritsu Org. Ind. Res. Lab., Japanese Kokai Tokkyo Koho JP 59 16,509, 1984; *Chem. Abstr.*, 100, 193259, 1984.

268. Ogawa, M. and Narushima, M., Japanese Kokai Tokkyo Koho JP 63 12,795, 1988; *Chem. Abstr.*, 108, 188747, 1988.

269. Farrar, D. and Flesher, P., European Patent Appl. EP 210,784, 1987; *Chem. Abstr.*, 106, 177,852, 1987.

270. McKague, A. B., Flocculating agents deriving from Kraft lignin, *J. Appl. Chem. Biotechnol.*, 24, 607, 1974.

271. De Groote, M. and Shen, K.-T., U.S. Patent 2,771,430, 1956; *Chem. Abstr.*, 51, 6996, 1957.

272. Kyoritsu Org. Ind. Res. Lab., Japanese Kokai Tokkyo Koho JP 58,153,506, 1983; *Chem. Abstr.*, 100, 86653, 1984.

273. McKinney, L. L., Setzkorn, E. A., and Uhing, E. H., U.S. Patent 2,717,263, 1955; *Chem. Abstr.*, 50, 7138, 1956.

274. Aign, V., Pusch, N., Wienkenhoever, M., Paulat, V., and Schubert, K., German Offen. DE 3,617,010, 1987; *Chem. Abstr.,* 106, 158047, 1987.

275. Makhmudov, T. M., Khidoyatov, K. K., and Khakimov, A. S., Synthesis, property study, and development of technology for making ampholyte surfactants based on aromatic hydroxy acids, in *Tr. 7th Mezhdunar, Kongr. Poverkhn Akt Veshchestvam,* 1976; *Chem. Abstr.,* 91, 59211, 1979.

276. Maksimov, A. V., Lazarev, V. A., Merkotun, Z. Y., Magar, N. G., Arsent'ev, V. A., and Pavlychev, V. N., U.S.S.R. Patent SU 1,592,048, 1990; *Chem. Abstr.,* 114, 86042, 1991.

277. Hartough, H. D., Dickert, J. J., and Meisel, S. L., U.S. Patent 2,647,117, 1953; *Chem. Abstr.,* 48, 8265, 1954.

278. Fechner, W. D., Polster, R., Kranz, J., and Hartmann, E., German Offen. DE 3,211,165, 1983; *Chem. Abstr.,* 100, 23599, 1984.

279. Akimova, I. V., Kolesova, N. R., Ponomarev, D. A., and Smetanina, S. S., Producing surfactants from wood-derived phenols, *Izv. Vyss. Uchebn. Zaved Lesn. Zh.,* 91, 1984; *Chem. Abstr.,* 100, 158581, 1984.

280. Leonte, M., Florea, T., Isbasciu, M., Pope, A., and Ionita, G., Romanian Patent RO 87,062, 1985; *Chem. Abstr.,* 105, 155097, 1986.

281. Shilling, P., U.S. Patent 4,786,720, 1988; *Chem. Abstr.,* 110, 233666, 1989.

282. Pied, J. P., German Offen. 2,011,192, 1970; *Chem. Abstr.,* 74, 4266, 1971.

283. Gross, H. and Seibt, H., German Offen. 2,312,910, 1974; *Chem. Abstr.,* 82, 59012, 1975.

284. Kuliev, A. M., Mamedov, F. N., Bairamova, A. G., Dzhafarov, A. S., and Alieva, N. M., Stabilization of polyethylene by Mannich bases, *Plast. Massy,* 42, 1976; *Chem. Abstr.,* 85, 161111, 1976.

285. Dexter, M., Knell, M., Klemchuk, P., and Stephen, J. F., U.S. Patent 4,116,930, 1978; *Chem. Abstr.,* 90, 55759, 1979.

286. Takeisa, M., Watanabe, T., Imamura, K., and Iwata, M., Japanese Patent 72 08,137, 1972; *Chem. Abstr.,* 77, 141200, 1972.

287. Vishnyakova, T. P., Golubeva, I. A., Gutnikova, L. P., Seregin, E. P., Prokudin, V. N., and Veselyanskaya, V. M., Phenolic antiozonants for rubber, *Khim. Tekhnol. Topl. Masel,* 39, 1980; *Chem. Abstr.,* 93, 75264, 1980.

288. Hanson, J. B., U.S. Patent 4,242,212 1980; *Chem. Abstr.,* 94, 177810, 1981.

289. Danilov, A. M., Selyagina, A. A., Demina, N. N., Mitusova, T. N., Perezhigina, I. Y., and Senekina, A. M., Stabilization of light gas oil from catalytic cracking, *Khim. Tekhnol. Topl. Masel,* 9, 1989; *Chem. Abstr.,* 112, 39419, 1990.

290. Parlman, R. M. and Burns, L. D., U.S. Patent 4,322,304, 1982; *Chem. Abstr.,* 97, 41279, 1982.

291. Nottes, O., German Offen. 2,104,470, 1972; *Chem. Abstr.,* 77, 126025, 1972.

292. Mizuch, K. G., Lapina, R. A., Kryukova, A. S., and Peksheva, N. V., U.S.S.R. Patent SU 205,212, 1982; *Chem. Abstr.,* 98, 199793, 1983.

293. Nikitenko, A. G., Lesman, B. I., Yushcenko, V. A., Lozinskii, M. O., Gevanzka, Y. I., Trikhleb, L. M., Markovskii, L. N., and Tmenov, D. M., U.S.S.R. Patent SU 973,525, 1982; *Chem. Abstr.,* 98, 73796, 1983.

294. Hisao, Y. and Kogiso, O., PCT Int. Appl. WO 82 03,880, 1982; *Chem. Abstr.,* 98, 181080, 1983.

295. Terada, Y., Ono K., Takagishi, H., Maeda, Y., Suguira, T., and Jinno, S., Japanese Kokai 74 80,399, 1974; *Chem. Abstr.,* 82, 74378, 1975.

296. Maier, L., U.S. Patent 3,359,266, 1967; *Chem. Abstr.,* 69, 19303, 1968.

297. Brown, S. H. and Crocker, R. E., U.S. Patent 3,868,329, 1975; *Chem. Abstr.,* 83, 8251, 1975.

298. Tatur, I. R., Yakovlev, D. A., Lazarev, V. A., Shestopalov, V. E., and Gilevich, G. G., Composition VNIINM-PAV-31/87 for preservation and hydraulic testing of heat-transfer and reservoir equipment, *Khim. Neft. Mashinostr.,* 38, 1989; *Chem. Abstr.,* 112, 80163, 1990.

299. Sano, H., Toyama, T., and Sasage, D., Japanese Kokai Tokkyo Koho JP 61,272,278, 1986; *Chem. Abstr.,* 107, 41790, 1987.

300. Orudzheva, I. M., Aliev, S. M., Efendiev, T. E., and Dzhafarov, Z. I., Study of some aminoderivatives of 2-mercaptopyridine as inhibitors of hydrogen saturation, *Korroz. Zashch. Neftegazov Promsti.,* 9, 1982; *Chem. Abstr.,* 98, 130275, 1983.

301. Allabergenov, K. D., Kurbanov, F. K., and Kuchkarov, A. B., Corrosion inhibiting effect of acetylenic aminoethers of substituted phenols, *Izv. Vyssh. Uchebn. Zaved., Khim. Khim. Tekhnol.,* 21, 478, 1978; *Chem. Abstr.,* 89, 179808, 1978.

302. Anderson, J. D. and Hayman, E. S., U.S. Patent 3,992,313, 1976; *Chem. Abstr.,* 87, 139496, 1977.

303. Semikolenov, G. F., Podobaev, N. I., and Voskresenskii, A. G., Chemical composition of the acid-corrosion inhibitor PKU [urotropine based condensation product], *Ingibitory Korroz. Met.,* 127, 1972.

304. Kuhn, K., East German Patent 77,236, 1970; *Chem. Abstr.,* 75, 99727, 1971.

305. Traise, T. P., Watson, R. W., and Little, R. Q., U.S. Patent 3,725,480, 1973; *Chem. Abstr.,* 80, 98204, 1974.

306. King, J. M., British Patent Appl. 2,019,850, 1979; *Chem. Abstr.,* 92, 149861, 1980.

307. Kaufman, B. J., U.S. Patent 4,490,155, 1984; *Chem. Abstr.,* 102, 134802, 1985.

308. Morris, J. R., U.S. Patent 2,736,707, 1956; *Chem. Abstr.,* 50, 8198, 1956.

309. Lindert, A. and Wolpert, S. M., PCT Int. Appl. WO 90 05,794, 1990; *Chem. Abstr.,* 113, 193575, 1990.

310. Speranza, G. P., Grigsby, R. A., Jr., and Yeakey, E. L., U.S. Patent 4,952,732, 1990; *Chem. Abstr.,* 114, 184973, 1991.

311. Worrel, C. J., Canadian Patent 966,309, 1975; *Chem. Abstr.,* 83, 150195, 1975.

312. Cahill, P. J. and Piasek, E. J., U.S. Patent 4,425,249, 1984; *Chem. Abstr.,* 100, 123864, 1984.

313. Lundberg, R. D. and Emert, J., European Patent Appl. EP 317,353, 1989; *Chem. Abstr.,* 111, 100176, 1989.

314. Gutierrez, A., Song, W. R., Lundberg, R. D., and Kleist, R. A., European Patent Appl., EP 356,010 1990; *Chem. Abstr.,* 112, 162049, 1990.

315. West, C. T., U.S. Patent 4,131,553, 1978; *Chem. Abstr.,* 90, 171324, 1979.

316. Smyser, G. L. and Cengel, J. A., European Patent Appl. EP 96,972, 1983; *Chem. Abstr.,* 100, 106372, 1984.

317. Lundberg, R. D. and Gutierrez, A., European Patent Appl. EP 302,643, 1989; *Chem. Abstr.,* 110, 234497, 1989.

318. Hart, W. P. and Liu, C. S., U.S. Patent 4,668,412, 1987; *Chem. Abstr.,* 107, 80812, 1987.

319. Fujii, H. and Hattori, T., Japanese Kokai 74 80,393, 1974; *Chem. Abstr.,* 82, 74379, 1975.

320. Mitchell, R. S., German Offen. 2,423,881, 1974; *Chem. Abstr.,* 82, 98141, 1974.

321. Hardy, T. A., U.S. Patent 4,083,897, 1978; *Chem. Abstr.,* 89, 109959, 1978.

322. Rotmaier, L. and Merten, R., European Patent Appl. EP 44,417, 1982; *Chem. Abstr.,* 96, 181294, 1982.

323. Boreiko, N. P., Vernov, P. A., Gizatullina, L. Y., Kurbatov, V. A., Lemaev, N. V., Liak-umovic, A. G., Sokovykh, L. I., and Tarasov, N. F., U.S.S.R. Patent 831,772, 1981; *Chem. Abstr.,* 95, 135610, 1981.

324. Albert, H. E., U.S. Patent 3,526,673, 1970; *Chem. Abstr.,* 73, 99842, 1970.

325. Stahly, E. E. and Lard, E. W., U.S. Patent 3,555,116, 1971; *Chem. Abstr.,* 74, 64991, 1971.

326. Vernov, P. A., Boreiko, N. P., Gizatullina, L. Y., Zuev, V. P., Ivanov, B. E., Kurbatov, V. A., Kirpichnikov, P. A., and Liakumovic, A. G., U.S.S.R. Patent 819,078, 1981; *Chem. Abstr.,* 95, 25870, 1981.

327. Birnbach, S., Oftring, A., Baur, R., Gousetis, C., and Trieselt, W., German Offen. DE 3,821,883, 1990; *Chem. Abstr.,* 113, 42829, 1990.

328. Mitchell, R. S., U.S. Patent 3,974,090, 1976; *Chem. Abstr.,* 85, 177612, 1976.

329. Redmore, D. and Paley, W. S., U.S. Patent 4,330,487, 1982; *Chem. Abstr.,* 97, 127817, 1982.

330. Hwa, C. M., Kelly, J. A., and Adhya, M., U.S. Patent 4,973,744, 1990.

331. Hwa, C. M., Kelly, J. A., and Adhya, M., U.S. Patent 4,977,292, 1990; Chem. Abstr., 114, 229150, 1991.

332. Quinlan, P. M., U.S. Patent 4,080,375, 1978; *Chem. Abstr.,* 89, 65122, 1978.

333. Grannen, E. A. and Robinson, L., Def. Publ. U.S. Pat. Off. T 917,001, 1973; *Chem. Abstr.,* 83, 60295, 1975.

334. *The Merck Index,* 11th ed., Merck & Co., Rahway, NJ, 1989.

335. *The Merck Index,* 9th ed., Merck & Co., Rahway, NJ, 1976.

336. Natarjan, P. N. and Burckhalter, J. H., α,α-Phenyl piperidyl analog of amodiaquine, *Eur. J. Med. Chem.,* 11, 89, 1976.

337. Burger, A., *Burger's Medicinal Chemistry,* 4th ed., Wolff, M. E., Ed., John Wiley & Sons, New York, 1980–1981, parts II and III.

338. Gall, M., Kamdar, B. V., Lipton, M. F., Chidester, C. G., and DuCamp, D. J., Mannich reactions of heterocycles with dimethyl(methylene)ammonium chloride: a high yield, one-step conversion of estazolam to adinazolam, *J. Heterocycl. Chem.,* 25, 1649, 1988.

339. Actor, P., Chow, A. W., Dutko, F. J., and McKinlay, M. A., Chemotherapeutics, in *Ullmann's Encyclopedia of Industrial Chemistry,* Vol. A6, 5th ed., Gerhartz, W., Ed., VCH, Weinheim, 1986, 173; see, in particular, pp. 195, 209.

340. Mattioda, G. and Christidis, Y., Glyoxylic acid, in *Ullmann's Encyclopedia of Industrial Chemistry,* Vol. A12, 5th ed., Gerhartz, W., Ed., VCH, Weinheim, 1989, 495–496.

341. Grepl, F., Minar, L., Kuapil, L., and Pospisil, J., Czechoslovakian Patent CS 267,606, 1990; *Chem. Abstr.,* 114, 163534, 1991.

342. Mrongovius, R., Neugebauer, M., and Rücker, G., Analgesic activity and metabolism in the mouse of morazone, famprofazone and related pyrazolones, *Eur. J. Med. Chem.,* 19, 160, 1984.

343. Kreighbaum, W., German Offen. DE 3,421,252, 1984; *Chem. Abstr.,* 103, 53952, 1985.

344. Flick, K., Frankus, E., and Friderichs, E., Untersuchungen zur chemischen Struktur und analgetischen Wirkung von phenylsubstituierten Aminomethylcyclohexanolen, *Arzneim. Forsch.,* 28, 107, 1978.

345. Jing, S., Lin, S. H., Che, Y. P., Wang, J. Y., and Wu, C. Y., Preparation of cinnarizine, *Hua Hsueh Tung Pao,* 341, 1980; *Chem. Abstr.,* 94, 103301, 1981.

346. Hasspacher, K., Bretz, U., and Schreyer, H., Antiallergic agents, in *Ullmann's Encyclopedia of Industrial Chemistry,* Vol. A2, 5th ed., Gerhartz, W., Ed., VCH, Weinheim, 1985, 419; see, in particular, p. 424.

347. Holbová, E., Mannichova reakcia 2-merkaptobenzotiazolu a jeho derivátov, *Chem. Listy,* 82, 943, 1988; *Chem. Abstr.,* 110, 114704, 1989.

348. Akhrem, A. A., Kokhomskaya, V. V., Krasovskaya, M. S., Stryzhakova, E. P., Vatulina, G. G., and Tuzhilkova, T. M., Synthesis and radioprotective properties of trans-2e-methyl-4-[3-(dialkylamino)propyn-1-yl] hexahydrothiochroman-4-ols, their sulfoxides and sulfones, *Vetsi Akad. Nauk BSSR Ser. Khim. Navuk,* 59, 1990; *Chem. Abstr.,* 113, 191102, 1990.

349. Katritzky, A. R., Rachwal, S., and Rachwal, B., The chemistry of benzotriazole. 3. The aminoalkylation of benzotriazoles, *J. Chem. Soc. Perkin Trans. 1,* 799, 1987.

350. Nelson, M. L., Park, B. H., Andrews, J. S., Georgian, V. A., Thomas, R. C., and Levy, S. B., Inhibition of the tetracycline efflux antiport protein by 13-thio-substituted 5-hydroxy-6-deoxytetracycline, *J. Med. Chem.,* 36, 370, 1993.

351. Csuk, R., Hönig, H. Weidmann, H., and Zimmerman, H. K., Aminoalkoholesther von Hydroxyboranen, 10. Tetracyclin-bor-Mannichbasen als potentielle Antitumorwirkstoffe, *Arch. Pharm.,* 317, 336, 1984.

352. Altman, J. and Julia, M., Réaction de Mannich de la pargyline avec des aminoacides, *Bull. Soc. Chim. Fr.,* 1427, 1973.

353. Huang, L. S., Wang, A., Ma, Z., Si, B. X., Synthesis of a new schistosomicide S 72014 and its homologs, *Yao Hsueh Hsueh Pao,* 346, 1980; *Chem. Abstr.,* 94, 121401, 1981.

354. Pitt, C. G., Bao, Y., Thompson, J., Wani, M. C., Rosenkrantz, H., and Metterville, J., Esters and lactones of phenolic aminocarboxylic acids: prodrugs for iron chelation, *J. Med. Chem.,* 29, 1231, 1986.

355. Suganuma, H. and Fuyimura, H., European Patent Appl. EP 330,186, 1989; *Chem. Abstr.,* 112, 76645, 1990.

356. Collin, G. and Höke, H., Indole, in *Ullmann's Encyclopedia of Industrial Chemistry,* Vol. A14, 5th ed., Gerhartz, W., Ed., VCH, Weinheim, 1989, 167–168.

357. Seki, I., Kitano, K., and Kondo, F., Japanese Kokai Tokkyo Koho 78 72,829, 1978; *Chem. Abstr.,* 90, 72167, 1979.

358. Maier, L. and Diel, P. J., Organic phosphorus compounds. 87. Some reactions of O-ethyl-2-chloroethylphosphonite, *Phosphorus, Sulfur, Silicon Relat. Elem.,* 45, 165, 1989.

359. Jain, K. R., Synthesis and insecticidal activity of some new Mannich bases from 5-chlorosalicylic acid hydrazide, *J. Inst. Chem. India,* 62, 73, 1990; *Chem. Abstr.,* 114, 61654, 1991.

360. Kovalenko, L. G., Viktorov-Nabokov, O. V., Skrynik, E. M., Ruban, E. M., Sholudchenko, L. L., Denisova, Z. A., Markina, V. V., and Bogdanova, E. N., Repellent properties of Mannich bases derived from hydroxy- and aminobenzoic acid esters, *Med. Parazitol. Parazit. Bolezni,* 52, 69, 1983; *Chem. Abstr.,* 99, 135492, 1983.

361. Kovalenko, L. G., Viktorov-Nabokov, O. V., Ruban, E. M., Skrynik, E. M., Denisova, Z. A., Dremòva, V. P., Markina, V. V., and Bogdanova, E. N., Repellent properties of Mannich bases derived from cresol and phenol for mosquitoes Aedes Aegypti L. and fleas Xenopsylla Cheopis Roths., *Med. Parazitol. Parazit. Bolezni,* 52, 46, 1983; *Chem. Abstr.,* 99, 18028, 1983.

362. Kovalenko, L. G., Viktorov-Nabokov, O. V., Ruban, E. M., Skrynik, E. M., Korneeva, L. A., Sichko, L. D., and Markina, V. V., Repellent effect of the Mannich bases of some phenol-methoxy derivatives on Aedes Aegypti mosquitoes and Xenopsylla Cheopis fleas, *Med. Parazitol. Parazit. Bolezni,* 68, 1989; *Chem. Abstr.,* 112, 50549, 1990.

363. Wollweber, D., Kraemer, W., Brandes, W., Dutzmann, S., and Paulus, W., German Offen. DE 4,004,035, 1991; *Chem. Abstr.,* 115, 183082, 1991.

364. Sen Gupta, A. K. and Gupta, A. A., Synthesis and biological evaluation of 1-arylaminomethyl-3-methyl-4-substituted benzylidene-5-pyrazolones. A new class of pesticides, *Bokin Bobai,* 8, 283, 1980; *Chem. Abstr.,* 94, 30629, 1981.

365. Maksudov, N. K., Tadzhieva, T. A., Alimov, E., and Maksumova, K. G., Fungicidal activity of some Mannich bases, Deposited doc., 1975 VINITI 2082–75; *Chem. Abstr.,* 87, 112830, 1977.

366. Ram, V. J. and Pandey, H. N., Synthesis of Mannich bases and sulfides derived from 5-(p-chlorophenyl)-1,3,4-oxadiazole-2-thione, *Agr. Biol. Chem.,* 37, 1465, 1973; *Chem. Abstr.,* 79, 92118, 1973.

367. Ghattas, A. B. A. G., Abdel-Rahman, M., El-Wassimy, M. T., and El-Saraf, G. A., Benzofurans. III. Synthesis and antifungal activity of some new Mannich bases and sulfides derived from (1,3,4-thiadiazolyl)- and (1,2,4-triazolyl)benzofurans, *Rev. Roum. Chim.,* 34, 1987, 1989; *Chem. Abstr.,* 114, 6381, 1991.

368. Holbova, E., Sidoova, E., Povazanec, F., Zemanova, M., and Drobnicova, I., Antimicrobially active 1(2-alkylthio-6-benzotriazolylaminomethyl)-5-(3-X^1–4-X^2-phenyl)-1,2,3,4-tetrazoles, *Chem. Pap.,* 44, 369, 1990; *Chem. Abstr.,* 113, 231260, 1990.

369. United States Rubber Co., British Patent 1,059,677, 1967; *Chem. Abstr.,* 67, 73602, 1967.

370. Schubart, R., Dithiocarbamic acid derivatives, in *Ullmann's Encyclopedia of Industrial Chemistry,* Vol. A9, 5th ed., Gerhartz, W., Ed., VCH, Weinheim, 1987, 1; see, in particular, pp. 12, 18.

371. Geigy, J. R., A.-G., French M. 6521 1969; *Chem. Abstr.,* 74, 88072, 1971.

372. Soroka, M., Polish Patent 107,789, 1980; *Chem. Abstr.,* 95, 43347, 1981.

373. Large, G. B. and Buren, L. L., U.S. Patent 4,341,549, 1982; *Chem. Abstr.,* 97, 198570, 1982.

374. Franz, J. E., U.S. Patent 4,471,131, 1984; *Chem. Abstr.,* 102, 7086, 1985.

375. Miller, W. H., Reitz, D. B., and Pulwer, M. J., European Patent Appl. EP 216,745 1987; *Chem. Abstr.,* 106, 214382, 1987.

376. Croxall, W. J. and Melamed, S., U.S. Patent 2,715,631, 1955; *Chem. Abstr.,* 50, 8742, 1956.

377. Kovalenko, L. G., Ruban, E. M., Skrynik, E. M., Viktorov-Nabokov, O. V., Korneeva, L. A., Dremova, V. P., and Markina, V. V., Repellent effect of Mannich bases derived from lactic acid esters, *Med. Parazitol. Parazit. Bolezni,* 44, 1987; *Chem. Abstr.,* 107, 192961, 1987.

378. Iordanova, A., Karparov, A., and Starkeva, M., Application of plant protoplasts for antiphytoviral means essay *in vitro, Dokl. Bolg. Akad. Nauk,* 43, 105, 1990; *Chem. Abstr.,* 115, 110401, 1991.

379. Metalsalts Corp., Netherlands Patent Appl. 6,600,061, 1966; *Chem. Abstr.,* 65, 17638, 1966.

380. Konkey, J. H., Relative toxicity of biostatic agents used in the pulp and paper industry, *Tappi,* 49, 124A, 1966; *Chem. Abstr.,* 65, 19241, 1966.

381. Baicu, T., Leonte, M., Jilaveanu, A., Georgescu, M., and Leonte, A., Romanian Patent 68,352, 1980; *Chem. Abstr.,* 95, 7088, 1981.

382. Shoeb, H. A., Tammam, G. H., Moharram, H. H., Korkor, M. I., and El-Amin, S. M., Synthetic schistosomicides. V. Synthesis of some antimonyl quinolines, *Egypt. J. Chem.,* 24, 201, 1981; *Chem. Abstr.,* 99, 38343, 1983.

383. Croxall, W. J. and Dawson, J. W., U.S. Patent 2,584,429, 1952; *Chem. Abstr.,* 46, 9605, 1952.

384. Beck, J. R., British Patent 1,405,862, 1975; *Chem. Abstr.,* 83, 193109, 1975.

385. Sandoz A. G., Japanese Kokai Tokkyo Koho JP 62,135,470, 1987; *Chem. Abstr.,* 107, 176070, 1987.

386. Sellet, L., U.S. Patent 3,655,619, 1972; *Chem. Abstr.,* 77, 20616, 1972.

387. Sellet, L., U.S. Patent 3,674,415, 1972; *Chem. Abstr.,* 77, 103363, 1972.

388. Seiko Chem. Ind., Hokuetsu Paper Mills, Japanese Kokai Tokkyo Koho JP 58,216,730, 1983; *Chem. Abstr.,* 100, 105419, 1984.

389. Anon., Concentrated emulsions of N-substituted carbamoyl polymers, *Res. Discl.,* 146, 43, 1976; *Chem. Abstr.,* 85, 79185, 1976.

390. Kyoritsu Organic Industrial Research Laboratory, Japanese Kokai Tokkyo Koho JP 58,127,719, 1983; *Chem. Abstr.,* 100, 36060, 1984.

391. Mikhailov, V. G. and Mikhailov, G. S., Synthesis and some properties of cationic arabinogalactan, *Khim. Drev.,* 70, 1987; *Chem. Abstr.,* 106, 139960, 1987.

392. Sellet, L., German Offen. 1,965,562, 1970; *Chem. Abstr.,* 73, 100165, 1970.

393. Fujiwara, H., Sekiya, M., and Suzuki, H., German Offen. 2,327,661, 1973; *Chem. Abstr.,* 80, 121730, 1974.

394. Guthrie, J. D., Pottle, M. S., and Marcavio, M. F., Attachment of compounds to aminized cotton fabric by the Mannich reaction, *Am. Dyest. Rep.*, 53, 19, 1964; *Chem. Abstr.*, 61, 8460, 1964.

395. Kirillova, M. N., Mel'nikov, B. N., Bolshakova, I. V., and Kurakin, E. N., *Otkrytiya, Izobret. Prom. Obratztsy Tavarnye Znaky*, 127, 1983; *Chem. Abstr.*, 99, 55026, 1983.

396. Baumann, H. P., U.S. Patent 4,582,649, 1986; *Chem. Abstr.*, 105, 99101, 1986.

397. Glen, H. M., U.S. Patent 2,527,323, 1950; *Chem. Abstr.*, 45, 2512, 1951.

398. Kryukova, A. S., Lapina, R. A., and Mizuch, K. G., U.S.S.R. Patent SU 176,290, 1984; *Chem. Abstr.*, 101, 90442, 1984.

399. Krsnak, F., Kodytek, V., Zadak, M., Jezek, Z., Zelenka, P., and Sip, M., Czechoslovakian Patent CS 255,211, 1988; *Chem. Abstr.*, 111, 79857, 1989.

400. Schoenberg, J. E., U.S. Patent 4,009,311, 1977; *Chem. Abstr.*, 86, 123253, 1977.

401. Takatsuji, M., Hirajima, S., Miyake, K., Nakai, T., Tokuda, M., and Kikuchi, K., Japanese Kokai Tokkyo Koho 79 25,944, 1979; *Chem. Abstr.*, 92, 24730, 1980.

402. Paar, W., Gmoser, J., and Hoenig, H., Austrian Patent AT 378,537, 1985; *Chem. Abstr.*, 104, 35608, 1986.

403. Paar, W., Hoenel, M., and Gmoser, J., Austrian Patent AT 382,160, 1987; *Chem. Abstr.*, 107, 41800, 1987.

404. Paar, W., European Patent Appl. EP 347,785, 1989; *Chem. Abstr.*, 112, 200736, 1990.

405. Lawrenz, D., Schupp, H., and Schwerzel, T., European Patent Appl. EP 351,612, 1988; *Chem. Abstr.*, 113, 8119, 1990.

406. Demmer, C. G. and Moss, N. S., A new development in water-based can coatings, *J. Oil Colour Chem. Assoc.*, 65, 249, 1982.

407. Waddill, H. G. and Speranza, G. P., U.S. Patent 4,581,421, 1986; *Chem. Abstr.*, 105, 25290, 1986.

408. Grigsby, R. A., Jr. and Speranza, G. P., U.S. Patent 4,714,750, 1987; *Chem. Abstr.*, 108, 132861, 1988.

409. Roos, L., German Offen. 3,041,223, 1981; *Chem. Abstr.*, 95, 159907, 1981.

410. Raue, R. and Corbett, J. F., Nitro and nitroso dyes, in *Ullmann's Encyclopedia of Industrial Chemistry*, Vol. A17, 5th ed., Gerhartz, W., Ed., VCH, Weinheim, 1991, 383; see, in particular, p. 385.

411. Hunter, B. A., South African Patent 68 04,029, 1968; *Chem. Abstr.*, 70, 116109, 1968.

412. Schubart, R., Eholzer, U., Roos, E., and Kempermann, T., German Offen. 2,227,338, 1973; *Chem. Abstr.*, 80, 109596, 1974.

413. Saidaliev, Z. G., Kadyrov, A., Safaev, A., and Islamova, D. V., Preparation of carbazyl-N-methyl-2-thiobenzothiazole, Deposited doc., 1974 VINITI 762–74; *Chem. Abstr.*, 86, 171312, 1977.

414. Kalennikov, E. A., Leshchenko, V. N., Dashevskaya, R. I., Tsyganova, L. V., Enokhovich, G. V., Garoyan, V. A., and Lakamtseva, I. D., U.S.S.R. Patent 763,391, 1980; *Chem. Abstr.*, 94, 16912, 1981.

415. Shenbor, M. I. and Azarov, A. S., Mannich reaction of 11-hydroxyfluoranthene, *Vopr. Khim. Khim. Tekhnol.*, 88, 40, 1988; *Chem. Abstr.*, 112, 98172, 1990.

416. Hechenbleikner, I., Hussar, J. F., Koeniger, F., and Bresser, E., German Offen. 2,059,916, 1971; *Chem. Abstr.*, 75, 110822, 1971.

417. Kline, R. H., German Offen. 2,417,981, 1974; *Chem. Abstr.*, 82, 87401, 1975.

418. Schloman, W. W., European Patent Appl. EP 174,262, 1986; *Chem. Abstr.*, 105, 7786, 1986.

419. Zimmerman, R. L. and Cuscurida, M., British Patent Appl. GB 2,087,911, 1982; *Chem. Abstr.*, 97, 73359, 1982.

420. Schoner, U., Noak, R., Stoll, J., and Mann, G., East German Patent DD 219,760, 1985; *Chem. Abstr.*, 104, 150035, 1986.

421. Rasberger, M., Rody, J., and Gugumus, F., German Offen. 2,654,058, 1977; *Chem. Abstr.*, 87, 102186, 1977.

422. Bredereck, H. and Bäder, E., U.S. Patent 2,750,357, 1956; *Chem. Abstr.*, 50, 15128, 1956.

423. Heraeus, W. C., British Patent 786,315, 1957; *Chem. Abstr.*, 52, 5031, 1958.

424. Paquin, A. M., German Patent 831,248, 1952; *Chem. Abstr.*, 52, 8620, 1958.

425. Laqua, A., Holtschmidt, U., and Schedlitzki, D., German Offen. 2,508,454, 1976; *Chem. Abstr.*, 85, 193521, 1976.

426. Kauffman, W. J., Observations on the synthesis and characterization of N,N',N''-tris(dimethylaminopropyl)hexa-hydro-s-triazine and isolable intermediates, *J. Heterocycl. Chem.*, 13, 409, 1975.

427. Bakirova, I. N., Zenitova, L. A., and Rozental, N. A., Study of the behaviour of poly(urethane isocyanurates) produced by using phenolic Mannich bases at elevated temperatures, *Svoistva Uretan. Elastom. M*, 62, 1984; *Chem. Abstr.*, 102, 186419, 1985.

428. Foucht, M. E., U.S. Patent 4,260,514, 1981; *Chem. Abstr.*, 95, 63432, 1981.

429. Rakhmatullina, G. M., Averko-Antonovich, L. A., and Kirpichnikov, P. A., Viscosimetric method of testing catalysts used in the formation of poly(thiourethanes), *Izv. Vyssh. Uchebn. Zaved. Khim. Khim. Tekhnol.*, 26, 735, 1983; *Chem. Abstr.*, 99, 124134, 1983.

430. Ashikhmina, L. I., Yamalieva, L. N., Nefed'ev, E. S., and Averko-Antonovich, L. A., Thiokol-epoxy composition vulcanized with a peroxide-amine system, *Kauch. Rezina*, 19, 1988; *Chem. Abstr.*, 110, 174911, 1989.

431. Nefed'ev, E. S., Ashikhmina, L. I., Ismaev, I. E., Kadirov, M. K., Averko-Antonovich, L. A., Il'yasov, A. V., and Khirpichnikov, P. A., Radical processes in curing the thiokol-epoxy compositions, *Dokl. Akad. Nauk SSSR*, 304, 1181, 1989; *Chem. Abstr.*, 111, 58844, 1989.

432. Dytyuk, L. T. and Samakaev, R. K., U.S.S.R. Patent 761,494, 1980; *Chem. Abstr.*, 93, 221485, 1980.

433. Oleinikov, A. N., U.S.S.R. Patent 829,637, 1981; *Chem. Abstr.*, 95, 43949, 1981.

434. Weers, J. J., Falkler, T. J., Duggan, G. G., and Garrecht, R. J., U.S. Patent 4,900,427, 1990; *Chem. Abstr.*, 112, 182786, 1990.

435. Roling, P. V. and Niu, J. H. Y., U.S. Patent 4,874,415, 1989; *Chem. Abstr.*, 111, 217104, 1989.

436. Mikryukova, N. K., Potapenko, A. P., Litvinenko, L. V., Matyash, L. G., Popovich, T. D., and Polevaya, V. I., U.S.S.R. Patent SU 1,152,613, 1985; *Chem. Abstr.*, 103, 89333, 1985.

437. Harle, O. L., Belgian Patent 872,543, 1979; *Chem. Abstr.*, 91, 94234, 1979.

438. Woitunik, D., Kuhn, K., Koeppert, G., Stoeffgen, R., and Wenzel, B., East German Patent 126,659, 1977; *Chem. Abstr.*, 88, 155614, 1978.

439. Stoldt, S. H. and Walsh, R. H., PCT Int. Appl. WO 87 01,720, 1987; *Chem. Abstr.*, 107, 25873, 1987.

440. Koch, F. W., Dorer, C. J., Evans, G. A., and Del Paggio, A. A., PCT Int. Appl. WO 87 01,721, 1987; *Chem. Abstr.*, 107, 10261, 1987.

441. Levin, A. Y., Ivanova, E. A., Monin, S. V., and Bogdanova, G. F., Study of the moisture resistance of alkylphenol additives, *Neftepererab. Neftekhim. Moscow*, 17, 1982; *Chem. Abstr.*, 97, 26053, 1982.

442. Gemmil, R. M., Jr. and Horodysky, A., U.S. Patent 4,396,517, 1983; *Chem. Abstr.*, 99, 107954, 1983.

443. Song, Y. S. and McCandless, H. A., U.S. Patent 4,663,392, 1987; *Chem. Abstr.*, 107, 42915, 1987.

444. Gutierrez, A., Kleist, R. A., and Bloch, R. A., European Patent Appl. EP 357,215, 1990; *Chem. Abstr.*, 112, 220189, 1990.

445. Nater, P., Kopernicky, I., Janosik, S., Stacho, D., Stanco, N., and Nadvornik, F., Czechoslovakian Patent CS 261,083, 1989; *Chem. Abstr.*, 112, 59523, 1990.

446. Ahmed, M. H., Ghuiba, F. M., Habib, O. M. O., and Garieb, H. K., Synthesis and evaluation of lubricating oil additives from local tar phenols, *Hung. J. Ind. Chem.*, 17, 311, 1989: *Chem. Abstr.*, 112, 121869, 1990.

447. Donner, J. P., Horodysky, A. G., and Keller, J. A., U.S. Patent 4,743,386, 1988; *Chem. Abstr.*, 109, 233984, 1988.

448. Dovgopoli, E. E., Eminov, E. A., Oberfel'd, M. S., et al., U.S.S.R. Patent 761,543, 1980; *Chem. Abstr.*, 94, 50010, 1981.

449. Konyushenko, V. P., Glavati, O. L., Bavika, N. P., and Nazarchuk, N. M., Preparation of adhesion additives for bitumes from unused alkylphenols, *Neftepererab. Neftekhim. Kiev*, 33, 10, 1987; *Chem. Abstr.*, 108, 225904, 1988.

450. Körbl, J. and Přibil, R., Some new metallochromic indicators of the Complexon type, *Chem. Ind.*, 233, 1957.

451. Tsirul'nikova, N. V., Temkina, V. Y., Yaroshenko, G. F., and Lastovskii, R. P., Synthesis of complexons of the hydroxyaryl series by the Mannich reaction, *Khim. Prom. Moscow*, 47, 502, 1971; *Chem. Abstr.*, 75, 88261, 1971.

452. Zhang, Z., Zhang, Z., and Yu, R., Complexons synthesized by Mannich reaction and applications in analytical chemistry, *Huoxue Shiji*, 6, 24, 1984; *Chem. Abstr.*, 101, 6697, 1984.

453. Minin, A. A. and Filippova, L. P., Improved process for obtaining alizarin complexon, *Uch. Zap. Permsk. Gos. Univ.*, 229, 208, 1970; *Chem. Abstr.*, 77, 151732, 1972.

454. Leonard, M. A. and Murray, G. T., Sulfonated alizarin fluorine blue. Improved reagent for the positive absorptiometric determination of the fluoride ion, *Analyst London*, 99, 645, 1974; *Chem. Abstr.*, 82, 92553, 1975.

455. Agency of Ind. Sci. and Tech., Japanese Kokai Tokkyo Koho JP 57,162 726, 1982; *Chem. Abstr.*, 98, 126868, 1983.

456. Horiguchi, S., Nakamura, M., and Nakajima, K., Japanese Patent 73 02,930, 1973; *Chem. Abstr.*, 80, 121738, 1974.

457. Kalopissis, G., Abegg, J. L., Ghilardi, G., and De Beaulieu, H. P., U.S. Patent 3,957,774, 1976; *Chem. Abstr.*, 85, 143118, 1976.

458. Jex Co., Ltd., Japanese Kokai Tokkyo Koho JP 82 25,813, 1982; *Chem. Abstr.,* 97, 78952, 1982.

459. Van Leusen, A. M. and Strating, J., p-Tolyl-sulfonyldiazomethane, *Org. Synth.,* 57, 95, 1977; *Chem. Abstr.,* 88, 104833, 1978. See also *Org. Synth.,* 57 102, 1977.

460. Boileau, J., Piteau, M., and Jacob, G., Synthesis of 1,3,5,5-tetranitrohexahydropyrimidine, *Propellants Explos. Pyrotech.,* 15, 38, 1990; *Chem. Abstr.,* 112, 182447, 1990.

461. Cichra, D. A. and Adolph, H. G., Synthesis of polynitro- and nitronitrosoperhydro-1,5-diazocines, *Synthesis,* 830, 1983.

462. Dunstan, I., Chemistry in the technology of explosives and propellants, *Chem. Brit.,* 7, 62, 1971.

463. Webb, W. P., U.S. Patent 2,823,515, 1958; *Chem. Abstr.,* 52, 7665, 1958.

464. Tonari, K., Ichimoto, I., Ueda, H., and Tatsumi, C., Mannich reaction of cyclotene. Novel synthesis of 3,5-dimethyl-2-cyclopenten-2-ol-1-one, *Nippon Nogei Kagaku Kaishi,* 44, 55, 1970; *Chem. Abstr.,* 72, 100105, 1970.

465. Sato, K. and Inoue, S., Japanese Kokai Tokkyo Koho 73 10,044, 1973; *Chem. Abstr.,* 78, 135750, 1973.

466. Ohashi, M., Takahashi, T., Inoue, S., and Sato, K., The Mannich reaction of alicyclic α-diketones. A novel synthesis of 2-hydroxy-3-methyl-2-cyclohexen-1-one, *Bull. Chem. Soc. Jpn.,* 48, 1892, 1975.

467. Amos, H. E., Evans, J. J., Himmelsbach, D. S., and Barton, F. E., *In vitro* stability, *in vivo* hydrolysis, and absorption of lysine and methionine from polymerized amino acid preparations, *J. Agric. Food Chem.,* 28, 1250, 1980; *Chem. Abstr.,* 93, 202990, 1980.

468. Leonte, M., Georgescu, M., Toma, G., Sinchievici, E., and Roncea, C., Romanian Patent 61,894, 1977; *Chem. Abstr.,* 91, 211070, 1979.

469. Eastman Kodak Co., French Patent 1,470,819, 1967; *Chem. Abstr.,* 67, 121359, 1967.

470. Caldwell, J. R. and Hill, E. H., U.S. Patent 3,312,665, 1967; *Chem. Abstr.,* 67, 12640, 1967.

471. Ishikawa, M., Kobayashi, K., and Koboshi, S., German Offen. DE 3,423,100, 1985; *Chem. Abstr.,* 103, 79391, 1985.

472. Pelton, R. H., Model cationic flocculant from Mannich reaction of polyacrylamide, *J. Polym. Sci. Polym. Lett. Ed.,* 22, 3955, 1984.

473. Takeuchi, T., Yasukawa, K., and Kitsugi, M., Japanese Kokai Tokkyo Koho JP 61,200,898, 1986; *Chem. Abstr.,* 106, 22855, 1987.

474. Racciato, J. S., Cottrell, I. W., and Shim, J. L., U.S. Patent 4,155,885, 1979; *Chem. Abstr.,* 91, 40348, 1979.

475. Matsuda, H., Kanki, K., Nakagawa, T., and Koike, M., Japanese Kokai Tokkyo Koho 75 67,287, 1975; *Chem. Abstr.,* 85, 83100, 1976.

476. Tomono, T., Hasegawa, E., and Tsuchida, E., Polyamine polymers from active hydrogen compounds, formaldehyde and amines, *J. Polym. Sci. Polym. Chem. Ed.,* 12, 953, 1974.

477. Falchetti, A., Giavarini, C., Moresi, M., and Sebastiani, E., Absorption of CO_2 in aqueous solutions: reaction kinetics of modified tetraethylenepentamine, *Ing. Chim. Ital.,* 17, 1, 1981; *Chim. Ind.,* 63, 1981.

478. Mazgarov, A. M., Vil'danov, A. F., Fakhriev, A. M., Sattarov, U. G., Neyaglov, A. V., and Latypova, M. M., U.S.S.R. Patent SU 1,150,007, 1985; *Chem. Abstr.,* 103, 24376, 1985.

479. Voronov, M. G. and Mikhailov, Z. I., Deodorization and utilization of methanethiol from industrial emissions, *Zh. Prikl. Khim. Leningrad,* 49, 2577, 1976; *Chem. Abstr.,* 86, 89105, 1977.

480. Levon, K., Meshitsuka, G., and Nekano, J., Acceleration of the beating of pulp by the use of modified lignosulfonate, *Mokuzai Gakkaishi,* 32, 1011, 1986; *Chem. Abstr.,* 106, 139978, 1987.

481. Craciun, D. C., Constantinescu, A., Socol, S., Cristian, M., Constantinescu, P., Ivascu, I., Cismaru, P., and Prodrom, R., Romanian Patent 55,110, 1973; *Chem. Abstr.,* 80, 85476, 1974.

Index

(s) indicates substrate in Mannich reaction; italic numbers refer to Tables